AF900745

# Metal Nanocluster Catalysis

From Fundamentals to Applications

*Manzhou Zhu* and *Man-Bo Li*

WILEY VCH

**Authors**

***Prof. Manzhou Zhu***
*Department of Chemistry*
*Anhui University*
*Hefei, 230601*
*P.R. China*

***Prof. Man-Bo Li***
*Institutes of Physical Science and Information Technology*
*Anhui University*
*Hefei, 230601*
*P.R. China*

**Cover Design:** Wiley
**Cover Image:** Courtesy of Manzhou Zhu, Man-Bo Li

**Typesetting:** Lumina Datamatics, Madurai, India
**Printing and Binding:** CPI Group (UK) Ltd, Croydon, CR0 4YY

**Library of Congress Card No.:** applied for

**British Library Cataloguing-in-Publication Data**
A catalogue record for this book is available from the British Library.

**Bibliographic information published by the Deutsche Nationalbibliothek**
The Deutsche Nationalbibliothek lists this publication in the Deutsche Nationalbibliografie; detailed bibliographic data are available on the Internet at http://dnb.d-nb.de.

**Print ISBN:** 9783527355266
**ePDF ISBN:** 9783527852604
**ePub ISBN:** 9783527852598
**oBook ISBN:** 9783527852611

© 2026 WILEY-VCH GmbH, Boschstraße 12, 69469 Weinheim, Germany

C9783527355266_240226

All rights reserved (including those of translation into other languages, text and data mining and training of artificial technologies or similar technologies). No part of this book may be reproduced in any form – by photoprinting, microfilm, or any other means – nor transmitted or translated into a machine language without written permission from the publishers. Registered names, trademarks, etc. used in this book, even when not specifically marked as such, are not to be considered unprotected by law.

The manufacturer's authorized representative according to the EU General Product Safety Regulation is WILEY-VCH GmbH, Boschstr. 12, 69469 Weinheim, Germany, e-mail: Product_Safety@wiley.com.

In view of ongoing research, equipment modifications, changes in governmental regulations, and the constant flow of information relating to the use of experimental reagents, equipment, and devices, the reader is urged to review and evaluate the information provided in the package insert or instructions for each chemical, piece of equipment, reagent, or device for, among other things, any changes in the instructions or indication of usage and for added warnings and precautions. While the publisher and the authors have used their best efforts in preparing this work, including a review of the content of the work, neither the publisher nor the authors make any representations or warranties with respect to the accuracy or completeness of the contents of this work and specifically disclaim all warranties, including without limitation any implied warranties of merchantability or fitness for a particular purpose. No warranty may be created or extended by sales representatives, written sales materials or promotional statements for this work. The fact that an organization, website, or product is referred to in this work as a citation and/or potential source of further information does not mean that the publisher and authors endorse the information or services the organization, website, or product may provide or recommendations it may make. This work is sold with the understanding that the publisher is not engaged in rendering professional services. The advice and strategies contained herein may not be suitable for your situation. You should consult with a specialist where appropriate. Further, readers should be aware that websites listed in this work may have changed or disappeared between when this work was written and when it is read. Neither the publisher nor authors shall be liable for any loss of profit or any other commercial damages, including but not limited to special, incidental, consequential, or other damages.

# Contents

# Preface

Metal nanoclusters are unique nanomaterials, bridging the gap between metal atoms (or metal chelates) and metal nanoparticles. They feature ultra-small size, atomically precise structure, and diversified metal–metal and metal–ligand interactions. Especially, as a metal catalyst, they combine the advantages of their homogeneous and heterogeneous metal analogs, demonstrating potentially high activity, unique selectivity, and recyclability. With the rapid development of synthetic chemistry of metal nanoclusters, they have been discovered as emerging catalysts in many catalytic processes, such as hydrogenation, oxidation, borylation, and cross-coupling reactions. Additionally, owing to their exceptional energy and electron transfer capability, they have been combined with photo-, electro-, and enzyme-catalysis, triggering efficient organic transformations.

With the development of metal nanocluster chemistry and their application as an emerging catalyst in a lot of efficient even new transformations, the book focusing on metal nanocluster catalysis is in high demand. Based on this situation, we presented this book entitled *Metal Nanocluster Catalysis: From Fundamentals to Applications*, systematically introducing the formation, structural features, catalytic properties, and mechanisms of this kind of emerging nanomaterial.

Chapter 1 gives a brief introduction of metal nanocluster catalysis from aspects of catalytic activity, selectivity, and mechanism. Chapters 2–7 cover three kinds of metal nanoclusters that have been widely employed in catalysis: (i) in situ generated metal nanoclusters, which are spontaneously formed by metal catalysts during the reaction; (ii) supported metal nanoclusters, which are obtained by loading the active metal nanoclusters on relatively robust supports, such as carbon nanomaterials (e.g. carbon nanotube, graphene), metallic oxides (e.g. $TiO_2$, $SiO_2$, $CeO_2$), etc.; (iii) organic ligand-protected metal nanoclusters, which have defined molecular formulas and atomically precise structures, and can be designed and synthesized by wet-chemical reduction methods and other recently developed methods. We introduce each metal nanocluster catalysis from their formation processes to their unique structures, and then to their catalytic performance and catalytic mechanisms. Apart from the introduction of the three kinds of metal nanoclusters and their metal catalytic performance, we describe their catalytic activities in photo-, electro-, and enzyme-mediated processes as well, which are included in Chapters 8–10. The practical applications of metal nanocluster catalysis, including large scale synthesis, recyclability, and industrial catalysis are presented in Chapter 11. Finally, comments on opportunities and challenges of metal nanocluster catalysis are listed at the end of the book, Chapter 12.

This book would be useful for the audiences to understand the current progress as well as the fundamental theory on metal nanocluster catalysis. It can be a handbook for the researchers who are going to start their research in this area, and can also be of value to chemists long involved in metal nanocluster in both academic and industrial research.

***Manzhou Zhu***
***Man-Bo Li***

# Acknowledgments

The financial support from the National Natural Science Foundation of China (U24A20480, 22422101, 22371002), the Anhui Provincial Natural Science Foundation (2508085ZD017), and the Ministry of Education are acknowledged.

# 1

# Introduction to Metal Nanocluster Catalysis

## 1.1 Catalysis and Metal Catalysts

Chemical reactions provide a way to create new molecules, which have been widely used in our lives. The energy difference between reactants and products gives driving force for the chemical reactions. However, in general, these reactions cannot happen directly because of the existence of intermediates (or transition states) which bear higher Gibbs energy than the reactants. To promote the smooth progress of reactions, catalysts are needed to significantly reduce the reaction energy barrier between the intermediate (or transition state) and the reactant (Figure 1.1). The catalysts-promoted chemical reactions can be called catalytic process, that is, catalysis. More than 80% of chemical reactions involve catalysis. Among of them, metal species-promoted catalytic processes play important roles in the construction of functional molecules [1–7], which are difficult to be obtained in other ways. Almost half of Nobel Prizes in Chemistry are directly or indirectly relevant to metal catalysis. Typical examples include the "click chemistry and biorthogonal chemistry" (2022), "palladium-catalyzed cross couplings in organic synthesis" (2010), "metathesis method in organic synthesis" (2005), "chirally catalyzed hydrogenation and oxidation reactions" (2001), etc.

Generally, metal catalysts activate molecules in three ways. First, σ-donation and π-backdonation process (Figure 1.2a). Taking the nitrogen gas ($N_2$) activation as an example, a number of transition metals have been found to be able to bind $N_2$ by the mode of σ-donation and π-backdonation [8]. This $N_2$ activation mode was found since the discovery of the Haber–Bosch process. This kind of catalytic activity of metals is attributed to their advantageous combination of unoccupied and occupied d orbitals, which are of appropriate energy and symmetry to synergistically accept electron density from and backdonate to $N_2$. The σ-donation and π-backdonation process highly weakens the nitrogen–nitrogen triple bond while simultaneously strengthening the metal–nitrogen bond, thus endowing metals with catalytic activity toward $N_2$ and other molecules such as alkenes, alkynes, etc.

*Metal Nanocluster Catalysis: From Fundamentals to Applications*, First Edition. Manzhou Zhu and Man-Bo Li. © 2026 Wiley-VCH GmbH. All rights reserved, including rights for text and data mining and training of artificial intelligence technologies or similar technologies. Published 2026 by Wiley-VCH GmbH.

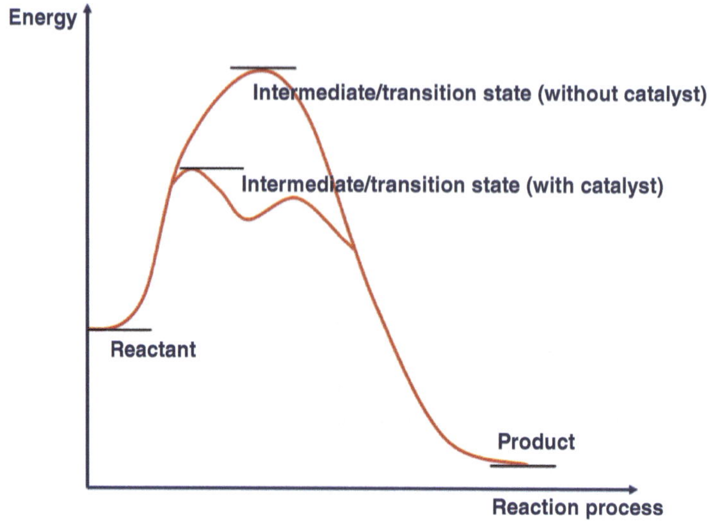


**Figure 1.1** Illustration of catalytic process.

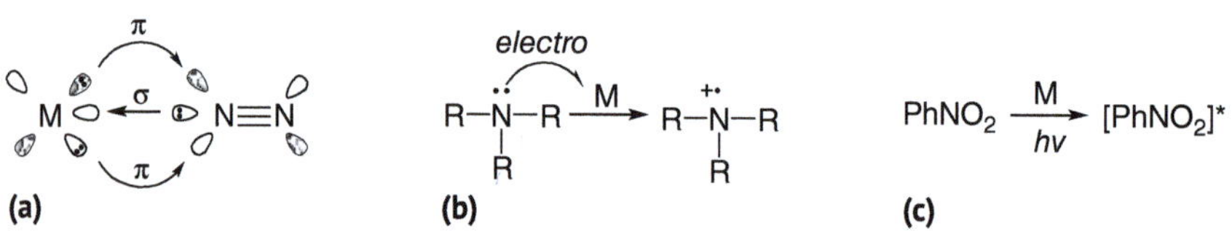


**Figure 1.2** Three catalytic modes of metal catalysts.

Second, electron transfer process (Figure 1.2b). Taking the trialkyl amine activation as an example, metals bearing abundant valence states are generally able to undergo electron transfer process. Under the conditions of electrolysis or light irradiation, one of the lone pair electrons on the trialkyl amine would be abstracted by the metal, thus generating the trialkyl amine radical cation [9]. This species is quite active and can be further transformed to carbon radicals or captured directly by other molecules. The as-formed radical intermediates abstract an electron from the reduced metal, giving M and finishing the catalytic cycle. Third, energy transfer process (Figure 1.2c). Taking the nitrobenzene activation as an example, photosensitive metal species are usually capable of energy transfer process. The photosensitive metal catalyst would be excited under light irradiation, giving the excited state metal catalyst. This active species contacts with nitrobenzene in the reaction system. Energy transfer between the excited metal catalyst and nitrobenzene would regenerate metal catalyst and produce activated nitrobenzene [10], which can be further transformed in the presence of other organic molecules. Sometimes the metal catalyst interacts with molecules and forms the metal-molecule complex that is more easily excited and undergoes intracomplex energy transfer process under light irradiation.

The three catalytic modes are involved in various metal catalysts including metal salts, metal-ligand chelates, metal nanoclusters, metal nanoparticles, metal–organic frameworks, etc. The content in the following sections and chapters will also focus on the three catalytic modes, introducing different reactions catalyzed by metals.

## 1.2 Single Metal Atom, Metal Nanocluster, and Metal Nanoparticle

From the size point of view, the common metal catalysts can be divided into the following three types: single metal atoms, metal nanoclusters and metal nanoparticles (Figure 1.3). For single metal atoms, their sizes range from 0.5 to 2.5 Å. For the roles of controlling catalytic activity and selectivity, the single metal atom can be surrounded by inorganic/organic carriers, counter ions, or organic ligands. Therefore, the true size of single metal atom catalysts is dramatically larger than the diameters of metal atoms, probably reaching to 1 nm. Metal nanoclusters composed of several metal atoms generally fall in the range of 1–3 nm. The bare metal nanoclusters are quite active because of their high surface energy. They are stabilized by either organic ligands or carriers, which help to reduce the surface energy and regulate catalytic activity and selectivity of the metal kernel as well. Generally speaking, the metal nanocluster is a group of metal atoms that are connected by metal–metal or metal–organic interactions [11, 12]. They have relatively blurring boundaries with metal nanoparticles. On the other hand, it should be clarified that the narrowly defined metal nanocluster is the metal aggregate possessing direct metal bonding interactions, well-defined composition and atomically precise structure [13–15]. Metal nanoparticle has been widely used currently in various subjects such as chemistry, materials, physics, biology, and information. It has a metal kernel with size ranging from several to tens and hundreds of nanometers, and even to micrometers. Inorganic or organic layers are surrounded on the surface to protect the metal kernel from aggregation or decomposition. Metal nanoclusters and nanoparticles have distinct physicochemical properties compared to single metal atoms. The 2023 Nobel Prizes in Chemistry "discovery and synthesis of quantum dots" displays the unique properties of this kind of material.

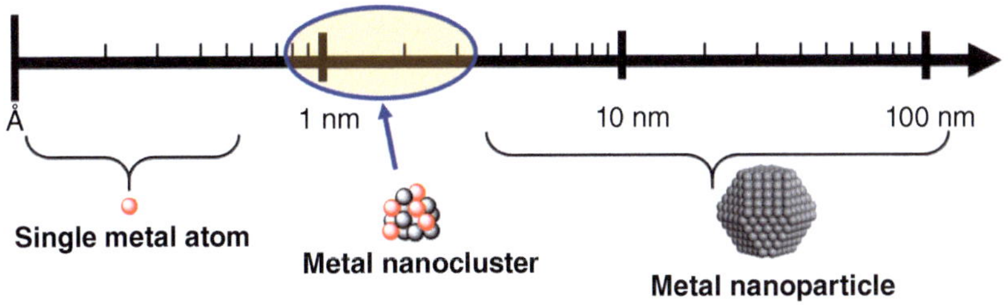


**Figure 1.3** The size of single metal atom, metal nanocluster, and metal nanoparticle.

From the catalytic property point of view, single metal atom has the most surface metal atoms, that is, the potential activation sites. Nowadays, the most frequently used metal catalysts are homogeneous single metal atom catalyst [16–19], such as metal salts and metal chelates. Each metal site contacts with reactants sufficiently under homogeneous reaction conditions, triggering high catalytic activity and selectivity. It should be noted that for certain metals, their homogeneous single atom forms with different valence states have significantly different catalytic activities. We are merely citing Pd catalysis as a typical example (Figure 1.4). The single $Pd^0$ atom catalyst such as $Pd(PPh_3)_4$ has been found to be highly active to the cross-coupling process [20]. The low-valence-state palladium would be oxidized by the aryl-halide, generating the $Pd^{II}$ intermediate via an elementary reaction of oxidative addition. This intermediate can be trapped by an organometallic reagent and transformed to the other $Pd^{II}$ intermediate via an elementary reaction of transmetallation. The as-obtained $Pd^{II}$ intermediate undergoes the third elementary reaction of reductive elimination, furnishing the final cross-coupling product and regenerating the $Pd^0$ catalyst. The single $Pd^{II}$ atom catalyst such as $PdCl_2$ demonstrates a dramatically distinct catalytic activity compared to the $Pd^0$ species. A typical example is the $Pd^{II}$-catalyzed Wacker process [21]. The high-valence-state palladium prefers coordinating with electron-rich compounds such as the alkene. The activated alkene would be attacked easily by a nucleophile such as $H_2O$, producing a $Pd^{II}$ intermediate via an elementary reaction of olefin insertion. The following elementary reaction of β-H elimination would give the aldehyde as the final product and produce $Pd^0$ species, which is oxidized to regenerate the $Pd^{II}$ catalyst. The significantly different catalytic processes shown in Figure 1.4 represent the catalytic difference of single metal atoms with different valence states, which have been widely applied in pharmaceutical industry and acetaldehyde production, respectively.

Apart from the homogeneous single metal atom catalysts, heterogeneous single metal atom catalysts [22–25] have shown exceptional catalytic performance in recent years. They are composed of single metal atoms that are anchored on different carriers to reduce the high surface energy. The carriers around metal atoms play the similar role with the organic ligands of metal chelates, on the one hand stabilizing

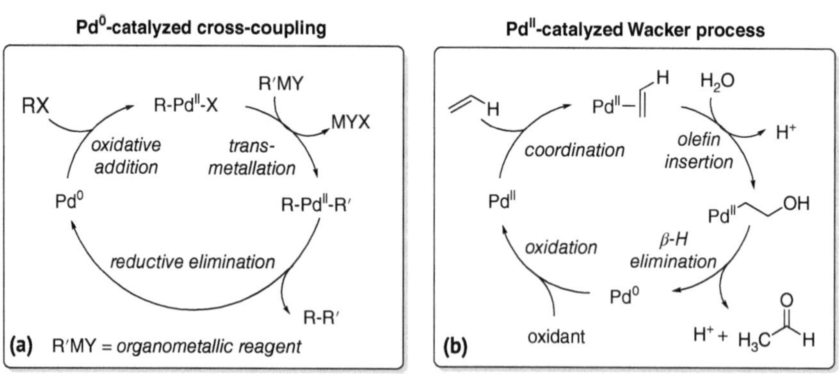


**Figure 1.4** Typical examples on single $Pd^0$ (a) and $Pd^{II}$ (b) atom catalysis.

the metal center and on the other hand facilitating control of catalytic activity and selectivity. Additionally, the heterogeneous single metal atom catalysts are readily recovered from the catalytic reactions by simple procedures such as centrifugation and then recycled, thus being highly beneficial for industrial applications.

Being different from the carrier-supported heterogeneous single metals, metal nanoparticles [26–28] are self-supported heterogeneous metal catalysts. They are usually composed of a metal kernel and an organic surface. Besides their advantage of recoverability and recyclability during catalysis, the catalytic synergy among different activation sites on the surface endows metal nanoparticles with different activity and selectivity compared to single metal atom catalysts. Molecular adsorption and desorption should be well considered for metal nanoparticle catalysis because both processes are rate-determining steps in many cases [29].

As a result of the significantly different catalytic mechanisms between metal nanoparticle catalysis and single metal atom catalysis, they exhibit distinct catalytic selectivity even if the metal and ligand types are the same, respectively. We still take the palladium catalysis as a typical example. $Pd^{II}$-catalyzed oxidative carbocyclization reactions constitute an efficient route to diverse cyclic carbon compounds. The group of Jan-E Bäckvall developed a series of $Pd^{II}$-catalyzed oxidative carbocyclizations of allenes via a key process of allenic $C(sp^3)$-H bond oxidation [30–33]. Recently, Jan-E Bäckvall and Man-Bo Li found that replacement of the single palladium atom catalyst ($Pd(OAc)_2$) by the palladium nanoparticle catalyst (1–2 nm palladium nanoparticles immobilized on amino-functionalized siliceous mesocellular foam) gave completely different selectivities for the catalyzed oxidative carbocyclizations (Figure 1.5) [34]. Specifically, for the oxidative carbocyclization-alkynylation of enallenes with a —OH substituent, the palladium nanoparticle gave the desired product with high yield and diastereoselectivity while $Pd(OAc)_2$ catalyzed the reaction and produced an oxidative carbocyclization product. A β-hydride elimination instead of alkynylation should be attributed to the different selectivity of $Pd(OAc)_2$ catalysis. Control experiments further revealed that the Pd-amine synergy on the surface of the palladium nanoparticle promoted the alkynylation of the palladium intermediates, thus affording the oxidative carbocyclization-alkynylation product with high efficiency (Figure 1.5a).

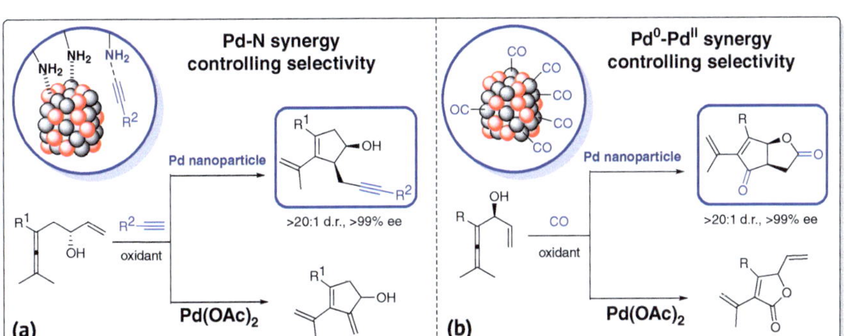


**Figure 1.5** Pd-N and Pd-Pd synergy for palladium nanoparticle catalysis.

Apart from the Pd-N synergy, the $Pd^0$-$Pd^{II}$ synergy on the surface of palladium nanoparticles also triggers selectivity switching of oxidative carbocyclization of allenes. As shown in Figure 1.5b, compared to $Pd^{II}$, Pd(0) atoms favor adsorption and coordination of CO, thus resulting in the aggregation of CO in the palladium nanoparticle catalyst. The local high CO concentration would promote the generation a subsequent carbocyclization to give the bicyclic compound as the final product (Figure 1.5b). In sharp contrast, $Pd(OAc)_2$-catalyzed reaction gave the monocyclic compound as the product because of the unfavored second carbonylation process. The distinct selectivity of metal nanoparticle-catalysis, combined with their recoverability and recyclability after catalysis, would be highly promising for their practical applications in industrial catalysis.

Being similar with metal nanoparticles, metal nanoclusters are metal aggregates stabilized by inorganic or organic species. As mentioned above, metal nanoclusters fall in the gap between single metal atoms and metal nanoparticles. They can be considered as the nascent states during the aggregation of metal atoms to metal nanoparticles. Therefore, they are usually more active than the stabilized metal nanoparticles, possessing promising catalytic potentials. Actually, in many metal-catalyzed transformations, the real catalysts are the in situ generated metal nanoclusters. They are formed either by bottom-up process from single metal atoms under homogeneous catalytic conditions or top-down process from metal nanoparticles under heterogeneous catalytic conditions. However, this kind of special species is highly active and difficult to be isolated and characterized. This part will be discussed in detail in the next chapter "In situ generated metal nanocluster catalysis" of the book. The main challenge of research of metal nanocluster catalysis is stabilize these highly active species while keeping them catalytically active. A stable and characterizable structure is the prerequisite for in-depth investigation of their unique catalytic activity, selectivity, and mechanism. Carriers or organic ligands are generally employed to support or protect the metal kernels, thus stabilizing metal nanoclusters for catalytic investigations. The stabilized metal nanoclusters that enable isolation and characterization after catalysis will be discussed in detail in Chapter 3 "Supported metal nanocluster catalysis" and Chapters 4–6 "Organic ligand protected metal nanocluster catalysis" of the book.

It should be noted that the single metal atom, metal nanocluster, and metal nanoparticle can be transformed reversibly in certain reaction conditions (Figure 1.6) [35, 36]. The aggregation of metal atoms would result in the generation of metal nanoclusters and even metal nanoparticles. On the other hand, metal leaching on

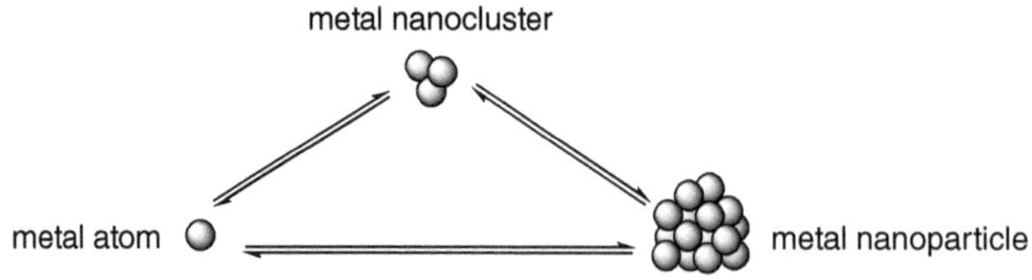


**Figure 1.6** Mutual transformation among the single metal atom, metal nanocluster, and metal nanoparticle during catalysis.

the surface of nanoparticles lead to the formation of metal atoms or metal nanoclusters in the reaction solutions. The idea of the aggregation and leaching processes has been discussed in many reports. For instance, the palladium leaching process (the formation of soluble metal species from palladium black) was discussed by Mizoroki in the early seventies, just after the first examples of Pd-catalyzed Mizoroki–Heck reaction had been published. The interconversion of various metal species during catalytic reactions have governed the development of the concept of a "Cocktail" of metal catalysts, which facilitate catalytic reactions such as the cross-coupling reaction, hydrogenation, C—H functionalization, etc. for a broad range of substrates. Meanwhile, the mutual transformation and coexistence of single metal atoms, metal nanoclusters, and metal nanoparticles would result in an ambiguous mechanism and raise the arguments on the topic of "What is the true catalyst?" In-depth mechanistic investigations including the kinetics will be helpful to clarify the true active species even though the catalytic reactions are dynamic and nothing can be static in catalysis.

## 1.3 Metal Nanocluster Catalysis: Activity

The highly catalytic activity of metal nanocluster catalysis would be the first feature comes into one's mind. In earlier research on metal nanocluster catalysis, it has been reported that metal nanoclusters demonstrated significantly higher catalytic activity than other metal species. For example, Huang and coworkers prepared gold nanocluster (<2.0 nm) deposited on alkaline-treated titanosilicalite by solid grinding with dimethyl gold acetylacetonate [37]. They further found that this nanocluster catalyst triggered the transformation from the propene to the propene epoxide with the metal time yield of 2.3 $g_{PO}g^{-1}{}_{Au}h^{-1}$. The propene conversion and selectivity were determined to be near 1.0% and above 50%, respectively. Although this catalytic performance is not sufficiently high for an industrial application, it is much better than those of general copper- and silver-based catalysts utilized for this reaction (Table 1.1).

In another report from Turner and coworkers [38], a sharp size threshold of nanogold catalysts was found for styrene oxidation. A series of nanogold with different sizes were synthesized and located on the titanium dioxide (110) single crystal. Only when the size falls below about 3.5 nm, the nanogold catalysts were catalytically active. Especially, the gold nanocluster composed of 55 gold atoms (~1.4 nm) was found to be an efficient and robust catalyst for the selective oxidation of styrene by dioxygen (Table 1.2). This typical case unambiguously indicates the high catalytic activity of metal nanoclusters compared to metal nanoparticles.

The high activity of metal nanoclusters was discovered not only in oxidation reactions but also in C—C cross-coupling reactions. Reetz and de Vries found that activated and nonactivated aryl bromides underwent smooth ligand-free Heck reactions [39], featuring low amounts of palladium salts such as $Pd(OAc)_2$ (ideally 0.01–0.1 mol%). Experimental results such as the in situ transmission electron microscopy (TEM) demonstrated that this industrially viable process appears to involve palladium

**Table 1.1** Oxidation of $C_3H_6$ with $O_2$ in the presence of $H_2O$ by gold catalysts.[a]

| Support | Au loading [wt%][b] | Pretreatment atmosphere,[c] temperature [K] | Au nanocluster diameter [nm][d] | Conv. of $C_3H_6$ [%] | Selectivity of propene epoxide [%] |
|---|---|---|---|---|---|
| Titanosilicalite | 0.19 | $H_2$, 423 | 4.6 | 0.07 | 0 |
| Titanosilicalite | 0.19 | $H_2$, 423 | 1.8 | 0.88 | 52 |
| Titanosilicalite | 9.0[e] | $H_2$, 473 | — | 0.47 | 57 |
| $\gamma$-$Al_2O_3$ | 0.86 | Air, 573 | 2.6 | 0.15 | 0 |
| $Al_2O_3$ (amorphous) | 0.19 | $H_2$, 423 | <2.0[f] | 0.15 | 0 |
| $TiO_2$ | 1.2 | Air, 573 | 2.9[f] | 0.03 | 0 |
| $TiO_2$ | 0.10 | Air, 573 | 1.8 | 0.05 | 0 |

a. Reaction conditions: catalyst 0.3 g; reaction temperature 473 K (for Au/$TiO_2$, 353 K); feed gas $C_3H_6/O_2/H_2O/Ar = 10/10/2/78$; space velocity 4000 $mLg^{-1}{}_{cat.}\,h^{-1}$. Data were taken under a steady state after at least 2-hour duration.
b. Actual Au loadings obtained by inductively coupled plasma (ICP) analysis.
c. $H_2$ (10 vol%) was diluted by 90 vol% Ar.
d. The diameters of gold nanoclusters determined by HAADF-STEM observation after catalytic tests.
e. Gold loading calculated from the gold content in the starting solution.
f. The diameters of gold nanoparticles or clusters determined by TEM observation after catalytic tests. Over Au/amorphous $Al_2O_3$, because no Au particles could be found by TEM, the diameter was estimated to be smaller than 2.0 nm.

**Table 1.2** Catalytic results of oxidation of styrene using $O_2$ alone for different Au catalysts.

| Catalyst | Au loading [wt%] | Au nanocluster/ particle diameter [nm] | Conversion [%] | Selectivity of styrene epoxide [%] |
|---|---|---|---|---|
| $Au_{55}$/BN | 0.63 | 1.6 | 19.2 | 14.0 |
| $Au_{55}/SiO_2$ | 0.67 | 1.5 | 25.8 | 12.0 |
| $Au_{55}/SiO_2$ recycled 1 | 0.67 | — | 21.4 | 23.7 |
| $Au_{55}/SiO_2$ recycled 2 | 0.67 | — | 15.9 | 27.1 |
| $Au/SiO_2$ | 6.35 | 3.0 | Trace | — |
| $Au/SiO_2$ | 0.60 | 4.0 | Trace | — |
| Au/C | 1.0 | 17.0 | No reaction | — |
| $Au/SiO_2$ | 5.0 | >30 | No reaction | — |

nanoclusters with the size of approximately 1.6 nm. Leyva-Perez and coworkers further investigated the active catalytic species in ligand-free C—C cross-coupling reactions [40], such as Heck, Sonogashira, Suzuki, and Stille coupling reactions. By using TEM together with dynamic light scattering (DLS) measurements, it was confirmed that the palladium atoms aggregated to form 2–10 nm palladium nanoparticles in a few minutes when the reaction was carried out. However, the reaction-induction period continued for approximately two additional hours after the formation of the palladium nanoparticles. The result unambiguously indicates that the palladium nanoparticles generated in situ are not the catalytically active metal species. Furthermore, they analyzed reaction samples before and after the induction time by electrospray ionization mass spectrometry (ESI-MS) and matrix-assisted laser desorption/ionization coupled to time-of-flight (MALDI-TOF) mass spectrometry. Results showed that the C—C coupling reactions initiated on when palladium nanoclusters with a mass below 500 Da appear. Based on the jellium model, significant emission signals (450 nm) were only observed in the range of 336–370 nm, which corresponds to small clusters with three and four metal atoms. This real catalytic species was observed to be partially stabilized by water in the reaction system, and demonstrated turnover frequencies (TOFs) between $10^5$ and $10^6$ molecules of product per atom of palladium per hour. Ellis and coworkers found the similar catalytic performance of in situ generated palladium nanoclusters and provided the direct evidence that the Suzuki coupling reaction can occur heterogeneously at the surface of polyvinylpyrrolidone (PVP)-stabilized palladium nanoclusters with the help of a combination of operando X-ray absorption spectroscopy (XAS), surface sensitive X-ray photoelectron spectroscopy (XPS), and detailed kinetic profiling [41].

Being different from the closest packing of metal atoms in bulk metal catalysts, for example, face-centered cubic (fcc) for gold, body-centered cubic (bcc) for ferrum and hexagonal close-packed (hcp) for cobalt, the packing style of metal atoms in metal nanoclusters are much more diverse. Therefore, not all metal nanoclusters are active in catalysis. The catalytic activities of metal nanoclusters are highly dependent on the metal atom packing style. A recent report from Ruan and coworkers indicates that palladium nanoclusters with the same composition $Pd_6(SR)_{12}$ but different metal atom packing styles exhibit different catalytic activities in Sonogashira coupling reaction and 4-nitrophenol reduction reaction (Figure 1.7) [42]. The $Pd_6(SR)_{12}$ nanocluster with tiara-like structure (Pd6-Tia) is more active than the $Pd_6(SR)_{12}$ nanocluster with octahedral structure (Pd6-Oct) during catalysis. Pd6-Tia displayed 100% selectivity and 90% catalytic yield for the Sonogashira coupling between iodobenzene and phenylacetylene, while Pd6-Oct catalyzed the reaction under the same

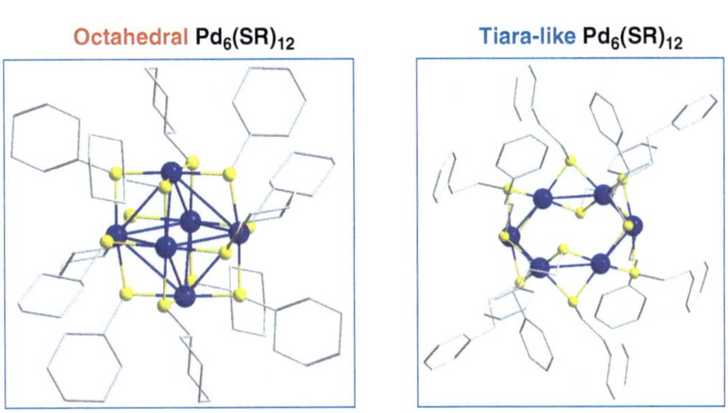


**Figure 1.7** $Pd_6$ nanoclusters with different metal atom packing styles.

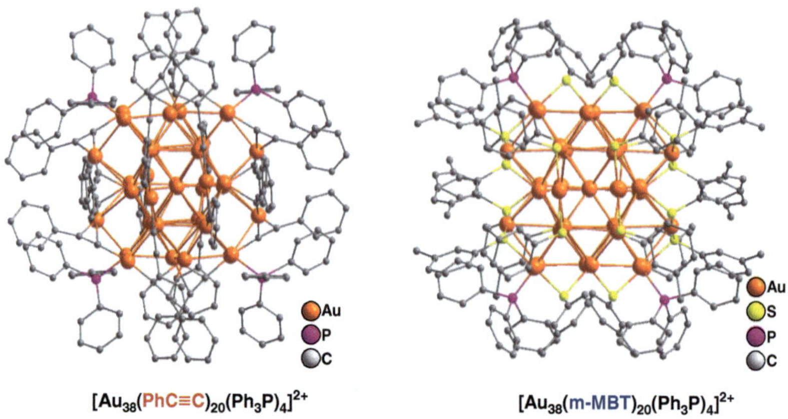


**Figure 1.8** $Au_{38}$ nanoclusters with different surrounding ligands.

conditions with only 24% yield. For the hydrogenation of 4-nitrophenol, the reaction rate constant with Pd6-Tia is significantly higher than that with Pd6-Oct.

The ligand surrounds the metal kernel is another key factor that impact the catalytic activity of metal nanoclusters. Wan and coworkers synthesized two isostructural $Au_{38}$ nanoclusters $[Au_{38}L_{20}(PPh_3)_4]^{2+}$, in which L is alkynyl and thiolate groups for the two nanoclusters, respectively (Figure 1.8). Both $Au_{38}$ nanoclusters possess an fcc gold kernel. However, they displayed completely different catalytic activities toward the semihydrogenation of alkynes [43]. $\left[Au_{38}(PhC{\equiv}C)_{20}(PPh_3)_4\right]^{2+}$ nanocluster with the alkyne ligand catalyzed the hydrogenation of terminal and internal alkynes, giving the corresponding alkenes with nearly 100% conversion and selectivity. In sharp contrast, $[Au_{38}(SR)_{20}(PPh_3)_4]^{2+}$ (SR = 3-methylbenzenethiol, m-MBT) nanocluster with the thiolate ligand was catalytically inert toward the terminal alkynes as well as the internal alkynes.

The potentially high and adjustable catalytic activities not only endow metal nanoclusters with interesting catalytic properties for fundamental research but also provide a kind of dominant metal catalyst that has promising catalytic performance in industrial applications. In the next chapters, we will discuss their diversified activities in various catalytic reactions.

## 1.4 Metal Nanocluster Catalysis: Selectivity

As we have mentioned in Section 1.2, metal nanoparticles present different catalytic selectivity compared with the single metal atom catalysts in some cases because of the metal–metal synergy on the surface of metal nanoparticles. Similarly, metal nanoclusters catalyzed reactions, giving different selectivity with single metal atoms as well. Pun and coworkers found an interesting phenomenon when they were studying the aerobic dehydrogenation of cyclohexanone by palladium catalysts (Figure 1.9) [44]. When $Pd(DMSO)_2(TFA)_2$ (DMSO = dimethylsulfoxide;

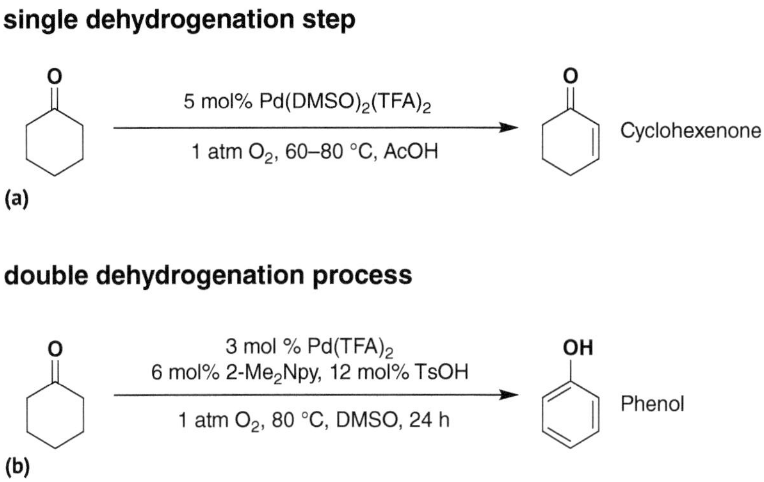


**Figure 1.9** Switched catalytic selectivity of Pd-catalyzed cyclohexanone dehydrogenation.

TFA = trifluoroacetate) was applied as the palladium catalyst, the reaction stopped at the first stage and gave the single dehydrogenation cyclohexenone as the final product. When $Pd(TFA)_2$/2-dimethylaminopyridine (2-$Me_2$Npy) was applied as the palladium catalyst, the reaction favored a second dehydrogenation and afforded the phenol as the final product. They envisioned that different catalytic species were involved with $Pd(DMSO)_2(TFA)_2$ and $Pd(TFA)_2$/2-$Me_2$Npy catalytic system, respectively. DLS, TEM, and kinetic studies all suggested that palladium nanoclusters were formed under the reaction conditions using $Pd(TFA)_2$/2-$Me_2$Npy. The as-obtained palladium nanoclusters are much active than the $Pd^{II}$ species in catalyzing the conversion from cyclohexanone to phenol, thus triggering the double dehydrogenation of cyclohexanone. Under the reaction conditions using $Pd(DMSO)_2(TFA)_2$, DMSO efficiently protects the $Pd^{II}$ species from aggregation [45] and holds the reaction at the single dehydrogenation stage. The distinct catalytic selectivity between the single palladium atoms and the palladium nanocluster should be ascribed to their different valence state, redox capability, and dehydrogenation pathways.

Similarly, Fricke and coworkers found an interesting experimental result [46], that is, aryl germanes are unreactive in single palladium atom $LnPd^0/LnPd^{II}$ catalysis and allow selective functionalization of established coupling partners in their presence. On the contrary, they display superior reactivity under palladium nanocluster conditions, outcompeting established coupling partners such as ArBPin and allowing air-tolerant, base-free, and orthogonal access to valuable and challenging diaryl motifs. The single palladium atoms and palladium nanoclusters conditions can be achieved, respectively, by using or not using phosphine ligands with palladium salts. Based on the different catalytic selectivity of single palladium atoms and palladium nanoclusters, they developed an orthogonal catalysis strategy with organogermanes (Figure 1.10).

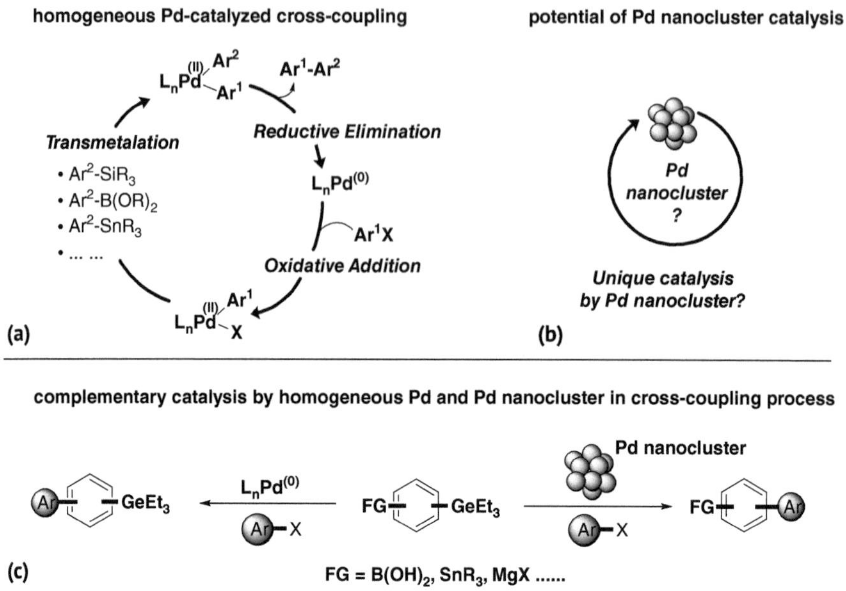


**Figure 1.10** Orthogonal catalysis strategy with organogermanes based on different catalytic selectivity of single palladium atoms and palladium nanoclusters.

On the one hand, the catalytic products present selectivity toward different metal species. On the other hand, the metal nanocluster catalysts are selective to the reactions. Lee and coworkers found that gold nanoclusters demonstrated selectivity toward the CO oxidation reaction [47]. They designed the experiment of room-temperature CO oxidation on planar model $Au_n/TiO_2$ catalysts prepared by deposition of size-selected $Au_n^+$ nanoclusters, in which $n = 1, 2, 3, 4, 7$. The catalytic activity of gold nanoclusters have been found to be strongly dependent on deposited nanocluster size, with substantial activity for $Au_n$ as small as three gold atoms.

No matter product selectivity or catalyst selectivity, metal nanocluster catalysis provide a route to control pathways of the reactions, thus furnishing valuable and desired products that are difficult to be achieved by other catalytic methods.

## 1.5 Metal Nanocluster Catalysis: Mechanism

Catalysis is a dynamic process, in which the single metal atoms, metal nanoclusters, and metal nanoparticles are all possibly involved. As the bridging point of the monoligated metal species (or metal salt) and the metal nanoparticle, metal nanocluster (especially with atomically precise structures) constitutes the key intermediate for clarifying the catalytic mechanism of metal catalysis. They can be used as model catalysts to reveal the correlation between catalytic performance and catalysts structures at the atomic scale, thus being a bridge between structure and properties as well. A review paper focusing on the model catalyst role of metal nanoclusters in catalysis is recommended, which illustrates the catalytic mechanisms in detail from theoretical and experimental perspectives [48].

We would like to take the copper-catalyzed azide-alkyne cycloaddition (CuAAC) reaction as a typical example to explain the importance of investigating the mechanism of metal catalysis by metal nanoclusters. CuAAC is one of the most famous click reactions developed by Karl Barry Sharpless, who has been awarded the 2022 Nobel Prize in Chemistry for the development of click chemistry and biorthogonal chemistry. It should be noted that this is the second time that Professor Karl Barry Sharpless was awarded the Nobel Prize in Chemistry. His first Nobel Prize came from the asymmetric catalysis in 2001. The well-recognized CuAAC reaction mechanism catalyzed by the homogeneous monoligated copper complex involves the activation of alkynes via carbon–hydrogen bond cleavage and the formation of a [R—C≡C—Cu] intermediate (Figure 1.11). The fact that internal alkynes are generally unreactive to azides in the presence of copper complexes further verifies this catalytic mechanism. On the other hand, it has been found that some heterogeneous copper catalysts were able to catalyze the AAC reaction with the internal alkyne as the substrate. As a result of the polydispersity and the imprecise structures of the heterogeneous copper catalysts, their catalytic mechanism is mysterious. Copper nanoclusters that fall into the critical point between copper atoms and copper nanoparticles provide an opportunity to understand the catalytic mechanism of CuAAC reactions under heterogeneous conditions. Fang and coworkers synthesized an $Au_4Cu_4$ nanocluster with the composition of $[Au_4Cu_4(dppm)_2(SAdm)_5]Br$ (dppm = bis(diphenylphosphino)methane; SAdm = adamantane-1-thiolate) [49]. This nanocluster has robust structure and its composition and structure can be well characterized by ESI-MS and single-crystal X-ray diffraction of (SCXRD). Zhu and coworkers found that $Au_4Cu_4$ nanocluster not only catalyzed the click reaction of azides and terminal alkynes but also smoothly catalyzed the click reaction between azides and internal alkynes. Based on the specific and repeatable ultraviolet and visible (UV–vis), ESI-MS, nuclear magnetic resonance (NMR), Fourier-transform infrared (FT-IR), and XPS spectra, they conducted the in-depth mechanism investigation on

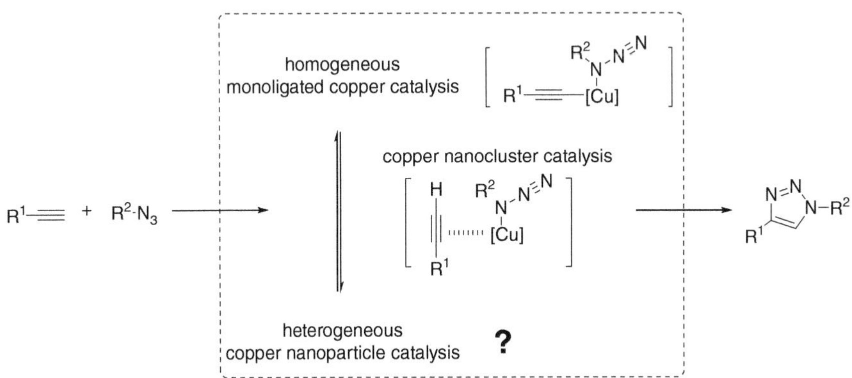


**Figure 1.11** Activation modes of CuAAC click reaction catalyzed by different types of metal species.

the $Au_4Cu_4$-catalyzed AAC reaction. A crucial intermediate [$Au_4Cu_4$-π-(R—C≡C—H] was observed by thin-layer chromatography (TLC) and further captured and structurally determined by ESI-MS. This intermediate demonstrates that $Au_4Cu_4$ activates alkynes by π-activation instead of deprotonation (Figure 1.11). This study reveals the alternative reaction pathway of CuAAC reaction under heterogeneous reaction conditions, with the copper metal nanocluster catalysis as a case study.

Another representative case is the metal-catalyzed hydrogenation reactions. Although there are many proposed mechanisms toward metal-catalyzed hydrogenations, the generally recognized key steps for homogeneous and heterogeneous catalytic conditions are the noncovalent ligand–substrate interaction and adsorption of substrates on the metal surface, respectively. For the monoligated metal species catalyzed hydrogenation, the molecular interactions between the unsaturated substrate, hydrogen gas, and the ligand are able to be well-defined by characterizations such as NMR, MS, and SCXRD because of the atomically precise structure of the catalyst [50–52]. For the heterogeneous catalysis, the adsorption of substrates and hydrogen gas, and the desorption of the hydrogenated product are generally considered to be the rate-determining steps (RDS). As a result of the relatively weak molecular interactions between surface ligand and the substrate, the control of selectivity under heterogeneous hydrogenation conditions is challenging. This problem is also partially attributed to the imprecise structure of the heterogeneous catalysts, which make it difficult to reveal the real processes during heterogeneous hydrogenation reactions [53, 54]. Metal nanoclusters with atomically precise structures would be ideal model catalysts for investigating the detailed catalytic processes of heterogeneous hydrogenation reactions. An early work from Zhu and coworkers revealed that the activation of C=O group at the icosahedral $Au_{13}$ core and the activation of hydrogen gas at the outer $Au_{12}$ surface are the key factors for the high efficiency of $Au_{25}(SR)_{18}$-catalyzed selective hydrogenation of α, β-unsaturated ketones and aldehydes [55]. This proposed catalytic model would be very helpful for the further structural design of catalytically active heterogeneous hydrogenation catalysts.

For the classic cross-coupling reaction, the catalytic mechanisms might be different as well by using monoligated metal complexes and metal nanoclusters as catalysts, respectively. For example, oxidative addition, transmetallation, and reductive elimination are the basic element reactions for the conventional $Pd^0$-catalyzed Suzuki reaction (Figure 1.12a). However, for palladium nanocluster that is composed of three palladium atoms, the catalytic process has been verified via a direct ligand exchange between the halogen atoms and the aryl group of the organoboronic acid (Figure 1.12b). The newly developed route of Suzuki reaction via palladium nanocluster catalysis is even more efficient than the typical monoligated palladium catalysis [56, 57]. Based on the previous reports, the heterogeneous palladium-catalyzed cross-coupling reactions possess significantly high turnover numbers under certain conditions. The results suggest that more efficient catalytic process might be involved during these heterogeneous metal-catalyzed cross-coupling reactions.

Apart from the distinct catalytic mechanisms involved in metal nanocluster catalysis, the methods for investigating the mechanisms of molecular homogeneous

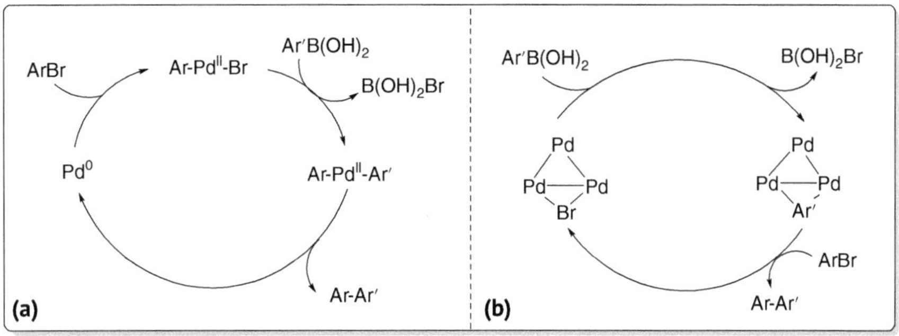


**Figure 1.12** (a) Conventional Pd$^0$-catalyzed Suzuki reaction. (b) Palladium nanocluster-catalyzed Suzuki reaction.

catalysis, heterogeneous catalysis and nanocluster catalysis are different as well. Detection and capture of intermediates, kinetic experiments, and simulation based on molecular structures are the powerful tools for revealing the catalytic mechanisms of molecular homogeneous catalysts. For heterogeneous catalysis, the first thing that should be confirmed is the real catalytic species because of the "cocktail"-type systems [58, 59] during heterogeneous catalysis as we mentioned in Section 1.2. After confirming the heterogeneous pathway, characterizations such as XPS and extended X-ray absorption fine structure (EXAFS) are usually employed for exploring the catalytic mechanism. Owing to the fact that metal nanoclusters combine the structural features of molecular homogeneous catalysts and heterogeneous catalysts, NMR, MS, SCXRD, XPS, EXAFS, etc. are all practical tools for revealing the catalytic processes of metal nanocluster catalysis. Nevertheless, clearly identifying the whole catalytic cycle of metal nanocluster catalysts is still a challenging task as a result of their dynamic transformations during catalysis compared to monoligated metal catalysts.

## References

**1** F. Meemken, A. Baiker, Recent Progress in Heterogeneous Asymmetric Hydrogenation of C=O and C=C Bonds on Supported Noble Metal Catalysts. *Chem. Rev.* **2017**, *117*, 11522–11569.

**2** A. Fürstner, P. W. Davies, Catalytic Carbophilic Activation: Catalysis by Platinum and Gold π Acids. *Angew. Chem. Int. Ed.* **2007**, *46*, 3410–3449; Katalytische carbophile Aktivierung: Platin- und Gold-p-Säuren als Katalysatoren. *Angew. Chem.* **2007**, *119*, 3478–3519.

**3** Ł. Banach, P. A. Guńka, J. Zachara, W. Buchowicz, Half-Sandwich Ni(II) Complexes [Ni(Cp)(X)(NHC)]: From an Underestimated Discovery to a New Chapter in Organonickel Chemistry. *Coord. Chem. Rev.* **2019**, *389*, 19–58.

4 J. Thongpaen, R. Manguin, O. Baslé, Chiral N-Heterocyclic Carbene Ligands Enable Asymmetric C—H Bond Functionalization. *Angew. Chem. Int. Ed.* **2020**, *59*, 2–12; *Angew. Chem.* **2020**, *132*, 10326–10335.

5 Y. Wang, H. Arandiyan, J. Scott, A. Bagheri, H. X. Dai, R. Amal, Recent Advances in Ordered Meso/Macroporous Metal Oxides for Heterogeneous Catalysis: A Review. *J. Mater. Chem. A.* **2017**, *5*, 8825–8846.

6 B. Su, J. F. Hartwig, Development of Chiral Ligands for the Transition-MetalCatalyzed Enantioselective Silylation and Borylation of C—H Bonds. *Angew. Chem. Int. Ed.* **2021**, *60*, 2–16; *Angew. Chem.* **2022**, *134*, e202113343.

7 Y. Dong, G. Goubert, M. N. Groves, J.-C. Lemay, B. Hammer, P. H. McBreen, Structure and Dynamics of Individual Diastereomeric Complexes on Platinum: Surface Studies Related to Heterogeneous Enantioselective Catalysis. *Acc. Chem. Res.* **2017**, *50*, 1163–1170.

8 M.-A. Légaré, G. Bélanger-Chabot, R. D. Dewhurst, E. Welz, I. Krummenacher, B. Engels, H. Braunschweig, Nitrogen Fixation and Reduction at Boron. *Science* **2018**, *359*, 896–900.

9 J. Twilton, M. Christensen, D. A. DiRocco, R. T. Ruck, I. W. Davies, D. W. C. MacMillan, Selective Hydrogen Atom Abstraction Through Induced Bond Polarization: Direct a-Arylation of Alcohols Through Photoredox, HAT, and Nickel Catalysis. *Angew. Chem. Int. Ed.* **2018**, *57*, 5369–5373; *Angew. Chem.* **2018**, *130*, 5467–5471.

10 A. Ruffoni, C. Hampton, M. Simonetti, D. Leonori, Photoexcited Nitroarenes for the Oxidative Cleavage of Alkenes. *Nature* **2022**, *610*, 81–86.

11 F. A. Cotton, Strong Homonuclear Metal-Metal Bonds. *Acc. Chem. Res.* **1969**, *2*, 240–247.

12 U. Schubert, Cluster-based Inorganic–Organic Hybrid Materials. *Chem. Soc. Rev.* **2011**, *40*, 575–582.

13 R. Jin, C. Zeng, M. Zhou, Y. Chen, Atomically Precise Colloidal Metal Nanoclusters and Nanoparticles: Fundamentals and Opportunities. *Chem. Rev.* **2016**, *116*, 10346–10413.

14 Q. Liu, X. Wang, X. Wang, Sub-1 nm Materials Chemistry: Challenges and Prospects. *J. Am. Chem. Soc.* **2024**, *146*, 26587–26602.

15 I. Chakraborty, T. Pradeep, Atomically Precise Clusters of Noble Metals: Emerging Link between Atoms and Nanoparticles. *Chem. Rev.* **2017**, *117*, 8208–8271.

16 D. J. Cole-Hamilton, Homogeneous Catalysis-New Approaches to Catalyst Separation, Recovery, and Recycling. *Science* **2003**, *299*, 1702–1706.

17 D. F. Chen, L. Z. Gong, Organo/Transition-Metal Combined Catalysis Rejuvenates Both in Asymmetric Synthesis. *J. Am. Chem. Soc.* **2022**, *144*, 2415–2437.

18 X. Huang, Q. Zhang, J. Lin, K. Harms, E. Meggers, Electricity-Driven Asymmetric Lewis Acid Catalysis. *Nat. Catal.* **2018**, *2*, 34–40.

19 D. A. Nicewicz, D. W. C. MacMillan, Merging Photoredox Catalysis with Organocatalysis: The Direct Asymmetric Alkylation of Aldehydes. *Science* **2008**, *322*, 77–80.

20 X. F. Wu, P. Anbarasan, H. Neumann, M. Beller, From Noble Metal to Nobel Prize: Palladium-Catalyzed Coupling Reactions as Key Methods in Organic Synthesis. *Angew. Chem. Int. Ed.* **2010**, *49*, 9047–9050; *Angew. Chem.* **2010**, *122*, 9231–9234.

21 J. Piera, J.-E. Bäckvall, Catalytic Oxidation of Organic Substrates by Molecular Oxygen and Hydrogen Peroxide by Multistep Electron Transfer—A Biomimetic Approach. *Angew. Chem. Int. Ed.* **2008**, *47*, 3506–3523; *Angew. Chem.* **2008**, *120*, 3558–3357.

22 B. Qiao, A. Wang, X. Yang, L. F. Allard, Z. Jiang, Y. Cui, J. Liu, J. Li, T. Zhang, Single-Atom Catalysis of Co Oxidation Using $Pt_1/FeO_x$. *Nat. Chem.* **2011**, *3*, 634–641.

23 L. N. Chen, K. P. Hou, Y. S. Liu, Z. Y. Qi, Q. Zheng, Y. H. Lu, J. Y. Chen, J. L. Chen, C. W. Pao, S. B. Wang, Y. B. Li, S. H. Xie, F. D. Liu, D. Prendergast, L. E. Klebanoff, V. Stavila, M. D. Allendorf, J. Guo, L. S. Zheng, J. Su, G. A. Somorjai, Efficient Hydrogen Production from Methanol Using a Single-Site $Pt_1/CeO_2$ Catalyst. *J. Am. Chem. Soc.* **2019**, *141*, 17995–17999.

24 Y. Zhang, S. Pang, Z. Wei, H. Jiao, X. Dai, H. Wang, F. Shi, Synthesis of a Molecularly Defined Single-Active Site Heterogeneous Catalyst for Selective Oxidation of N-Heterocycles. *Nat. Commun.* **2018**, *9*, 1465.

25 J. L. Sun, H. Jiang, P. H. Dixneuf, M. Zhang, Multicomponent Reductive Coupling for Selective Access to Functional γ-Lactams by a Single-Atom Cobalt Catalyst. *J. Am. Chem. Soc.* **2024**, *146*, 11289–11298.

26 P. Ryabchuk, G. Agostini, M.-M. Pohl, H. Lund, A. Agapova, H. Junge, K. Junge, M. Beller, Intermetallic Nickel Silicide Nanocatalyst—A Non-noble Metal–based General Hydrogenation Catalyst. *Sci. Adv.* **2018**, *4*, eaat0761.

27 A. Balanta, C. Godard, C. Claver, Pd Nanoparticles for C—C Coupling Reactions. *Chem. Soc. Rev.* **2011**, *40*, 4973–4985.

28 A. T. Bell, The Impact of Nanoscience on Heterogeneous Catalysis. *Science* **2003**, *299*, 1688–1691.

29 A. Bergmann, B. Roldan Cuenya, Operando Insights into Nanoparticle Transformations During Catalysis. *ACS Catal.* **2019**, *9*, 10020–10043.

30 C. Zhu, B. Yang, T. Jiang, J.-E. Bäckvall, Olefin-Directed Palladium-Catalyzed Regio- and Stereoselective Oxidative Arylation of Allenes. *Angew. Chem. Int. Ed.* **2015**, *54*, 9066–9069; *Angew. Chem.* **2015**, *127*, 9194–9197.

31 J. Piera, A. Persson, X. Caldentey, J.-E. Bäckvall, Water as Nucleophile in Palladium-Catalyzed Oxidative Carbohydroxylation of Allene-Substituted Conjugated Dienes. *J. Am. Chem. Soc.* **2007**, *129*, 14120–14121.

32 J. Liu, A. Ricke, B. Yang, J.-E. Bäckvall, Efficient Palladium-Catalyzed Aerobic Arylative Carbocyclization of Enallenynes. *Angew. Chem. Int. Ed.* **2018**, *57*, 16842–16846; *Angew. Chem.* **2018**, *130*, 17084–17088.

33 B. Yang, Y. Qiu, J.-E. Bäckvall, Control of Selectivity in Palladium(II)-Catalyzed Oxidative Transformations of Allenes. *Acc. Chem. Res.* **2018**, *51*, 1520–1531.

34 M.-B. Li, A. K. Inge, D. Posevins, K. P. J. Gustafson, Y. Qiu, J.-E. Bäckvall, Chemodivergent and Diastereoselective Synthesis of Gamma-Lactones and Gamma-Lactams: A Heterogeneous Palladium-Catalyzed Oxidative Tandem Process. *J. Am. Chem. Soc.* **2018**, *140*, 14604–14608.

35 A. M. Trzeciak, A. W. Augustyniak, The Role of Palladium Nanoparticles in Catalytic C—C Cross-Coupling Reactions. *Coord. Chem. Rev.* **2019**, *384*, 1–20.

36 A. H. M. de Vries, J. M. C. A. Mulders, J. H. M. Mommers, H. J. W. Henderickx, J. G. de Vries, Homeopathic Ligand-Free Palladium as a Catalyst in the Heck Reaction. A Comparison with a Palladacycle. *Org. Lett.* **2003**, *5*, 3285–3288.

**37** J. Huang, T. Akita, J. Faye, T. Fujitani, T. Takei, M. Haruta, Propene Epoxidation with Dioxygen Catalyzed by Gold Clusters. *Angew. Chem. Int. Ed.* **2009**, *48*, 7862–7866; *Angew. Chem.* **2009**, *121*, 8002–8006.

**38** M. Turner, V. B. Golovko, O. P. Vaughan, P. Abdulkin, A. Berenguer-Murcia, M. S. Tikhov, B. F. Johnson, R. M. Lambert, Selective Oxidation with Dioxygen by Gold Nanoparticle Catalysts Derived from 55-Atom Clusters. *Nature* **2008**, *454*, 981–983.

**39** M. T. Reetz, J. G. de Vries, Ligand-Free Heck Reactions Using Low Pd-Loading. *Chem. Commun.* **2004**, *14*, 1559–1563.

**40** A. Leyva-Perez, J. Oliver-Meseguer, P. Rubio-Marques, A. Corma, Water-Stabilized Three- and Four-Atom Palladium Clusters as Highly Active Catalytic Species in Ligand-Free C—C Cross-Coupling Reactions. *Angew. Chem. Int. Ed.* **2013**, *52*, 11554–11559; *Angew. Chem.* **2013**, *125*, 11768–11773.

**41** P. J. Ellis, I. J. Fairlamb, S. F. Hackett, K. Wilson, A. F. Lee, Evidence for the Surface-Catalyzed Suzuki-Miyaura Reaction over Palladium Nanoparticles: An Operando XAS Study. *Angew. Chem. Int. Ed.* **2010**, *49*, 1820–1824; *Angew. Chem.* **2010**, *122*, 1864–1868.

**42** C. Ruan, C. Meng, K. Wang, Y. Li, H. Xiang, H. Yan, W. W. Xu, M. Zhou, C. Yao, Octahedral Vs Tiara-Like $Pd_6(SR)_{12}$ Clusters. *Nano. Lett.* **2024**, *24*, 13054–13060.

**43** X.-K. Wan, J.-Q. Wang, Z.-A. Nan, Q.-M. Wang, Ligand Effects in Catalysis By Atomically Precise Gold Nanoclusters. *Sci. Adv.* **2017**, *3*, e1701823.

**44** D. Pun, T. Diao, S. S. Stahl, Aerobic Dehydrogenation of Cyclohexanone to Phenol Catalyzed by $Pd(TFA)_2$/2-Dimethylaminopyridine: Evidence for the Role of Pd Nanoparticles. *J. Am. Chem. Soc.* **2013**, *135*, 8213–8221.

**45** D. Wang, A. B. Weinstein, P. B. White, S. S. Stahl, Ligand-Promoted Palladium-Catalyzed Aerobic Oxidation Reactions. *Chem. Rev.* **2018**, *118*, 2636–2679.

**46** C. Fricke, G. J. Sherborne, I. Funes-Ardoiz, E. Senol, S. Guven, F. Schoenebeck, Orthogonal Nanoparticle Catalysis with Organogermanes. *Angew. Chem. Int. Ed.* **2019**, *58*, 17788–17795; *Angew. Chem.* **2019**, *131*, 17952–17959.

**47** S. Lee, C. Fan, T. Wu, S. L. Anderson, CO Oxidation on $Au_n/TiO_2$ Catalysts Produced by Size-Selected Cluster Deposition. *J. Am. Chem. Soc.* **2004**, *126*, 5682–5683.

**48** Y. Du, H. Sheng, D. Astruc, M. Zhu, Atomically Precise Noble Metal Nanoclusters as Efficient Catalysts: A Bridge between Structure and Properties. *Chem. Rev.* **2020**, *120*, 526–622.

**49** Y. Fang, K. Bao, P. Zhang, H. Sheng, Y. Yun, S. X. Hu, D. Astruc, M. Zhu, Insight into the Mechanism of the CuAAC Reaction by Capturing the Crucial $Au_4Cu_4$-$\pi$-Alkyne Intermediate. *J. Am. Chem. Soc.* **2021**, *143*, 1768–1772.

**50** J. H. Xie, S. F. Zhu, Q. L. Zhou, Transition Metal-Catalyzed Enantioselective Hydrogenation of Enamines and Imines. *Chem. Rev.* **2011**, *111*, 1713–1760.

**51** L. C. Misal Castro, H. Li, J.-B. Sortais, C. Darcel, When Iron Met Phosphines: A Happy Marriage for Reduction Catalysis. *Green Chem.* **2015**, *17*, 2283–2303.

**52** R. H. Morris, Exploiting Metal-Ligand Bifunctional Reactions in the Design of Iron Asymmetric Hydrogenation Catalysts. *Acc. Chem. Res.* **2015**, *48*, 1494–1502.

**53** F. Zaera, Regio-, Stereo-, and Enantioselectivity in Hydrocarbon Conversion on Metal Surfaces. *Acc. Chem. Res.* **2009**, *42*, 1152–1160.
**54** M. Heitbaum, F. Glorius, I. Escher, Asymmetric Heterogeneous Catalysis. *Angew. Chem. Int. Ed.* **2006**, *45*, 4732–4762; *Angew. Chem.* **2006**, *118*, 2794–2797.
**55** Y. Zhu, H. Qian, B. A. Drake, R. Jin, Atomically Precise $Au_{25}(SR)_{18}$ Nanoparticles as Catalysts for the Selective Hydrogenation of α,β-Unsaturated Ketones and Aldehydes. *Angew. Chem. Int. Ed.* **2010**, *49*, 1295–1298; *Angew. Chem.* **2010**, *122*, 1317–1320.
**56** F. Fu, J. Xiang, H. Chemg, L. Cheng, H. Chong, S. Wang, P. Li, S. Wei, M. Zhu, Y. Li, A Robust and Efficient $Pd_3$ Cluster Catalyst for the Suzuki Reaction and Its Odd Mechanism. *ACS Catal.* **2017**, *7*, 1860–1867.
**57** N. Jeddi, N. W. J. Scott, I. J. S. Fairlamb, Well-Defined $Pd_n$ Clusters for Cross-Coupling and Hydrogenation Catalysis: New Opportunities for Catalyst Design. *ACS Catal.* **2022**, *12*, 11615–11638.
**58** D. B. Eremin, V. P. Ananikov, Understanding Active Species in Catalytic Transformations: From Molecular Catalysis to Nanoparticles, Leaching, “Cocktails” of Catalysts and Dynamic Systems. *Coord. Chem. Rev.* **2017**, *346*, 2–19.
**59** R. H. Crabtree, Resolving Heterogeneity Problems and Impurity Artifacts in Operationally Homogeneous Transition Metal Catalysts. *Chem. Rev.* **2012**, *112*, 1536–1554.

# 2

# In Situ Generated Metal Nanocluster Catalysis

In situ generated metal nanoclusters under different catalytic conditions have been discovered and seriously taken several decades ago. The special and highly active metal catalysts form in situ either via the combination of homogeneous metal monomeric species, or by leaching from a heterogeneous metal nanoparticle. Although various metal species generally coexist under these complicated reaction conditions (the "cocktail"-type systems mentioned in Chapter 1), the "naked" metal nanoclusters that are generated in situ have been identified to play important roles in catalyzing the transformation of substrates. This key point has been confirmed by elegant experiments designed. The in situ generated metal nanoclusters are highly active in catalysis, while only kinetically stable and difficult to be isolated. The thermodynamically stable metal nanoclusters that are supported by carriers or stabilized by organic ligands will be discussed in detail in Chapters 3 and 4. In this chapter, we focus on the metal nanoclusters that are in situ formed under various reaction conditions of hydrogenation, oxidation, borylation, and cross-coupling reactions.

## 2.1 Catalytic Hydrogenation

Hydrogenation reaction is one of typical catalytic systems that involve the in situ generated metal nanocluster catalysts because the reducing condition favors aggregation of metal atoms. Widegren and Finke reviewed this kind of soluble transition metal nanoclusters in 2003 [1], which were observed frequently in transition metal-catalyzed arene hydrogenation reactions. Arene hydrogenation is a very important research area in catalysis. Sabatier, who was awarded the 1912 Nobel Prize in Chemistry, cooperated with Senderens a century ago and studied the hydrogenation of benzene using the finely divided nickel as the catalyst. This might be the earliest report on arene hydrogenation reaction catalyzed by metals. The research on hydrogenation of benzene to cyclohexane is still very active to this day because of the industrial application of the hydrogenated products. For example, cyclohexane, generated from the hydrogenation of benzene, has

*Metal Nanocluster Catalysis: From Fundamentals to Applications*, First Edition. Manzhou Zhu and Man-Bo Li. © 2026 Wiley-VCH GmbH. All rights reserved, including rights for text and data mining and training of artificial intelligence technologies or similar technologies. Published 2026 by Wiley-VCH GmbH.

been used in the production of adipic acid, which is a precursor to nylon. Arene hydrogenation reaction is generally performed with heterogeneous metal catalysts, such as the rhodium supported on $Al_2O_3$, Raney nickel, etc. A few single metal complexes are capable of catalyzing the reaction, and typically this kind of homogeneous metal catalyst has been considered to have poor catalytic activity toward the arene hydrogenation.

The in situ generated metal nanoclusters have been discovered under homogeneous as well as heterogeneous hydrogenation reaction conditions. They provide advantages over traditional metal nanoparticle heterogeneous catalysts on solid supports in the following aspects. First, they have been found to be more active than the corresponding supported metal nanoparticle catalysts. This is reasonable because the traditional heterogeneous catalysts are typically prepared at high temperature, causing annealing to the most stable but probably not the most active surface structures. In sharp contrast, metal nanoclusters are generated in situ under mild conditions, leading to a kinetically controlled surface structure that is highly active in catalysis. Second, the generated metal nanoclusters have been found to be more selective than the corresponding traditional heterogeneous metal catalysts for some hydrogenation reactions. Schmid and coworkers studied the hydrogenation of 2-hexyne and found that the metal nanoclusters favor the semi-hydrogenation of the alkyne while the selectivity of metal nanoparticles is much lower [2]. Finally, and the most importantly, the in situ generated metal nanoclusters are soluble. The solubility allows the use of simple but efficient characterization methods such as NMR, high-performance liquid chromatography (HPLC), SCXRD, and ESI-MS to conduct the kinetic studies, mechanism studies, etc. This is very important for the further optimization of the catalytic conditions.

Two disadvantages are also listed here before we systematically introduce the in situ generated metal nanocluster-catalyzed hydrogenation reactions. The first one is the short lifetime of the as-obtained metal nanoclusters. They are not thermodynamically stable and are readily aggregated or decomposed under the reaction conditions. Furthermore, the presence of sulfur in substrates would poison the metal nanoclusters and result in their deactivation. The second disadvantage is the separation problem of the in situ generated metal nanoclusters, which would limit their further industrial applications. Nevertheless, this problem is relatively surmountable by the use of multiphase reaction system or immobilization technology.

In the early days, there is still problem of distinguishing the in situ generated metal nanoclusters from the homogeneous single metal complex catalysts [3]. One of the typical examples is the arene hydrogenation catalyzed by $RhCl_3$ and $[(C_8H_{17})_3NCH_3]Cl$. Initially, it was believed that the true metal catalyst was the $[(C_8H_{17})_3NCH_3]^+[RhCl_4]^-$ ion pair. Later work using TEM identified the presence of the in situ generated Rh nanoclusters. Moreover, modern characterizations plus powerful solution-phase kinetic studies unambitiously indicated that the as-obtained Rh nanoclusters were the true catalysts. The problem of distinguishing metal nanocluster catalysts from the homogenous single atom catalysts is pervasive and not limited to the hydrogenation reactions [4–6]. However, this kind of misjudgment has now been greatly avoided with the development of advanced characterization techniques.

Widegren and Finke summarized typical reports of hydrogenation reactions catalyzed by the in situ generated metal nanoclusters. We added representative and recent examples in it and the corresponding results are listed in Table 2.1.

**Table 2.1** Summary of monocyclic arene hydrogenation with soluble nanocluster catalysts.

| | Authors (Publication date) | System | Results |
|---|---|---|---|
| 1 | K. R. Januszkiewicz, H. Alper (1983) | This is an important paper because much of the other work in the area is based directly upon it. The precatalyst is $[RhCl(1,5\text{-hexadiene})]_2$ and a tetraalkylammonium halide. Several substrates were tried, including benzene, *o*-xylene, *p*-methylanisole and phenol. The conditions are room temperature and 1 atm $H_2$; biphasic benzene/water or hexane/water is the reaction medium. The authors do not probe whether or not the catalyst is homogeneous or heterogeneous, but in a 1984 paper by the same authors they suggest that the actual catalyst is "a highly active form of colloidal rhodium," and reference "unpublished results." Use of variations of this catalyst system (see below) by later groups, who recognize its colloidal nature, makes it fairly certain that the catalyst described in this paper is indeed colloidal | Up to 100 TTO are demonstrated for some substrates; pH is shown to be an important variable. No activity is given (i.e. the authors do not say how long their hydrogenation experiments last) so the catalytic results cannot be compared to later work. Only the *cis* product is observed in the hydrogenation of 2-methylanisole (92% yield) |
| 2 | J. Foise, R. Kershaw, K. Dwight, A. Wold (1985) | In this study, the electrochemical reduction of benzene is combined with direct catalytic hydrogenation of benzene. Platinum and ruthenium colloids are used as the hydrogenation catalysts in solution and are prepared from the decomposition of Pt and Ru sulfite acid solutions. The conditions are room temperature and 1.7 atm $H_2$. The solvent is ethanol mixed with HMPA | Only the ruthenium colloids are active. However, the colloids are poorly characterized by today's standards and it is not possible to calculate an activity or lifetime for the hydrogenation |

*(Continued)*

**Table 2.1** (Continued)

| | Authors (Publication date) | System | Results |
|---|---|---|---|
| 3 | Z. Duan, M. J. Hampden Smith, A. P. Sylwester (1992) | $[(1,5\text{-COD})RhH]_4$ is synthesized from $[(1,5\text{-COD})RhCl]_2$ and used as a precursor for 2 nm sized colloids. NMR monitoring of the loss of $[(1,5\text{-COD})RhH]_4$ and the evolution of cyclooctane shows that hydrogenation activity is still seen after all of the $[(1,5\text{-COD})RhH]_4$ is gone, leading the authors to conclude that the nanoclusters are the (primary) active species. Cyclohexane-$d_{12}$ is sometimes used as a solvent. Toluene is the primary substrate, but benzene and pyridine are also used. The conditions are room temperature and 1 atm $H_2$ | A lifetime of at least 127 TTO is demonstrated, though this seems to be based on substrate/$[(1,5\text{-COD})RhH]_4$, not on substrate/Rh atom. The authors calculate an activity of $1.83 \times 10^{-3}$ molecules of toluene/(Rh atom·s), which is compared to a literature value of benzene hydrogenation with $Rh/Al_2O_3$ of $9.6 \times 10^{-4}$ molecules of benzene/(Rh atom·s). Assuming a crystallite size of 2 nm, the authors calculate an activity of $2.3 \times 10^{-3}$ molecules of toluene/(Rh surface atom·s). TEM shows the presence of agglomerated 2 nm sized nanoclusters |
| 4 | P. Drognat Landre´, M. Lemaire, D. Richard, P. Gallezot (1993) | The precatalyst is $RhCl_3 \cdot 3H_2O$ in the presence of Aliquat 336 (tricaprylylmethylammonium chloride) and/or trioctylamine. Dibenzo-18-crown-6 ether is the only substrate. The conditions are room temperature and 1–50 atm $H_2$; biphasic $CH_2Cl_2$/water is the solvent | The authors demonstrate the hydrogenation of about 20 mol of crown ether per mole of Rh in <1 h. TEM shows the presence of nanoclusters in the 2–3 nm size range. At higher pressure, the stereoselectivity increases to a 95/5 ratio of the *syn/anti* isomers of the dicyclohexyl-18-crown-6 ether |
| 5 | K. Nasar, F. Fache, M. Lemaire, J. C. Beziat, M. Besson, P. Gallezot (1994) | The precatalyst is $RhCl_3 \cdot 3H_2O$ in the presence of Aliquat 336 (tricaprylylmethylammonium chloride), trioctylamine or dioctylcyclohexylethylamine. The substrates are 2-methylanisole (for which the *cis/trans* hydrogenation selectivity is investigated) and *o*-cresol derivatives (for which the enantioselective hydrogenation is investigated). The conditions are room temperature and 1–50 atm $H_2$; biphasic $CH_2Cl_2$/water is the solvent | The authors demonstrate 40 TTO in 24 h for the hydrogenation of 2-methylanisole. About 10% hydrogenolysis of the methoxy group of 2-methylanisole is observed. No activity is observed in the absence of water, but this may be due to the low solubility of $RhCl_3 \cdot 3H_2O$ in the organic phase. About 5% ee is observed in the enantioselective reductions using a chiral amine as the nanocluster stabilizer. TEM shows the presence of nanoclusters in the 2–3 nm size range |

| | | | |
|---|---|---|---|
| 6 | P. Drognat Landré, D. Richard, M. Draye, P. Gallezot, M. Lemaire (1994) | The authors survey colloidal catalysts based on various metal chlorides ($RhCl_3$, $RuCl_3$, $NiCl_2$, $PdCl_2$, $IrCl_3$, $K_2PtCl_6$) with tertiary amines or Aliquat-336 (tricaprylylmethylammonium chloride) as the stabilizer. They also survey several supported heterogeneous catalysts (e.g. $Ru/Al_2O_3$, Rh/C, $Rh/SrTiO_3$). Dibenzo-18-crown-6 ether is the only substrate. The reaction conditions are varied from 25 to 60 °C and 1–50 atm $H_2$; the solvent is biphasic $CH_2Cl_2$/water | The Rh colloids are by far the most active, exhibiting the hydrogenation of 20 mol of crown ether per mol of Rh in 42 min at 5 MPa $H_2$. The authors observe hydrogenolysis of the crown ether, the main reaction pathway under certain conditions. An increase of the hydrogen pressure to 5 MPa leads to a *cis–syn–cis/cis–anti–cis* isomer ratio of 95/5. TEM shows the presence of colloids. Also, a modified "Maitlis test" (a reaction rate comparison before and after filtration through a graphite powder) points to nanoclusters as the true catalyst |
| 7 | F. Fache, S. Lehuede, M. Lemaire (1995) | The authors use $RuCl_3 \cdot 3H_2O$ (and $RhCl_3 \cdot 3H_2O$) in the presence of trioctylamine as the precursor to their colloidal catalyst. Several mono- and di-substituted benzene derivatives are hydrogenated (containing a variety of different functional groups), including 2-methylanisole and 2-methylbenzoate. The conditions are room temperature and 50 atm $H_2$; the solvent is methanol/water | The authors appear to demonstrate about 40 TTO in 1 h for the most easily hydrogenated substrates (the data are not completely clear). The authors say that no hydrogenolysis is observed. The authors show that the activity of $Ru/Al_2O_3$ is 5- or 10-fold lower (on a per-metal-atom basis) than the colloidal catalyst. *Cis/trans* selectivities up to 60 are observed in the hydrogenation of di-substituted arenes. No attempt is made to characterize the putative nanocluster catalyst. It is noted that the rate is slower without water |
| 8 | B. R. James, Y. Wang, T. Q. Hu (1996) | Much of this work is based on the catalyst system developed by Januszkiewicz and Alper (entry 1 in this table). $[RhCl(\eta^4\text{-}1,5\text{-hexadiene})]_2$, $[RhCl(\eta^4\text{-}1,5\text{-cyclooctadiene})]_2$, $RhCl_3 \cdot 3H_2O$, $Rh_6(CO)_{16}$ and $[Rh(OC_6H_5)(\eta^4\text{-}1,5\text{-cyclooctadiene})]_2$ are used as nanocluster precursors; tetrabutylammonium hydrogen sulfate is the nanocluster stabilizer. The substrates are 4-propylphenol, 2-methoxy-4-propylphenol and 2,6-dimethoxy-4-propylphenol. The reaction is performed in a biphasic (aqueous/organic) medium at 20 °C and 1 atm $H_2$ | The initial red or yellow solutions darken within 0.5 h to give the catalytically active solutions; over longer periods, bulk Rh metal precipitates with a resulting loss of activity. The formation of Rh metal, and a comparison to literature colloids (noting the necessity of the tetrabutylammonium hydrogen sulfate), are given as evidence that nanoclusters are the real catalyst. The TTO and TOF cannot be determined from the information given, but appear to be in the same range as later papers published by these researchers |

(Continued)

**Table 2.1** (Continued)

| | Authors (Publication date) | System | Results |
|---|---|---|---|
| 9 | T. Q. Hu, B. R. James, S. J. Rettig, C.-L. Lee (1997) | This catalyst system is based on that of Januszkiewicz and Alper (entry 1 in this table). $[RhCl(\eta^4\text{-1,5-hexadiene})]_2$ is the nanocluster precursor. A tetrabutylammonium salt is the nanocluster-stabilizing agent. The substrates are lignin model compounds: 4-propylphenol, 2-methoxy-4-propylphenol, and 2,6-dimethoxy-4-propylphenol. The reaction conditions include room temperature and 1–13.6 atm $H_2$; the solvent is biphasic hexane/water | The authors demonstrate about 50 TTO in about 50 h at 1 atm $H_2$. The reaction time decreases to about 6 h at 13.6 atm $H_2$. No nanocluster characterization is done. The authors simply say, probably correctly, that "the active Rh catalyst is probably present in a colloidal form: as the hydrogenation proceeds, metal particles accumulate and these are not active for subsequent hydrogenation reactions." The all-*cis* diastereomer is obtained selectively when the phenolic hydroxyl group is protected as a methyl ether or when a model compound possessing two methoxy substituents adjacent to the phenolic hydroxyl group is used |
| 10 | T. Q. Hu, B. R. James, C.-L. Lee (1997) | A water-soluble, polymer-stabilized, colloidal Rh catalyst is prepared by reducing $RhCl_3{\cdot}3H_2O$ with ethanol in the presence of polyvinylpyrrolidone and triethylamine. The substrates are benzyl acetone, 4-propylphenol, 2-methoxy-4-propylphenol, and 1,2-dimethoxy-4-propylbenzene. The reaction conditions include 25 °C and 1 atm $H_2$; monophasic, aqueous ethanol is the solvent | The authors demonstrate about 50 TTO for the hydrogenation of benzyl acetone in <43 h. About 20% hydrogenolysis is observed for compounds with methoxy groups. The authors attempt to use small, *neutral* organic molecules to stabilize their colloids, but with little success. The polymer-stabilized nanoclusters are prepared using a literature method, and no further characterization of the catalyst is done |
| 11 | T. Q. Hu, B. R. James, C.-L. Lee (1997) | This work is based on that of Januszkiewicz and Alper (entry 1 in this table). $[RhCl(\eta^4\text{-1,5-hexadiene})]_2$ is the primary catalyst precursor. A tetrabutylammonium salt is used as the nanocluster-stabilizing agent. The authors describe the hydrogenation of benzyl acetone, 4-propylphenol, eugenol, 1,2-dimethoxy-4-propylbenzene, and 2,6-dimethoxy-4-propylphenol under mild conditions (25 °C, 1 atm $H_2$) in a biphasic (water/hexane) medium at pH 7.5 | The authors demonstrate about 50 TTO in <24 h for the hydrogenation of benzyl acetone. Moderate to high diastereoselectivities are observed for the hydrogenated products; only the all-*cis* diastereomer is obtained for substrates where the phenolic hydroxyl group is protected as a methyl ether or where the substrate possesses two methoxy substituents adjacent to the phenolic hydroxyl group. $[RhCl(\eta^4\text{-1,5-cyclooctadiene})]_2$, $RhCl_3{\cdot}3H_2O$ and $RuCl_3{\cdot}3H_2O$ are found to be less active precatalysts |

| | | | |
|---|---|---|---|
| 12 | K. S. Weddle, J. D. Aiken III, R. G. Finke (1998) | This paper re-examines a putative "homogeneous" arene hydrogenation catalyst developed by others. Several experiments (TEM, reaction kinetics, Hg poisoning, H/D exchange) provide definitive evidence for Rh(0) nanoclusters being the true catalyst when using $RhCl_3{\cdot}3H_2O$ and trioctylamine and Aliquat 336 as the catalyst precursor. Benzene is the only substrate used in this paper. Reaction conditions include 30 °C and 0.9 atm $H_2$; the reactions are done in biphasic dichloroethane/water or in monophasic THF | A total of 40 TTO are demonstrated in an experiment where the catalyst is re-used once (i.e. after doing 20 TTO in about 2 h, more benzene is added and another 20 TTO are demonstrated in an additional 4 h). Additional notes from the previous paper: "In general, hydrogenation of benzenes did not give partially hydrogenated products. Cyclohexene derivatives were isolated when sterically hindered compounds were reduced." The authors observed water effect that a minimum of 2 equiv. of water (in addition to the waters of hydration on the $RhCl_3{\cdot}3H_2O$) are necessary for activity; H/D exchange between water and substrate occurs under catalytic conditions |
| 13 | B. R. James, Y. Wang, C. S. Alexander, T. Q. Hu (1998) | The nanoclusters are synthesized from RhCl3·3H2O, $[RhCl(diene)]_2$ or $RuCl_3/(n\text{-}C_8H_{17})_3N$. A Ziegler-Natta catalyst and a Bennett-type Ru catalyst are also tried. The nanocluster stabilizers are tetraalkylammonium salts, $R_4NX$ (R = alkyl, X = Cl, Br or $HSO_4$). The substrates are 4-propylphenols (used as lignin model compounds) and "milled wood lignin." Reaction conditions include 1–50 atm $H_2$ and 20–100 °C; biphasic (aqueous/organic) solvent systems are used | This paper discusses the effects of water in detail. D-labeling studies show that some of the H in the product comes from water. Hydrogenolysis of the methoxy group is observed. The authors demonstrate up to 300 TTO in 24 h for the hydrogenation of 2-methoxy-4-propylphenol with $RuCl_3/(n\text{-}C_8H_{17})_3N$ as the catalyst. Some *cis/trans* selectivity is observed |
| 14 | J. Schulz, A. Roucoux, H. Patin (1999) | What separates this system from several of the others is the use of *hydroxy*alkylammonium bromide salts (which are highly water soluble) as the nanocluster stabilizer. The nanoclusters are synthesized by reducing $RhCl_3{\cdot}3H_2O$ with $NaBH_4$ in dilute aqueous solutions of the hydroxyalkylammonium salts. The organic phase apparently consists of just the substrate and products. Various mono- and di-substituted arenes are hydrogenated including benzene, toluene, anisole and xylene. The conditions include 20 °C and 1 atm $H_2$; the reaction solutions are biphasic (aqueous/organic) | The nanoclusters have an average diameter of 3.6 nm by TEM. The authors demonstrate 1000 TTO for anisole hydrogenation in 47 h. Some *cis/trans* selectivity is observed for the hydrogenation of multiply substituted arenes. No hydrogenolysis products are seen |

*(Continued)*

**Table 2.1** (Continued)

| | Authors (Publication date) | System | Results |
|---|---|---|---|
| 15 | R.-W. Albach, M. Jautelat (1999) | In this patent, nanoclusters are synthesized from a number of different precursors [$RuCl_3$, $RhCl_3$, $PdCl_2$, $NiBr_2$, $Pd(OAc)_2$]. The sulfobetaine 3-(dodecyldimethylammonium) propanesulfonate was used as a nanocluster stabilizer. Substrates include isopropylbenzene and benzene. The reaction conditions include 50–180 °C and 10–400 atm $H_2$; the reaction solutions are biphasic (aqueous/organic) | The authors appear to demonstrate 200 TTO for the hydrogenation of benzene and isopropyl benzene |
| 16 | J. Schulz, A. Roucoux, H. Patin (2000) | This is the follow-up full paper to these authors' 1999 Chem. Commun. paper (see entry 14 in this table). The nanoclusters are synthesized by reducing $RhCl_3 \cdot 3H_2O$ with $NaBH_4$ in dilute aqueous solutions of the hydroxyalkylammonium salts. Various mono- and di-substituted arenes are hydrogenated including benzene, toluene, anisole and xylene. The conditions include 20 °C and 1 atm $H_2$; the reaction solutions are biphasic (aqueous/organic) | The nanoclusters are 2–2.5 nm in diameter by TEM. The metal particle nature of the catalyst is confirmed with a Hg(0) poisoning experiment. No hydrogenolysis products are formed. The activity of the nanocluster catalyst for anisole hydrogenation is compared with the activity of Hirai's polymer-stabilized Rh nanoclusters and with 5% Rh/C, all under identical conditions. The authors observe a TOF of 20 $h^{-1}$ for Rh/C, of 47 $h^{-1}$ for the PVP stabilized nanoclusters, and of 60 $h^{-1}$ for the nanoclusters. The Rh nanoclusters give 2000 TTO for anisole hydrogenation in 37 h. The maximum TOF for the same experiment is 188 mol $H_2$/(mol Rh·h). *Cis/trans* selectivities up to 99:1 are observed for the hydrogenation of di-substituted benzenes |

| | | | |
|---|---|---|---|
| 17 | R. J. Bonilla, P. G. Jessop, B. R. James (2000) | $[RhCl(1,5\text{-}COD)]_2$ is the nanocluster precursor and tetrabutylammonium hydrogen sulfate is the surfactant and nanocluster stabilizer. Several arenes are used as substrates, including lignin model compounds, anisole, phenol and p-xylene. The reaction conditions include 36 °C and 10 atm $H_2$. The reactions are performed in a biphasic aqueous/supercritical ethane reaction medium | The authors demonstrate as many as 110 TTO in 62 h for the hydrogenation of benzyl alcohol. Some *cis/trans* selectivity is observed for the hydrogenation of multiply substituted arenes. The nanoclusters are not characterized |
| 18 | J. A. Widegren, R. G. Finke | The nanocluster catalyst is formed in situ by reducing $[Bu_4N]_5Na_3[(1,5\text{-}COD)Rh{\cdot}P_2W_{15}Nb_3O_{62}]$ with $H_2$ in a monophasic propylene carbonate solution, which gives polyoxoanion- and tetrabutylammonium-stabilized Rh(0) nanoclusters. Anisole is the only substrate. The reaction conditions include 22 °C and 3.7 atm $H_2$; propylene carbonate is the solvent | With 10 equiv. (vs. Rh) of $HBF_4{\cdot}Et_2O$ added, the authors demonstrate 2600 TTO for anisole hydrogenation in 144 h. The Rh(0) nanoclusters are shown by TEM to have an average diameter of between ~4 and 6 nm, depending on the nanocluster formation conditions. A combination of hydrogenation kinetics and the kinetics and stoichiometry of nanocluster formation provide excellent evidence that the nanoclusters are the true catalyst. These nanoclusters also display an unprecedented selectivity for the partial hydrogenation of anisole to 1-methoxycyclohexene, with initial selectivities of ~30% and overall yields up to 8% |

Based on the brief introduction in the table, the following conclusions can be made: (i) Rh- and Ru-based nanoclusters that are generated in situ have been frequently observed in catalytic hydrogenation reactions; (ii) two in situ formed metal nanocluster catalyst systems developed by Roucoux research group and Finke research group stand out from the rest in terms of catalytic activity and lifetime; (iii) selective hydrogenation of arenes for the synthesis of all-*cis* diastereomer of substituted cyclohexanes is a promising area for the metal nanocluster catalysis compared to the homogeneous single metal complex catalysis and heterogeneous metal nanoparticle catalysis.

A recent work by Kim and coworkers gives us a unique perspective concerning the formation process of metal nanoclusters during hydrogenation reactions and their catalytic features [7]. They conducted in-depth investigation on the homogeneous rhodium complex-catalyzed hydrogenation of N-heteroarenes (Figure 2.1). They designed a single Rh complex catalyst $[(\eta^5\text{-}C_5Me_5)Rh(N{-}C)H]$(N—C= 2-phenylpyridinyl, abbreviated as ppy). This homogeneous Rh catalyst was used to

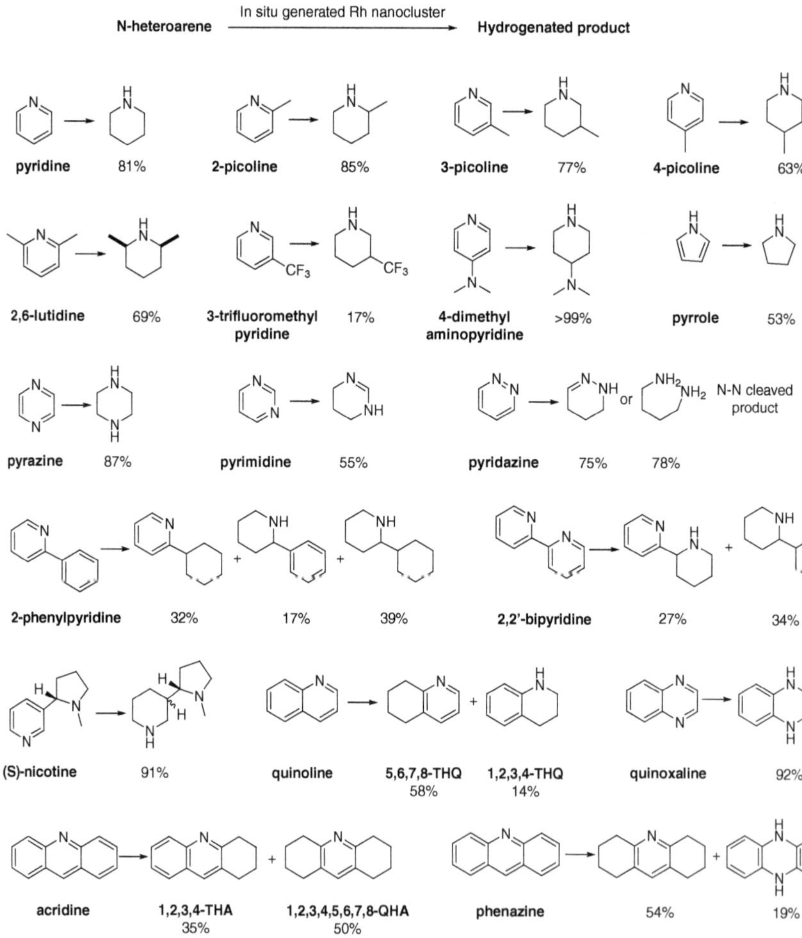


**Figure 2.1** In situ generated Rh nanocluster-catalyzed arene hydrogenation.

catalyze the hydrogenation reaction of N-heteroarenes, which is a challenging task by using other catalysts. This catalyst showed unexpectedly high activity toward hydrogenation of unsubstituted N-heteroarenes such as pyridine, pyrrole, pyrazine, 3-picoline, etc., many of which has not been previously hydrogenated. Unusual chemoselectivity was also observed in the hydrogenation reactions of polyaromatic N-heteroarenes. For example, quinoline and acridine are partially hydrogenated by other known catalysts, giving 1,2,3,4-tetrahydroquinoline and 1,10-dihydroacridine, respectively. However, Kim and coworkers found that 5,6,7,8-tetrahydroquinoline and 1,2,3,4,5,6,7,8-octahydroacridine were obtained as the major products in moderate yields, respectively. The unique catalytic activity and selectivity stimulate the authors to reveal the catalytic process of the homogeneous Rh-catalyzed hydrogenation reactions. With utilization of a combined characterization technologies such as NMR spectroscopy, X-ray diffraction, neutron diffraction, etc., and detailed kinetic studies, they found that the real catalysts should be multi-metallic Rh nanoclusters generated in situ and depicted the whole evolution of the metal species during the hydrogenation reaction. Notably, the catalytically active Rh nanoclusters is also structurally sensitive, and favorable to undergo aggregation to give the catalytically inert $[(\eta^5\text{-}C_5Me_5)_4Rh_5H_7]$ nanocluster and finally Rh nanoparticles, which are catalytically active but less reactive than the most active multimetallic Rh nanoclusters.

## 2.2 Catalytic Oxidation

Palladium-catalyzed oxidation constitutes an efficient way to functional molecules with high atom and step economy [8–12]. As shown in Figure 2.2 of the catalytic cycle, reduction of pd(II) to Pd(0) accompanies with the oxidation of the substrate. In many cases, the active Pd(0) species has aggregated to palladium black before reoxidation of Pd(0) by oxidants [13]. The as-generated palladium black is generally catalytically inert, thus resulting in the palladium deactivation problem in Pd-catalyzed oxidation reactions. On the other hand, the appropriate aggregation of the homogeneous single Pd complex would lead to the formation

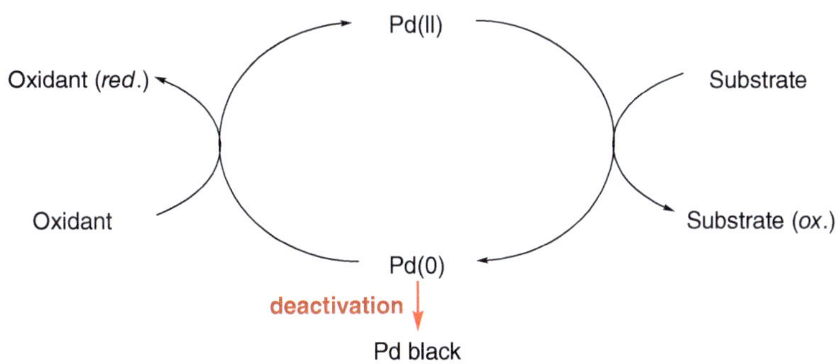


**Figure 2.2** Palladium deactivation in homogeneous Pd-catalyzed oxidation reactions.

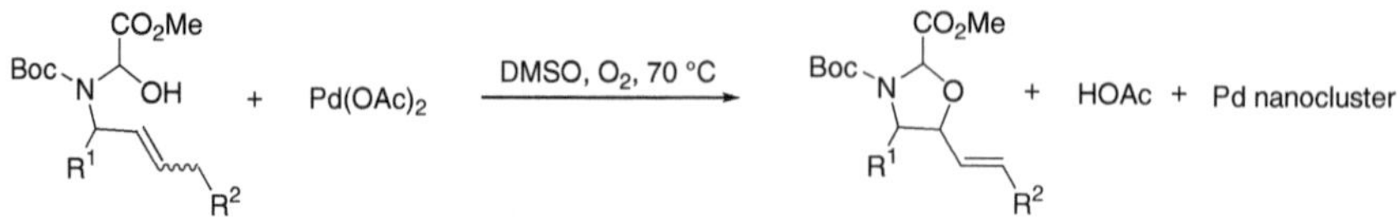


**Figure 2.3** Generation of Pd nanocluster during the Wacker like reaction.

of ultrasmall palladium nanoclusters, which is even more active than the precatalyst. In this section, we are discussing in situ generated metal nanoclusters during oxidation reactions with homogeneous Pd-catalyzed oxidation as the typical example.

As early as 30 years ago, van Benthem and coworkers have found that small (2.2–3.0 nm) palladium nanoclusters formed in situ in the Wacker like reaction when they used $Pd(OAc)_2$ as the catalyst and dimethylsulfoxide (DMSO) as the solvent at 70 °C under oxygen atmosphere (Figure 2.3) [14]. Bright- and dark-field TEM images of the reaction system clearly showed that palladium nanoclusters were formed with the reaction process. The researchers even isolated the as-obtained nanoclusters after two turnovers as a black powder, and then tested its catalytic activity. Results showed that the palladium nanoclusters were still catalytically active after isolation. Elemental analysis indicated that DMSO molecules were adsorbed on the surface of the palladium nanoclusters and the ratio of the number of DMSO molecules per palladium atom was identified to be ranging from 2.5 to 3.0/10. The stabilization of DMSO toward the palladium kernel might be the reason that the as-isolated palladium nanoclusters can retain their catalytic activity. IR spectrum showing strong adsorption bands at 1023 and 1127 $cm^{-1}$ (S=O) further indicated that DMSO was coordinated through both O and S atoms.

In 2010, Mifsud and coworkers investigated the oxidation reaction from secondary alcohols to ketones [15]. Different kinds of bidentate nitrogen ligands were used with $Pd(OAc)_2$ in the presence of water as the solvent. Detailed analysis by TEM indicated that the in situ generated palladium nanoclusters were the real catalytic species. Notably, they found that the steric and electronic effect of ligands highly affect the formation of palladium nanoclusters during the reaction. For example, quantity of palladium nanocluster was observed when neocuproine was used as the ligand, while the reaction kept homogeneous and none of nanoclusters were observed when bathophenanthroline disulfonate was employed as the ligand. With the combination of single $Pd^{II}$ complex and neocuproine, they realized a series of highly efficient aerobic oxidation of secondary alcohols, which are actually catalyzed by the in situ generated palladium nanoclusters.

Pun and coworkers described structural evolution and catalytic performance of the in situ generated palladium nanoclusters in detail [16]. They studied the aerobic oxidation of cyclohexanone to cyclohexenone and finally to phenol by $Pd(TFA)_2$/ligand catalytic system. This is a typical example and can be shared herein for understanding the in situ generated metal nanocluster catalyzed oxidation reactions. Initially, they observed that different kinds of $Pd(TFA)_2$/ ligand combinations lead to different products. Cyclohexenone was isolated as the major product by using $Pd(TFA)_2$/DMSO catalytic system (Figure 2.4), while phenol was detected as the major product in the presence of $Pd(TFA)_2$ and 2-dimethylaminopyridine (2-$Me_2$Npy). They investigated the mechanism of $Pd(TFA)_2$/DMSO catalyzed oxidative dehydrogenation of cyclohexanone, and the corresponding results support a catalytic cycle shown in Figure 2.5. It is a homogeneous Pd catalytic cycle. A turnover-limiting α C—H bond activation and subsequential β-H elimination are verified as key steps for the formation of cyclohexenone. DMSO has minimal influence on the rate of the Pd-catalyzed oxidative dehydrogenation of cyclohexanone. However, it plays very important roles in preventing the homogeneous palladium catalyst from decomposition and generation of palladium black. On the other hand, DMSO inhibits the conversion of cyclohexenone to phenol efficiently, thereby providing the basis for the highly selective formation of cyclohexenone.

In contrast to the homogeneous process of $Pd(TFA)_2$/DMSO catalysis, the oxidative dehydrogenation reaction of cyclohexanone catalyzed by $Pd(TFA)_2$/2-$Me_2$Npy is much more complicated. The kinetic studies show that cyclohexenone is the

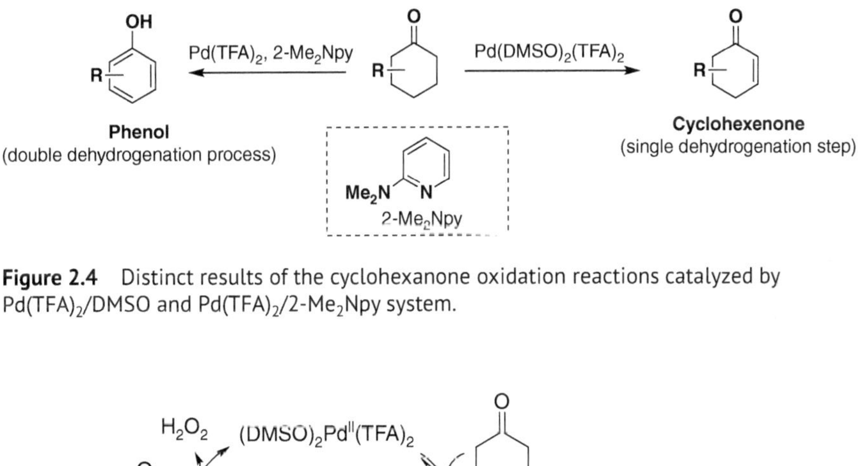


**Figure 2.4** Distinct results of the cyclohexanone oxidation reactions catalyzed by $Pd(TFA)_2$/DMSO and $Pd(TFA)_2$/2-$Me_2$Npy system.

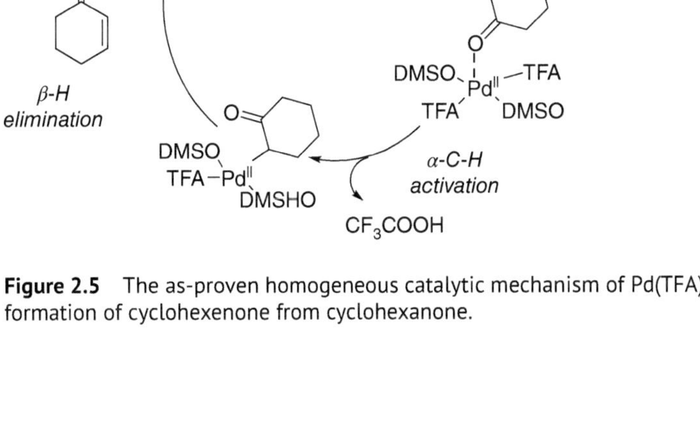


**Figure 2.5** The as-proven homogeneous catalytic mechanism of $Pd(TFA)_2$/DMSO catalyzed formation of cyclohexenone from cyclohexanone.

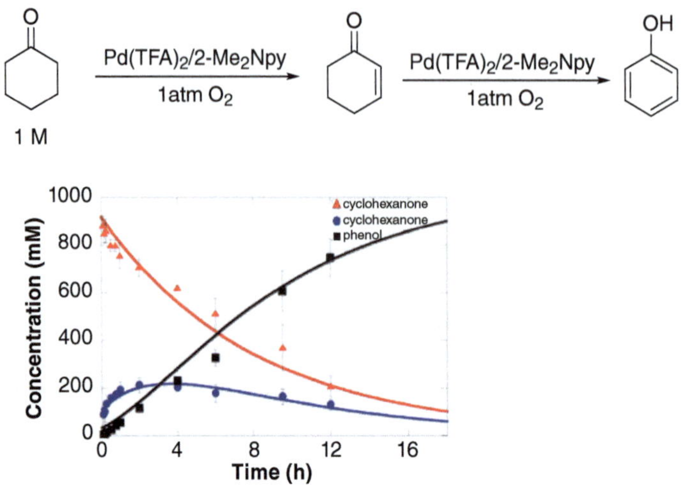


**Figure 2.6** Kinetic studies of the transformation from cyclohexanone to phenol catalyzed by $Pd(TFA)_2$/2-$Me_2$Npy system. *Source:* [16]/with permission of American Chemical Society.

intermediate during the transformation from cyclohexanone to phenol (Figure 2.6). Meanwhile, hot filtration demonstrates that only the soluble component of the reaction system is catalytically active. Further, Hg poisoning tests indicate that homogeneous $Pd^{II}$ is not the real catalyst, suggesting the existence of soluble palladium nanoclusters during the catalytic oxidations. Based on these experimental results, Pun et al. conducted characterizations such as TEM and DLS (Figure 2.7), which unambitiously indicate that palladium nanoclusters are formed in situ and act as the real catalyst for the further transformation from cyclohexenone to phenol. To further confirm this conclusion, they prepared neocuproine-stabilized palladium nanoparticles. These particles were tested as catalysts in the oxidative dehydrogenation of cyclohexanone and led to a 64% yield of phenol finally. This elegant work shows the detailed formation and catalysis of in situ generated metal nanoclusters in oxidation reactions, and suggests a broad application of metal nanoclusters in aerobic dehydrogenation reactions.

It should be pointed out that the formation and propagation of palladium nanoclusters during homogeneous Pd-catalyzed oxidation reactions is not an isolated case but a common phenomenon. Baumann and coworkers investigated quantity of palladium catalytic systems of C—H oxidative functionalization [17]. Their results indicate that palladium nanoclusters are formed in most of C—H oxidative functionalization reactions within seconds or one to two hours, depending on the reaction conditions. This experimental phenomenon is especially common when polar solvents such as N,N-Dimethylformamide (DMF), DMSO and acetic acid were used, probably due to the fact that the polar solvents are capable of stabilizing the metal kernel efficiently. However, it cannot simply be concluded that the in situ generated palladium nanoclusters are moribund forms of the active catalyst phase. For some

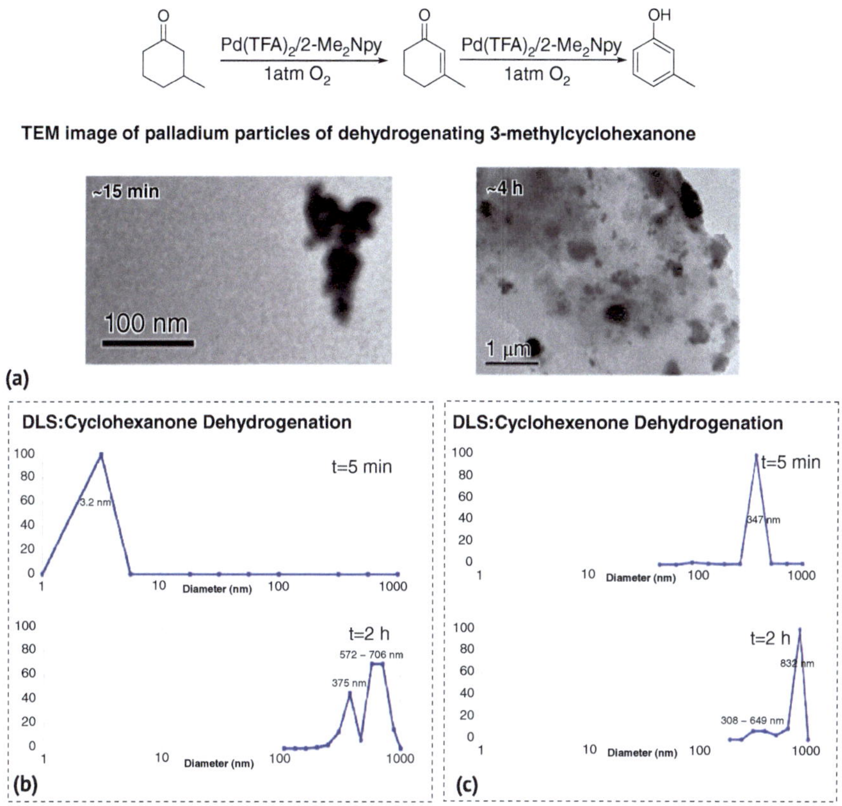


**Figure 2.7** TEM and DLS analysis of Pd(TFA)$_2$/2-Me$_2$Npy catalyzed oxidation of cyclohexanone. *Source:* Reproduced with permission from [16]/American Chemical Society.

of the oxidation reactions they examined in their work, the as-obtained palladium nanoclusters are catalytically active, while the catalyst activity is commensurate with the homogeneous Pd$^{II}$ complex (e.g. Pd(OAc)$_2$) in several cases. Nevertheless, the generation of metal nanoclusters and their potential catalytic activity should be considered and carefully examined when we are studying metal-catalyzed oxidation reactions. Meanwhile, designing of structurally stable and catalytically durable metal nanoclusters for highly efficient and unique selective oxidations would be an interesting research field for chemists and material scientists.

## 2.3 Catalytic Borylation

Boron compounds are very useful chemicals that are widely applied in medicinal chemistry [18–21]. Meanwhile, they are easily functionalized, thus providing a versatile platform for diversified chemical derivatizations. Hydroboration of unsaturated hydrocarbons constitute an efficient route to boron-functional compounds [22–25]. Considerable efforts have been made in this research area based on organic ligand-functionalized homogeneous or heterogeneous transition metal catalysis, which enables hydroboration via M-B intermediates. Although these mature and effective methods have been used for industrials, there is still a large room for the improvement of the activity and selectivity.

Very recently, some elegant work on the highly efficient and selective hydroboration of unsaturated bonds have been reported [26–30]. The reaction conditions are quite mild and simple. Only trace of metal salt was added and none of organic ligand was needed. These kinds of concise system show even much higher efficiency than the sophisticatedly designed catalytic system involving metals, ligands, and additives. Further investigation on some catalytic processes have verified the participation of in situ generated metal nanoclusters, which are highly active and easily deactivated after their isolation from the reaction mixtures. In the following, we will discuss a typical example on that.

Li and coworkers published a work entitled "Selective hydroboration of unsaturated bonds by an easily accessible heterotopic cobalt catalyst" in 2021 [31]. In this work, they developed a concise but efficient catalytic system for the hydroboration reactions, in which the sophisticated ligands or hazardous activators were not involved at all. Their reaction conditions and typical substrate scope are shown in Figure 2.8. Specifically, the combination of catalytic amounts of cobalt iodine ($CoI_2$) and potassium *t*-butoxide ($KO^tBu$) is enough to trigger the reaction under tetrahydrofuran (THF) at 30 °C. In most cases, only 0.5 mol% of $CoI_2$ is essential to deliver the hydroborated products in >80% yields. Moreover, excellent Markovnikov regioselectivity (>40/1) that are rarely approached with base-metal catalysis was observed for the reactions, highlighting the efficiency and practicality of this method. Under the developed reaction conditions, not only the alkenes but also nitriles are available substrates, undergoing hydroboration reactions smoothly and affording the corresponding secondary boronic esters and dihydroboration products, respectively. The dihydroboration products are readily to be derivatized without purification, giving ammonium salts with high efficiency. Most of the one-pot two-step reactions finished in 0.5 hours and the overall yields reach to 98%.

Kinetic studies were conducted by the authors to under the highly efficient hydroboration reaction. The results showed that no obvious induction period was involved in the reaction, and the yield of the hydroborated product reached 61% in five minutes! However, the reaction rate was slowed down obviously after that point, demonstrating the high activity and easy deactivation features of the catalytically active species. Addition of more equivalents of the starting materials to the reaction system at 5.0 minutes resulted in no obvious increasement of the yield of the hydroborated product, further confirming the conclusion that most of the in situ generated active catalysts deactivated in the first few minutes. They further conducted kinetic poisoning studies by employing three kinds of reagents. First, Hg had almost no effect toward the reaction rate, indicating that the real

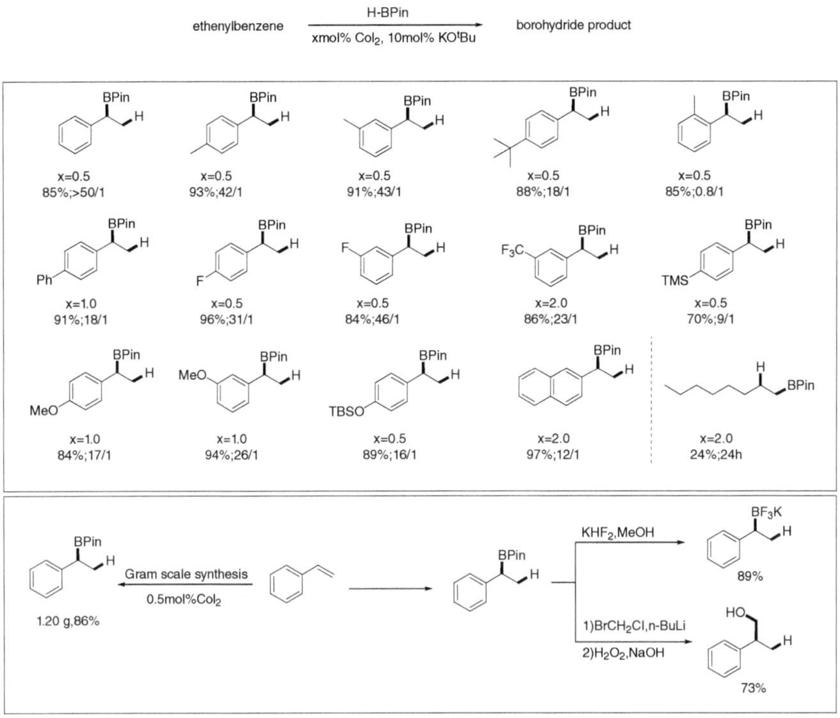


**Figure 2.8** $CoI_2$ and $KO^tBu$ catalyzed hydroboration of unsaturated bonds.

catalyst should be some soluble species. Second, addition of substoichiometric trimethylphosphite ($P(OMe)_3$) to the reaction mixture under standard conditions at 1.0 minute led to significant inhibition of the catalytic turnover, further verifying the soluble feature of the catalytically active species. Third, the selective homogeneous catalyst poison dibenzo[a,e]cyclo-octatetraene (dct) showed no effect for the formation of the hydroborated product, suggesting a heterotopic mechanism that involves initial reduction of $CoI_2$ and rapid aggregation of the resulting low-valent Co species to Co nanoclusters that are act as the real catalysts. Apart from the kinetic studies, more evidence was acquired concerning the in situ generated and catalytically active Co nanoclusters, including the observation of a $[Co_3\text{-HBpin-styrene-H}]^+$ peak in ESI-MS, cobalt nanoparticles with small sizes (~2.5 nm) in TEM and the rapid aggregation of the catalytically active species into catalytically inert larger particles when the reaction system was exposed to air. Although the exact mechanism of Co nanocluster-catalyzed hydroboration reaction is still unclear, this work provides an opportunity to manifest the high activity (and further practical application) of the in situ generated metal nanoclusters under borylations and prefigures the intriguing metal–metal interactions of metal nanoclusters in catalysis (Figure 2.9).

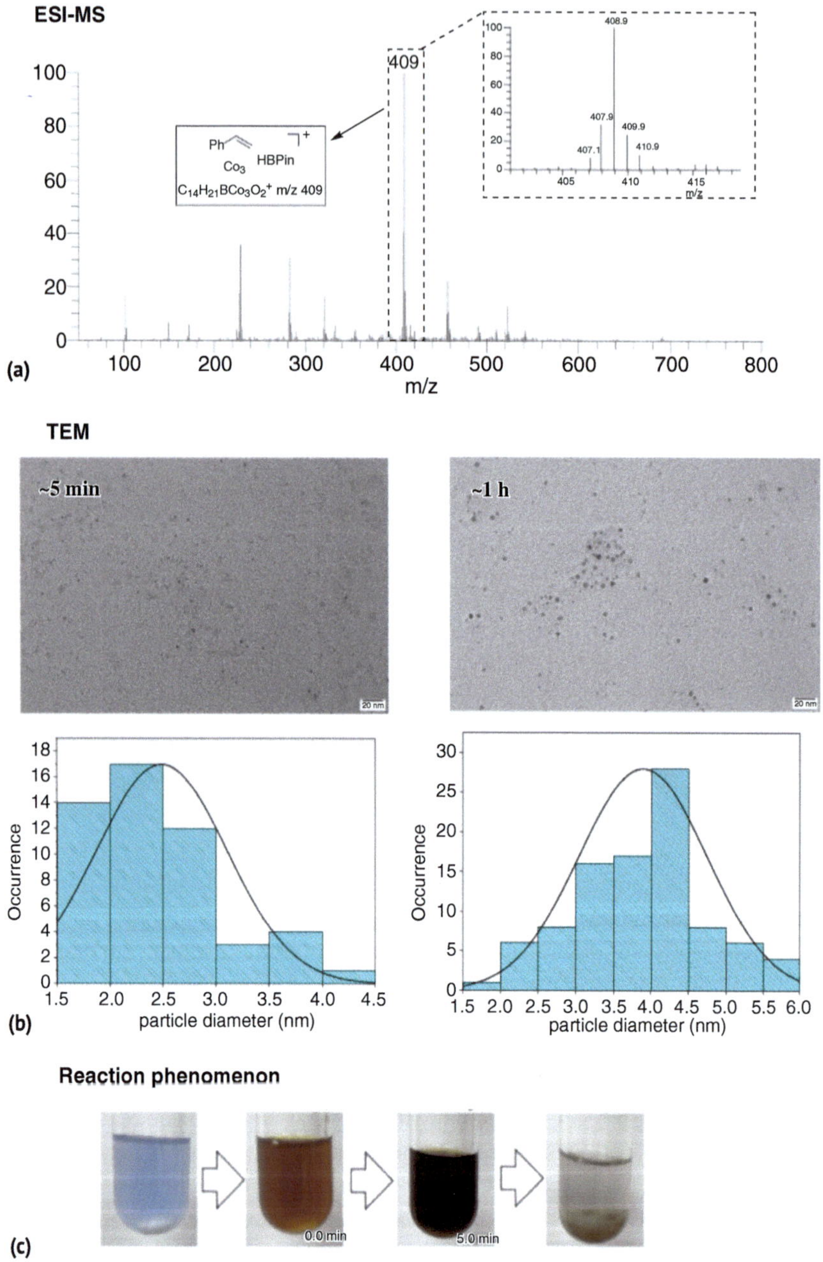


**Figure 2.9** Evidence of the in situ generated Co nanoclusters during hydroboration reactions. *Source:* [31]/Springer Nature/CC BY 4.0.

## 2.4 Catalytic Cross-coupling Reactions

Metal-catalyzed cross-coupling reactions are ubiquitous in organic synthesis. Especially, Pd-catalyzed cross carbon–carbon coupling reactions, including Suzuki-Miyaura, Negishi, Sonogashira, Kumada-Corriu, Stille, Hiyama, and Heck reactions are among the most useful reactions [32–35]. The as-obtained coupling products are widely used as intermediates for polymers, natural products, bioactive compounds, etc. The most common catalysts employed for these reactions are palladium complexes accompanied by organic ligands (e.g. phosphines, N-heterocyclic carbenes). From the aspect of green chemistry and industrial application, these organic ligands are air- and moisture-sensitive, toxic, expensive, and involve multistep syntheses. Meanwhile, these homogeneous palladium catalysts are difficult to be separated from the final product. Therefore, alternative palladium catalysts, that is, the so-called "ligandless" heterogeneous palladium, attract continuous attention. The usage of palladium nanoparticle may be a good choice for industrial application, but whether the catalytic process occurs with the heterogeneous palladium nanoparticle or the leached palladium nanoclusters are still unknown in many cases. From the fundamental research point of view, it is crucial to identify the real catalytic species. In this section, we are going to discuss the in situ generated metal nanoclusters from heterogeneous metal catalysts that are used for cross-coupling reactions. The generation of metal nanoclusters from homogeneous metal complexes will be discussed as well in some cases.

The research group of Gadi Rothenberg investigated the leached palladium species in Pd-catalyzed cross carbon–carbon coupling reactions by a well-designed reactor [36]. They built a special U-tube permeation cell membrane reactor, in which two stainless steel compartments are separated by a membrane. The membrane is designed with ~5-nm pores that only allow the diffusion of palladium species below 5 nm size. A traditional Sonogashira reaction was tested in their experiments. As shown in Figure 2.10, the substrates and bases of the reaction were placed on the one side (side A) of the reactor, while the prepared palladium nanoparticles were placed on the other side (side B) of the reactor. The minimum size of the prepared palladium nanoparticles was 12 nm and their average size is 14.5 nm. As a result of the existence of membrane with 5 nm pores that separates two sides of the reactor, the palladium nanoparticles could not cross between the two sides. The reaction was performed in DMF at 70 °C. After three days, 72% yield of the coupling product were observed on side A. This experiment suggests that leached palladium species (<5 nm) cross the membrane from side B to side A and catalyze the Sonogashira reaction. Meanwhile, 9% yield of the product was observed as well on side B. Thus, the possibility that the substrates and bases diffused from side A to side B and are catalyzed by the palladium nanoparticles cannot be excluded.

Based on the analysis, they designed another cross carbon–carbon coupling reaction, in which insoluble bases were used. This is a typical Heck reaction, as shown in Figure 2.11. The substrates were introduced on both sides of the reactor

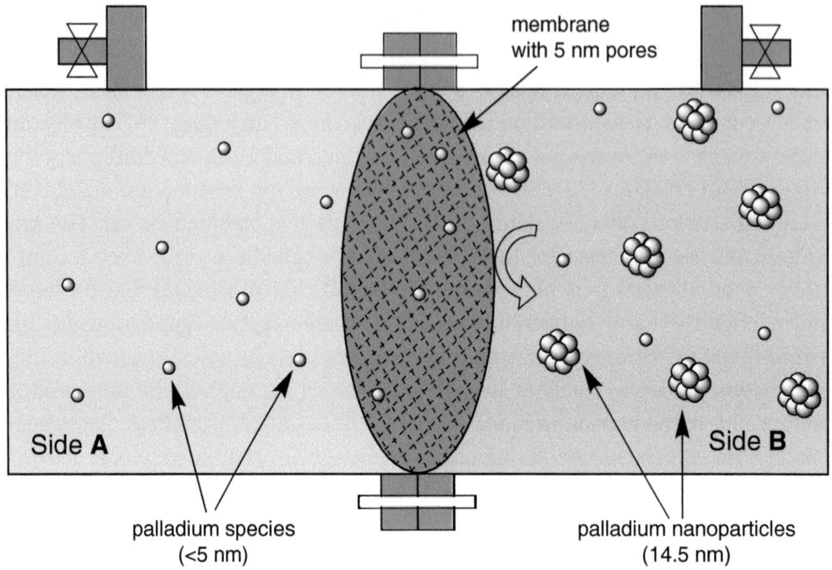


**Figure 2.10** The special U-tube permeation cell membrane reactor designed for cross-coupling reactions.

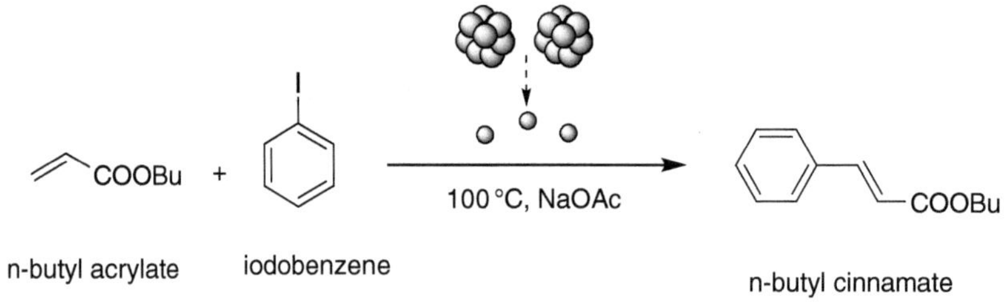


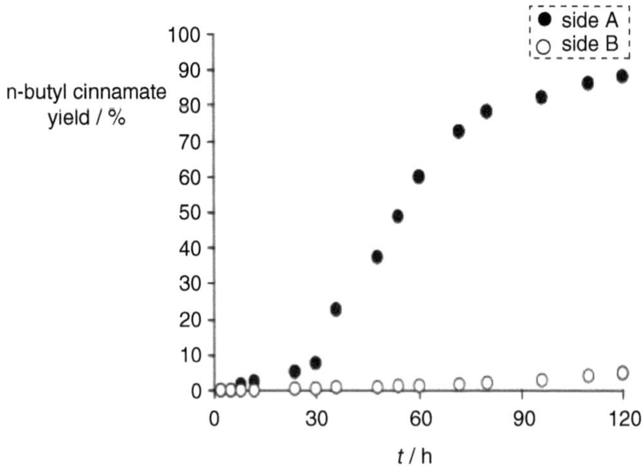


**Figure 2.11** Heck reaction with insoluble base tested in the special U-tube permeation cell membrane reactor.

to avoid concentration gradients. The insoluble base (NaOAc) and palladium nanoparticles were placed on side A and side B, respectively, of the reactor. If the cross carbon–carbon coupling product was observed, leached palladium species should be confirmed because the base is necessary for closing the catalytic cycle. Finally, they observed 88% yield of the product on side A. This result demonstrates that the as-obtained product should be attributed to the catalysis of the leached palladium species, probably the $Pd^0$ nanoclusters or $Pd^{II}$ atom that both are less than 5 nm. Notably, 4.9% yield of the coupling product was observed as well ion side B, which should be ascribed to the diffusion of the product from side A to side B. The smart design of the special U-tube permeation cell membrane reactor permits the testing of real catalysts in Pd-catalyzed cross-coupling reactions, and the in situ generated palladium nanoclusters and leached palladium atoms have been identified as contributing to catalytically active species.

The palladium nanoparticles in the above mentioned work were prepared by reducing $Pd(NO_3)_2$ with tetraoctylammonium glycolate (TOAG). For ligand-free palladium catalyzed cross-coupling reactions this kind of in situ formation of palladium nanoparticles and their subsequential evolution into palladium nanoclusters are common, despite the fact that the catalytically active species of ligated-palladium catalysis are generally considered to be the relatively stable single palladium atom protected by the organic ligand. Thus, although the reaction conditions of ligand-free palladium catalyzed cross-coupling reactions are more concise than that of the ligated-palladium catalyzed ones, more complicated mechanistic pictures usually hide behind the well-known and simple experimental procedures [37, 38]. Kashin and Ananikov summarized the cocktail-type systems of ligand-free palladium catalyzed cross carbon–carbon and carbon–heteroatom coupling reactions [39]. This picture has been described in Section 1.2 and Figure 1.6 of this book. No matter initiated from $Pd^0$ or $Pd^{II}$, recent mechanism investigations have clearly pointed out the complexity of the catalytically active sites interconversions under the reaction mixture and the diversity of palladium-containing species in catalytic cross-coupling reactions (Figure 2.12). Such interconversion processes occur in the case of in situ generated palladium catalysts and in the case of application of palladium nanoparticles as catalyst precursors.

Experimentally, the capture of the catalytically palladium nanoclusters is quite difficult. Reetz and Westermann investigated the phosphane-free palladium-catalyzed cross-coupling reactions by TEM in the year of 2000 [40]. They compared the kinetics of the Heck reaction with the formation process of palladium nanoclusters/nanoparticles. Their studies point to the involvement of the in situ generated nano-sized palladium colloids in three preparatively significant phosphine-free catalyst systems based on simple palladium salts, that is, $Pd(OAc)_2$/DMG (DMG means N,N-dimethylglycine), $Pd(OAc)_2/(n\text{-}Bu)_4N^+Cl^-$ and $Pd(OAc)_2$. It has been confirmed that these palladium colloids or their precursors nanoclusters catalyzed the Heck reaction and gave the corresponding coupling products. They also point out that other phosphine-free Heck reactions and related carbon–carbon bond formation processes in the presence of palladium salts without special ligands are also catalyzed by the nanosized palladium species.

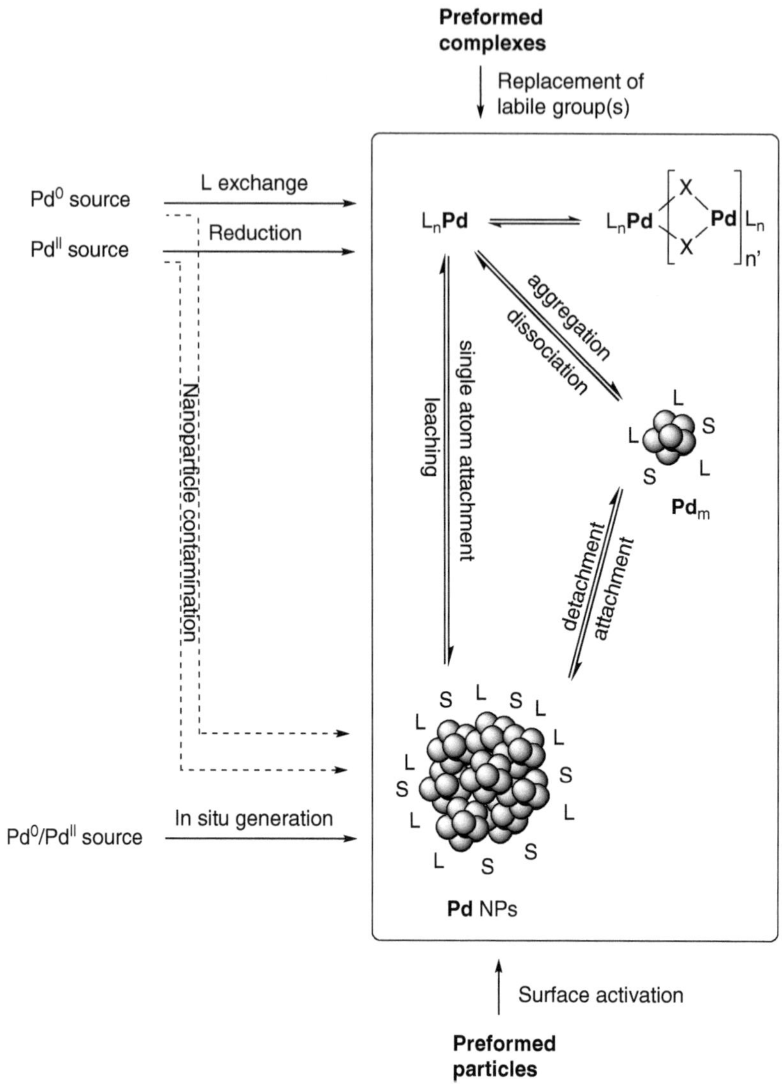


**Figure 2.12** The complicated pathways of ligand-free palladium catalyzed cross-coupling reactions and the in situ generated palladium nanoclusters.

Although the complexity of the "cocktail-of-species" model of ligand-free palladium catalyzed cross-coupling reactions results in the experimental capture of the catalytically active palladium nanoclusters being difficult, Polynski and Ananikov still find the theoretical validity by detailed computational study of leaching/capture pathways and analysis of related stabilization energies [41]. They investigated the Pd-catalyzed cross-coupling reactions involving ArX reagents by modeling some key pathways and proposed the formation and evolution of "cocktail"-type system during the reaction process. It should be noted that the starting materials ArX are also involved in the evolution of the palladium catalysts. Rational models such as $[ArPdX]_2$, $[Pd_2X_4]$, $[ArPdX]_4$ have been found as the key intermediates under the reaction conditions (Figure 2.12). These intermediates themselves are possible active nanoclusters that are able to efficiently catalyze the corresponding cross-coupling reactions.

Based on the experimental results from de Vries research group [42] and Fernandez research group [43], high temperature and base are the two crucial factors that impact the formation of palladium nanoclusters and their catalytic activity in ligand-free palladium catalyzed cross carbon–carbon coupling reactions. de Vries summarized the palladium acetate-catalyzed Heck reaction at temperature higher than 120 °C and found that almost all investigated reactions involve the formation of palladium nanoclusters, which are $Pd^0$ soluble colloids [42]. The aryl halides continuously attack the palladium atoms on the surface of the palladium nanoclusters at high temperature, thus forming soluble anionic complexes and protecting the as-obtained palladium catalysts. After completion of a cross-coupling cycle, the palladium catalysts fall back to the original palladium nanoclusters and engage in the next oxidative addition process. It is also found that the as-generated palladium nanocluster catalyst at high temperature performed better than the palladium acetate and other palladium complexes in the activation of aryl chlorides. This experimental phenomenon should be ascribed to the high catalytic activity of in situ generated palladium nanoclusters.

Generally, oxidative addition has been considered to be the rate determine step in metal-catalyzed cross-coupling reactions. Unexpectedly, Fernandez and coworkers revealed that base plays very important roles in altering the catalytic energy barrier [43]. Based on a combined experimental and computational study, they found that the oxidative addition is not the kinetically relevant step any more for the cross-coupling reactions catalyzed by palladium or platinum nanoclusters. The exposed metal atoms on the surface of the palladium or platinum nanoclusters cooperate to enable an extremely easy oxidative addition of aryl halides, even aryl chlorides, and allow the base to bifurcate the coupling products. Specifically, amines and carbonates trigger the Heck reaction and Sonogashira reaction, respectively, catalyzed by the palladium nanocluster, while phosphates and acetates favor the Suzuki reaction catalyzed by the palladium nanocluster and cross-coupling reactions catalyzed by the platinum nanocluster, respectively. The based-controlled effect arises from overcoming the energetic barrier of the oxidative addition and translating the rate determine step of the cross-coupling reactions to the later steps where the base actively participates. This kind of unexpected base effect discovered in metal nanocluster catalysis also provides new opportunities for the design of cross-coupling reactions in organic transformations.

The in situ generated metal nanoclusters are even employed for activating inert compounds and participating in asymmetric catalysis. A very recent work reported by Liu research group revealed the formation of palladium nanoclusters during Pd/Cu synergistic catalysis and their excellent catalytic activity toward the oxidation of alkynes [44]. Nonproteinogenic $\alpha,\alpha$-disubstituted $\alpha$-amino acids constitute useful building blocks in organic synthesis and new drugs and bioactive products. Their synthetic approach high depends on the metal-catalyzed asymmetric allylic alkylation of aldimine esters and allylic precursors such as the allylic alcohols, allylic halides, etc. A chiral copper molecular catalyst is responsible for activating the imino ester via formation of a rigid metalated azomethine ylide and controlling the stereoselectivity of the intermediates. Meanwhile, a palladium molecular catalyst is generally employed to activate the allylic precursors and form the active π-allylpalladium intermediate that reacts with the copper-activated imino ester (Figure 2.13). Considering that those allylic precursors such as the allylic alcohols and allylic halides are not easily accessible synthons, the bench-stable and easily accessible alkynes have attracted great attention of chemists. Concerning this synthetic route, the activation of alkynes would be a challenging task because the commonly used molecular palladium catalysts in the catalytic system are catalytically inert toward alkynes. Liu and coworkers unexpectedly found that under certain reaction conditions, palladium nanocluster formed in situ and was able to activate the alkyne, triggering its coupling with the aldimine ester to generate nonproteinogenic α, α-disubstituted α-amino acids. The palladium nanoclusters were identified under the reaction conditions of $Pd(dba)_2$ (dba = dibenzalacetone), AcOH, KOAc, the copper catalyst, the phosphine ligand and the solvent at 80 °C. The combination of kinetic studies, DLS, TEM, and poisoning experiments unambitiously indicates that activation of the alkyne is initiated by the in situ generated palladium nanoclusters. The palladium nanocluster catalysis cooperates very well with the molecular copper catalysis, furnishing the desired amino acids in high yields with excellent enantioselectivities.

In summary, we systematically introduce the metastable metal nanoclusters formed in situ during catalysis. Although they are catalytically active and able to initiate the reactions that are hardly catalyzed by metal complexes, they are generally instable. Isolation, characterization and recycling of these in situ generated metal nanoclusters are difficult. However, their superior activity motivates chemists and material scientists to stabilize them and create highly efficient metal catalysts. In the next two chapters, we will introduce two widely used methods for stabilizing metal nanoclusters, that is, immobilization and organic ligand protection.

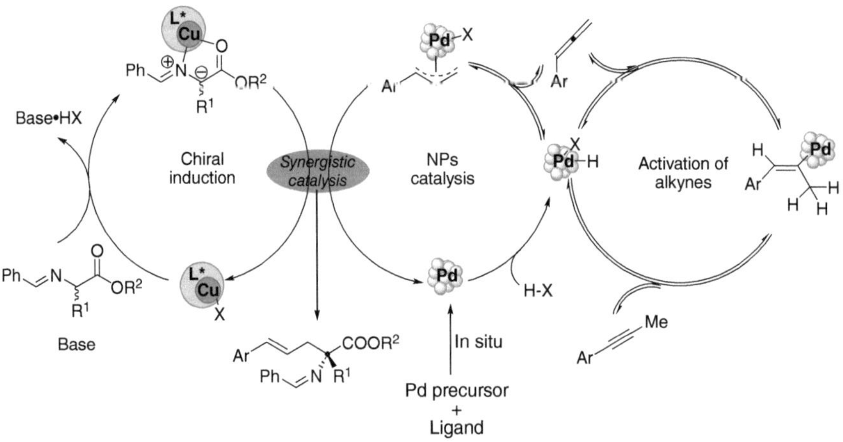


**Figure 2.13** Asymmetric catalysis triggered by synergy of homogeneous Cu molecular catalyst and palladium nanocluster catalyst. *Source:* [44]/with permission of American Chemical Society.

## References


1 J. A. Widegren, R. G. Finke, A Review of Soluble Transition-metal Nanoclusters as Arene Hydrogenation Catalysts. *J. Mol. Catal. A: Chem.* **2003**, *191*, 187–207.
2 G. Schmid, V. Maihack, F. Lantermann, S. Peschel, Ligand-stabilized Metal Clusters and Colloids: Properties and Applications. *J. Chem. Soc., Dalton Trans.* **1996**, *5*, 589–595.
3 J. A. Widegren, R. G. Finke, A Review of the Problem of Distinguishing True Homogeneous Catalysis from Soluble or Other Metal-particle Heterogeneous Catalysis under Reducing Conditions. *J. Mol. Catal. A: Chem.* **2003**, *198*, 317–341.
4 P. J. Dyson, Arene Hydrogenation by Homogeneous Catalysts: Fact or Fiction? *Dalton Trans.* **2003**, *15*, 2964–2974.
5 D. Gärtner, S. Sandl, A. Jacobi von Wangelin, Homogeneous vs. Heterogeneous: Mechanistic Insights Into Iron Group Metal-catalyzed Reductions from Poisoning Experiments. *Catal. Sci. Technol.* **2020**, *10*, 3502–3514.
6 L. Liu, D. M. Meira, R. Arenal, P. Concepcion, A. V. Puga, A. Corma, Determination of the Evolution of Heterogeneous Single Metal Atoms and Nanoclusters Under Reaction Conditions: Which Are the Working Catalytic Sites? *ACS Catal.* **2019**, *9*, 10626–10639.
7 S. Kim, F. Loose, M. J. Bezdek, X. Wang, P. J. Chirik, Hydrogenation of N-Heteroarenes Using Rhodium Precatalysts: Reductive Elimination Leads to Formation of Multimetallic Clusters. *J. Am. Chem. Soc.* **2019**, *141*, 17900–17908.
8 X. F. Wu, H. Neumann, M. Beller, Palladium-catalyzed Oxidative Carbonylation Reactions. *ChemSusChem* **2013**, *6*, 229–241.
9 X. Jiang, J. Zhang, S. Ma, Iron Catalysis for Room-temperature Aerobic Oxidation of Alcohols to Carboxylic Acids. *J. Am. Chem. Soc.* **2016**, *138*, 8344–8347.
10 J. He, M. Wasa, K. S. L. Chan, Q. Shao, J. Q. Yu, Palladium-catalyzed Transformations of Alkyl C—H Bonds. *Chem. Rev.* **2017**, *117*, 8754–8786.
11 G. Yin, X. Mu, G. Liu, Palladium(II)-catalyzed Oxidative Difunctionalization of Alkenes: Bond Forming at a High-valent Palladium Center. *Acc. Chem. Res.* **2016**, *49*, 2413–2423.
12 Y. Liu, B. Shi, Z. Liu, R. Gao, C. Huang, H. Alhumade, S. Wang, X. Qi, A. Lei, Time-resolved EPR Revealed the Formation, Structure, and Reactivity of N-centered Radicals in an Electrochemical C($sp^3$)-H Arylation Reaction. *J. Am. Chem. Soc.* **2021**, *143*, 20863–20872.
13 M.-B. Li, J. E. Backvall, Efficient Heterogeneous Palladium Catalysts in Oxidative Cascade Reactions. *Acc. Chem. Res.* **2021**, *54*, 2275–2286.
14 R. A. T. M. van Benthem, H. Hiemstra, P. W. N. M. van Leeuwen, J. W. Geus, W. N. Speckamp, Sulfoxide-stabilized Giant Palladium Clusters in Catalyzed Oxidations. *Angew. Chem. Int.: Ed. Engl.* **1995**, *34*, 457–460.
15 M. Mifsud, K. V. Parkhomenko, I. W. C. E. Arends, R. A. Sheldon, Pd Nanoparticles as Catalysts for Green and Sustainable Oxidation of Functionalized Alcohols in Aqueous Media. *Tetrahedron* **2010**, *66*, 1040–1044.
16 D. Pun, T. Diao, S. S. Stahl, Aerobic Dehydrogenation of Cyclohexanone to Phenol Catalyzed by $Pd(TFA)_2$/2-Dimethylaminopyridine: Evidence for the Role of Pd Nanoparticles. *J. Am. Chem. Soc.* **2013**, *135*, 8213–8221.

**17** C. G. Baumann, S. De Ornellas, J. P. Reeds, T. E. Storr, T. J. Williams, I. J.S. Fairlamb, Formation and Propagation of Well-defined Pd Nanoparticles (PdNPs) During C—H Bond Functionalization of Heteroarenes: Are Nanoparticles a Moribund Form of Pd or an Active Catalytic Species? *Tetrahedron* **2014**, *70*, 6174–6187.

**18** G. J. Irvine, M. J. G. Lesley, T. B. Marder, N. C. Norman, C. R. Rice, E. G. Robins, W. R. Roper, G. R. Whittell, L. J. Wright, Transition Metal-boryl Compounds: Synthesis, Reactivity, and Structure. *Chem. Rev.* **1998**, *98*, 2685–2722.

**19** R. Barbeyron, E. Benedetti, J. Cossy, J.-J. Vasseur, S. Arseniyadis, M. Smietana, Recent Developments in Alkyne Borylations. *Tetrahedron* **2014**, *70*, 8431–8452.

**20** K. Yang, Q. Song, Transition-metal-free Regioselective Synthesis of Alkylboronates From Arylacetylenes and Vinyl Arenes. *Green Chem.* **2016**, *18*, 932–936.

**21** T. J. Mazzacano, N. P. Mankad, Dehydrogenative Borylation and Silylation of Styrenes Catalyzed by Copper-carbenes. *ACS Catal.* **2016**, *7*, 146–149.

**22** S. Liu, X. Zeng, B. Xu, (E)-Alkene Synthesis via Nano-copper/Homogeneous Palladium Co-catalysis and Selectivity Amplification. *Asian J. Org. Chem.* **2017**, *6*, 507–511.

**23** I. Beletskaya, C. Moberg, Element—Element Additions to Unsaturated Carbon—Carbon Bonds Catalyzed by Transition Metal Complexes. *Chem. Rev.* **2006**, *106*, 2320–2354.

**24** S. Jin, K. Liu, S. Wang, Q. Song, Enantioselective Cobalt-catalyzed Cascade Hydrosilylation and Hydroboration of Alkynes to Access Enantioenriched 1,1-Silylboryl Alkanes. *J. Am. Chem. Soc.* **2021**, *143*, 13124–13134.

**25** M. Zhong, Y. Gagne, T. O. Hope, X. Pannecoucke, M. Frenette, P. Jubault, T. Poisson, Copper-photocatalyzed Hydroboration of Alkynes and Alkenes. *Angew. Chem. Int. Ed.* **2021**, *60*, 14498–14503; *Angew. Chem.* **2021**, *133*, 14619–14624.

**26** W.-H. Li, J. Yang, H. Jing, J. Zhang, Y. Wang, J. Li, J. Zhao, D. Wang, Y. Li, Creating High Regioselectivity by Electronic Metal-support Interaction of a Single-atomic-site Catalyst. *J. Am. Chem. Soc.* **2021**, *143*, 15453–15461.

**27** M. Kidonakis, M. Stratakis, Gold-nanoparticle-catalyzed Mild Diboration and Indirect Silaboration of Alkynes Without the Use of Silylboranes. *Eur. J. Org. Chem.* **2017**, *2017*, 4265–4271.

**28** H.-Y. Tsai, M. Madasu, M. H. Huang, Polyhedral $Cu_2O$ Crystals for Diverse Aryl Alkyne Hydroboration Reactions. *Chem. Eur. J.* **2019**, *25*, 1300–1303.

**29** Y. You, S. Ge, Cobalt-catalyzed One-pot Asymmetric Difunctionalization of Alkynes to Access Chiral Gem-(Borylsilyl)Alkanes. *Angew. Chem. Int. Ed.* **2021**, *60*, 20684–20688; *Angew. Chem.* **2021**, *133*, 20852–20856.

**30** X.-Y. Yang, Y. Li, S. A. Pullarkat, A One-pot Diastereoselective Self Assembly of C-Stereogenic Copper(I) Diphosphine Clusters. *Inorg. Chem.* **2014**, *53*, 10232–10239.

**31** C. Li, S. Song, Y. Li, C. Xu, Q. Luo, Y. Guo, X. Wang, Selective Hydroboration of Unsaturated Bonds by an Easily Accessible Heterotopic Cobalt Catalyst. *Nat. Commun.* **2021**, *12*, 3813.

32 C. Liu, H. Zhang, W. Shi, A. Lei, Bond Formations Between Two Nucleophiles: Transition Metal Catalyzed Oxidative Cross-coupling Reactions. *Chem. Rev.* **2011**, *111*, 1780–1824.

33 B. P. Fors, D. A. Watson, M. R. Biscoe, S. L. Buchwald, A Highly Active Catalyst for Pd-catalyzed Amination Reactions: Cross-coupling Reactions Using Aryl Mesylates and the Highly Selective Monoarylation of Primary Amines Using Aryl Chlorides. *J. Am. Chem. Soc.* **2008**, *130*, 13552–13554.

34 X. Shang, Z. Q. Liu, Transition Metal-catalyzed $C_{Vinyl}$—$C_{Vinyl}$ Bond Formation Via Double $C_{Vinyl}$—H Bond Activation. *Chem. Soc. Rev.* **2013**, *42*, 3253–3260.

35 M. J. Ardolino, J. P. Morken, Construction of 1,5-Enynes by Stereospecific Pd-Catalyzed Allyl-propargyl Cross-couplings. *J. Am. Chem. Soc.* **2012**, *134*, 8770–8773.

36 M. B. Thathagar, J. E. ten Elshof, G. Rothenberg, Pd Nanoclusters in C—C Coupling Reactions: Proof of Leaching. *Angew. Chem. Int. Ed.* **2006**, *45*, 2886–2890; *Angew. Chem.* **2006**, *118*, 2952–2956.

37 R. Narayanan, C. Tabor, M. A. El-Sayed, Can the Observed Changes in the Size or Shape of a Colloidal Nanocatalyst Reveal the Nanocatalysis Mechanism Type: Homogeneous or Heterogeneous? *Top. Catal.* **2008**, *48*, 60–74.

38 K. Uno, T. Itoh, H. Sato, G. C. Lloyd-Jones, Y. Orito, Direct Observation of Palladium Leaching from Pd/C by a Simple Method: X-ray Absorption Spectroscopy of Heterogeneous Mixtures. *ACS Omega* **2023**, *8*, 21787–21792.

39 A. S. Kashin, V. P. Ananikov, Catalytic C—C and C—Heteroatom Bond Formation Reactions: In Situ Generated or Preformed Catalysts? Complicated Mechanistic Picture Behind Well-known Experimental Procedures. *J. Org. Chem.* **2013**, *78*, 11117–11125.

40 M. T. Reetz, E. Westermann, Phosphane-free Palladium-catalyzed Coupling Reactions: The Decisive Role of Pd Nanoparticles. *Angew. Chem. Int. Ed.* **2000**, *39*, 165–168.

41 M. V. Polynski, V. P. Ananikov, Modeling Key Pathways Proposed for the Formation and Evolution of "Cocktail"-type Systems in Pd-catalyzed Reactions Involving ArX Reagents. *ACS Catal.* **2019**, *9*, 3991–4005.

42 J. G. de Vries, A Unifying Mechanism for All High-temperature Heck Reactions. The Role of Palladium Colloids and Anionic Species. *Dalton Trans.* **2006**, 421–429.

43 E. Fernandez, M. A. Rivero-Crespo, I. Dominguez, P. Rubio-Marques, J. Oliver-Meseguer, L. Liu, M. Cabrero-Antonino, R. Gavara, J. C. Hernandez-Garrido, M. Boronat, A. Leyva-Perez, A. Corma, Base-Controlled Heck, Suzuki, and Sonogashira Reactions Catalyzed by Ligand-free Platinum or Palladium Single Atom and Sub-nanometer Clusters. *J. Am. Chem. Soc.* **2019**, *141*, 1928–1940.

44 Y. Liu, H. Chen, X. Wang, Synergistic Homogeneous Asymmetric Cu Catalysis with Pd Nanoparticle Catalysis in Stereoselective Coupling of Alkynes With Aldimine Esters. *J. Am. Chem. Soc.* **2024**, *146*, 28427–28436.

# 3

# Immobilized Metal Nanocluster Catalysis

## 3.1 Carbon Nanomaterial Immobilized Metal Nanocluster

### 3.1.1 Features of Carbon Nanomaterials as Support

Bearing three kinds of hybridization modes, that is, sp, $sp^2$, and $sp^3$, carbon atoms provide the base unit to form a variety of carbon isomers, which can be obtained under different hybridization forms and rich combinations of carbon atoms arrangement. Especially, in the scale of nanometer, carbon nanomaterials feature high porosity and excellent dispersity, thus being ideal supports for metal nanoclusters. Additionally, the structures and properties of carbon nanomaterials can be readily tailored by surface modification, defect modulation, interfacial engineering, and many other methods to match with the size, type, and structure of metal nanoclusters. Apart from the structural features, carbon nanomaterials are inexpensive and easy to synthesize, and have the advantages of being stable and lightweight. They have been widely employed as supports for metal nanoclusters.

Carbon nanomaterials-immobilized metal nanoclusters can be synthesized in situ by mixing the carbon nanomaterials and metal precursors under typical reaction conditions of metal nanocluster synthesis. In some cases, even the carbon nanomaterials are synthesized in situ with metal nanoclusters. On the other hand, carbon nanomaterials-immobilized metal nanoclusters can be prepared indirectly by mixing the presynthesized carbon nanomaterials and metal nanoclusters. The choice of synthetic approaches depends on the size, structure, metal–support interactions, etc., that we need. We do not discuss the complex situations in details in the book. To mainly focus on the catalytic features of carbon nanomaterials-immobilized metal nanoclusters, we select some current works as typical examples and manifest the role of carbon nanomaterials in regulating the stability, activity, and selectivity of metal nanoclusters.

*Metal Nanocluster Catalysis: From Fundamentals to Applications*, First Edition. Manzhou Zhu and Man-Bo Li. © 2026 Wiley-VCH GmbH. All rights reserved, including rights for text and data mining and training of artificial intelligence technologies or similar technologies. Published 2026 by Wiley-VCH GmbH.

Hasegawa and coworkers reported that a carbon nanomaterials-immobilized Rh–Ru bimetallic nanocluster [1], optimized from a series of carbon-immobilized mono- and bimetallic nanocluster catalysts they synthesized, shows high catalytic activity and selectivity toward cross-dehydrogenative coupling reactions of arenes and carboxylic acids (Figure 3.1). They synthesized carbon nanomaterials-immobilized monometallic nanocluster catalysts such as Rh, Ru, Ir, Pd, Au, Ag, and Pt, and bimetallic nanocluster catalysts such as Rh–Ru, Rh–Ir, Rh–Pt, Pt–Ru, and Ir–Ru by depositing the as-prepared metal nanoclusters on carbon supports. The nanocarbon supports well protect the metal nanoclusters from aggregation or decomposition, affording metal nanocluster catalysts with a narrow size distribution and an average diameter of 1.2 nm. These nanocluster catalysts were sent for the cross-dehydrogenative coupling reactions between arenes and carboxylic acids, which constitute a straightforward route to aryl esters from nonactivated substrates. Homogeneous metal catalysts, such as Rh, Ru, Pd, Au, and Co complexes, have been developed for this kind of coupling reaction, while expensive and toxic oxidants, such as phenyliodine(III) diacetate, silver salts, and persulfate, are generally required. Inspired by the previous reports that the carbon nanomaterials and heterogeneous metal catalysts are beneficial for adsorption and direct activation of molecular oxygen, Hasegawa and coworkers envisioned the possibility of employing the carbon nanomaterials-immobilized metal nanoclusters for catalyzing cross-dehydrogenative coupling reactions of arenes and carboxylic acids in the presence of $O_2$ as the sole oxidant.

The carbon nanomaterials were found to be good supports for the metal nanoclusters because the immobilized nanocluster catalysts kept their compositions and

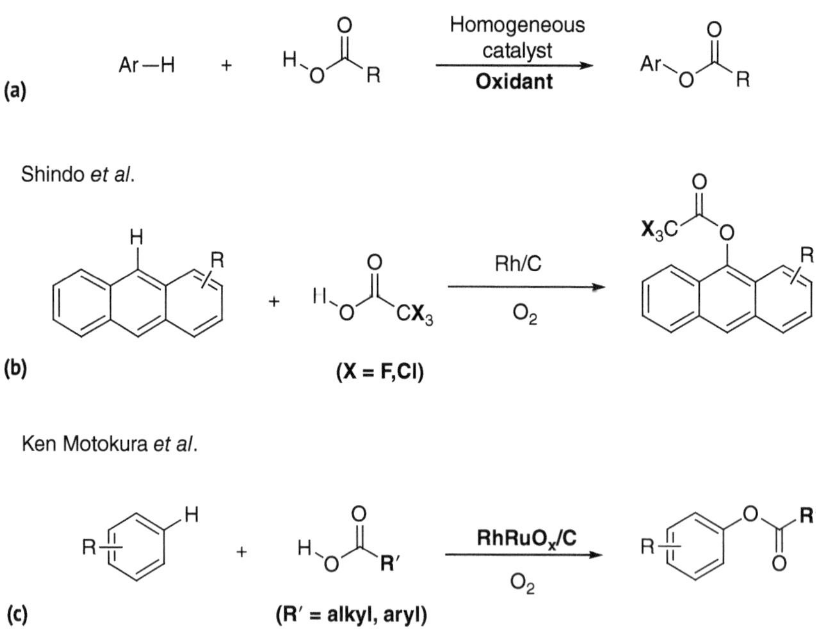


**Figure 3.1** Cross-dehydrogenative coupling reactions of arenes and carboxylic acids using carbon nanomaterials-immobilized metal nanoclusters as the catalysts.

structures during catalysis and none of leached metal species were observed. After screening catalytic performance of monometallic and bimetallic metal nanocluster catalysts, the Rh–Ru bimetallic nanocluster immobilized on carbon nanomaterial was picked up as an efficient catalyst for the cross-dehydrogenative coupling of benzene and acetic acid. Kinetic isotope and substituent effect experiments show that the carbon–hydrogen bond cleavage of benzene is the rate-determining step, which proceeds at the Rh center of the Rh–Ru bimetallic nanocluster catalyst. Further mechanism studies indicate that the Ru part of the catalyst also plays important roles in catalysis because it indirectly enhances the electrophilicity of the Rh activation center by decreasing the negative charges of the oxygen atoms that coordinate on the Rh. Based on the highly stable and active Rh–Ru bimetallic nanocluster catalyst immobilized on carbon nanomaterials, they realized effective cross-dehydrogenative couplings of various arenes and carboxylic acids using oxygen gas as the sole oxidant.

### 3.1.2 Nano-diamond as Support

Different kinds of carbon nanomaterials were designed, synthesized, and applied for immobilizing specific metal nanoclusters and enabling their efficient catalysis toward certain reactions. Jin and coworkers synthesized nano-diamond immobilized Ir nanoclusters [2]. The nano-diamond is a type of diamond structured nanocarbon material, which has a size distribution of 5–10 nm. The nano-diamond immobilized Ir nanoclusters were synthesized in situ by mixing the nano-diamond and iridium precursors. The loading of Ir on the carbon nano-diamond is adjustable from 0.05 to 3.8 wt%. The nano-diamond was found to be beneficial for stabilizing the Ir nanoclusters with a size of 1–2 nm and coordination number of 9.7. This kind of Ir nanoclusters immobilized on nano-diamond were able to efficiently catalyze the partial oxidation of ethane to valuable $C_2$ chemicals in the presence of $O_2$ and CO. CO plays critical role in maintaining the low-valence state of iridium that is required for ethane adsorption during the reaction. This is clearly revealed by the in situ XPS experiments. The immobilized Ir nanoclusters catalyze the transformation of ethane, affording $C_2$ chemicals such as $CH_3CH_2OH$, $CH_3CHO$, $CH_3COOH$, and $CH_3CH_2COOH$, with $CH_3COOH$ as the major products. Notably, the nano-diamond immobilized Ir nanoclusters demonstrated much higher catalytic activity than the atomically dispersed Ir atoms immobilized on nano-diamond. This might be ascribed to the synergistic effect between adjacent Ir atoms and the interactions between Ir nanoclusters and the nano-diamond support.

### 3.1.3 Graphene as Support

In spite of the rapid development of zero-dimensional, one-dimensional, two-dimensional, and three-dimensional carbon nanomaterials, such as carbon quantum dots, carbon nanotubes, and graphynes, the graphene [3] is still the most commonly used carbon nanomaterials for immobilizing metal nanoclusters for catalysis. It was experimentally isolated in 2004 and quickly became a hot topic in catalysis as a result of its intrinsic carrier mobility at room temperature and perfect chemical and thermal stability. Other advantages of graphene include high adsorption capacity, excellent electrical conductivity, and easiness of modification. There is a good review paper systematically summarizing the synthesis, properties, and applications of graphene-immobilized metal catalysts in heterogeneous catalysis. The synthetic approach (including microwave-assisted method, photochemical method, atomic layer deposition method, etc.) and catalyzed reactions (oxidation, hydrogenation, dehydrogenation, cross-coupling, etc.) of graphene-immobilized metal catalysts, including graphene-immobilized metal nanocluster catalysts, are analyzed in details.

Among the investigated reactions catalyzed by carbon nanomaterials-immobilized metal nanoclusters, electrochemical conversion is a unique catalog, manifesting the structural features of carbon nanomaterials as well as metal nanoclusters for efficient catalysis. On the one hand, metal nanoclusters are composed of several metal atoms. The delocalized electrons in metal nanoclusters act as an "electron pool," which is beneficial for the electron transfer between the metal nanoclusters and substrates. Thus, metal nanoclusters are ideal catalysts for electrochemical conversions. On the other hand, graphene features excellent electrical conductivity, which provides a fast diffusion channel for electrons and reduces the hindrance of electron movement. This is also highly desirable for electrochemical conversions. The combination of metal nanocluster and graphene provides potential for efficiently reducing the activation energy required for the electrochemical reactions and significantly increasing the rate of electrochemical reactions. In addition, the graphene is able to protect the catalytically active metal nanoclusters from deactivation during the electrochemical process. We select four electrochemical conversion reactions, that is, hydroxylation reaction (HOR), oxygen reduction reaction (ORR), carbon dioxide reduction reaction ($CO_2$RR), and nitrogen reduction reaction (NRR) to show the catalytic performance of graphene-immobilized metal nanoclusters in the following.

HOR is an energy storage technology that converts hydrogen into electricity. This is of great research significance to alleviate the energy and environmental crisis of the current year. Due to that a small amount of CO always accompanies with $H_2$ in the current industrial process, how to avoid the deactivation of metal catalysts in the presence of CO is a crucial challenge. For example, Pt-based catalysts are favorable for the HOR because of its excellent electrocatalytic property, while the high binding energy between Pt and CO results in the deactivation of the corresponding catalysts. In a current work, Pt–Ni nanoclusters were designed, synthesized and immobilized on graphene [4]. Owing to the ability of Ni to adsorb hydroxyl groups at a low potential, the Pt–Ni nanoclusters demonstrate strong resistance to CO toxicity. Meanwhile, the high specific surface area of the nanoclusters and the participation of well-conductive and robust graphene endow the graphene-immobilized Pt–Ni nanoclusters with high catalytic activity and durability for HOR. The performance of the nanocluster catalyst is significantly better than the commercialized Pt/C catalyst. After 1000 endurance tests, the catalytic activity of Pt–Ni/graphene decreased by 36%, which is better than that of Pt/C (72.2%). The mass-specific activity of Pt–Ni/graphene was 826.1 A/g, which is 2.58 times higher than that of Pt/C.

ORR is another important electrochemical technology [5–7], and has attracted great research interest in electrochemistry field. Due to its sluggish kinetics, it is regarded as the potential-limiting step in proton-exchange membrane fuel cells, which are one of the most promising renewable energy conversion devices to address the ever-grown energy crisis and environmental concerns. Additionally, it is important in water electrolysis, corrosion, diverse industrial processes, and so on. Based on the considerable effort from chemists and material scientists, Pt has been recognized as an excellent catalyst for ORR, while commercially available Pt catalyst cannot satisfy the need of research and do not fully demonstrate the catalytic potential of Pt in ORR. Reducing the size of Pt catalysts constitutes an efficient strategy to improve their catalytic activity. A good practice by Shin and coworkers demonstrates that when graphene and ferrum (Fe) present in Pt nanoclusters [8], the electronic structure of Pt is modulated and its geometric structure is stabilized, leading to highly enhanced activity of Pt in ORR. This kind of graphene-immobilized Pt–Fe alloy nanocluster was synthesized by carbonizing hyper-crosslinked polystyrene particles with Fe precursors, doping nitrogen species, depositing Pt precursors, then reducing them. Pt–Fe alloy nanoclusters with average size of 1.5 nm are well stabilized by the graphene support. This catalyst remains stable even after annealing at 800 °C. On the other hand, graphene-immobilized Pt–Fe alloy nanoclusters are highly active for ORR. Despite the fact that the metal nanoclusters are surrounded by the catalytically inert graphene, substrates such as proton, oxygen molecule, $Br^-$, formic acid, and methanol are able to diffuse through the carbon support to reach the Pt activation sites. When it was applied in a cathode of proton-exchange membrane fuel cells with a small Pt loading of 0.05 $mg/cm^2$, it presented high current density at 0.65 V. After 30,000 cycles, the current density was barely changed. Therefore, the metal nanocluster catalyst demonstrates high mass activity and exceptional durability for the ORR, contrasting with the lower catalytic activity and significant decrease of current density observed in commercially available Pt/C and Pt–Fe/C catalysts.

$CO_2$RR represents one of the important routes to reducing atmospheric $CO_2$ emissions. In the meantime, it constitutes an effective method to obtain high added value single carbon (e.g. CO, $CH_4$, $CH_3OH$) and multicarbon products ($C_2H_4$, $C_2H_6$, $CH_3CH_2OH$) in terms of electrochemical devices. Nevertheless, reduction of $CO_2$ is a highly challenging task, because it possesses extremely high thermal stability and chemical inertness. The average bond energy of the carbon-oxygen bonds in $CO_2$ are about 750 kJ/mol. High temperature, high pressure and high energy are required to overcome the potential barriers of the direct reduction of $CO_2$. On the other hand, the introduction of catalysts would significantly reduce the activation energy required for the reduction of $CO_2$ and increase the electrochemical reaction rate, which is crucial for improving the efficiency of electrochemical devices. Among the employed catalysts, the two-dimensional material graphene-immobilized metal nanoclusters demonstrated attractive performance (Figure 3.2). The $CO_2$RR process with metal catalysts generally includes the following three steps: (i) $CO_2$ is dissolved and diffused in the electrolyte and chemically adsorbed on the metal activation sites of the cathode catalysts. (ii) The carbon–oxygen bond cleavage and the carbon–hydrogen bond formation occur by electron transfers or proton transfers. Different

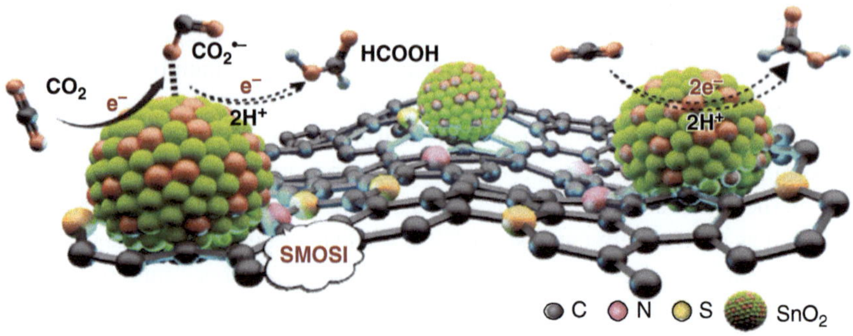


**Figure 3.2** Illustration of the interactions between metal nanoclusters and the graphene. *Source:* [9]/with permission of Royal Society of Chemistry.

intermediates are then generated by dimerization, trimerization, hydrogenation, or dehydration. (iii) Product configuration is rearranged followed by the desorption of the final products from the surface of the electrode. All of these steps occur on the surface of the catalyst. The ultra-small size endows metal nanoclusters abundant activation sites on the surface for catalysis and the single-atom-thick hexagonal honeycomb structure of graphene provides quantity of anchoring sites for the metal nanoclusters, thereby enabling the metal nanoclusters stably anchored to the top sites, bridge sites, and center sites of the carbon atoms and improving the dispersion performance of the as-immobilized metal nanoclusters. The features of graphene-immobilized metal nanoclusters match well with the characteristics of $CO_2RR$.

In 2024, Chen and coworkers have used the dual constraints of carbon defects and micropores to develop a simple impregnation-calcination method and prepared a graphene-immobilized Cu nanocluster catalyst [9], featuring an average size of 1 nm and about 10 atoms supported on defect-rich graphene. The as-prepared catalyst shows high catalytic activity and structural stability toward the selective $CO_2RR$. The d-centers of Cu nanoclusters enhance the adsorption strength of activated H (H*) and CO (CO*) species, thereby promoting the formation of the final product $CH_4$. Additionally, the strong metal–support interaction between Cu nanoclusters and defective carbon also reduces the $\Delta GH^*$, facilitating the second steps of $CO_2RR$ process mentioned above. It should also be noteworthy that density functional theory (DFT) calculations can offer relatively stable nanocluster structures in advances, thus providing certain guidance for the experimental synthesis of metal nanocluster-immobilized in graphene in some cases.

NRR is a chemical process that converts $N_2$ to $NH_3$. The conventional method for NRR is called Haber–Bosch process that is quite important in industrial applications, while it is energy intensive and accompanied by significant greenhouse gas emissions. Designing catalysts for the synthesis of ammonia from nitrogen gas with high efficiency and low energy consumption is highly demanded while challenging. Metal nanoclusters are able to create low coordination sites and also reduce bonding strength of starting materials and intermediates, thus being potentially effective catalyst for NRR process. Theoretical studies also indicate the high catalytic performance of metal nanoclusters in NRR. For example, Cui and coworkers investigated the $Y_3$ and $Nb_3$ nanoclusters-immobilized on graphene for the conversion from $N_2$ to $NH_3$ [10]. Calculation results demonstrate that both nanocluster catalysts are structurally stable and favor the adsorption of $N_2$, resulting in a strong force between nanoclusters and the nitrogen gas molecules and a fast charge transfer rate. The nitrogen–nitrogen bond lengths of $N_2$ adsorbed on the nanocluster catalyst are calculated to be ranging from 1.39 to 1.43 Å, suggesting that the bond strength of $N_2$ is significantly weaken on the surface of $Y_3$ or $Nb_3$ nanoclusters-immobilized on graphene. Experimental results unambitiously indicate that metal nanocluster-catalyzed NRR is a feasible route. Yao and coworkers prepared atomically precise $Au_4Pt_2(SR)_8$ and $Au_4Pd_2(SR)_8$ nanoclusters that are immobilized on defective graphene [11]. Atomic force microscopy (AFM) images showed that the as-prepared bimetallic nanoclusters were uniformly distributed on the surface of graphene. Especially, the graphene-immobilized $Au_4Pt_2(SR)_8$ nanocluster was found to be highly active for $N_2$ fixation. Electrochemical process from $N_2$ to $NH_3$ was well established based on the metal nanocluster catalysis, featuring high catalytic activity and selectivity.

### 3.1.4 Nitrogen-doped Carbon Nanomaterials as Support

Nitrogen doping in carbon nanomaterials provides composition diversity as well as structural modification way toward carbon support. More importantly, nitrogen-doped carbon nanomaterials possess abundant loading sites for metal nanoclusters and affect their steric and electronic properties efficiently. Liu and coworkers applied nitrogen-doped carbon nanomaterials to immobilize Fe–Co bimetallic nanoclusters with the well-defined composition of $FeCoN_6(OH)$ [12]. This catalyst showed high activity as well as high selectivity toward alcohol dehydrogenation reactions (Figure 3.3). Compared to the single-atom catalysts (Fe or Co) that are immobilized in the same support, Fe–Co nanocluster catalyst displayed higher activity, producing the desired

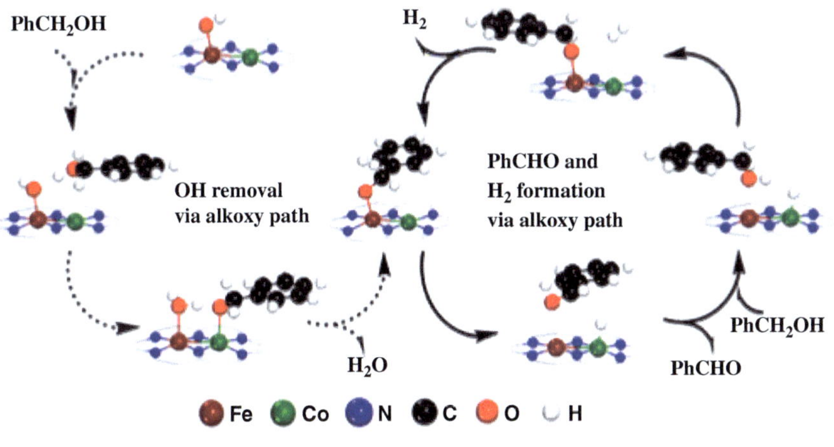


**Figure 3.3** Nitrogen-doped carbon nanomaterials-immobilized Fe–Co nanocluster-catalyzed alcohol dehydrogenation reactions. *Source:* [12]/with permission of American Chemical Society.

products (aldehyde) in up to 98% yields. Computational studies combined with kinetic investigations show that the alcohol dehydrogenation undergoes a completely different route in the presence of the nitrogen-doped carbon nanomaterials-immobilized Fe–Co nanocluster catalyst, compared with the catalytic processes by Fe or Co single-atom catalysts. The reaction catalyzed by the Fe–Co bimetallic nanoclusters features a much lower activation energy (30.1 vs. 38.1 kJ/mol by Fe single-atom catalyst or 42.0 kJ/mol by Co single-atom catalyst) hydroxyalkyl path, in which the C—H bond activation of alcohols occurs firstly.

Nitrogen-doped carbon nanomaterials are also employed to stabilize highly active palladium nanoclusters. It has been reported that porous nitrogen-doped carbon supports are able to stabilize palladium nanoclusters within 2 nm [13]. The strong chemical interactions between the palladium nanoclusters and the nitrogen atoms on the support effectively suppress the palladium aggregation during nitrogen fixing process at 400 °C in the presence of hydrogen gas. Meanwhile, the nitrogen-doped carbon nanomaterials-immobilized palladium nanoclusters exhibited a high turnover frequency (TOF) of 1092 $h^{-1}$ for the semi-hydrogenation of phenylacetylene under mild reaction conditions, along with a high selectivity (>93%) to styrene. Apart from the hydrogenation reaction, the palladium nanoclusters immobilized on porous nitrogen-doped carbon support demonstrate high catalytic activity for the hydrogen release from formic acid, with TOF up to 2365 $h^{-1}$. The nitrogen content in the porous nitrogen-doped carbon nanomaterials is as high as 16.87% and surface area of the material is as large as 1544 $m^2/g$ [14]. The high nitrogen content and large surface area protect the nanocluster from agglomeration and lead to uniformly distribution of the palladium nanoclusters. Thus, the porous nitrogen-doped carbon nanomaterials-immobilized palladium nanocluster catalyst showed robust structure and excellent catalytic activity even after 10 recycles. Being similar with the nitrogen doping method, sulfur can also be doped in the carbon nanomaterials [15], thereby helping to stabilize ultra-small metal nanoclusters for catalysis. As a representative, Yin and his research group doped sulfur in a carbon matrix and realized the stabilization of a series of metal nanoclusters, including Pt, Ir, Ru, Rh, and Os. These nanoclusters immobilized on the sulfur-doped carbon materials retain their structures and compositions even at high temperature up to 700 °C. The enhanced adhesion strength between metal nanoclusters and the sulfur-doped carbon support was discovered, greatly retarding the diffusion and migration of metal atoms in metal nanoclusters. Owing to their high stability, these metal nanocluster catalysts were applied in propane dehydrogenation reaction at 550 °C, showing promising catalytic performance.

## 3.2 Metal or Metallic Oxide Immobilized Metal Nanocluster

Up to now, metal and metal oxide are the most commonly used support for metal nanoclusters and efficiently protect and cooperate with metal nanoclusters in catalysis. The catalytic synergy between metal nanoclusters and their metal/metal oxide support is much more significant than the carbon nanomaterials-immobilized metal nanoclusters. In other words, the metal/metal oxide support on the one hand protect the metal nanoclusters from decomposition or aggregation in catalytic process, on the other hand and more importantly cooperate with metal nanoclusters for substrates activation. In this section, we dissect the structural features, catalytic properties and structural-property correlation of metal/metallic oxide-immobilized metal nanoclusters in reduction, oxidation, and photocatalytic reactions, which cover 90% of the catalytic processes catalyzed by metal nanoclusters.

### 3.2.1 Metal/Metallic Oxide Immobilized Metal Nanoclusters for Reduction

As we mentioned in Section 3.1.3, $CO_2$RR reaction is one of the most important reduction reactions widely applied in industrials. Metal/metallic oxide-immobilized metal nanoclusters show promising performance in the reduction. Park and coworkers synthesized a porous silver-immobilized copper nanocluster catalyst [16], which features ethanol as the final product. The catalyst was prepared by pulsed Cu electrodeposition on well-ordered Ag inverse opal structures. The Cu/Ag atomic percentage on the surface of the catalyst was determined to be 6.0/15.7 based on XPS analysis and the size of copper nanoclusters on the porous silver are less than 1.0 nm. Compared to the major production of ethylene by normal copper nanoparticles, this porous silver-immobilized copper nanocluster catalyst gives ethanol as the major product. This unique catalytic performance has been attributed to the following two aspects based on the operando shell-isolated nanoparticle-enhanced Raman spectroscopy and other in situ characterizations. (i) The ultra-small size of copper nanoclusters favors the adsorption of in situ generated CO and carbonaceous intermediates. (ii) Abundant Ag/Cu boundaries and adjacent spacing between copper nanoclusters and the support silver enhance the surface migration of CO, further facilitating the aggregation of CO and carbonaceous intermediates on the copper nanoclusters. Thus, the essential coverage of CO and carbonaceous intermediates on Cu triggers the production of ethanol.

With the similar copper nanoclusters as the activation sites while immobilized on zeolitic imidazolate framework-8, the selectivity of $CO_2$RR is switched to give methanol as the major product [17]. The copper nanoclusters are in situ generated in the zeolitic imidazolate framework-8 from $Cu(NO_3)_2$. The as-obtained nanoclusters are then captured by the support and stabilized. It should be noted that the immobilized nanoclusters are quite robust and kept the composition and structure even at high temperature (523 K) for over 150 hours. This catalyst enables thermal catalysis and features the reduction of $CO_2$ to methanol in high efficiency. Based on the authors' investigation, the high activity and stability of the catalyst should be ascribed to the following reasons. First, highly uniform distribution of active copper sites in a Zn-based MOF material provides sufficient adsorption sites for $CO_2$ and $H_2$. Second, the framework and the —OH groups remaining in the catalyst after Cu reduction and $CO_2$ transformation efficiently stabilize the copper nanoclusters activation sites. Third, the constituents of the Zn–O–Cu interface formed through significant metal–support interactions also act as activation sites for strongly adsorbing $CO_2$.

By employing Cu–Au alloyed nanoclusters immobilized on ZnO, the reaction temperature for the conversion from $CO_2$ to methanol can be reduced to 473 K. Mosrati and coworkers prepared this kind of catalyst [18]. The Au was proven to be alloyed in the core of the copper nanoclusters. The outer surface of the Cu–Au alloyed nanoclusters is mainly Cu which are probably alloying with the support Zn. The existence of Au has a pronounced effect on the electronic properties of the surface-exposed Cu in the ZnO-immobilized Cu–Au alloyed nanoclusters, by which means enhancing the catalytic activity for $CO_2$RR in water and selectivity toward methanol. Their study revealed that the presence of Au helped to stabilize the formation of nanoclusters in the small and narrow size range and reduce their catalytic temperature.

Apart from the $CO_2$RR, the metal-metal oxide-immobilized metal nanoclusters show high catalytic activity toward the reduction of nitrate. Very recently, Liu and coworkers synthesized a silver nanocluster catalyst immobilized on MXene [19]. The silver nanoclusters were prepared by a co-reduction method with the precursors of $AgNO_3$ and thiolate, and have well-defined composition of $Ag_9(SR)_8H_8$. These $Ag_9$ nanoclusters are evenly distributed on the prepared single-layered ultrathin $Ti_3C_2$ MXene material, with mean size of 1.16 nm. This catalyst enables a tandem catalytic process for nitrate reduction, significantly increasing the selectivity and Faradaic efficiency of $NH_3$. Herein, the MXene support not only plays roles in stabilizing the $Ag_9$ nanoclusters but also participates in the catalysis. The $Ag_9$ nanoclusters promote the conversion of $NO_3^-$ to $NO_2^-$, and the intermediate $NO_2^-$ is further transformed to $NH_3$ on MXene.

By immobilizing osmium nanoclusters in palladium metallenes, Prabhu and coworkers developed a highly efficient ORR catalyst [20]. A facile solvothermal reduction methodology was employed to prepare high-density Os nanoclusters on the surface of Pd metallenes. The loading content of Os can be modified through composition control (changing Os precursor content) and determined by quantitative energy-dispersive X-ray spectroscopy (EDX). They found that $Os_{5\%}$@Pd catalyst, consisting of $Os_{18}$ nanoclusters immobilized on palladium metallenes' surface, showed optimal alkaline HER and ORR catalytic performance with a low overpotential and a superior TOF value. The corresponding value is about 20 times higher than that by pure Pd metallenes and commercially available Pt/C catalyst. Their investigation reveals that the strong electronic interactions and charge redistribution from Os to Pd atoms lead to the formation of distinct active centers with appropriate d-band centers, which should be attributed to the enhanced ORR performance. Meanwhile, the downshift in Pd d-band center induced by Os immobilization weakens the interactions of oxygenated species with Pd surface sites, thus also providing optimized energetics along the ORR reaction pathway.

Metallic oxide immobilized metal nanoclusters can also be used for selective hydrogenation of aldehydes. Shirayama and coworkers synthesized $CeO_2$-immobilized Au nanoclusters [21]. They found that this catalyst featured selective hydrogenation of aldehydes with easily reducible functional groups, such as alkenes, alkynes, and halogens unchanged when syngas was used as the reducing reagent (Figure 3.4). The unique selectivity originates from the site blocking of the catalyst by CO in the

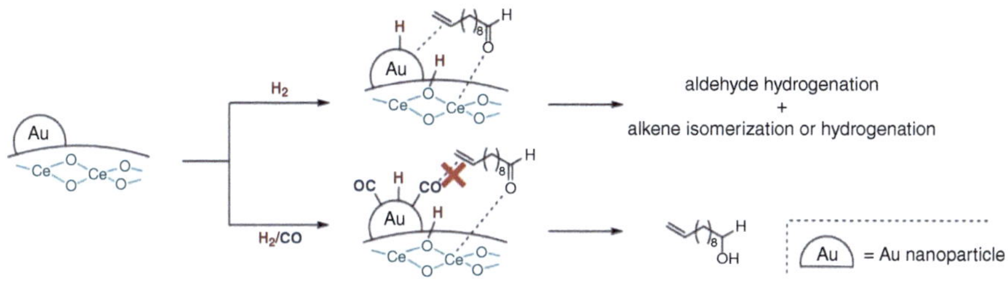


**Figure 3.4** $CeO_2$-immobilized Au nanoclusters catalyzed selective hydrogenation of aldehydes with syngas as the reducing reagent. *Source:* [21]/with permission of American Chemical Society.

syngas, thus significantly inhibiting the side reactions. When pure hydrogen gas was used as the reducing reagent, the substrates that demonstrated good selectivity under syngas conditions suffered over-reduction or dehalogenation.

### 3.2.2 Metal/Metallic Oxide Immobilized Metal Nanoclusters for Oxidation

CO oxidation reaction and $CO_2$ reduction reaction are a pair of counter reactions, and are equally important. CO is a highly toxic pollutant gas and a major contributor to the worsening air quality, mainly produced by incomplete combustion of fossil fuels. Palladium has been proven to be highly active toward oxidation of CO, while it is expensive and suffers from deactivation and poisoning. The deactivation mechanism is ascribed to the strong adsorption of CO to Pd. Once CO adsorbs on the surface of Pd catalysts, the $O_2$ dissociation sites are blocked and there is no more catalytic activation site for the CO oxidation reaction. An elegant strategy developed by Song and coworkers well overcome the above limitations [22]. Zeolite was used as support in this work, which is crystalline aluminosilicate with well-defined pore structures and high surface areas. Very small $Pd_4$ nanoclusters were immobilized in zeolite, on the one hand inhibiting the aggregation of Pd activation sites during catalysis thus reducing the dosage of expensive Pd catalyst, and on the other hand alleviating the CO poisoning process because of the significantly improved affinity of $O_2$ on molecular palladium nanoclusters compared to metallic palladium nanoparticles. Their theoretical and experimental results all suggest that the Pd nanoclusters immobilized in zeolite show a very small amount of CO adsorbed during the flow of CO and $O_2$ mixture, which should be attributed to the molecular feature and different coordination environment of palladium nanoclusters compared to palladium nanoparticles. The as-prepared Pd nanocluster catalysts immobilized in zeolite exhibit 85% conversion of CO oxidation at 90 °C, representing a quite promising results concerning CO oxidation under low temperature. In sharp contrast, palladium nanoparticles only exhibit less than 20% conversion even at 120 °C (Figure 3.5).

Serna research group investigated the catalytic performance of Pt nanoclusters and single-site Pt catalysts immobilized in chabazite zeolite in oxidation reactions [23].

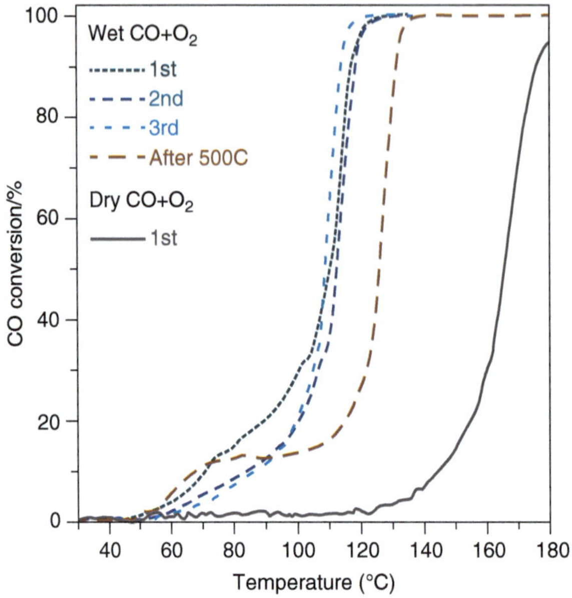


**Figure 3.5** Different catalytic activity of palladium nanoclusters immobilized in zeolite and palladium nanoparticles in CO oxidation reaction. *Source:* [22]/with permission of American Chemical Society.

Their results revealed the different activity of these two catalysts in methane oxidation and CO oxidation. Pt nanoclusters that are small than 1 nm are catalytically active toward the activation, scrambling and desorption of two $O_2$ molecules at moderate temperature, while the single-site Pt and Pt nanoclusters that are larger than 1 nm are inaccessible. Oxidation of methane at high temperature is attributed to stable single-site Pt catalyst even for Pt nanoclusters, which are transferred to the single-site Pt catalysts finally. The C—H cleavage of methane is determined to be less sensitive to changes in metal core compared to its effect on $O_2$ scrambling. In sharp contrast to the methane oxidation, metal nanoclusters are dominant in CO oxidation, compared to the much lower activity of single-site Pt catalysts.

Actually, methane oxidation and CO oxidation are the two oxidation reactions that are most studied for platinum group metals. Gashnikova research group used a $CeO_2$ nano-island confinement strategy to stabilize ultra-small palladium nanoclusters on $Al_2O_3$ support for highly efficient CO oxidation (Figure 3.6a) [24]. The preferential location of palladium nanoclusters on $CeO_2$ islands leads to a high local surface palladium concentration and low palladium loading. This strategy on the one hand overcomes the palladium nanoclusters deactivation issue and on the other hand reduces the amount of palladium catalysts used. Meanwhile, the Pd nanoclusters immobilized on $CeO_2$–$Al_2O_3$ nano-island demonstrate high activity toward CO oxidation at low temperature (100 °C). Yu research group investigated the palladium-support interaction-dependent catalytic activity of metal nanoclusters (Figure 3.6b) [25]. They found that $PdO_x$ nanoclusters anchored on defective alumina support ($Al_2O_{3\text{-}x}$) showed much higher activity in methane oxidation reaction,

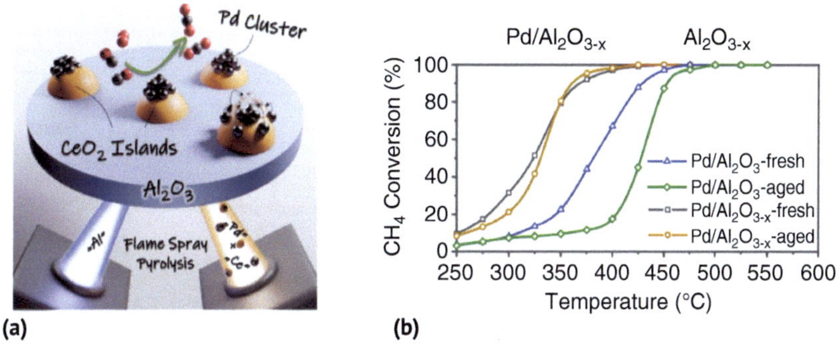


**Figure 3.6** (a) $CeO_2$ nano-island-immobilized Pd nanoclusters for catalytic CO oxidation. *Source:* [24]/John Wiley & Sons/CC BY 4.0. (b) Defective alumina-immobilized Pd nanoclusters for catalytic methane oxidation. *Source:* [25]/Springer Nature/CC BY 4.0.

compared to the similar palladium catalyst immobilized on nondefective alumina support ($Al_2O_3$). The stronger Pd nanocluster–support interactions would facilitate the high dispersion and high content of $Pd^{2+}$ oxidation state of the Pd catalyst, and resist the deactivation of the catalyst by carbonates, bicarbonate, and water.

Organic ligand-protected atomically precise metal nanoclusters are also involved in CO oxidation reaction. Wang and coworkers reported $MnO_2$-immobilized atomically precise $Au_{25}(SR)_{18}$ nanocluster for catalytic CO oxidation at very low temperature (−70 °C) [26]. The $Au_{25}(SR)_{18}$ nanocluster (SR = $SC_2H_4Ph$) and α-$MnO_2$ nanowire were synthesized by the wet-chemical reduction and the hydrothermal methods, respectively, and then mixed, stirred, and pretreatment under high temperature to prepare the $MnO_2$-immobilized $Au_{25}(SR)_{18}$ nanocluster catalysts. The experimental results reveal that the $Au_{25}(SR)_{18}$ nanoclusters are relatively stable and remain their composition and structure even after pretreatment at 250 °C. Higher pretreatment temperatures result in the aggregation of $Au_{25}(SR)_{18}$ nanoclusters. The $MnO_2$-immobilized $Au_{25}(SR)_{18}$ nanocluster catalysts pretreated at 250 °C exhibits extremely high activity for CO oxidation at ultra-low temperature, achieving 100% conversion at −50 °C and 50% conversion at −70 °C. The specific reaction rate reached as high as $1.52mol_{CO} \cdot h^{-1} \cdot g_{Au}^{-1}$ at −70 °C. In situ modulation excitation spectroscopy (MES) studies revealed that the on-top CO adsorption on the $Au_{25}(SR)_{18}$ nanocluster is crucial in the CO catalytic oxidation process. Meanwhile, the surface ligands of the nanocluster are also involved in the catalysis and affect the negative apparent activation energy at intermediate temperatures, thus providing a reference basis to avoid the negative apparent activation energy.

Masuda and coworkers prepared layered double hydroxide (LDH)-immobilized $Au_{25}(p\text{MBA})_{18}$ and $Au_{102}(p\text{MBA})_{44}$ nanocluster catalysts ($p$MBA = $p$-mercaptobenzoic acid) [27]. The electrostatic interaction between the deprotonated $p$MBA and cationic LDH is the key for stabilizing both kinds of nanoclusters on the support (Figure 3.7). It was found that LDH-immobilized $Au_{102}(p\text{MBA})_{44}$ nanoclusters showed much higher catalytic activity (22 times higher) than DH-immobilized $Au_{25}(p\text{MBA})_{18}$ nanoclusters in catalytic oxidation of 5-hydroxymethylfurfural

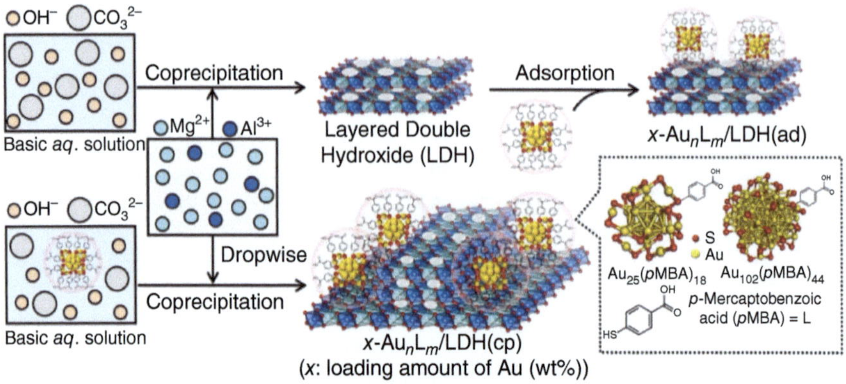


**Figure 3.7** $Au_{25}$ and $Au_{102}$ nanocluster catalysts immobilized on LDH for catalytic 5-hydroxymethylfurfural oxidation. *Source:* [27]/with permission of American Chemical Society.

(Figure 3.7). The 5-hydroxymethylfurfural was sequentially converted to 5-hydroxymethyl-2-furancarboxylic acid, 5-formyl-2-furancarboxylic acid, and 2,5-furandicarboxylic acid. Based on the atomically precise structures of both nanocluster catalysts and control experiments on the pressure-dependent rate constant of the $O_2$ and the substrate-dependent activation barrier, the higher catalytic activity of $Au_{102}(pMBA)_{44}$ nanocluster catalyst is ascribed to the size-specific electronic structure, that is, hydride abstraction from the adsorbed alkoxide is accelerated by the electron transfer from $Au_{102}(pMBA)_{44}$ (denoted as $Au_{102}$ because some of ligands have been removed during the catalyst preparation process) and the higher affinity of hydride to $Au_{102}$.

### 3.2.3 Metal/Metallic Oxide Immobilized Metal Nanoclusters for Photocatalysis

Metal and metallic oxide supports possess good light absorption and conversion capacity, thus cooperating with metal nanoclusters and demonstrating photo catalytic performance. We have one chapter (Chapter 8) specially discussing metal nanocluster photocatalysis. In this part, focusing on the interactions between metal/metallic oxide support and the metal nanoclusters, we give a brief introduction on the photo catalytic property of metal nanoclusters-immobilized on metal/metallic oxide.

There is a very typical example concerning metal/metallic oxide immobilized metal nanoclusters for efficient photocatalysis [28]. A composite photocatalyst $(Cu_1Pd_2)$@PCN-222(Co) consisting of $Cu_1Pd_2$ nanoclusters and porphyrin-based metal–organic frameworks (MOFs) PCN-222(Co) was designed and synthesized by Fu research team. The PCN-222(Co) is a porphyrin-based MOF material with Co single-site activation centers, which is fabricated by the reaction of $ZrCl_4$ and CoTCPP. $Cu_1Pd_2$ nanoclusters are confined in the MOF material (Figure 3.8). This composite catalyst demonstrates high efficiency in converting $CO_2$, iodobenzene,

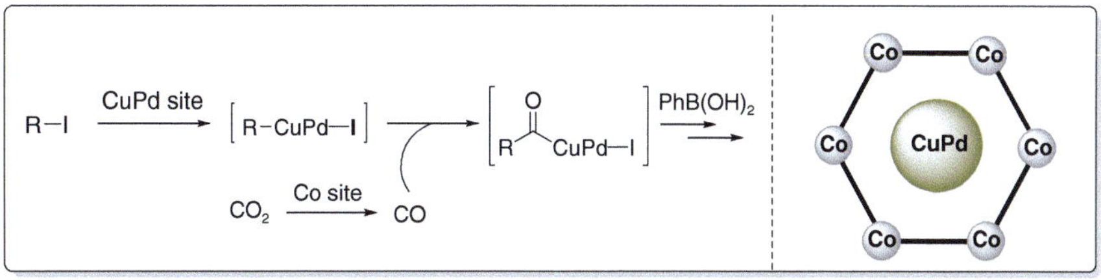


**Figure 3.8** $(Cu_1Pd_2)$@PCN-222(Co) composite catalyst for highly efficient photocatalysis of $CO_2$, iodobenzene, and phenylboronic acid.

and phenylboronic acid to benzophenone. During the catalysis, the porphyrin part of this composite catalyst acts as a photosensor, which is exited under light irradiation for the electron-hole separation. The Co single-site part that accepts electrons from porphyrin then plays roles in photoreduction of $CO_2$ to CO. The $Cu_1Pd_2$ nanocluster part activates iodobenzene and phenylboronic acid, producing the phenyl-Pd-phenyl intermediate that is inserted by CO and finally affording benzophenone (Figure 3.8). It should be noted that lack of any part in this composite catalyst results in low activity or even switched products, such as the simple cross-coupling product Ph–Ph. It is the strong interaction and precision collaboration between metal nanoclusters and MOF support that contribute to the sequential process of oxidative addition, CO insertion, transmetalation, and reductive elimination elementary reactions under light irradiation and the effective production of benzophenone.

Apart from the photocatalytic cascade reaction involving $CO_2$ reduction, metal/metallic oxide immobilized metal nanoclusters exhibit high performance in methane and $O_2$ activation reactions. A recent report from Wang research group indicates that compared to Ag, Cu, Pt, Pd, Ni, and Ru nanoclusters, Au nanoclusters that immobilized on ZnO nanoparticles show the highest activity and selectivity toward the oxidative coupling of methane [29]. The ZnO-immobilized Au nanocluster catalyst with 1 wt% of Au content delivers the remarkable $C_2$–$C_4$ hydrocarbon production rate of 683 $\mu mol \cdot g^{-1} \cdot h^{-1}$ under light irradiation. This catalyst also showcases excellent structural and catalytic stability under light over two days. The synergy between ZnO support and the Au nanocluster is crucial for the high performance of the hybrid catalyst. The $Zn^+$–$O^-$ sites act as photosensors, enabling methane excitation to methyl intermediates, while Au nanoclusters participate in the next step as a relay for the coupling of activated methane. The metal-activated methane interaction controls the selectivity of the oxidative coupling of methane, and the strong d–σ orbital hybridization of Au reduces the metal–C–H bond angles and steric hindrance, thereby showing higher activity and selectivity than the other six metal nanoclusters immobilized on ZnO.

By immobilization gold nanoclusters on $TiO_2$, Chen and coworkers prepared $Au_{25}(PPh_3)_{10}Cl_2(SC_3H_6SiO_3)_5/TiO_2$ nanocluster composite catalysts [30]. The golds nanoclusters connect with the $TiO_2$ support by the Si–O and Ti–O chemical interactions, endowing the whole catalytic system with high stability during catalysis. Moreover, the efficient electron transfer between the gold nanoclusters and the $TiO_2$ support triggers the reduction of $O_2$ to produce $O_2^{\bullet-}$, which are response to the subsequential selective oxidation of amine (Figure 3.9). Specifically, the combined $Au_{25}(PPh_3)_{10}Cl_2(SC_3H_6SiO_3)_5/TiO_2$ catalytic system catalyze the oxidation of benzylic amine under blue LED irradiation, giving the desired imine product with TOF up to 1522 $h^{-1}$. By using $TiO_2$ as the support, the same research group synthesized a $Au_{12}Cu_{13}(PPh_3)_{10}I_7/TiO_2$ catalytic system and realized a highly efficient photocatalytic methanol oxidation [31]. Being different from the $Au_{25}(PPh_3)_{10}Cl_2(SC_3H_6SiO_3)_5/TiO_2$ catalytic system mentioned above, the gold-copper alloyed nanoclusters connect with the $TiO_2$ support by physical interaction. Nonetheless, this catalytic system is robust and catalyzes the oxidation of methanol, giving methyl formate with 92% conversion and 84% chemoselectivity. The structural stability and catalytic durability of $Au_{12}Cu_{13}(PPh_3)_{10}I_7/TiO_2$ are attributed to the Au–Cu synergistic effect of the nanocluster itself and the interaction between nanoclusters and the $TiO_2$ support.

The metal/metallic oxide immobilized metal nanoclusters were also applied for the rhodamine B photodegradation [32], featuring high photostability during long-time light illumination conditions. Multifunctional branched poly-ethylenimine was utilized as a surface charge modifying, reducing and stabilizing agent to prepare $SiO_2$ sphere-immobilized gold nanoclusters protected by glutathione. A sequential coating of thickness controlled $TiO_2$ shells further improve the photostability and photocatalytic efficiency of the composite material. The high photostability of the functionalized $SiO_2$ sphere-immobilized gold nanoclusters is attributed to the high reduction ability of branched poly-ethylenimine, by which the organic ligand

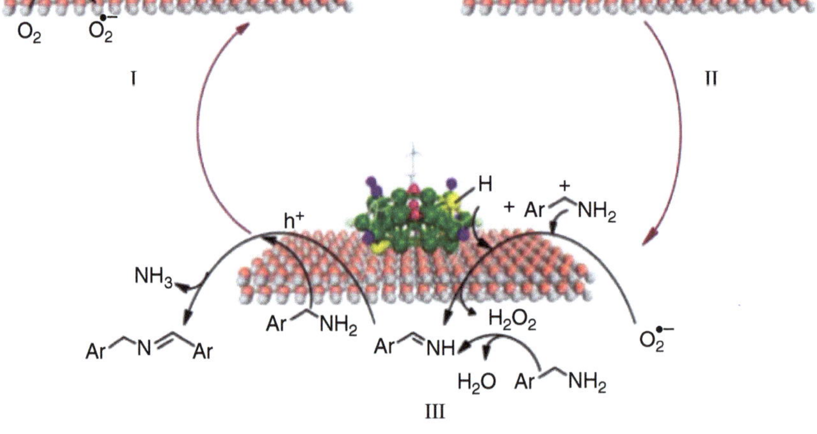


**Figure 3.9** $Au_{25}(PPh_3)_{10}Cl_2(SC_3H_6SiO_3)_5/TiO_2$ catalytic system for photocatalytic oxidation of benzylic amine. *Source:* [30]/with permission of American Chemical Society.

oxidation process over glutathione-protected gold nanoclusters can be effectively inhibited, and this in turn leads to the improvement of durability. Moreover, the sequential coating of thickness tunable $TiO_2$ shells for the construction of core–shell $SiO_2$-immobilized gold nanoclusters has been demonstrated to significantly enhance the photoactivity by regulating the photoelectrochemical and adsorption properties of the nanostructures synergistically.

To prepare atomically precise and photo-catalytically active metal nanoclusters immobilized on metallic oxide support, Kawawaki and coworkers systematically investigated the ligand-desorption process of 2-phenylethanethiolated $Au_{25}$ nanoclusters during calcination using five experimental techniques [33]. Their results revealed that the ligand desorption process consists of three steps, that is, the ligand dissociation on the surface of the $Au_{25}(SR)_{18}$ nanoclusters, the adsorption of the generated compounds on the metallic oxide support, and the desorption of the compounds from the support. They compared the ligand desorption process of two kinds of immobilized metal nanocluster catalysts: $Au_{25}(SR)_{18}/BaLa_4Ti_4O_{15}$ and $Au_{25}(SR)_{18}/Cr(OH)_3/BaLa_4Ti_4O_{15}$. For the former catalyst, the dissociation of S–C and Au–S bonds of $Au_{25}(SR)_{18}$ occurs firstly during its calcination. The desorbed compounds are then adsorbed onto the support. With the increment of the temperature, most of the compounds on the support are desorbed as styrene or $CO_2$. Although most of the thiolate ligands can be removed from the surface of $Au_{25}(SR)_{18}$ nanocluster by calcination at 300 °C and maintaining the size of $Au_{25}$, some organic compounds and thiol oxides still remain on the $BaLa_4Ti_4O_{15}$ support even at 500 °C. For the latter catalyst, the desorption process is similar with the former catalyst, while the $Au_{25}(SR)_{18}$ nanoclusters loaded on the $Cr(OH)_3/BaLa_4Ti_4O_{15}$ support aggregate seriously with time. Light irradiation soon after calcination has been confirmed to be efficient to prevent the aggregation, and the average size of nanoclusters is 1.5 nm, indicating the successful suppression of the aggregation process. The as-obtained metallic oxide-immobilized Au nanocluster catalysts all demonstrate high activity and stability toward water-splitting reaction.

## 3.3 Polymer or Supermolecule Immobilized Metal Nanocluster

Polymers and supermolecules possess multidentate coordination sites and huge surface area. They stabilize metal nanoclusters by ligation of surface metal atoms as well as physically occupying space around the metal nanocluster. Nanoclusters of various metals, such as Au, Ag, Pd, Pt, Ir, etc., have been prepared by immobilization with polymers or supermolecules. Several factors have been verified to affect the size, composition and structure of polymers/supermolecules-immobilized metal nanoclusters. First, chain length of polymers plays important roles in the size of the as-obtained metal nanoclusters. For example, although the stabilization is independent of the quantity of polymers employed, it has been shown that the polymer chain length has significant influence on the size of the Ni nanoclusters obtained [34]. Moreover, we can expect that polymers with longer chains would attach to more nanoclusters than the polymers with shorter chains [35]. Second, many polymers contain adventitious water, leading to compositionally ill-defined metal nanoclusters that consist of $H_2O$, $OH^-$, etc. on their surface [36]. The corresponding studies have shown that the amount of $H_2O$ presenting during the formation process of polymers/supermolecules-immobilized metal nanoclusters influence both the stabilization and formation of $Ir_n$ nanoclusters. The removal of the water by drying polymer solutions over activated molecular sieves for more than 12 hours would help to solve this problem. Third, the metal halide nanocluster precursors and boron-containing reductants significantly impact the composition of the as-obtained metal nanoclusters as a result of the interactions between halides/B-containing compounds and the polymers/supermolecules protectors [37].

There are five kinds of polymers that are widely utilized for stabilizing metal nanoclusters, that is, poly(vinylpyrrolidone) (PVP) [38–40], poly(vinyl alcohol) (PVA) [41], poly(methyl methacrylate) (PMMA) [42], poly(styrene) (PS) [43], and poly(bis(ethoxy)phosphazene) (PBEP) [44]. Among these polymers, PVP is mostly used. However, these PVP-immobilized metal nanoclusters are frequently prepared in water beginning with chlorinated precursors, such as $HAuCl_4$ and $RhCl_3$, leading once again, to the possibility of halide stabilization and compositionally ill-defined nanocluster surface as mentioned above. Additionally, as in photo-type PVP-immobilized metal nanoclusters, an alcohol is frequently utilized as both solvent and reductant. This protocol again leads to compositionally ill-defined nanocluster surface which may be coordinated by alkoxy molecules or possibly the resultant oxidized carbonyl products, such as $H_2CO$. The study of PVA-immobilized metal nanoclusters could trace back to the year of 1941 by Rampino and coworkers, who developed PVA-immobilized Pd [45] and Pt [46] nanoclusters. Since then, Au and Ag nanoclusters immobilized by PVA have also been reported. PVA polymers have the potential to be both reductant and nanocluster stabilizer during the synthesis of metal nanoclusters. Being similar with PVP-immobilized metal nanoclusters, compositionally poorly defined metal nanoclusters are the rule to date in putative PVA-immobilized metal nanoclusters. For the other three kinds of polymers, PMMA is most similar to PVP in providing a modest level of immobilization. PS and PBEP have been found to be ineffective for the formation and immobilization of catalytically active $Ir_n$ nanoclusters.

Although the chemical and physical interactions between the surface of metal nanoclusters and the polymers/supermolecules occupy the surface metal atoms of metal nanoclusters as a result of the steric blocking [47, 48], polymers or supermolecules-immobilized metal nanoclusters have been shown to be active in several catalytic reactions. For example, PVP-immobilized metal nanoclusters demonstrate interesting activity in hydrogenation [49], Suzuki coupling reaction [50, 51], and Heck coupling reactions [52]. PVA-immobilized Pt nanoclusters show high catalytic efficiency in oxidizing methyl viologen during the light-induced hydrogen evolution from water as part of photo redox process for solar cells [53].

There is a kind of special polymer, that is, siloxane polymer that has been used to stabilize metal nanoclusters [54]. Conventional examples include poly(methylhydrosiloxane) (PMHS) [55] and poly(octadecylsiloxane) (PODS) [56].

The feature of this kind of polymer is that their silanol moieties are capable of reducing metal precursors. Thus, reducing agents are generally not essential, resulting in the advantage of simple reaction conditions. Au, Pt, Pd, and Ag nanoclusters were reported based on the siloxane polymer immobilization [57]. The authors claimed that the Si–H bonds play important roles in reducing the organometallic precursors, while this advantage is relatively overshadowed by the ill-characterized nature of the oxidized product and stabilizer as well as by the extensive metal nanocluster agglomeration shown by TEM. Branched polymers are also capable of stabilizing metal nanocluster [58]. Although they generally have a less well-defined and uncontrolled structure, branched polymers such as polyglycerol [59] and polyaramids [60] have been reported to yield Pd nanoclusters that are stable in solutions for several months. Especially, the branched polymer-immobilized Pd nanoclusters were capable of catalyzing hydrogenation of cyclohexene at a rate much faster than the commercial supported Pd/activated charcoal catalyst. In some cases, aggregation of metal nanoclusters immobilized by polyaramids or polyglycerols was observed by TEM and SEM.

As mentioned above, one of the drawbacks by using polymers or supermolecules to stabilize metal nanoclusters is the ill-defined structures of the as-obtained nanoclusters [61], but there are still elegant works that prepared atomically precise metal nanoclusters immobilized by polymers or supermolecules for efficient catalysis. We herein give two typical examples: a polymer-immobilized metal nanocluster and a supermolecule-immobilized metal nanocluster.

The research group of Wang made major contribution to the polymer-immobilized metal nanoclusters with atomically precise structures. The breakthrough comes in 2022 [62], showcasing the high stability and catalytic activity of poly-$Cu_{14}$cba, poly-$Cu_6Au_6$cbz, and poly-$Cu_6Ag_4$cbz nanoclusters. These polymer-immobilized metal nanoclusters were prepared by electropolymerization of carbazole ligands-protected metal nanoclusters. Initially, three atomically precise metal nanoclusters were synthesized by a conventional wet chemical reduction method. Their molecular formulas are determined to be $Cu_{14}(cba)_6(S^tBu)_8$ (abbreviated as $Cu_{14}$cba), $Cu_6Au_6(cbz)_6(S^tBu)_6$ (abbreviated as $Cu_6Au_6$cbz), and $Cu_6Ag_4(cbz)_{10}(PPN)_2$ (abbreviated as $Cu_6Ag_4$cbz). Herein, Hcba represents 5-(9*H*-carbazol-9-yl)pentanoic acid, Hcbz represents 9-(prop-2-yn-1-yl)-9*H*-carbazole, and PPN represents 2,6-bis(diphenylphosphino)pyridine. Subsequentially, based on the polymeric active sites on the carbazole molecules (Figure 3.10), all of the three nanoclusters were polymerized into hybrid materials by the electropolymerization method, and the as-obtained polymer-immobilized metal nanoclusters are denoted as poly-$Cu_{14}$cba, poly-$Cu_6Au_6$cbz, and poly-$Cu_6Ag_4$cbz. Notably, the electropolymerization occurs at the indium-doped tin oxide glass electrode and these polymer-immobilized metal nanoclusters are formed on the electrode as a poly-metal nanocluster membrane after ten cycles of the electropolymerization process. Characterizations including UV–vis, FT-IR, EXAFS, TEM, etc. verify that the metal nanoclusters immobilized on the polymers basically remain their original chemical composition and structure.

Further investigations revealed that the polymer immobilization not only improves the stability of $Cu_{14}$cba, $Cu_6Au_6$cbz, and $Cu_6Ag_4$cbz nanoclusters catalyst by eliminating their efficiency deactivation but also facilitates electron transport to

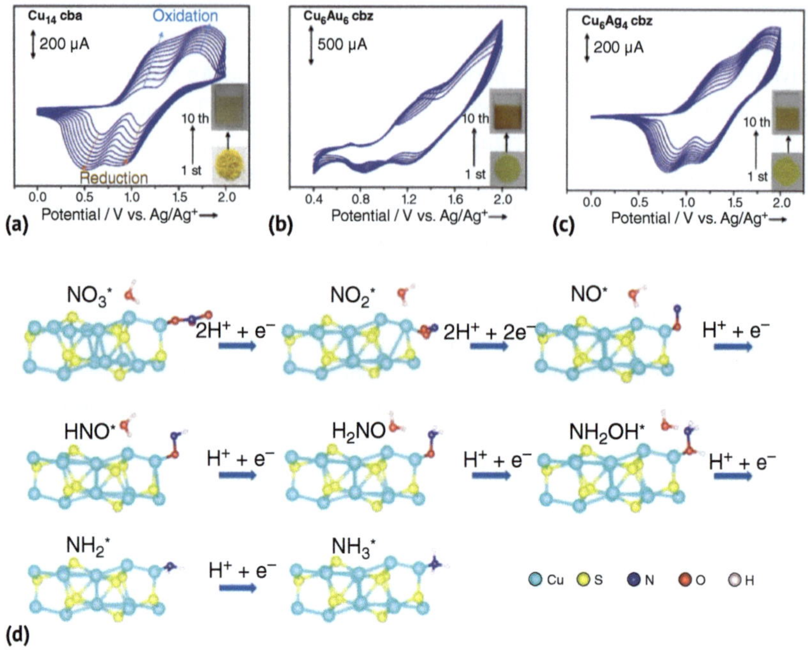


**Figure 3.10** Electropolymerization method for the preparation of polymer-immobilized metal nanoclusters and their improved electrocatalytic performance. *Source:* Reproduced with permission from [62]/John Wiley & Sons.

the metal nanocluster catalytic sites as a result of the polymerized extended network. Thus, the poly-$Cu_{14}$cba, poly-$Cu_6Au_6$cbz, and poly-$Cu_6Ag_4$cbz nanocluster-based catalysts show excellent electroreduction $NO_3^-$ to $NH_3$ performance (Figure 3.10). As a proof-of-concept, poly-$Cu_{14}$cba was evaluated by the authors as an electrocatalyst for reducing nitrate to ammonia, exhibiting fourfold ammonia yield rate and twofold Faraday efficiency enhancement compared to that of the isolated $Cu_{14}$cba crystals.

By applying the similar strategy, Wang and coworkers successfully combined the single atom site catalyst and metal nanocluster catalyst in a polymer [63]. They firstly synthesized carbazole ligand-modified $[Au_8(dppp)_4(CzPA)_2]^{2+}$ nanocluster (HCzPA = 9-(4-ethynylphenyl)carbazole)) and carbazole ligand-chelated single-atom metal complex M-DCP (M = Fe, Co, Ni, Cu, Zn; DCP = 3,8-di(9H-carbazol-9-yl)-1,10-phenanthroline). In the next step, by using the electropolymerization method, they prepared uniformly dispersed metal nanoclusters modified with single-atom sites that were loaded in a poly-carbazole matrix. Owing to the fact that functionalization of single-atom sites enables altering the electronic structures of the $Au_8$ nanoclusters, the polymer-immobilized $Au_8$ nanocluster-single atom site composite catalysts amplify their electrocatalytic reduction performance of $CO_2$ to CO by 18.07-fold, compared to the isolated $Au_8$ nanocluster catalyst. Systematic characterization and theoretical modeling revealed that the rearrangements of the electronic structure of metal nanoclusters not only strengthen the adsorption of the key intermediate COOH* but also establish a favorable reaction pathway for the $CO_2$ reduction reaction.

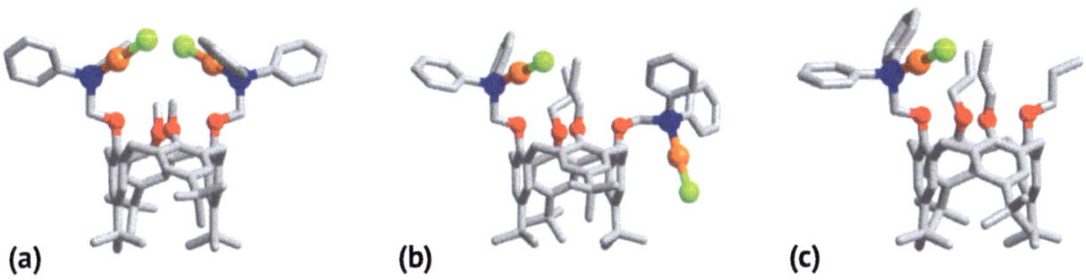


**Figure 3.11** Au nanoclusters confined in calix[4]arene supermolecules with different structures.

The improvement of stability of the metal nanocluster catalysts was also observed due to inhibition of nanocluster aggregation by the polymerization process.

The typical example concerning the supermolecule-immobilized atomically precise metal nanoclusters is achieved by using calix[4]arene macrocycles as the stabilizer [64]. It was found that the macrocycles with different structures lead to gold nanoclusters confined in the calix[4]arene with different sizes (Figure 3.11). Specifically, by using *tert*-butyl-calix[4]arene-$(OMe)_2(OCH_2PPh_2)_2$ as the supermolecule stabilizer, a $[Au_{11}L_2Cl_3]^{2+}$ nanocluster (L = calixarene phosphine ligand) was identified to be confined inside the supermolecules. It was further confirmed by UV–vis, ESI-MS, high-angle annular dark-field scanning transmission electron microscopy (HAADF-STEM), and XPS. Most importantly, it was found that the accessibility of these supermolecule-immobilized metal nanoclusters verified with each other, mimicking amino acid-surrounded enzymes. By employing gold-affinity and photoluminescent 2-naphthalenethiol as the probe, the authors found that 25% of all the gold atoms within the 0.9 nm-sized $Au_{11}$ nanocluster bind to the chemisorption probe 2-naphthalenethiol, while the accessibility dramatically decreases with the calix[4]arene-immobilized 1.1 nm-sized gold nanocluster and the calix[4]arene-immobilized 4 nm-sized gold nanoparticle. The authors also synthesized $PPh_3$-protected 0.9 nm-sized $Au_{11}$ nanoclusters, such as $Au_{11}(PPh_3)_7(SCN)_3$ and $Au_{11}(PPh_3)_8Cl_2$. Both nanoclusters demonstrate an entire lack of 2-naphthalenethiol binding, indicating there is none of exchange of thiols for phosphines in $Au_{11}$-type nanoclusters. The high accessibility of calix[4]arene-immobilized 0.9 nm-sized $Au_{11}$ nanocluster would be valuable for its activation of organic substrates and further application in catalysis. After this work, the research groups of Hassinen [65], Guan [66, 67], Wang [68, 69], etc. have developed various of atomically precise metal nanoclusters protected by calix[4]arene supermolecules and discovered their catalytic properties.

## 3.4 Dendrimer Immobilized Metal Nanocluster

Dendrimer is a kind of special and uniform polymer with the connectivity of fractal trees and generally radial symmetry. It has been utilized for stabilizing metal nanoclusters since 1998, in which year Zhao and coworkers reported the synthesis of copper nanoclusters immobilized in poly(amidoamine) (PAMAM) dendrimers [70]. Dendrimers have a precisely branched and well controlled structure. The multivalent interactions between the dendrimer support and the metal nanoclusters inside efficiently protect the as-generated metal nanoclusters from aggregation or leaching. Meanwhile, the weak nature of individual interactions with surface sites enables the metal nanoclusters catalytic activity. The dendrimers employed in metal nanocluster immobilization generally are nitrogen-based, such as PAMAM, poly(propylene imine) (PPI), etc. [71]. It has been demonstrated by spectrophotometric titration that the number of metal precursor atoms adsorbed in dendrimer corresponds to the number of interior amine sites. The most attractive advantage by employing dendrimers as the stabilizers is their good control on the size, composition and structure of metal nanoclusters immobilized inside (Figure 3.12).

After the pioneer work from Richard M. Crooks, there are various dendrimer-immobilized metal nanoclusters that have been prepared and reported [72–80]. Catalysis employing dendrimer-immobilized metal nanoclusters is followed since then and several research groups investigated the cross-coupling reactions [81–86], such as Heck and Suzuki reactions catalyzed by Pd nanoclusters encapsulated in PAMAM and PPI dendrimers. Promising results such as 100% selectivity in Heck reaction between iodobenzene and *n*-butyl acrylate at 90 °C were achieved based on this kind of catalyst [87]. As a representative, Ogasawara and kato prepared ~2.0 nm-sized Pd nanoclusters immobilized in PAMAM dendrimer by a polymerization-induced phase separation technique [88]. The Pd nanoclusters formed in situ with the formation of dendrimer and the addition of reductant is not essential. Catalytic performance of the dendrimer-immobilized Pd nanoclusters was

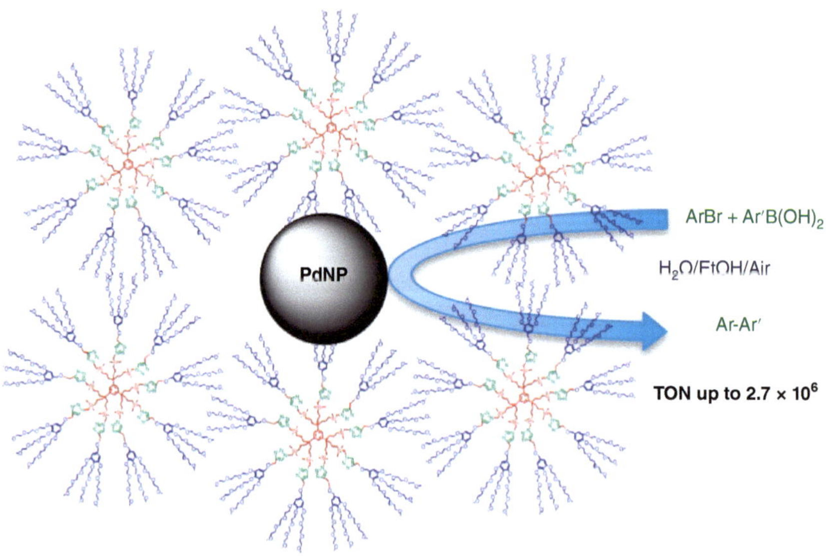


**Figure 3.12** Dendrimer as the support for immobilization of metal nanoclusters. *Source:* [71]/with permission of American Chemical Society.

studied in a heterogeneous Suzuki coupling reaction of para-bromoacetophenone and phenylboronic acid using water as the solvent. A 1 mol% of polymer gave the coupling product para-acetylbiphenyl in >90% yield in two hours. The TON of the reaction was determined to be 85,000 and the dendrimer-immobilized Pd nanocluster catalyst is recoverable and recyclable without obvious loss of catalytic activity and Pd content based on the ICP analysis.

Whereas commercially available dendrimers such as PAMAM and PPI are usually employed as the stabilizer for metal nanoclusters (i.e. Pd nanoclusters), the design of ferrocenyl group functionalized triazoyl dendrimers by click reaction allows evaluation of metal content that immobilized on the dendrimer (Figure 3.13) [89, 90]. Ornelas and coworkers contributes to this research area as early as 2007. This kind of recognition can be realized owing to the ferrocenyl group attached to triazoyl dendrimer branch termini, and can be used for precisely control of the size and structure of as-obtained metal nanoclusters. For example, the electrochemical titration indicates a 1/1 of $Pd^{II}$ precursor/triazole branch stoichiometry. After reduction of the $Pd^{II}$ precursor to $Pd^0$ nanoclusters, the forecasted number of palladium atoms and the size of the Pd nanoclusters immobilized in the dendrimer are able to be verified by TEM and other characterizations. Dendrimers containing 27 and 81 ferrocenyl triazole termini lead to immobilized Pd nanoclusters with 1.3 and 1.6 nm, respectively. Interestingly, smaller dendrimer containing nine ferrocenyl triazole termini results in larger Pd nanoclusters (2.8 nm) instead, because several dendrimers surround one Pd nanocluster rather than several Pd nanoclusters encapsulated inside on dendrimer [91]. These three different dendrimer-immobilized Pd nanoclusters were employed for Suzuki reaction, and it was found that they were equally efficient. The catalytic activity was independent of the size of the Pd nanoclusters, whether they are encapsulated inside the larger dendrimers or surrounded by smaller dendrimers. By using the dendrimer prepared from click reaction as the stabilizer, the Pd nanoclusters show extremely high catalytic activity under low Pd loading conditions. With 1 ppm of Pd content, a TON of 540,000 was obtained at room temperature although the catalysis is relatively slow [92]. High Pd concentration did not give quantitative coupling conversion due to the atom quenching by the Pd nanoclusters in analogy to the high temperature Heck "homeopathic" mechanism [93]. When the terminal position of the dendrimer is sulfonated, they are water-soluble and can be used to immobilize Pd nanoclusters for catalyzing Suzuki reaction under aqueous conditions. The corresponding investigation demonstrates that the TON and TOF are up to 10,000 and 1500 $h^{-1}$, respectively [94]. The value is higher than the results catalyzed by Pd nanoclusters immobilized in common sulfonate polymers [95] that are prepared by click reaction as well.

Triazolyl linkages combined with tri(ethylene glycol) termini in Percec-type tripodal dendrons are utilized for designing water-soluble and catalytically active dendrimers [96, 97]. $Pd^{II}$ precursors such as $K_2PdCl_4$ coordinated inside the dendrimer containing 27 tri(ethylene glycol) termini and 9 triazole linkers can be reduced to give ultra-small Pd nanoclusters, with the size of 1.4 nm and approximately

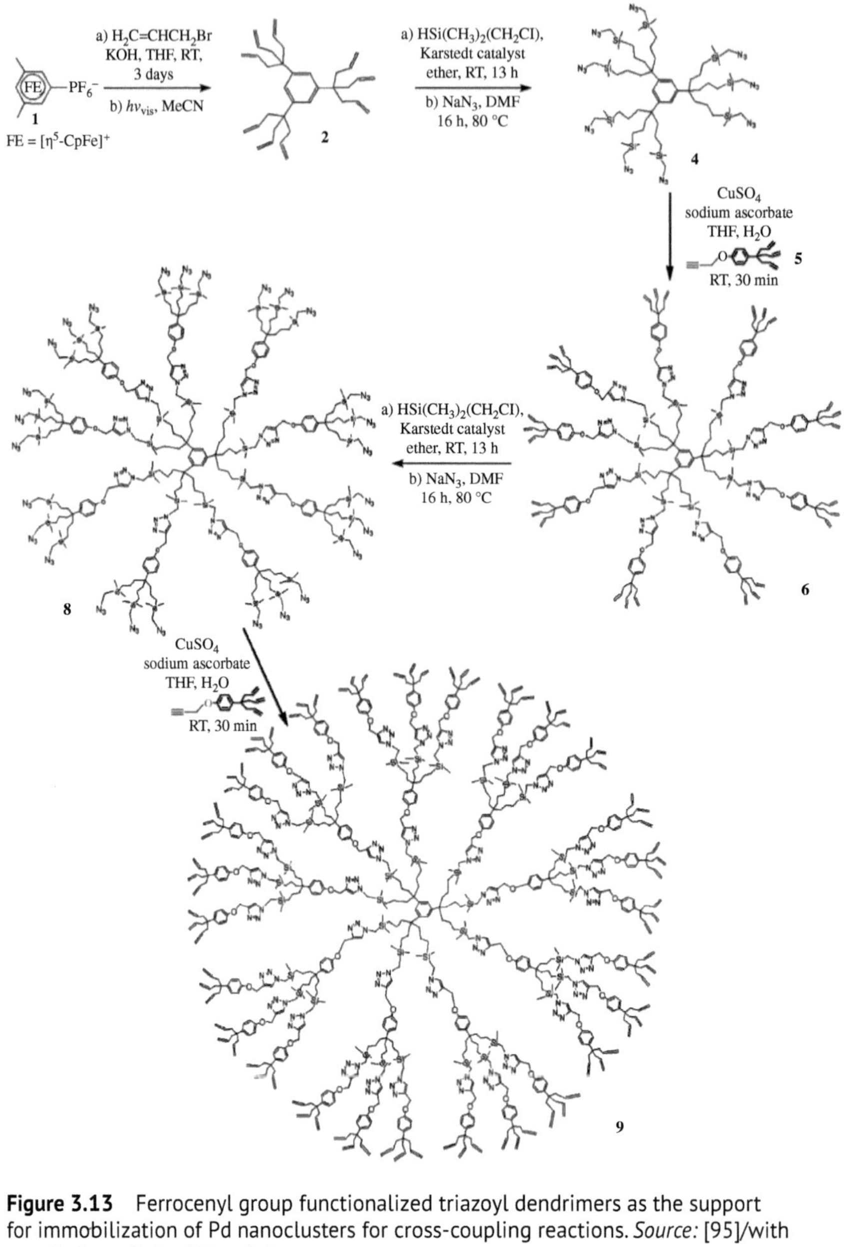


**Figure 3.13** Ferrocenyl group functionalized triazoyl dendrimers as the support for immobilization of Pd nanoclusters for cross-coupling reactions. *Source:* [95]/with permission of John Wiley & Sons.

100 palladium atoms. This kind of catalyst is able to trigger Suzuki reaction of inactivated bromoarenes with phenylboronic acids at 80 °C with TON reaching to one million [98]. For more challenging substrate such as chloroarenes, the result is less impressive due to the decomposition of catalyst at temperature above 100 °C. It also catalyzes efficient Sonogashira reaction at 100 °C with the amount of Pd/substrate being 0.0001/1. For obvious comparison of the dendrimer-immobilized

Pd nanoclusters with the other catalysts in cross-coupling reactions, we list the results in Table 3.1, which was summarized by Christophe Deraedt and Didier Astruc [93].

Further adsorption of dendrimer-immobilized metal nanoclusters onto mesoporous silica provides a new adventure for developing highly robust and efficient catalysts. The dense multivalency, shape persistence, and structural uniformity of dendrimers, combined with the robustness and high specific surface area of mesoporous silica, endow the as-obtained metal nanocluster catalysts with both structural stability and catalytic activity. Therefore, they demonstrate unique performance in some challenging transformations that otherwise cannot be accomplished by other heterogeneous catalytic system.

Huang and coworkers synthesized this kind of mesoporous silica/dendrimer-immobilized metal nanocluster catalyst for the first time [99]. PAMAM dendrimer was applied for encapsulation of metal precursors (e.g. Rh, Pt, Pd, Au). An excess

**Table 3.1** Comparison of the catalytic performance of dendrimer-immobilized Pd nanoclusters with the other Pd catalysts.

| Catalyst (or precatalyst) | Pd cluster (nm) | R | Pd loading, TON |
|---|---|---|---|
| $Pd(OAc)_2$ | — | H | 5 ppm, 200,000 |
| $Pd(OAc)_2$ | — | H | 0.9 ppm, 94,444 |
| $Pd(OAc)_2$+n-$Bu_4NBr$ | — | S | 50 ppm, 16,600 |
| Pd/C | 2.4 | H | 50 ppm, 18,000 |
| $Pd(OAc)_2$/G3-P3 | 2.0 | S | 10 ppm, 85,000 |
| $Pd(OAc)_2$/porous | 2.0–5.0 | S | 10 ppm, 95,000 |
| $K_2PdCl_4$, PVP | 10 | S | 8 ppm, 117,500 |
| $(NH_4)_2PdCl_4$/enzyme | 4.9 | S | 0.28 ppm, 3,570,000 |
| $PdCl_2$/imidazole | 3.0 | S | 10 ppm, 100,000 |
| $H_2PdCl_4$/SBA-15+TBAB | 6.4 | S | 6 ppm, 121,666 |
| $K_2PdCl$/PPPI | — | H | (3–5) × $10^4$ ppm, low |
| $K_2PdCl_4$/G4-OH | — | H | 25 ppm, 36,800 |
| $K_2PdCl_4$/G4-OH | 1.4 | S | 1.5 × $10^4$ ppm, low |
| $K_2PdCl_4$/G4-OH | 3.2 | S | 550 ppm, 1771 |
| $K_2PdCl_4$/G4-OH | 1.6 | St | 3 × $10^3$ ppm, low |
| $Pd(OAc)_2$/G0-9Fc | 2.8 | S | 1 ppm, 540,000 |
| $K_2 PdCl_4$/G0-9$SO_3^-$ | 2.3 | S | 100 ppm, 9200 |
| $K_2 PdCl_4$/G1-27$SO_3^-$ | 2.8 | S | 100 ppm, 9400 |
| $K_2PdCl_4$/G0-27TEG | 1.4 | S | 0.3 ppm, 2,700,000 |
| $K_2PdCl_4$/G1-81TEG | 2.7 | S | 1 ppm, 390,000 |

R, reaction (S, Suzuki; H, Heck; St, Stille; and N, Negishi).

of freshly prepared $NaBH_4$ solution was added subsequently for reducing the metal precursors and generating metal nanoclusters. The solution was purified by dialysis in cellulose dialysis tubing against an excess of deionized water. After that, mesoporous silica such as SBA-15 [100] was employed for the final immobilization via sonication. The hydrogen bonding between the terminal hydroxyl groups on the dendrimer and the surface of the $SiO_2$ support plays important role for the immobilization between SBA-15 and dendrimer-encapsulated metal nanoclusters. The interaction efficiently prevents the dendrimer-encapsulated metal nanoclusters from detachment even under liquid phase catalytic conditions and high temperature. A variety of techniques, including TEM, XPS, ICP, EXAFS, and NEXAFS, can be used to characterize this kind of heterogeneous catalyst before and after catalysis [101]. Sufficient information supports the fact that none of detectable metal was leached from the mesoporous silica/dendrimer-immobilized metal nanocluster catalyst during catalysis. Meanwhile, the revealed tunable oxidation states of the metal atoms on the surface render the catalyst effective in various transformations.

The mesoporous silica/dendrimer-immobilized metal nanoclusters have proven to be promising heterogeneous catalyst in different kinds of reactions. Ye and coworkers made great progress [102]. They developed π-bond activation reactions (e.g. hydroalkoxylation of alkynes) [103], C–C bond activation reactions (e.g. ring opening of cyclopropane) [104], Hayashi-Ito aldol reactions [105], cascade cyclopropanation rearrangement reactions [100], cascade dihydropyran formation reactions [106], and hydrogenative isomerization reactions [107] by the catalyst (Figure 3.14). These catalytic reactions have been previously the realm of homogeneous catalysis, while the metal nanoclusters immobilized in mesoporous silica/dendrimer support not only demonstrate comparable catalytic performance to homogeneous metal catalysts (Figure 3.14a) [103], but also trigger transformations that otherwise cannot be achieved by conventional homogeneous metals (Figure 3.14b) [104].

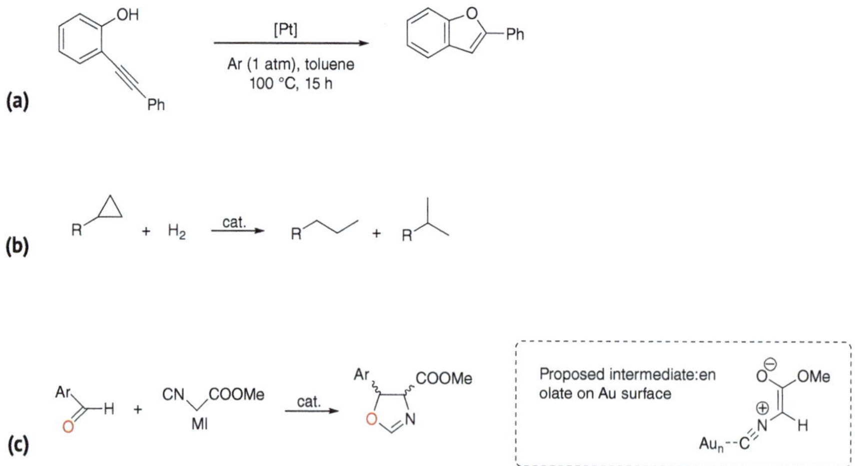


**Figure 3.14** Mesoporous silica/dendrimer-immobilized metal nanoclusters-catalyzed efficient organic transformations with the alkyne hydroalkoxylation reaction (a), cyclopropane ring opening reaction (b), and Hayashi-Ito aldol reaction (c) as representatives.

Moreover, the heterogeneous metal nanocluster catalyst is able to control the cis/trans selectivity of the aldol reaction by adjusting pore size of the mesoporous silica support (Figure 3.14c) [105]. Of course, recyclability would be another advantage of this kind of heterogeneous catalyst. It has been confirmed by three-phase tests and other control experiments that these transformations happen under heterogeneous conditions and the catalyst can be recycled without loss of catalytic activity and metal leaching.

With the aim of taking full advantage of the heterogeneous nature of the mesoporous silica/dendrimer-immobilized metal nanocluster catalyst and facilitating a broad spectrum of complex chemical transformations, Huang and coworkers also studied various reactions in a fixed bed plug reactor in a continuous flow mode. Full conversion from reactants to the desired product was observed in the π-bond activation reaction [108]. Thanks to the flow reactor, separation of the catalyst from the reaction medium or any further purification of the product is obviated. Apart from that, more appealing results were obtained. In the cyclopropanation-rearrangement cascade reaction of propargyl pivalate with the enyne, the unique activity and selectivity was achieved by the mesoporous silica/dendrimer-immobilized metal nanocluster catalyst [100]. Homogeneous $AuCl_3$ catalyst gave 40% yield of the primary product of cyclopropane with a cis/trans ration of 3/1, and 10% yield of the secondary products formed by the rearrangement of the cis-cyclopropane. In sharp contrast, the metal nanocluster catalyst immobilized in mesoporous silica/dendrimer in the batch mode catalyzed the reaction, affording the primary product of cyclopropane with a much higher cis/trans ration of 12/1 and only a small amount of the secondary products. By further modifying the flow rate and reaction time of the flow reactor, 18% yield of the exclusive cis-cyclopropane (cis/trans = 100/1) or 72% yield of the rearranged secondary product can be selectively obtained, respectively (Figure 3.15). This activity and selectivity control should be attributed to the precisely controlling of all reaction parameters of the flow reactors, based on the silica/dendrimer immobilization technique for metal nanoclusters.

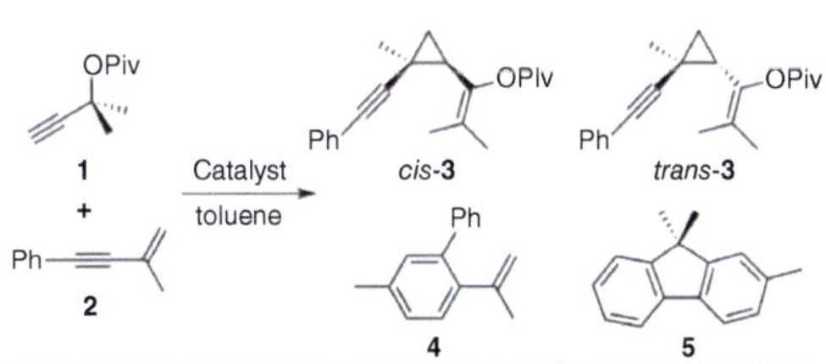


| Catalyst | % Conversion | % 3 (cis:trans) | % 4 + 5 (4:5) |
|---|---|---|---|
| $AuCl_3$* | 50 | 40 (3:1) | 10 (3:2) |
| Au-G4OH batch*,† | 40 | 38 (12:1) | 2 (1:1) |
| Au-G4OH flow (5 mL/h)‡ | 20 | 18 (100:0) | 2 (100:0) |
| Au-G4OH flow (0.1 mL/h)‡ | 72 | 0 | 72 |

(a)

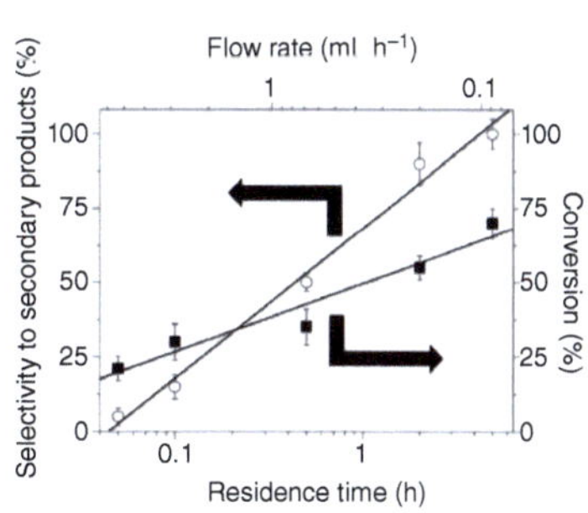


(b)

**Figure 3.15** Flow reaction based on mesoporous silica/dendrimer-immobilized metal nanocluster catalyst. *Source:* [102]/with permission of American Chemical Society.

The flow reactor can be designed to be big for industrial application, while can also be designed to be small enough for the detection of intermediates during catalysis. Based on the micrometer-sized synchrotron-based IR and X-ray beams with a spatial resolution of 15 μm, spectral mapping of the organic phase and metal nanocluster catalyst during multistep formation of products in a flow microreactor has been achieved (Figure 3.16) [106]. The reaction between substituted propargyl vinyl ether and alcohol was selected as the model reaction for the investigation. Au nanoclusters-immobilized in SBA-15/dendrimer was used as the catalyst. In situ synchrotron-sourced IR microspectroscopy detected that an allenic aldehyde existed before the formation of the final product dihydropyran. In-depth investigation on the catalyst revealed that the oxidation states of Au were altered along the formation of the allenic aldehyde intermediate, thus affording a deeper mechanism understanding of the reaction. Other applications of the flow reactor include the gas reactions catalyzed by the mesoporous silica/dendrimer-immobilized metal nanoclusters at temperatures ranging from 150 to 250 °C. A Pt-based SBA-15/dendrimer catalyst showed high efficiency for the hydrogenation reaction of the methylcyclopentane at about 200 °C (Figure 3.16) [106]. The TOF of the reaction was determined to be 334 $h^{-1}$, much higher than other catalysts reported. Moreover, up to 99.6% selectivity was achieved for the ring opening isomerization reaction at high conversion of 94%. Rh and Pd-based SBA-15/dendrimer catalysts in the fixed bed then brought switched selectivity. Notably, the temperature above 250 °C results in the aggregation of the metal nanoclusters inside dendrimer and deactivation of the corresponding catalyst.

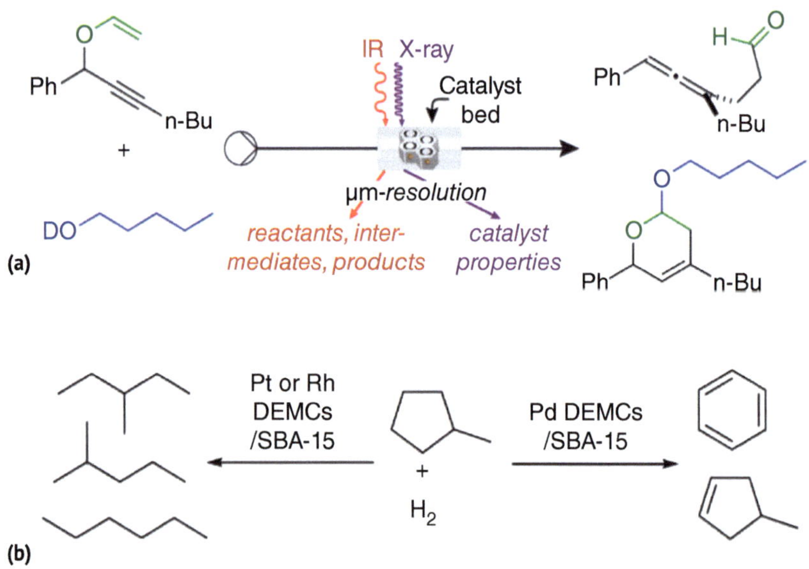


**Figure 3.16** Intermediates detection and gas flow reactor based on mesoporous silica/dendrimer-immobilized metal nanocluster catalyst. *Source:* [102]/with permission of American Chemical Society.

## 3.5 Biomolecule Immobilized Metal Nanoclusters

In the nature, metal–metal bonding nanoclusters that are surrounded by organism widely exist as enzymes and other functional systems. For example, photo-system II is very important in plants, cyanobacteria and algae, acting as the catalyst for splitting of water and generating $O_2$, protons, and electrons. Crystallographic structures reveal that a $Mn_3CaO_4$ cubane nanocluster is the key structure of the oxygen-evolving center in photo-system II [109]. This nanocluster is coordinated with four water molecules, one imidazole molecule and six carboxylate groups of the amino acid residues of the photo-system II polypeptides. Computational and spectroscopic results have given the detailed information concerning the reaction intermediates and catalytic mechanisms. For mimicking the functions of the natural nanoclusters and the demand of green and environmentally friendly chemistry, considerable effort has been devoted to the synthesis and applications of the biomolecule immobilized metal nanoclusters.

There is increasing interest in biomolecules for templating the growth of a variety of inorganic nanomaterials 20 years ago [110]. Three kinds of biomolecules have been developed and outlined. First, crystalline surface layers of bacterial cells with a regular distribution of physicochemical affinity sites at the surface of protein have demonstrated interesting roles in the fabrication of highly oriented semiconductor and metal nanocluster arrays. Second, ferritin and ferritin-like protein cages with nanometer size have been utilized as the size-constrained reaction environments for the encapsulation of inorganic materials. Third, biomolecules with linear morphology such as viruses, microtubules, and lipid nanotubes have employed for creation of one-dimensional nanoparticles and tubular nanostructures. Being similar with the strategy of creating the inorganic materials, proteins, peptides, and DNA are the most common biomolecules utilized for the metal nanoclusters' immobilization. Further deposition of the biomolecule immobilized metal nanoclusters has also been developed to improve their stability and property and for in-depth investigation of the metal–support interactions. For example, Christensen and coworkers applied a variety of X-ray spectroscopic techniques to investigate the biofunctionalized metal nanoclusters supported on titanium and acquire better understanding of the metal–thiolate interactions between the metal kernel and the biomolecular surface [111]. Au, Ag, and Pd metal nanoclusters immobilized in biomolecules such as (2-mercaptopropionyl)glycine and bovine serum albumin were prepared. SEM and XPS results revealed that the immobilization of metal nanoclusters with varying model biomolecule systems have a significant effect on both surface coverage and organization. Based on the study, it was also anticipated that the immobilization of biofunctionalized Au nanoclusters onto titanium would be helpful to improve osteointegration of Ti-based implants in orthopedics, whereas the introduction of Ag metal nanoclusters would add antibacterial functionality to titanium.

For the catalytic activity of biomolecule-immobilized metal nanoclusters, the enzymatic activity is the first one that comes to one's mind [112–114]. It has been shown that the combination of metal nanoclusters and enzymes results in decrease of the enzyme activity in some cases and also leads to the catalytic activity improvement in other cases. For example, bovine pancreatic α-chymotrypsin has been employed to immobilize silver nanoclusters. Compared to the naked enzyme, the metal nanocluster functionalized enzyme demonstrated slower catalytic kinetics by two to three times [115]. The retarded activity of the enzyme-immobilized Ag nanoclusters is ascribed to the manifestation of the dynamic rigidity of the bovine pancreatic α-chymotrypsin upon the attachment of the Ag nanoclusters. On the other hand, enzymes such as glucose oxidase and horseradish peroxidase were used to immobilize metal nanoclusters under physiological conditions. Owing to the fact that glucose oxidase and horseradish peroxidase are readily combined to $O_2$ and $H_2O_2$, respectively, these biomolecule-immobilized metal nanoclusters can be used to catalyze the oxidation of a variety of organic molecules [116].

Nevertheless, as an impressive artificial enzyme, the biomolecule-immobilized metal nanocluster has been developed and attracted particular attention. Their peroxidase (POD)-like, superoxide dismutase (SOD)-like, glucose oxidase (GOD)-like, catalase-like (CAT), nuclease-like, and protease-like activities endow this kind of artificial enzyme with wide application scope. Compared to the natural enzymes that are highly sensitive to environment changes and readily denatured by harsh conditions, the artificial enzymes, that is, biomolecule-immobilized metal nanoclusters are low-cost, easy to be prepared and more stable. Moreover, they are capable of catalyzing chemical and biochemical reactions with superior catalytic activity and high substrate specificity as well.

Bovine serum albumin-immobilized gold nanoclusters have been found to exhibit prominent intrinsic POD-like activity. They were able to catalyze the oxidation of a conventional peroxidase substrate 3,3′,5,5′-tetramethylbenzidine in the presence of $H_2O_2$ to generate a blue-colored oxidized product [117]. Thus, they have been applied as a sensitive and selective colorimetric sensors and probes for biological detection. For example, bovine serum albumin-immobilized gold nanoclusters that are loaded in liposomes have been developed as a robust biosensor based on their prominent POD-like activity for the colorimetric detection of human epidermal growth factor receptor 2 [118].

SOD is a metal holoenzyme and widely exists in plants, animals, and microbes [119, 120]. It converts superoxide anions into hydrogen peroxide and oxygen, which are then converted into water by other enzymes [121, 122]. It has been proposed that gold nanoclusters can adsorb superoxide anions and trigger the further electron transfer process, thus catalyzing the denaturation of superoxide anions. In other words, the gold kernel of the gold nanoclusters acts as an electron storage pool during the process. Glutathione-immobilized gold nanoclusters with red luminescence were fabricated in HeLa cells with cationic polymeric chitosan as the cross-linker and stabilizer [123]. The material has both SOD- and POD-like activities, accelerating the conversion from superoxide anions to $H_2O_2$ and then hydroxyl radicals. The enzymic activity of this material was effectively activated under acidic conditions, exhibiting an excellent killing effect on tumor cells but with low toxicity to normal cells.

Gold nanoclusters that are immobilized in lysozyme and curcumin showed impressive GOD-like activity and reactive oxygen species-clearing capacity [124], enabling the efficient removal of excess glucose without producing additional $H_2O_2$ and thereby alleviating the oxidative stress. This material acts as a multifunctional dressing for regulating the regeneration cascade and promoting the healing of infected diabetic wound in vivo. The mechanism investigation indicates that in the catalytic process, glucose and oxygen gas co-adsorb on the surface of the gold nanoclusters that are immobilized in lysozyme and curcumin, which would promote the transformation of two hydrogen atoms from glucose to oxygen gas, thus completing the catalytic cycle [125].

CAT-like activity of enzyme is regarded as an essential part of the regulation of reactive oxygen species. Several biomolecule-immobilized gold nanoclusters have been found to have both POD-like and CAT-like activities [126]. It was demonstrated that the CAT-like activity still remained under physiological conditions, while the POD-like activity was lost under the same conditions [127]. In addition to the aforementioned enzymic activities, biomolecule-immobilized metal nanoclusters have other biomedical applications based on their enzymic activity. For example, catalytically active gold nanoclusters with atomic precision have been fabricated for noninvasive early intervention of neurotrauma [128]. The gold nanoclusters alleviated the oxidative stress and decreased the excessive superoxide anions and $H_2O_2$ in vitro and in vivo. They could also promote the wound healing of brain trauma and relieve inflammation through inhibiting the activation of astrocytes and microglia via noninvasive administration to decrease superoxide anions and $H_2O_2$ in brain tissue. Another interesting example is the super-charged gold nanoclusters immobilized in a nano-platform of bioactive glass nanoparticles that demonstrate highly enhanced enzymic activity for catalytic antibacterial action [126].

Biomolecule-immobilized metal nanoclusters generally demonstrate promising photoluminescent intensity and peak emission wavelengths that range across the visible and near-IR spectrum [129]. The high photoconversion efficiency, on the other hand, endows them photocatalytic activity toward the generation of reactive oxygen species. Thus, these biomolecule-immobilized metal nanoclusters have become the attractive antibacterial agents to chemists and material scientists. For instance, DNA-immobilized silver nanoclusters with wide color palette, low toxicity, and high photoconversion efficiency have found practical application in antibacterial activity [130]. Actually, the use of the silver plate and spoon has known to be more hygienic from the Greco-Roman period. Metallic silver materials have also been utilized extensively to fight infections and control spoilage since the ancient times. Nowadays, we know that the direct usage of silver ion or metallic silver as antibacterial agents is not advisable as it may lead to adverse cytotoxic effects. Biomolecule-immobilized silver nanoclusters then become a growing subject of interest for antibacterial applications because of their ultra-small size, high local concentration of metal ions that are assembled on the surface of metal nanoclusters. More importantly, the inherent toxicity of the metal nanoclusters is significantly reduced after functionalization with biomolecules. As mentioned above, several biomolecules, including proteins, peptides, and plant extracts, are known to be potent to microorganisms. The integration of these biomolecules with metal nanoclusters immobilized inside would lead to synergistically antibacterial effects. In the following paragraphs, we list several biomolecule-immobilized metal nanoclusters as representatives to show their antibacterial activities.

Lysozyme-immobilized gold nanoclusters have been used to target notorious pathogenic bacteria, including pandrug-resistant *Acinetobacter baumannii*, vancomycin-resistant *Enterococcus*, etc. due to the synergy between the gold nanoclusters and the lysozyme support, which can damage bacterial cell walls by attacking the peptidoglycans [131]. Díez and coworkers fabricated a hybrid material of photoluminescent Ag nanoclusters and native cellulose nanofibers. This material showed high efficacy for bacteria killing, probably due to the large surface-to-volume ratio of the native cellulose nanofibers that can further improve the antibacterial efficacy of ultra-small Ag nanoclusters [132].

Yuan and coworkers developed some high-quality glutathione-immobilized Ag nanoclusters, which showed superior bactericidal efficacy against *Pseudomonas aeruginosa* [133]. They also developed gold nanoclusters immobilized by 6-mercaptohexanoic acid via precisely adjusting the nanoclusters' size [134]. The resulted ultra-small gold nanoclusters materials exhibited efficient wide-spectrum antimicrobial activity, and the capability of killing both gram-positive and gram-negative bacteria via intracellular reactive oxygen species that were generated by the catalysis of gold nanoclusters.

Cysteine can also be used as a simple biomolecule for the immobilization of gold nanoclusters. The as-obtained material possesses orange-red-emitting fluorescence, high aqueous solubility and excellent biocompatibility. The photoluminescent intensity of cysteine-immobilized gold nanoclusters decreased in the presence of *Escherichia coli* [135]. The metabolism of cysteine-immobilized gold nanoclusters by *E. coli* accelerated the generation of intracellular reactive oxygen species to kill *E. coli*. Thus, it is suggested that the cysteine-immobilized gold nanoclusters could be practically applied for bacterial detection and antimicrobial agents.

Deoxyribonuclease (DNase)-immobilized gold nanoclusters are able to kill the gram-positive and gram-negative bacteria. The extracellular polymeric substance matrix was decomposed by DNase and then the defenseless bacteria were exposed to photodynamic therapy and photothermal therapy by DNase-immobilized gold nanoclusters under 808 nm laser irradiation. About 90% of bacteria were killed during the process [136].

Bovine serum albumin, which has been mentioned above, is able to immobilize cationic Au–Ag-alloyed nanoclusters that could induce severe disruption of outer membrane metabolism due to strong electrostatic interactions and increased reactive oxygen species accumulation leading bacteria to demise [137]. It has been demonstrated that photoluminescent biomolecule-immobilized gold nanoclusters had great potential for antimicrobial agents. In addition, novel gold nanoclusters functionalized with quaternary ammonium salts have also been developed for antibiotic agent targeting multidrug-positive bacteria both in vitro and in vivo [138].

## 3.6 New Types of Supports Immobilized Metal Nanoclusters

With the development of organic-inorganic hybrid materials, new types of supports such as MOFs and polyoxometalates (POMs) have been employed to immobilize metal nanoclusters for efficient catalysis. These new types of supports feature abundant chemical/physical linking sites and well-defined composition and structures, thus being beneficial to further stabilize sensitive metal nanoclusters and generating well-defined composite catalysts.

A presentative work comes from the research group of Yun, who successfully synthesized a robust MOF@nanocluster@MOF composite catalyst [139]. This kind of sandwich catalysts was prepared by four steps (Figure 3.17a). First, ZIF-8 (a typical MOF) and cysteine-stabilized $Au_{25}$ nanocluster $Au_{25}$(L-Cys)$_{18}$ were synthesized and carefully characterized. Second, both precursors were evenly dispersed in aqueous solution and the $Au_{25}$(L-Cys)$_{18}$ solution was slowly poured into the ZIF-8 solution. Physical impregnation resulted in the generation of a composite $Au_{25}$(L-Cys)$_{18}$/ZIF-8 nanomaterial. Third, $Au_{25}$(L-Cys)$_{18}$/ZIF-8 was uniformly dispersed in methanol and a solution of $Co(NO_3)_2$ with specific concentration was added, resulting in a $Co^{2+}$-coated $Au_{25}$(L-Cys)$_{18}$/ZIF-8 composite materials by the coordination of $Co^{2+}$ with the COOH groups on the surface of $Au_{25}$(L-Cys)$_{18}$. Finally, a solution of 2-methylimidazole (2-MeIm) was slowly added into the above system and the sandwich composite ZIF-8@$Au_{25}$/ZIF-67 catalyst was obtained by gentle agitation at room temperature for 24 hours. The thickness of the ZIF-67 shell can be well controlled by regulating the concentration of $Co^{2+}$ added, leading to ZIF-8@$Au_{25}$/ZIF-67 catalysts with ZIF-67 thickness of 2, 8, 12, 18, and 25 nm. These composite sandwich catalysts were well characterized by XRD, XPS, FTIR, and TEM. Especially, TEM results showed that $Au_{25}$(L-Cys)$_{18}$ nanoclusters were evenly distributed in the interlayer of the ZIF-8@$Au_{25}$@ZIF-67.

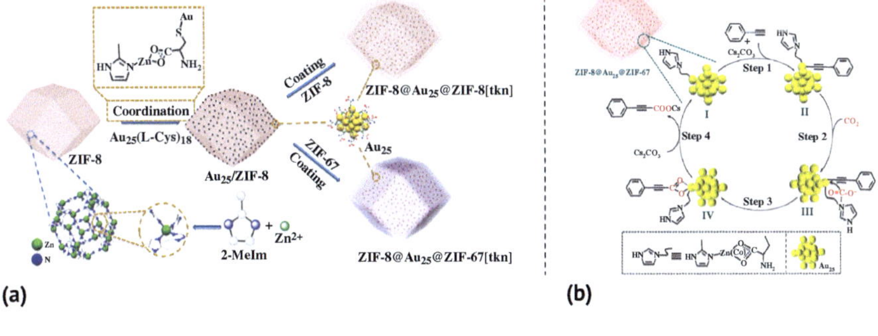


**Figure 3.17** (a) Construction of ZIF-8@$Au_{25}$@ZIF-67 sandwich catalysts. (b) ZIF-8@$Au_{25}$@ZIF-67-catalyzed highly efficient carboxylation of terminal alkynes. *Source:* [139]/ with permission of American Chemical Society.

Further catalytic performance investigations revealed that the composite sandwich ZIF-8@$Au_{25}$@ZIF-67 catalyst showed the best activity, compared to ZIF-8@$Au_{25}$@ZIF-8, $Au_{25}$@ZIF-8, ZIF-8@ZIF-67, ZIF-8, $Au_{25}$(L-Cys)$_{18}$, and the simple mixture of $Au_{25}$(L-Cys)$_{18}$ and ZIF-8 in the carboxylation reaction of terminal alkynes. Moreover, the thickness of the ZIF-67 shell significantly affects the catalytic activity of the composite sandwich ZIF-8@$Au_{25}$@ZIF-67 catalysts, and the one with ZIF-67 shell thickness of 12 nm gave the desired product alkyne acid in 99% yield. The shell thickness effect suggests that the ZIF-67 shell may play more important roles beyond the mass transfer and stabilizing roles. Time-programmed desorption (TPD) analysis of carbon dioxide ($CO_2$) revealed that the ZIF-67 shell was able to capture and activate the $CO_2$ molecules. A synergistic catalytic mechanism was proposed based on the structural feature and catalytic performance of this composite sandwich ZIF-8@$Au_{25}$@ZIF-67 catalyst (Figure 3.17b). This robust catalyst was applied to catalyze the reduction of 4-nitrophenol, demonstrating high activity and good recyclability as well.

A simple combination of metal nanoclusters and MOF also trigger the formation of efficient catalyst for photocatalytic hydrogen production. Wang and coworkers synthesized a series of $MAg_{24}$@UiO-66-$NH_2$ (M = Ag, Pd, Pt, Au) composite catalysts [140], which showed improved catalytic activity than the independent $MAg_{24}$ nanoclusters or UiO-66-$NH_2$ (Figure 3.18). Among them, the $AuAg_{24}$@UiO-66-$NH_2$ catalyst exhibited the best performance, producing $H_2$ under light irradiation with the rate up to 3.6 mmol $g^{-1}$ $h^{-1}$. These composite catalysts were prepared at room temperature by a self-assembly strategy to avoid the structural destruction of the $MAg_{24}$ nanoclusters under the common solvothermal conditions. Specifically, the $MAg_{24}$ nanoclusters are negatively charged and the presynthesized $Zr_6$-oxo clusters

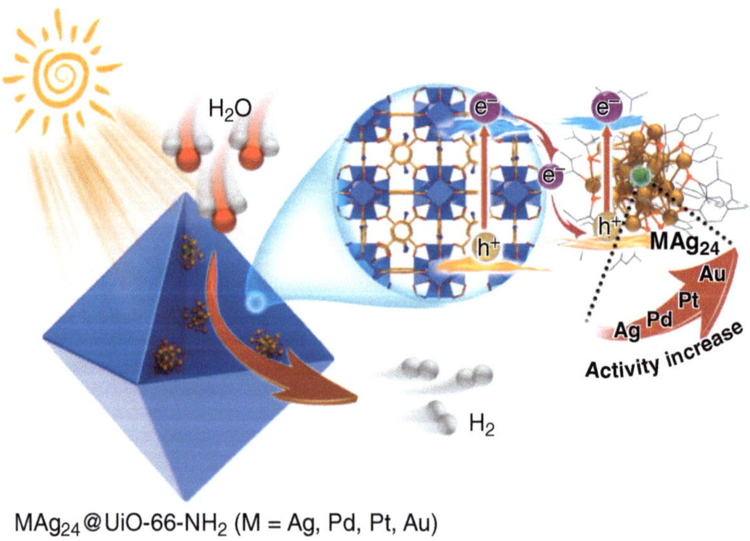


**Figure 3.18** $MAg_{24}$@UiO-66-$NH_2$ composite catalysts for photocatalytic hydrogen production. *Source:* [140]/with permission of John Wiley & Sons.

are positively charged. Therefore, $MAg_{24}$ nanoclusters can be electrostatically attracted onto the surface of $Zr_6$-oxo clusters in the presence of 2-aminoterephthalic acid ($NH_2$-BDC). A further MOF growth process would afford the desired $MAg_{24}$@UiO-66-$NH_2$ catalysts. These composite catalysts were well characterized by XPS, XRD, TEM, EDS and femtosecond time-resolved transient absorption (fs-TA), confirming the excellent dispersion of $MAg_{24}$ in UiO-66-$NH_2$. The analysis on XPS and transient absorption spectra indicated that the encapsulation of $MAg_{24}$ in UiO-66-$NH_2$ constituted a Z-scheme heterojunction system with suppressed charge recombination. In the composite catalytic system, the rate of photo-induced production of $H_2$ is closely associated with both charge transfer efficiency and the electronic state of the silver activation sites on the surface, which account for the high photocatalytic activity of the $MAg_{24}$@UiO-66-$NH_2$ catalysts. On the other hand, the MOF highly improves the stability of the nanoclusters, endowing the composite catalysts with good recyclability.

It should be noted that the bottom-up construction of UiO-66 MOF lorica on the surface of metal nanoclusters can be extended for different cluster structures. Apart from the M-Ag doped nanoclusters mentioned above, You and coworkers synthesized three structurally similar gold nanocluster, namely $Au_{38}I$, $Au_{38}S$, and $Au_{38}IS$, and realized the bottom-up construction of UiO-66-H, UiO-66-COOH, and UiO-66-$NH_2$ on their surface [141]. The as-obtained composite $Au_{38}$@UiO-66 catalysts showed significantly improved catalytic reduction activity than the $Au_{38}I$, $Au_{38}S$, $Au_{38}IS$, and UiO-66 (Figure 3.19). More interestingly, the growth of MOF triggers the partial removal of the surface ligands of these metal nanoclusters, and the order of thiolate removal is consistent with the order of catalytic activity ($Au_{38}S$@UiO-66 > $Au_{38}IS$@UiO-66 > $Au_{38}I$@UiO-66). The solubility of the metal

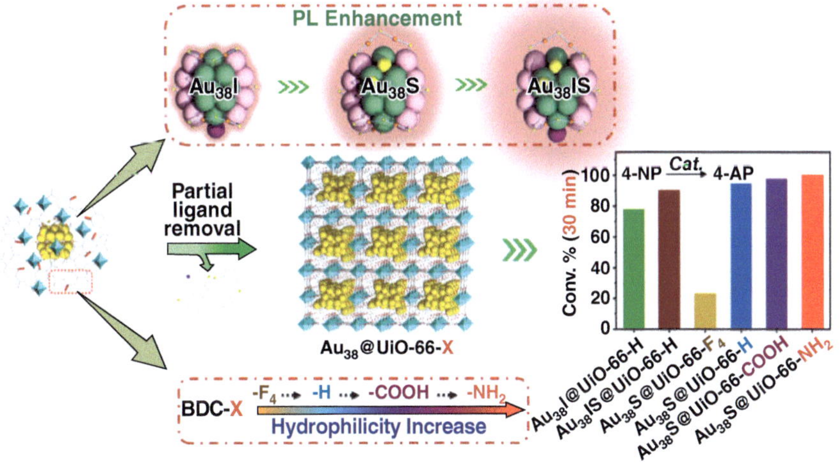


**Figure 3.19** Bottom-up construction of UiO-66 MOF lorica on the surface of metal nanoclusters for regulating solubility and catalytic performance. *Source:* [141]/with permission of American Chemical Society.

nanoclusters was also improved with the construction of MOF on their surface. For example, the hydrophilicity of $Au_{38}S$@UiO-66-$NH_2$ is better than $Au_{38}S$@UiO-66-COOH, which is also better than $Au_{38}S$@UiO-66-H. This strategy provides a new dimension for turning the catalytic performance of metal nanoclusters under hydrophilic conditions.

The above three examples present the encapsulation of metal nanoclusters in MOF materials, affording MOF-immobilized metal nanocluster catalysts. Actually, immobilization of metal nanoclusters can be achieved by different methods. For example, Huang and coworkers synthesized a series of functional ligands-protected metal nanoclusters and applied them as the structural unit to construct MOF materials [142, 143]. These metal nanocluster-based MOFs showed significantly improved stability than the organic ligand-protected metal nanoclusters and other intriguing properties. For example, using 4,4′-bipyridine (bpy) protected $Ag_{12}$ nanocluster as the structural units, they constructed rigid MOFs by the interconnection between $Ag_{12}$ nanoclusters and revealed their atomically precise structures [142]. The as-obtained $Ag_{12}$-based MOFs showed interesting photoluminescence. In a following work, they constructed $Cu_4$-based MOFs by using the functional ligands with –COOH groups (Figure 3.20) [143].

Mu and coworkers also utilized the interconnection between functional metal nanoclusters and synthesized metal nanocluster-based frameworks [144]. Specifically, they used thiacalix[4] arene and alkynes as the mixed ligand to stabilize a series of copper nanoclusters, including $Cu_9$, $Cu_{13}$, $Cu_{17}$, $Cu_{22}$, etc. These copper nanoclusters were found to be assembled in the presence of active anions such as $O^{2-}$, $Cl^-$, etc. to afford self-assembly secondary structures. Typical examples include $Cu_{38}$ and $Cu_{43}$ cluster structures. The newly formed high-nuclear cluster secondary structures showed highly improved stability, and exhibited catalytic activity toward

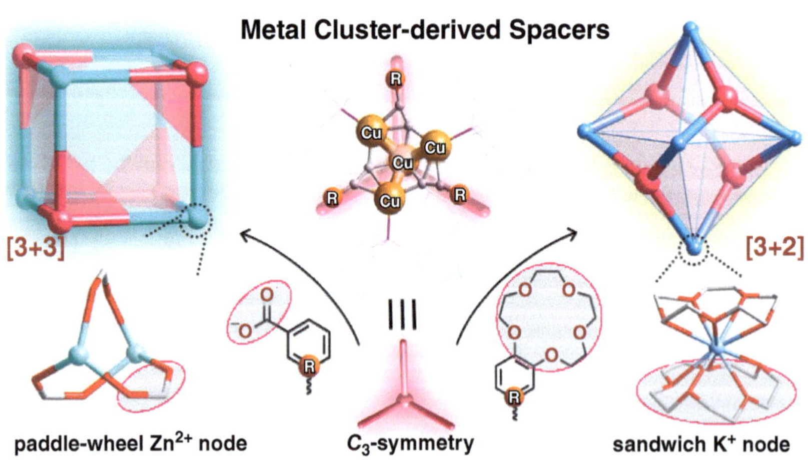


**Figure 3.20** Metal nanocluster-based MOFs. *Source:* [143]/with permission of American Chemical Society.

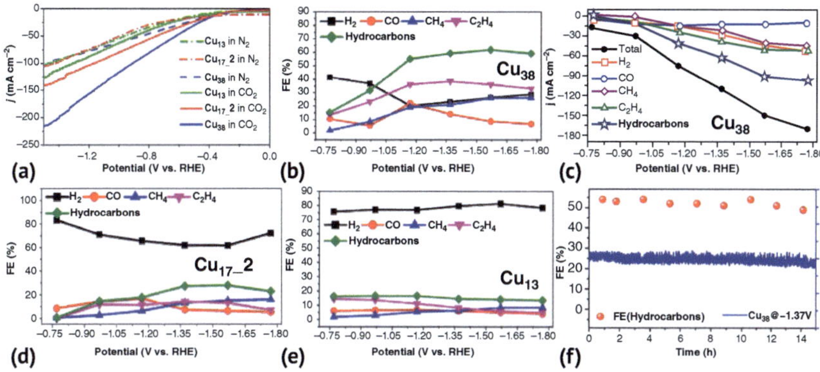


**Figure 3.21** Metal nanocluster-based frameworks for catalyzing electroreduction of $CO_2$ toward hydrocarbons. (a) LSV curves of $Cu_{13}$, $Cu_{17}$, and $Cu_{38}$ in $N_2$ or $CO_2$ saturated 1 M KOH. (b) FEs of production at different working potentials for $Cu_{38}$. (c) Partial current density for $Cu_{38}$. (d) FEs of production at different working potentials for $Cu_{17}$. (e) FEs of production at different working potentials for $Cu_{13}$. (f) Stability test of $Cu_{38}$ @ −1.37 V vs. RHE for 15 hours. *Source:* [144]/with permission of American Chemical Society.

the electrochemical conversion of $CO_2$ reduction to hydrocarbons (Figure 3.21). For example, the $Cu_{38}$ demonstrated an impressive Faradic efficiency of 62.01% for hydrocarbons at −1.57 V in $CO_2$ electroreduction, with catalytic selectivity of 34.03% for ethane and 27.98% for methane.

POM is another kind of famous organic–inorganic hybrid material and has gained considerable application development in various research field. Efficient $O_2$ activation was achieved by integration of POMs and silver nanoclusters. Kamachi research group reported an elegant work [145], in which an $Au_8Ag_{26}$ alloyed nanocluster was confined in a ring-shaped $[P_8W_{48}O_{184}]^{40-}$ POM (Figure 3.22). This is a long-pursued while challenging task that POM is applied as the surface ligand for atomically precise metal nanoclusters. The same research group achieved the synthesis of $[P_8W_{48}O_{184}]^{40-}$ POM-immobilized $Ag_{30}$ nanoclusters in the year of 2023 [146]. Compared to the [POM-$Ag_{30}$] nanocluster, the [POM-$Au_8Ag_{26}$] nanocluster demonstrates substantially enhanced structural stability and photocatalytic performance for the activation of $O_2$ to $^1O_2$. Based on the photocatalytic activity of the $[P_8W_{48}O_{184}]^{40-}$ POM-immobilized $Au_8Ag_{26}$ nanocluster, a variety of substrates, including α-terpinene, 1,3-cyclohexadiene, thioanisole, and benzylamine were effectively oxidized to ascaridole, 2,3-dioxabicyclo[2.2.2]octane, methylphenylsulfoxide, and *N*-benzylidenebenzylamine, respectively, in high yields.

Feng and coworkers applied lacunary POM to stabilize silver nanoclusters in the presence of organic ligands as the co-stabilizers [147, 148]. A $Ag_{24}$ nanoclusters was obtained when $[SiW_9O_{34}]^{10-}$ was employed as the POM stabilizer and 1,4-Bis(diphenylphosphino)butane) (DPPB) was employed as the organic ligand stabilizer [147]. By replacing DPPB with 1,3,5-benzenetricarboxylic acid ($H_3BTC$), a $Ag_{27}$ was obtained instead as the final product. Notably, the POM stabilizers of

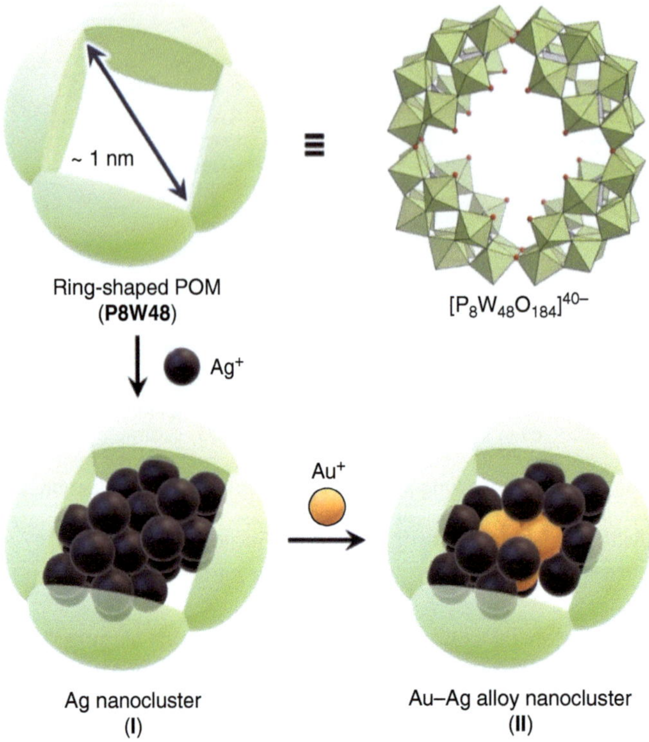


**Figure 3.22** $[P_8W_{48}O_{184}]^{40-}$ POM-immobilized $Au_8Ag_{26}$ nanocluster for photocatalytic oxidation reactions. *Source:* [145]/John Wiley & Sons/CC BY 4.0.

both silver nanoclusters were identified as three $Si_2W_{18}O_{66}$. That is, the POM grows with the synthesis of the silver nanoclusters. Femtosecond/nanosecond transient absorption spectroscopy revealed the rapid charge transfer between the silver kernel and the outer POM stabilizer of both $Ag_{24}(Si_2W_{18}O_{66})_3$ and $Ag_{27}(Si_2W_{18}O_{66})_3$ nanoclusters, which is beneficial for their catalytic $CO_2$ reduction to HCOOH under light irradiation (Figure 3.23). By applying the same strategy, Guo-Yu Yang, Hongjin Lv et al. synthesized two silver nanocluster isomers [148]. Both nanoclusters were stabilized by bowl-like $[Sb_3W_{30}]$ POM, and two $[Sb_3W_{30}]$ stabilize one $Ag_{14}$ kernel in both cases. The $Ag_{14}(Sb_3W_{30})_2$-1 and $Ag_{14}(Sb_3W_{30})_2$-2 nanoclusters have the completely same compositions while different structures, resulting in their distinct electronic structures as well. Thus, $Ag_{14}(Sb_3W_{30})_2$-1 and $Ag_{14}(Sb_3W_{30})_2$-2 isomers demonstrate different physicochemical properties, including distinct performance in photocatalytic hydrogen evolution. The metal nanocluster isomers show intriguing phenomena. We will discuss this kind of unique material in detail in Chapters 5 and 6.

The last example we present in this section is the POM nanocluster self-assembled super structures. In this case, negatively charged POMs are connected with each other to afford nanowires, tetragonal nanosheets, rectangular nanosheets, etc. with

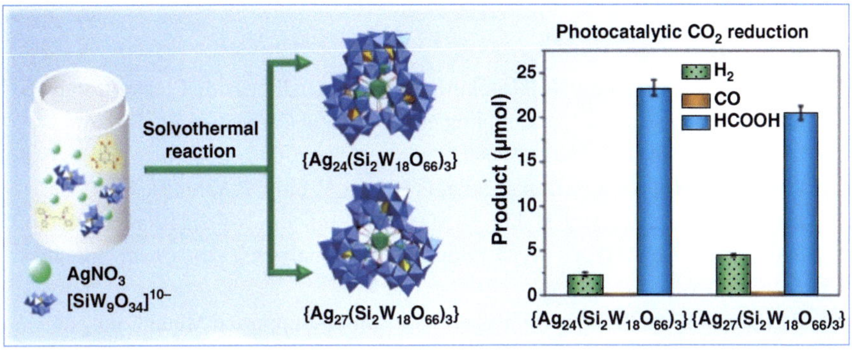


**Figure 3.23** [$Si_2W_{18}O_{66}$] POM-immobilized $Ag_{24}$ and $Ag_{27}$ nanoclusters for photocatalytic $CO_2$ reduction. *Source:* [147]/with permission of John Wiley & Sons.

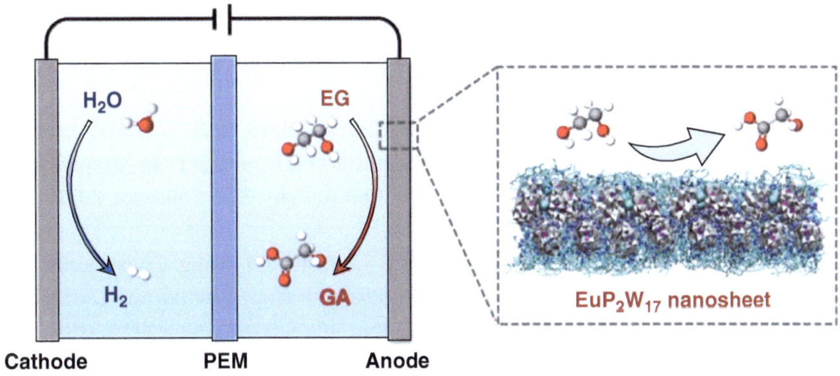


**Figure 3.24** Self-immobilized $MP_2W_{17}O_{61}$ POM clusters for electrocatalytic ethylene glycol oxidation reaction. *Source:* [149]/with permission of American Chemical Society.

cations ($K^+$, $CAT^+$, $TBA^+$) as the bridging units, where CTAB and TBAB represent cetyltrimethylammonium bromide and tetrabutylammonium bromide [149]. Specifically, Li and coworkers synthesized about 20 POM cluster-assembled superstructures built by two types of $MP_2W_{17}O_{61}$ (M = La-Lu). The successful controlled assembly of POM clusters manifests the advantages of POM as structural units for the construction of materials at the subnanometer scale. Especially, one of assembled materials, $EuP_2W_{17}$ nanosheet exhibited high Faradaic efficiency of 90.2% and selectivity of 87.3% for glycolic acid in the electrocatalytic ethylene glycol oxidation reaction (Figure 3.24). This catalytic activity is significantly higher than the isolated POM cluster catalysts. This work provides a new vision for the construction of catalytically active POM-immobilized materials apart from the POM-immobilized coin metal nanoclusters.

## References

1 S. Hasegawa, K. Harano, K. Motokura, RhRu Bimetallic Oxide Cluster Catalysts for Cross-Dehydrogenative Coupling of Arenes and Carboxylic Acids. *J. Am. Chem. Soc.* **2024**, *146*, 19059–19069.

2 R. Jin, M. Peng, A. Li, Y. Deng, Z. Jia, F. Huang, Y. Ling, F. Yang, H. Fu, J. Xie, X. Han, D. Xiao, Z. Jiang, H. Liu, D. Ma, Low Temperature Oxidation of Ethane to Oxygenates by Oxygen over Iridium-Cluster Catalysts. *J. Am. Chem. Soc.* **2019**, *141*, 18921–18925.

3 F. Huang, H. Liu, D. Su, Graphitized Nanocarbon-Supported Metal Catalysts: Synthesis, Properties, and Applications in Heterogeneous Catalysis. *Sci. China Mater.* **2017**, *60*, 1149–1167.

4 Z. Song, H. Shi, L. Liu, M. Zeng, S. Wang, X. Wu, Z. Xu, Application of 1D/2D Carbon Material Supported Metal Nanoclusters for Electrochemical Conversion. *Catal. Sci. Technol.* **2024**, *14*, 1462–1479.

5 H. Tsunoyama, A. Ohnuma, K. Takahashi, A. Velloth, M. Ehara, N. Ichikuni, M. Tabuchi, A. Nakajima, Enhanced Oxygen Reduction Activity of Platinum Subnanocluster Catalysts through Charge Redistribution. *Chem. Commun.* **2019**, *55*, 12603–12606.

6 J. N. Tiwari, K. Nath, S. Kumar, R. N. Tiwari, K. C. Kemp, N. H. Le, D. H. Youn, J. S. Lee, K. S. Kim, Stable Platinum Nanoclusters on Genomic DNA-Graphene Oxide with a High Oxygen Reduction Reaction Activity. *Nat. Commun.* **2013**, *4*, 2221.

7 A. Han, W. Sun, X. Wan, D. Cai, X. Wang, F. Li, J. Shui, D. Wang, Construction of $Co_4$ Atomic Clusters to Enable Fe-$N_4$ Motifs with Highly Active and Durable Oxygen Reduction Performance. *Angew. Chem. Int. Ed.* **2023**, *62*, e202303185; *Angew. Chem.* **2023**, *135*, e202303185.

8 S. Shin, E. Lee, J. Nam, J. Kwon, Y. Choi, B. J. Kim, H. C. Ham, H. Lee, Carbon-Embedded Pt Alloy Cluster Catalysts for Proton Exchange Membrane Fuel Cells. *Adv. Energy Mater.* **2024**, *14*, 2400599.

9 S. Chen, H. Zhu, T. Li, P. Liu, C. Wu, S. Jia, Y. Li, B. Suo, Applications of Metal Nanoclusters Supported on the Two-Dimensional Material Graphene in Electrocatalytic Carbon Dioxide Reduction. *Phys. Chem. Chem. Phys.* **2024**, *26*, 26647–26676.

10 C. Cui, H. Zhang, R. Cheng, B. Huang, Z. Luo, On the Nature of Three-Atom Metal Cluster Catalysis for $N_2$ Reduction to Ammonia. *ACS Catal.* **2022**, *12*, 14964–14975.

11 C. Yao, N. Guo, S. Xi, C. Q. Xu, W. Liu, X. Zhao, J. Li, H. Fang, J. Su, Z. Chen, H. Yan, Z. Qiu, P. Lyu, C. Chen, H. Xu, X. Peng, X. Li, B. Liu, C. Su, S. J. Pennycook, C. J. Sun, J. Li, C. Zhang, Y. Du, J. Lu, Atomically-Precise Dopant-Controlled Single Cluster Catalysis for Electrochemical Nitrogen Reduction. *Nat. Commun.* **2020**, *11*, 4389.

12 C. Liu, T. Li, X. Dai, J. Zhao, D. He, G. Li, B. Wang, X. Cui, Catalytic Activity Enhancement on Alcohol Dehydrogenation Via Directing Reaction Pathways from Single- to Double-Atom Catalysis. *J. Am. Chem. Soc.* **2022**, *144*, 4913–4924.

**13** L. Wang, P. Yin, L.-L. Zhang, S.-C. Shen, S.-L. Xu, P. Chen, H.-W. Liang, Nitrogen-Fixing of Ultrasmall Pd-Based Bimetallic Nanoclusters on Carbon Supports. *J. Catal.* **2020**, *389*, 297–304.

**14** Z. Duan, Q. Lei, W. Wu, H. Chen, 2-Dimensional Metal-Organic Framework Derived N-Doped Carbon Immobilized Pd Nanoparticles for Hydrogen Release from Formic Acid. *Int. J. Hydrogen Energy* **2023**, *48*, 7708–7718.

**15** P. Yin, X. Luo, Y. Ma, S. Q. Chu, S. Chen, X. Zheng, J. Lu, X. J. Wu, H. W. Liang, Sulfur Stabilizing Metal Nanoclusters on Carbon at High Temperatures. *Nat. Commun.* **2021**, *12*, 3135.

**16** J. Park, C. Jeong, M. Na, Y. Oh, K.-S. Lee, Y. Yang, H. R. Byon, Subnanometer Cu Clusters on Porous Ag Enhancing Ethanol Production in Electrochemical $CO_2$ Reduction. *ACS Catal.* **2024**, *14*, 3198–3207.

**17** V. K. Velisoju, J. L. Cerrillo, R. Ahmad, H. O. Mohamed, Y. Attada, Q. Cheng, X. Yao, L. Zheng, O. Shekhah, S. Telalovic, J. Narciso, L. Cavallo, Y. Han, M. Eddaoudi, E. V. Ramos-Fernández, P. Castaño, Copper Nanoparticles Encapsulated in Zeolitic Imidazolate Framework-8 as a Stable and Selective $CO_2$ Hydrogenation Catalyst. *Nat. Commun.* **2024**, *15*, 2045.

**18** J. Mosrati, T. Ishida, H. Mac, M. Al-Yusufi, T. Honma, M. Parliniska-Wojtan, Y. Kobayashi, A. Klyushin, T. Murayama, A. M. Abdel-Mageed, Low-Temperature Hydrogenation of $CO_2$ to Methanol in Water on ZnO-Supported CuAu Nanoalloys. *Angew. Chem. Int. Ed.* **2023**, *62*, e202311340; *Angew. Chem.* **2023**, *135*, e202311340.

**19** L. Liu, S. J. Zheng, H. Chen, J. Cai, S. Q. Zang, Tandem Nitrate-to-Ammonia Conversion on Atomically Precise Silver Nanocluster/Mxene Electrocatalyst. *Angew. Chem. Int. Ed.* **2024**, *63*, e202316910; *Angew. Chem.* **2024**, *136*, e202316910.

**20** P. Prabhu, V. H. Do, T. Yoshida, Y. Zhou, H. Ariga-Miwa, T. Kaneko, T. Uruga, Y. Iwasawa, J. M. Lee, Subnanometric Osmium Clusters Confined on Palladium Metallenes for Enhanced Hydrogen Evolution and Oxygen Reduction Catalysis. *ACS Nano* **2024**, *18*, 9942–9957.

**21** K. Shirayama, X. Jin, K. Nozaki, Selective Hydrogenation of Aldehydes under Syngas Using $CeO_2$-Supported Au Nanoparticle Catalyst. *J. Am. Chem. Soc.* **2024**, *146*, 14086–14094.

**22** I. Song, I. Z. Koleva, H. A. Aleksandrov, L. Chen, J. Heo, D. Li, Y. Wang, J. Szanyi, K. Khivantsev, Ultrasmall Pd Clusters in Fer Zeolite Alleviate CO Poisoning for Effective Low-Temperature Carbon Monoxide Oxidation. *J. Am. Chem. Soc.* **2023**, *145*, 27493–27499.

**23** P. Serna, A. Rodriguez-Fernandez, S. Yacob, C. Kliewer, M. Moliner, A. Corma, Single-Site vs. Cluster Catalysis in High Temperature Oxidations. *Angew. Chem. Int. Ed.* **2021**, *60*, 15954–15962; *Angew. Chem.* **2021**, *133*, 16090–16098.

**24** D. Gashnikova, F. Maurer, E. Sauter, S. Bernart, J. Jelic, P. Dolcet, C. B. Maliakkal, Y. Wang, C. Woll, F. Studt, C. Kubel, M. Casapu, J. D. Grunwaldt, Highly Active Oxidation Catalysts through Confining Pd Clusters on $CeO_2$ Nano-Islands. *Angew. Chem. Int. Ed.* **2024**, *63*, e202408511; *Angew. Chem.* **2024**, *136*, e202408511.

**25** X. Yu, N. S. Genz, R. G. Mendes, X. Ye, F. Meirer, M. Nachtegaal, M. Monai, B. M. Weckhuysen, Anchoring $PdO_X$ Clusters on Defective Alumina for Improved Catalytic Methane Oxidation. *Nat. Commun.* **2024**, *15*, 6494.

**26** X. Wang, J. Zhao, H. Eliasson, R. Erni, A. Ziarati, S. McKeown Walker, T. Burgi, Very Low Temperature CO Oxidation over Atomically Precise $Au_{25}$ Nanoclusters on $MnO_2$. *J. Am. Chem. Soc.* **2023**, *145*, 27273–27281.
**27** S. Masuda, T. Tsukuda, Partially Thiolated $Au_n$ (N = 25, 102) Clusters on Layered Double Hydroxides Anchored by Electrostatic Interactions: Size Effect on 5-Hydroxymethylfurfural Oxidation Catalysis. *ACS Catal.* **2023**, *13*, 16179–16187.
**28** S. Fu, S. Yao, S. Guo, G. C. Guo, W. Yuan, T. B. Lu, Z. M. Zhang, Feeding Carbonylation with $CO_2$ Via the Synergy of Single-Site/Nanocluster Catalysts in a Photosensitizing MOF. *J. Am. Chem. Soc.* **2021**, *143*, 20792–20801.
**29** P. Wang, R. Shi, Y. Zhao, Z. Li, J. Zhao, J. Zhao, G. I. N. Waterhouse, L. Z. Wu, T. Zhang, Selective Photocatalytic Oxidative Coupling of Methane Via Regulating Methyl Intermediates over Metal/Zno Nanoparticles. *Angew. Chem. Int. Ed.* **2023**, *62*, e202304301; *Angew. Chem.* **2023**, *135*, e202304301.
**30** H. Chen, C. Liu, M. Wang, C. Zhang, N. Luo, Y. Wang, H. Abroshan, G. Li, F. Wang, Visible Light Gold Nanocluster Photocatalyst: Selective Aerobic Oxidation of Amines to Imines. *ACS Catal.* **2017**, *7*, 3632–3638.
**31** Y. Zhang, J. Zhang, Z. Li, Z. Qin, S. Sharma, G. Li, Atomically Precise Copper Dopants in Metal Clusters Boost Up Stability, Fluorescence, and Photocatalytic Activity. *Commun. Chem.* **2023**, *6*, 24.
**32** B. Weng, K. Q. Lu, Z. Tang, H. M. Chen, Y. J. Xu, Stabilizing Ultrasmall Au Clusters for Enhanced Photoredox Catalysis. *Nat. Commun.* **2018**, *9*, 1543.
**33** T. Kawawaki, Y. Kataoka, M. Hirata, Y. Akinaga, R. Takahata, K. Wakamatsu, Y. Fujiki, M. Kataoka, S. Kikkawa, A. S. Alotabi, S. Hossain, D. J. Osborn, T. Teranishi, G. G. Andersson, G. F. Metha, S. Yamazoe, Y. Negishi, Creation of High-Performance Heterogeneous Photocatalysts by Controlling Ligand Desorption and Particle Size of Gold Nanocluster. *Angew. Chem. Int. Ed.* **2021**, *60*, 21340–21350; *Angew. Chem.* **2021**, *133*, 21510–21520.
**34** T. Ould Ely, C. Amiens, B. Chaudret, Synthesis of Nickel Nanoparticles. Influence of Aggregation Induced by Modification of Poly(vinylpyrrolidone) Chain Length on Their Magnetic Properties. *Chem. Mater.* **1999**, *11*, 526–529.
**35** N. Toshima, Y. Shiraishi, T. Teranishi, M. Miyake, T. Tominaga, H. Watanabe, W. Brijoux, H. Bönnemann, G. Schmid, Various Ligand-Stabilized Metal Nanoclusters as Homogeneous and Heterogeneous Catalysts in the Liquid Phase. *Appl. Organomet. Chem.* **2001**, *15*, 178–196.
**36** L. Starkey Ott, B. J. Hornstein, R. G. Finke, A Test of the Transition-Metal Nanocluster Formation and Stabilization Ability of the Most Common Polymeric Stabilizer, Poly(vinylpyrrolidone), as Well as Four Other Polymeric Protectants. *Langmuir* **2006**, *22*, 9357–9367.
**37** I. Pastoriza-Santos, L. M. Liz-Marzán, Formation of PVP-Protected Metal Nanoparticles in DMF. *Langmuir* **2002**, 18, 2888–2894.
**38** Y. Li, X. M. Hong, D. M. Collard, M. A. El-Sayed, Suzuki Cross-Coupling Reactions Catalyzed by Palladium Nanoparticles in Aqueous Solution. *Org. Lett.* **2000**, *2*, 2385–2388.

**39** J. U. Köhler, J. S. Bradley, A Kinetic Probe of the Effect of a Stabilizing Polymer on a Colloidal Catalyst: Accelerated Enantioselective Hydrogenation of Ethyl Pyruvate Catalyzed by Poly(vinylpyrrolidone)-Stabilized Platinum Colloid. *Langmuir* **1998**, *14*, 2730–2735.
**40** W.-Y. Yu, H.-F. Liu, Q. Tao, Modification of Metal Cations to Metal Clusters in Liquid Medium. *Chem. Commun.* **1996**, 1773–1774.
**41** R. Patakfalvi, Z. Virányi, I. Dékány, Kinetics of Silver Nanoparticle Growth in Aqueous Polymer Solutions. *Colloid Polym. Sci.* **2004**, *283*, 299–305.
**42** S. King, K. Hyunh, R. Tannenbaum, Kinetics of Nucleation, Growth, and Stabilization of Cobalt Oxide Nanoclusters. *J. Phys. Chem. B.* **2003**, *107*, 12097–12104.
**43** E. Tadd, A. Zeno, M. Zubris, N. Dan, R. Tannenbaum, Adsorption and Polymer Film Formation on Metal Nanoclusters. *Macromolecules* **2003**, *36*, 6497–6502.
**44** A. Spitaleri, P. Pertici, N. Scalera, G. Vitulli, M. Hoang, T. W. Turney, M. Gleria, Supported Ruthenium Nanoparticles on Polyorganophosphazenes: Preparation, Structural and Catalytic Studies. *Inorg. Chim. Acta.* **2003**, *352*, 61–71.
**45** L. D. Rampino, F. F. Nord, Applicability of Palladium Synthetic High Polymer Catalysts. *J. Am. Chem. Soc.* **1941**, *63*, 3268.
**46** L. D. Rampino, F. F. Nord, Preparation of Palladium and Platinum Synthetic High Polymer Catalysts and the Relationship between Particle Size and Rate of Hydrogenation. *J. Am. Chem. Soc.* **1941**, *63*, 2745–2749.
**47** W. Hoogsteen, L. G. J. Fokkink, Polymer-Stabilized Pd Sols:Kinetics of Sol Formation and Stabilization Mechanism. *J. Colloid Interf. Sci.* **1995**, *175*, 12–26.
**48** J. Ramos, A. Milán, F. Palacio, Production of Magnetic Nanoparticles in A Polyvinylpyridine Matrix. *Polymer* **2000**, *41*, 8461–8464.
**49** Y. Huang, Y. Li, J. Hu, P. Cheng, H. Chen, R. Li, X. Li, C. W. Yip, A. S. C. Chan, Enantioselective Hydrogenation of Ethyl Pyruvate Catalyzed by Polyvinylpyrrolidone-Stabilized and Supported Rhodium Nanocluster. *J. Mol. Catal. A: Chem.* **2002**, *189*, 219–224.
**50** R. Narayanan, M. A. El-Sayed, Effect of Colloidal Catalysis on the Nanoparticle Size Distribution: Dendrimer-Pd vs PVP-Pd Nanoparticles Catalyzing the Suzuki Coupling Reaction. *J. Phys. Chem. B* **2004**, *108*, 8572–8580.
**51** R. Narayanan, M. A. El-Sayed, Effect of Catalysis on the Stability of Metallic Nanoparticles: Suzuki Reaction Catalyzed by PVP-Palladium Nanoparticles. *J. Am. Chem. Soc.* **2003**, *125*, 8340–8347.
**52** J. L. Bars, U. Specht, J. S. Bradley, D. G. Blackmond, A Catalytic Probe of the Surface of Colloidal Palladium Particles Using Heck Coupling Reactions. *Langmuir* **1999**, *15*, 7621–7625.
**53** J. Kiwi, M. Graetzel, Role of Metal Cocatalysts in the Photocatalytic Production of Hydrogen from Water Revisited. *J. Am. Chem. Soc.* **1979**, *101*, 7214–7217.
**54** B. P. S. Chauhan, J. S. Rathore, T. Bandoo, "Polysiloxane-Pd" Nanocomposites as Recyclable Chemoselective Hydrogenation Catalysts. *J. Am. Chem. Soc.* **2004**, *126*, 8493–8500.

**55** B. P. S. Chauhan, J. S. Rathore, M. Chauhan, A. Krawicz, Synthesis of Polysiloxane Stabilized Palladium Colloids and Evidence of Their Participation in Silaesterification Reactions. *J. Am. Chem. Soc.* **2003**, *125*, 2876–2877.

**56** L. M. Bronstein, D. M. Chernyshov, P. M. Valetsky, E. A. Wilder, R. J. Spontak, Metal Nanoparticles Grown in the Nanostructured Matrix of Poly(octadecylsiloxane). *Langmuir* **2000**, *16*, 8221–8225.

**57** B. P. S. Chauhan, R. Sardar, Self-Assembled Stable Silver Nanoclusters and Nanonecklace Formation: Poly(methylhydrosiloxane)-Mediated One-Pot Route to Organosols. *Macromolecules* **2004**, *37*, 5136–5139.

**58** D. Tabuani, O. Monticelli, A. Chincarini, C. Bianchini, F. Vizza, S. Moneti, S. Russo, Palladium Nanoparticles Supported on Hyperbranched Aramids: Synthesis, Characterization, and Some Applications in the Hydrogenation of Unsaturated Substrates. *Macromolecules* **2003**, *36*, 4294–4301.

**59** D. Tabuani, O. Monticelli, H. Komber, S. Russo, Preparation and Characterisation of Pd Nanoclusters in Hyperbranched Aramid Templates to Be Used in Homogeneous Catalysis. *Macromol. Chem. Phys.* **2003**, *204*, 1576–1583.

**60** S. Mecking, R. Thomann, H. Frey, A. Sunder, Preparation of Catalytically Active Palladium Nanoclusters in Compartments of Amphiphilic Hyperbranched Polyglycerols. *Macromolecules* **2000**, *33*, 3958–3960.

**61** L. S. Ott, R. G. Finke, Transition-Metal Nanocluster Stabilization for Catalysis: A Critical Review of Ranking Methods and Putative Stabilizers. *Coord. Chem. Rev.* **2007**, *251*, 1075–1100.

**62** Y. M. Wang, J. Cai, Q. Y. Wang, Y. Li, Z. Han, S. Li, C. H. Gong, S. Wang, S. Q. Zang, T. C. W. Mak, Electropolymerization of Metal Clusters Establishing a Versatile Platform for Enhanced Catalysis Performance. *Angew. Chem. Int. Ed.* **2022**, *61*, e202114538; *Angew. Chem.* **2022**, *134*, e202114538.

**63** Y. M. Wang, F. Q. Yan, Q. Y. Wang, C. X. Du, L. Y. Wang, B. Li, S. Wang, S. Q. Zang, Single-Atom Tailored Atomically-Precise Nanoclusters for Enhanced Electrochemical Reduction of $CO_2$-to-CO Activity. *Nat. Commun.* **2024**, *15*, 1843.

**64** N. de Silva, J. M. Ha, A. Solovyov, M. M. Nigra, I. Ogino, S. W. Yeh, K. A. Durkin, A. Katz, A Bioinspired Approach for Controlling Accessibility in Calix[4]Arene-Bound Metal Cluster Catalysts. *Nat. Chem.* **2010**, *2*, 1062–1068.

**65** J. Hassinen, P. Pulkkinen, E. Kalenius, T. Pradeep, H. Tenhu, H. Hakkinen, R. H. Ras, Mixed-Monolayer-Protected Au25 Clusters with Bulky Calix[4]Arene Functionalities. *J. Phys. Chem. Lett.* **2014**, *5*, 585–589.

**66** Z.-J. Guan, J.-L. Zeng, Z.-A. Nan, X.-K. Wan, Y.-M. Lin, Q.-M. Wang, Thiacalix[4]arene: New Protection for Metal Nanoclusters. *Sci. Adv.* **2016**, *2*, e1600323.

**67** Z.-J. Guan, F. Hu, S.-F. Yuan, Z.-A. Nan, Y.-M. Lin, Q.-M. Wang, The Stability Enhancement Factor Beyond Eight-Electron Shell Closure in Thiacalix[4]Arene-Protected Silver Clusters. *Chem. Sci.* **2019**, *10*, 3360–3365.

**68** Z. Wang, H.-F. Su, Y.-W. Gong, Q.-P. Qu, Y.-F. Bi, C.-H. Tung, D. Sun, L.-S. Zheng, A Hierarchically Assembled 88-Nuclei Silver-Thiacalix[4]Arene Nanocluster. *Nat. Commun.* **2020**, *11*, 308.

**69** Z. Wang, H.-F. Su, L.-P. Zhang, J.-M. Dou, C.-H. Tung, D. Sun, L. Zheng, Stepwise Assembly of $Ag_{42}$ Nanocalices Based on a $Mo^{VI}$-Anchored Thiacalix[4]Arene Metalloligand. *ACS Nano.* **2022**, *16*, 4500–4507.

**70** M. Zhao, L. Sun, R. M. Crooks, Preparation of Cu Nanoclusters within Dendrimer Templates. *J. Am. Chem. Soc.* **1998**, *120*, 4877–4878.

**71** R. M. Crooks, M. Zhao, L. Sun, V. Chechik, L. K. Yeung, Dendrimer-Encapsulated Metal Nanoparticles: Synthesis, Characterization, and Applications to Catalysis. *Acc. Chem. Res.* **2001**, *34*, 181–190.

**72** M. Zhao, R. M. Crooks, Intradendrimer Exchange of Metal Nanoparticles. *Chem. Mater.* **1999**, *11*, 3379–3385.

**73** V. Chechik, R. M. Crooks, Dendrimer-Encapsulated Pd Nanoparticles as Fluorous Phase-Soluble Catalysts. *J. Am. Chem. Soc.* **2000**, *122*, 1243–1244.

**74** R. W. J. Scott, H. Ye, R. R. Henriquez, R. M. Crooks, Synthesis, Characterization, and Stability of Dendrimer-Encapsulated Palladium Nanoparticles. *Chem. Mater.* **2003**, *15*, 3873–3878.

**75** Y. Niu, R. M. Crooks, Preparation of Dendrimer-Encapsulated Metal Nanoparticles Using Organic Solvents. *Chem. Mater.* **2003**, *15*, 3463–3467.

**76** R. W. J. Scott, O. M. Wilson, R. M. Crooks, Synthesis, Characterization, and Applications of Dendrimer-Encapsulated Nanoparticles. *J. Phys. Chem. B.* **2005**, *109*, 692–704.

**77** J. C. Garcia-Martinez, R. W. J. Scott, R. M. Crooks, Extraction of Monodisperse Palladium Nanoparticles from Dendrimer Templates. *J. Am. Chem. Soc.* **2003**, *125*, 11190–11191.

**78** Y. Luo, Y. Li, X. Jia, H. Yang, L. Yang, Q. Zhou, Y. Wei, Fabrication of Ag Nanoparticle–Encapsulating Multilayer Films Based on PAMAM Dendrimers with Covalent Interlayer Linkages. *J. Appl. Polym. Sci.* **2003**, *89*, 1515–1519.

**79** L. Balogh, D. A. Tomalia, Poly(Amidoamine) Dendrimer-Templated Nanocomposites. 1. Synthesis of Zerovalent Copper Nanoclusters. *J. Am. Chem. Soc.* **1998**, *120*, 7355–7356.

**80** P. N. Floriano, C. O. Noble, J. M. Schoonmaker, E. D. Poliakoff, R. L. McCarley, Cu(0) Nanoclusters Derived from Poly(propylene imine) Dendrimer Complexes of Cu(II). *J. Am. Chem. Soc.* **2001**, *123*, 10545–10553.

**81** J. Lemo, K. Heuzé, D. Astruc, Synthesis and Catalytic Activity of DAB-Dendrimer Encapsulated Pd Nanoparticles for the Suzuki Coupling Reaction. *Inorg. Chim. Acta.* **2006**, *359*, 4909–4911.

**82** V. S. Myers, M. G. Weir, E. V. Carino, D. F. Yancey, S. Pande, R. M. Crooks, Dendrimer-Encapsulated Nanoparticles: New Synthetic and Characterization Methods and Catalytic Applications. *Chem. Sci.* **2011**, *2*, 1632–1646.

**83** L. K. Yeung, R. M. Crooks, Heck Heterocoupling within a Dendritic Nanoreactor. *Nano Lett.* **2001**, *1*, 14–17.

**84** L. K. Yeung, C. T. Lee Jr, K. P. Johnston, R. M. Crooks, Catalysis in Supercritical $CO_2$ Using Dendrimer-Encapsulated Palladium Nanoparticles. *Chem. Commun.* **2001**, *21*, 2290–2291.

**85** E. H. Rahim, F. S. Kamounah, J. Frederiksen, J. B. Christensen, Heck Reactions Catalyzed by PAMAM-Dendrimer Encapsulated Pd(0) Nanoparticles. *Nano Lett.* **2001**, *1*, 499–501.

**86** M. Pittelkow, K. Moth-Poulsen, U. Boas, J. B. Christensen, Poly(amidoamine)-Dendrimer-Stabilized Pd(0) Nanoparticles as a Catalyst for the Suzuki Reaction. *Langmuir* **2003**, *19*, 7682–7684.

**87** Y. M. Yamada, S. M. Sarkar, Y. Uozumi, Self-Assembled Poly(Imidazole-Palladium): Highly Active, Reusable Catalyst at Parts Per Million to Parts Per Billion Levels. *J. Am. Chem. Soc.* **2012**, *134*, 3190–3198.

**88** S. Ogasawara, S. Kato, Palladium Nanoparticles Captured in Microporous Polymers: A Tailor-Made Catalyst for Heterogeneous Carbon Cross-Coupling Reactions. *J. Am. Chem. Soc.* **2010**, *132*, 4608–4613.

**89** C. Ornelas, J. Ruiz Aranzaes, E. Cloutet, S. Alves, D. Astruc, Click Assembly of 1,2,3-Triazole-Linked Dendrimers, Including Ferrocenyl Dendrimers, Which Sense Both Oxo Anions and Metal Cations. *Angew. Chem. Int. Ed.* **2007**, *46*, 872–877; *Angew. Chem.* **2007**, *119*, 890–895.

**90** C. Ornelas, L. Salmon, J. R. Aranzaes, D. Astruc, Catalytically Efficient Palladium Nanoparticles Stabilized by "Click" Ferrocenyl Dendrimers. *Chem. Commun.* **2007**, *46*, 4946–4948.

**91** A. K. Diallo, C. Ornelas, L. Salmon, J. Ruiz Aranzaes, D. Astruc, "Homeopathic" Catalytic Activity and Atom-Leaching Mechanism in Miyaura-Suzuki Reactions under Ambient Conditions with Precise Dendrimer-Stabilized Pd Nanoparticles. *Angew. Chem. Int. Ed.* **2007**, *46*, 8644–8648; *Angew. Chem.* **2007**, *119*, 8798–8802.

**92** C. Ornelas, J. R. Aranzaes, L. Salmon, D. Astruc, "Click" Dendrimers: Synthesis, Redox Sensing of $Pd(OAc)_2$, and Remarkable Catalytic Hydrogenation Activity of Precise Pd Nanoparticles Stabilized by 1,2,3-Triazole-Containing Dendrimers. *Chem. Eur. J.* **2008**, *14*, 50–64.

**93** C. Deraedt, D. Astruc, "Homeopathic" Palladium Nanoparticle Catalysis of Cross Carbon-Carbon Coupling Reactions. *Acc. Chem. Res.* **2014**, *47*, 494–503.

**94** C. Ornelas, J. Ruiz, L. Salmon, D. Astruc, Sulphonated "Click" Dendrimer-Stabilized Palladium Nanoparticles as Highly Efficient Catalysts for Olefin Hydrogenation and Suzuki Coupling Reactions under Ambient Conditions in Aqueous Media. *Adv. Synth. Catal.* **2008**, *350*, 837–845.

**95** C. Ornelas, A. K. Diallo, J. Ruiz, D. Astruc, "Click" Polymer-Supported Palladium Nanoparticles as Highly Efficient Catalysts for Olefin Hydrogenation and Suzuki Coupling Reactions under Ambient Conditions. *Adv. Synth. Catal.* **2009**, *351*, 2147–2154.

**96** V. Percec, G. Johansson, G. Ungar, J. Zhou, Fluorophobic Effect Induces the Self-Assembly of Semifluorinated Tapered Monodendrons Containing Crown Ethers into Supramolecular Columnar Dendrimers Which Exhibit a Homeotropic Hexagonal Columnar Liquid Crystalline Phase. *J. Am. Chem. Soc.* **1996**, *118*, 9855–9866.

**97** A. K. Diallo, E. Boisselier, L. Liang, J. Ruiz, D. Astruc, Dendrimer-Induced Molecular Catalysis in Water: The Example of Olefin Metathesis. *Chemistry* **2010**, *16*, 11832–11835.

**98** C. Deraedt, L. Salmon, L. Etienne, J. Ruiz, D. Astruc, "Click" Dendrimers as Efficient Nanoreactors in Aqueous Solvent: Pd Nanoparticle Stabilization for Sub-Ppm Pd Catalysis of Suzuki-Miyaura Reactions of Aryl Bromides. *Chem. Commun.* **2013**, *49*, 8169–8171.

**99** W. Huang, J. N. Kuhn, C.-K. Tsung, Y. Zhang, S. E. Habas, P. Yang, G. A. Somorjai, Dendrimer Templated Synthesis of One Nanometer Rh and Pt Particles Supported on Mesoporous Silica: Catalytic Activity for Ethylene and Pyrrole Hydrogenation. *Nano Lett.* **2008**, *8*, 2027–2034.

**100** D. Zhao, J. Feng, Q. Huo, N. Melosh, G. H. Fredrickson, B. F. Chmelka, G. D. Stucky, Triblock Copolymer Syntheses of Mesoporous Silica with Periodic 50 to 300 Angstrom Pores. *Science* **1998**, *279*, 548–552.

**101** Y. Li, J. H. Liu, C. A. Witham, W. Huang, M. A. Marcus, S. C. Fakra, P. Alayoglu, Z. Zhu, C. M. Thompson, A. Arjun, K. Lee, E. Gross, F. D. Toste, G. A. Somorjai, A Pt-Cluster-Based Heterogeneous Catalyst for Homogeneous Catalytic Reactions: X-Ray Absorption Spectroscopy and Reaction Kinetic Studies of Their Activity and Stability against Leaching. *J. Am. Chem. Soc.* **2011**, *133*, 13527–13533.

**102** R. Ye, A. V. Zhukhovitskiy, C. V. Deraedt, F. D. Toste, G. A. Somorjai, Supported Dendrimer-Encapsulated Metal Clusters: Toward Heterogenizing Homogeneous Catalysts. *Acc. Chem. Res.* **2017**, *50*, 1894–1901.

**103** C. A. Witham, W. Huang, C.-K. Tsung, J. N. Kuhn, G. A. Somorjai, F. D. Toste, Converting Homogeneous to Heterogeneous in Electrophilic Catalysis Using Monodisperse Metal Nanoparticles. *Nat. Chem.* **2010**, *2*, 36–41.

**104** R. Ye, B. Yuan, J. Zhao, W. T. Ralston, C. Y. Wu, E. Unel Barin, F. D. Toste, G. A. Somorjai, Metal Nanoparticles Catalyzed Selective Carbon-Carbon Bond Activation in the Liquid Phase. *J. Am. Chem. Soc.* **2016**, *138*, 8533–8537.

**105** R. Ye, J. Zhao, B. Yuan, W. C. Liu, J. Rodrigues De Araujo, F. F. Faucher, M. Chang, C. V. Deraedt, F. D. Toste, G. A. Somorjai, New Insights into Aldol Reactions of Methyl Isocyanoacetate Catalyzed by Heterogenized Homogeneous Catalysts. *Nano. Lett.* **2017**, *17*, 584–589.

**106** E. Gross, X. Z. Shu, S. Alayoglu, H. A. Bechtel, M. C. Martin, F. D. Toste, G. A. Somorjai, In Situ Ir and X-Ray High Spatial-Resolution Microspectroscopy Measurements of Multistep Organic Transformation in Flow Microreactor Catalyzed by Au Nanoclusters. *J. Am. Chem. Soc.* **2014**, *136*, 3624–3629.

**107** C. Deraedt, G. Melaet, W. T. Ralston, R. Ye, G. A. Somorjai, Platinum and Other Transition Metal Nanoclusters (Pd, Rh) Stabilized by Pamam Dendrimer as Excellent Heterogeneous Catalysts: Application to the Methylcyclopentane (MCP) Hydrogenative Isomerization. *Nano. Lett.* **2017**, *17*, 1853–1862.

**108** W. Huang, J. H.-C. Liu, P. Alayoglu, Y. Li, C. A. Witham, C.-K. Tsung, F. Dean Toste, G. A. Somorjai, Highly Active Heterogeneous Palladium Nanoparticle Catalysts for Homogeneous Electrophilic Reactions in Solution and the Utilization of a Continuous Flow Reactor. *J. Am. Chem. Soc.* **2010**, *132*, 16771–16773.

**109** C. Zhang, C. Chen, H. Dong, J.-R. Shen, H. Dau, J. Zhao, A Synthetic $Mn_4Ca$-Cluster Mimicking the Oxygen-Evolving Center of Photosynthesis. *Science* **2015**, *348*, 690–693.

**110** T. Ueno, M. Suzuki, T. Goto, T. Matsumoto, K. Nagayama, Y. Watanabe, Size-Selective Olefin Hydrogenation by a Pd Nanocluster Provided in an Apo-Ferritin Cage. *Angew. Chem. Int. Ed.* **2004**, *43*, 2527–2530; *Angew. Chem.* **2004**, *116*, 2581–2584.

**111** S. L. Christensen, A. Chatt, P. Zhang, Biomolecule-Coated Metal Nanoparticles on Titanium. *Langmuir* **2012**, *28*, 2979–2985.

**112** Z. Wang, Y. Shao, Z. Zhu, J. Wang, X. Gao, J. Xie, Y. Wang, Q. Wu, Y. Shen, Y. Ding, Novel Gold Nanozyme Regulation Strategies Facilitate Analytes Detection. *Coord. Chem. Rev.* **2023**, *495*, 215369.

**113** F. Wen, Y. Dong, L. Feng, S. Wang, S. Zhang, X. Zhang, Horseradish Peroxidase Functionalized Fluorescent Gold Nanoclusters for Hydrogen Peroxide Sensing. *Anal. Chem.* **2011**, *83*, 1193–1196.

**114** X. Xia, Y. Long, J. Wang, Glucose Oxidase-Functionalized Fluorescent Gold Nanoclusters as Probes for Glucose. *Anal. Chim. Acta.* **2013**, *772*, 81–86.

**115** S. S. Narayanan, S. K. Pal, Structural and Functional Characterization of Luminescent Silver-Protein Nanobioconjugates. *J. Phys. Chem. C.* **2008**, *112*, 4874–4879.

**116** N. Goswami, K. Zheng, J. Xie, Bio-NCs - the Marriage of Ultrasmall Metal Nanoclusters with Biomolecules. *Nanoscale* **2014**, *6*, 13328–13347.

**117** S. Abarghoei, N. Fakhri, Y. S. Borghei, M. Hosseini, M. R. Ganjali, A Colorimetric Paper Sensor for Citrate as Biomarker for Early Stage Detection of Prostate Cancer Based on Peroxidase-Like Activity of Cysteine-Capped Gold Nanoclusters. *Spectrochim. Acta A Mol. Biomol. Spectrosc.* **2019**, *210*, 251–259.

**118** H. Wei, E. Wang, Nanomaterials with Enzyme-Like Characteristics (Nanozymes): Next-Generation Artificial Enzymes. *Chem. Soc. Rev.* **2013**, *42*, 6060–6093.

**119** X. Shen, Z. Wang, X. J. Gao, X. Gao, Reaction Mechanisms and Kinetics of Nanozymes: Insights from Theory and Computation. *Adv. Mater.* **2024**, *36*, e2211151.

**120** M. Liang, X. Yan, Nanozymes: From New Concepts, Mechanisms, and Standards to Applications. *Acc. Chem. Res.* **2019**, *52*, 2190–2200.

**121** Z. Wang, J. Wu, J. J. Zheng, X. Shen, L. Yan, H. Wei, X. Gao, Y. Zhao, Accelerated Discovery of Superoxide-Dismutase Nanozymes Via High-Throughput Computational Screening. *Nat. Commun.* **2021**, *12*, 6866.

**122** X. Shen, W. Liu, X. Gao, Z. Lu, X. Wu, X. Gao, Mechanisms of Oxidase and Superoxide Dismutation-Like Activities of Gold, Silver, Platinum, and Palladium, and Their Alloys: A General Way to the Activation of Molecular Oxygen. *J. Am. Chem. Soc.* **2015**, *137*, 15882–15891.

**123** W. Mi, S. Tang, S. Guo, H. Li, N. Shao, In Situ Synthesis of Red Fluorescent Gold Nanoclusters with Enzyme-Like Activity for Oxidative Stress Amplification in Chemodynamic Therapy. *Chinese Chem. Lett.* **2022**, *33*, 1331–1336.

**124** T. Wang, Z. Xu, M. Wen, N. Li, L. Zhang, Y. Xue, L. Shang, Multifunctional Gold Clusterzymes with Distinct Glucose Depletion and Macrophage Reprogramming Capability Towards Regulating the Regeneration Cascade. *Chem. Eng. J.* **2024**, *482*, 149068.

**125** A. R. Deshmukh, H. Aloui, B. S. Kim, Novel Biogenic Gold Nanoparticles Catalyzing Multienzyme Cascade Reaction: Glucose Oxidase and Peroxidase Mimicking Activity. *Chem. Eng. J.* **2021**, *421*, 127859.

**126** Z. Xu, T. Wang, Y. Li, M. Wen, K. Liang, C. Ruan, L. Zhang, Y. Xue, L. Shang, Nanozyme-Engineered Bioglass through Supercharged Interface for Enhanced Anti-Infection and Fibroblast Regulation. *Adv. Funct. Mater.* **2022**, *33*, 2209438.

**127** C. P. Liu, T. H. Wu, Y. L. Lin, C. Y. Liu, S. Wang, S. Y. Lin, Tailoring Enzyme-Like Activities of Gold Nanoclusters by Polymeric Tertiary Amines for Protecting Neurons against Oxidative Stress. *Small* **2016**, *12*, 4127–4135.

**128** Y. Zhang, S. Sun, H. Liu, Q. Ren, W. Hao, Q. Xin, J. Xu, H. Wang, X. D. Zhang, Catalytically Active Gold Clusters with Atomic Precision for Noninvasive Early Intervention of Neurotrauma. *J. Nanobiotechnol.* **2021**, *19*, 319.

**129** Y. Xie, W. Zheng, X. Jiang, Near-Infrared Light-Activated Phototherapy by Gold Nanoclusters for Dispersing Biofilms. *ACS Appl. Mater. Interfaces.* **2020**, *12*, 9041–9049.

**130** E. Gwinn, D. Schultz, S. M. Copp, S. Swasey, DNA-Protected Silver Clusters for Nanophotonics. *Nanomaterials* **2015**, *5*, 180–207.

**131** W. Y. Chen, J. Y. Lin, W. J. Chen, L. Luo, E. Wei-Guang Diau, Y. C. Chen, Functional Gold Nanoclusters as Antimicrobial Agents for Antibiotic-Resistant Bacteria. *Nanomedicine* **2010**, *5*, 755–764.

**132** I. Diez, P. Eronen, M. Osterberg, M. B. Linder, O. Ikkala, R. H. Ras, Functionalization of Nanofibrillated Cellulose with Silver Nanoclusters: Fluorescence and Antibacterial Activity. *Macromol. Biosci.* **2011**, *11*, 1185–1191.

**133** X. Yuan, M. I. Setyawati, A. S. Tan, C. N. Ong, D. T. Leong, J. Xie, Highly Luminescent Silver Nanoclusters with Tunable Emissions: Cyclic Reduction-Decomposition Synthesis and Antimicrobial Properties. *NPG Asia Mater.* **2013**, *5*, e39.

**134** K. Zheng, M. I. Setyawati, D. T. Leong, J. Xie, Antimicrobial Gold Nanoclusters. *ACS Nano.* **2017**, *11*, 6904–6910.

**135** T.-K. Chang, T.-M. Cheng, H.-L. Chu, S.-H. Tan, J.-C. Kuo, P.-H. Hsu, C.-Y. Su, H.-M. Chen, C.-M. Lee, T.-R. Kuo, Metabolic Mechanism Investigation of Antibacterial Active Cysteine-Conjugated Gold Nanoclusters in *Escherichia coli*. *ACS Sustainable Chem. Eng.* **2019**, *7*, 15479–15486.

**136** J. Qiu, F. Ahmad, J. Ma, Y. Sun, Y. Liu, Y. Xiao, L. Xu, T. Shu, X. Zhang, From Synthesis to Applications of Biomolecule-Protected Luminescent Gold Nanoclusters. *Anal. Bioanal. Chem.* **2024**, *416*, 3923–3944.

**137** Y. Li, S. Qu, Y. Xue, L. Zhang, L. Shang, Cationic Antibacterial Metal Nanoclusters with Traceable Capability for Fluorescent Imaging the Nano-Bio Interactions. *Nano. Res.* **2022**, *16*, 999–1008.

**138** Y. Xie, Y. Liu, J. Yang, Y. Liu, F. Hu, K. Zhu, X. Jiang, Gold Nanoclusters for Targeting Methicillin-Resistant *Staphylococcus aureus* in Vivo. *Angew. Chem. Int. Ed.* **2018**, *57*, 3958–3962; *Angew. Chem.* **2018**, *130*, 4022–4026.

**139** Y. Yun, H. Sheng, K. Bao, L. Xu, Y. Zhang, D. Astruc, M. Zhu, Design and Remarkable Efficiency of the Robust Sandwich Cluster Composite Nanocatalysts ZIF-8@$Au_{25}$@ZIF-67. *J. Am. Chem. Soc.* **2020**, *142*, 4126–4130.

**140** H. Wang, X. Zhang, W. Zhang, M. Zhou, H.-L. Jiang, Heteroatom-Doped $Ag_{25}$ Nanoclusters Encapsulated in Metal-Organic Frameworks for Photocatalytic Hydrogen Production. *Angew. Chem. Int. Ed.* **2024**, *63*, e202401443.

**141** Q. You, H. Wang, Y. Zhao, W. Fan, W. Gu, H.-L. Jiang, Z. Wu, Bottom-Up Construction of Metal-Organic Framework Loricae on Metal Nanoclusters with Consecutive Single Nonmetal Atom Tuning for Tailored Catalysis. *J. Am. Chem. Soc.* **2024**, *146*, 9026–9035.

**142** R.-W. Huang, Y.-S. Wei, X.-Y. Dong, X.-H. Wu, C.-X. Du, S.-Q. Zang, T. C. W. Mak, Hypersensitive Dual-Function Luminescence Switching of a Silver-Chalcogenolate Cluster-Based Metal-Organic Framework. *Nat. Chem.* **2017**, *9*, 689–697.

**143** Y. Jin, X.-Y. Dong, C. Zhang, S. Li, S.-Q. Zang, T. C. W. Mak, Predesigned Cluster-Based Spacers for Versatile Luminescent Metallacages. *J. Am. Chem. Soc.* **2023**, *145*, 13514–13519.

**144** W.-L. Mu, X.-Z. Cong, X. Chen, P. Xia, Q. Liu, L. Wang, J. Yan, C. Liu, Hierarchical Assembly of High-Nuclearity Copper(I) Alkynide Nanoclusters: Highly Effective $CO_2$ Electroreduction Catalyst toward Hydrocarbons. *J. Am. Chem. Soc.* **2024**, *146*, 28131–28140.

**145** M. Kamachi, K. Yonesato, T. Okazaki, D. Yanai, S. Kikkawa, S. Yamazoe, R. Ishikawa, N. Shibata, Y. Ikuhara, K. Yamaguchi, K. Suzuki, Synthesis of a Gold-Silver Alloy Nanocluster within a Ring-Shaped Polyoxometalate and Its Photocatalytic Property. *Angew. Chem. Int. Ed.* **2024**, *63*, e202408358; *Angew. Chem.* **2024**, *136*, e202408358.

**146** K. Yonesato, D. Yanai, S. Yamazoe, D. Yokogawa, T. Kikuchi, K. Yamaguchi, K. Suzuki, Surface-Exposed Silver Nanoclusters Inside Molecular Metal Oxide Cavities. *Nat. Chem.* **2023**, *15*, 940–947.

**147** Y. Feng, F. Fu, L. Zeng, M. Zhao, X. Xin, J. Liang, M. Zhou, X. Fang, H. Lv, G.-Y. Yang, Atomically Precise Silver Clusters Stabilized by Lacunary Polyoxometalates with Photocatalytic $CO_2$ Reduction Activity. *Angew. Chem. Int. Ed.* **2024**, *63*, e202317341.

**148** M. Chi, J. Wei, Z. Zhao, X. Liu, W. Sun, Y. Feng, H. Lv, G.-Y. Yang, Structural Isomerism of $\{Ag_{14}\}^{10+}$ Nanocluster Encapsulated by Bowl-Like Polyoxometalates. *Angew. Chem. Int. Ed.* **2025**, *64*, e202424499.

**149** Z. Li, H. Li, Y. Lin, W. Shi, X. Wang, Cluster-Cation Pairs Mediated Assembly of Subnanometer Polyoxometalates Superstructures. *J. Am. Chem. Soc.* **2024**, *146*, 28874–28884.

# 4

# Synthesis and Characterizations of Organic Ligand-protected Metal Nanoclusters

Compared to immobilized metal nanoclusters on supports, organic ligand-protected metal nanoclusters feature ultra-small size, atomically precise structures, and switchable homogeneous/heterogeneous catalytic modes. Considerable progress has been made in this research area since the pioneering work in structural determination of $Au_{102}(SC_6H_4COOH)_{44}$ [1] and $Au_{25}(SC_2H_4Ph)_{18}$ [2] by single-crystal X-ray crystallography almost 20 years ago. They have attracted extensive interest of researchers in synthetic chemistry and catalysis because of their available structures at atomic level, which provide basis for investigating structure–property correlation of nanomaterials. This is a long-pursued while challenging goal in metal nanocluster catalysis research field. Additionally, their ultra-small size is beneficial for acquiring high catalytic activity, and the diverse and regulable geometric and electronic structures provide the possibility of activating various types of substrates and controlling the corresponding reaction selectivity. Organic ligands play important roles in stabilizing the metal kernel and cooperating with the metal activation sites to trigger the catalysis. Thus, the selection of organic ligands for metal nanocluster stabilization and catalysis is crucial. We divide this topic into four parts, and the representative organic ligands such as the thiol, phosphine, alkyne, *N*-heterocyclic carbene, amidine, pincer ligand, and hydride are sequentially introduced to manifest their important roles in regulating the structure and catalytic property of metal nanoclusters in Chapters 4–6. Isostructural metal nanocluster catalysis and metal nanocluster asymmetric catalysis are also described in Chapter 7.

## 4.1 Synthetic Methods Toward Organic Ligand-protected Metal Nanoclusters

We start from the synthesis, characterization, and atomically precise structure of organic ligand-protected metal nanoclusters in this chapter. The most straightforward approach to organic ligand-protected metal nanoclusters is the in situ synthesis. The

*Metal Nanocluster Catalysis: From Fundamentals to Applications*, First Edition. Manzhou Zhu and Man-Bo Li. © 2026 Wiley-VCH GmbH. All rights reserved, including rights for text and data mining and training of artificial intelligence technologies or similar technologies. Published 2026 by Wiley-VCH GmbH.

prototype of this method could date back to the year of 1994, when Brust et al. developed a two-phase synthetic method for the preparation of ultra-small gold nanoparticles (Figure 4.1) [3]. The protocol consists of four steps. First, the Au part $[AuCl_4]^-$ of gold complex $HAuCl_4$ is transferred from the aqueous solution to the organic phase in the presence of tetraoctylammonium bromide (TOAB). Partial bromide anions in TOAB are carried into the aqueous solution instead for charge conservation. Second, the ligand exchange between the chloride of $[AuCl_4]^-$ and bromide of TOAB happens, giving $[AuBr_4]^-$ and tetraoctylammonium chloride. Third, the $Au^{III}$ species of $[AuBr_4]^-$ is preliminarily reduced to $Au^I$ with the addition of thiols. Finally, $Au^I$ is further reduced by strong reductants such as $NaBH_4$ to produce metal nanoparticles.

This method was later named as Brust–Schiffrin method and attracted extensive attention of chemists and material scientists because it enables preparation of highly stable and ultra-small (1–3 nm) metal nanoparticles that have wide applications. Inspired by the Brust–Schiffrin method, researchers developed modified procedures to synthesize metal nanoclusters with well-defined structures [4–9]. For example, Zhu and coworkers established a kinetically controlled method and synthesized an atomically precise Au nanocluster in quite high yield (40% based on Au atom) [8, 9]. The composition of the Au nanocluster was determined to be $Au_{25}(SR)_{18}$ by mass spectrometry and thermogravimetric analyses. In a following work, they further revealed the molecular structure of this nanocluster, which has been a milestone work in the research area [2]. The key point of the kinetically controlled method is controlling the kinetics of the formation of $Au^I$–SR intermediates (Figure 4.2). The aggregation state of the $Au^I$–SR intermediate is crucial for the subsequent growth of atomically precise Au nanoclusters and low temperature (0 °C) and slow stirring

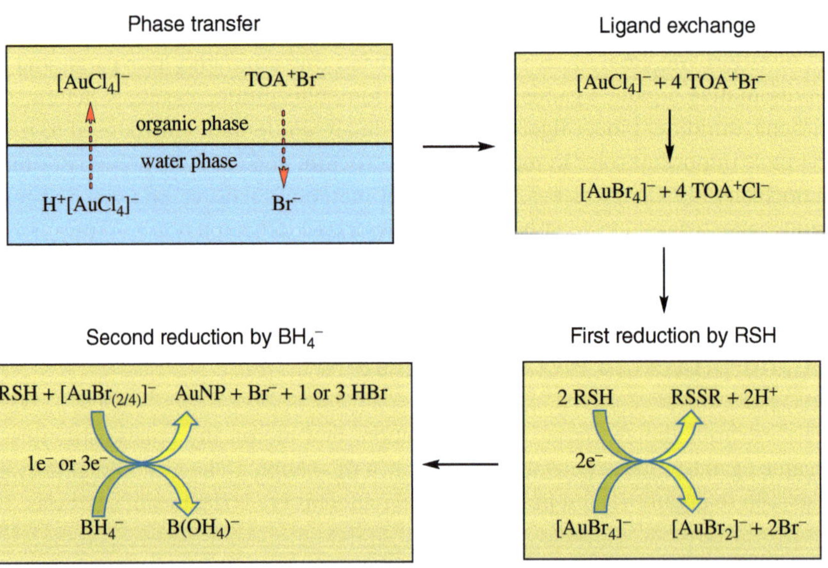


**Figure 4.1** Four steps involved in the two-phase synthetic method developed by Brust et al. [3].

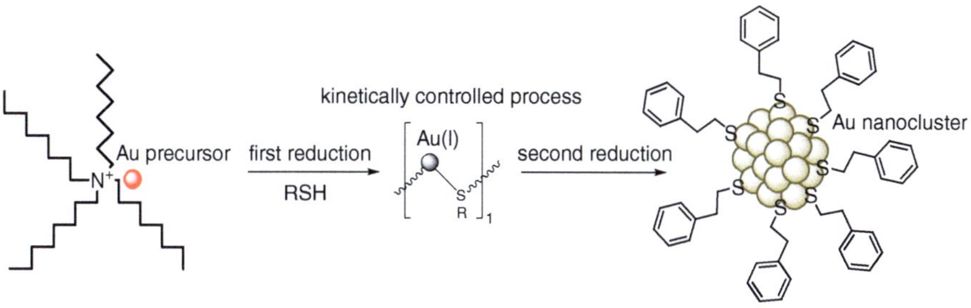


**Figure 4.2** Kinetically controlled method for the construction of $Au_{25}(SR)_{18}$ nanocluster in high yield. *Source:* [2]/with permission of American Chemical Society.

conditions are helpful for creating the beneficial $Au^{I}$-SR species that favors the further transformation into $Au_{25}(SR)_{18}$.

The in situ synthesis has become the most widely used method for the construction of organic ligand-protected Au nanoclusters. By slightly modifying the metal precursors, the organic ligands and the reductants, various metal nanoclusters with the other metal kernels such as Ag, Cu and alloyed metals (e.g. Au–Ag, Au–Cu, Ag–Cu, Au–Cd, Au–Ni) and the other organic surface such as the phosphines, alkynes, and carbenes were successfully prepared. Table 4.1 summarizes the typical metal nanoclusters that were obtained by the in situ synthetic method. The composition- and structure-diversity provides the basis for the further investigation on the property and practical application of organic ligand-protected metal nanoclusters.

Apart from the in situ synthesis, other methods such as the metal-exchange, antigalvanic reduction, ligand-exchange, photo-induced synthesis, and intercluster reaction have been developed by different research groups and constitute important supplement for the synthesis of organic ligand-protected atomically precise metal nanoclusters. We give a brief introduction to these synthetic methods in the following paragraphs.

The metal-exchange method [10] is based on a parent metal nanocluster, which is converted to another metal nanocluster or several metal nanoclusters in the presence of a metal source (e.g. metal ions and metal–ligand chelates). This method focuses on the reconstruction of the metal kernel in metal nanoclusters, resulting in three kinds of outcomes (Figure 4.3): (i) replacement or addition of metal atom(s) while keeping the geometric structure of metal kernel basically unchanged, (ii) foreign metals alloying in metal kernel with significantly changed geometric structure, and (iii) foreign metals-induced structural transformation without metal alloying.

For the first category, Choi and coworkers investigated the reactivity of $[Au_{25}(SCH_2CH_2Ph)_{18}]^-$ nanocluster with several metal ions [11], such as $Ag^+$, $Cu^{2+}$, and $Pb^{2+}$. Among the metal ions, $Ag^+$ triggered the transformation of $[Au_{25}(SCH_2CH_2Ph)_{18}]^-$ nanocluster, affording several new nanoclusters $[Au_{25-x}Ag_x(SCH_2CH_2Ph)_{18}]^-$, in which the Au sites are replaced by Ag atoms. However, the atomically precise structures of these doped nanoclusters were not obtained on that time. Wang and coworkers proposed the metal-exchange method for the synthesis of a series of doped nanoclusters [12].

**Table 4.1** Typical examples of organic ligand-protected metal nanoclusters prepared via in situ synthesis.

| Entry | Nanocluster molecular formula | Publication | Authors |
|---|---|---|---|
| 1 | $[Au_{25}(SCH_2CH_2Ph)_{18}]TOA$ | *J. Am. Chem. Soc.* **2008**, *130*, 5883–5885. | Manzhou Zhu Christine M. Aikens, Frederick J. Hollander, George C. Schatz, Rongchao Jin |
| 2 | $Au_{18}(S\text{-}Adm)_8(SbPh_3)_4Br_2$ | *J. Am. Chem. Soc.* **2022**, *144*, 478–484. | Justin B. Patty, Shana Havenridge, Dylan Tietje-Mckinney, Maxime A. Siegler, Kundan K. Singh, Roumina Hajy Hosseini, Mohamed Ghabin, Christine M. Aikens, and Anindita Das |
| 3 | $[(NHC\text{-}S)_4Au_6]Br_2$ | *Chem. Commun.* **2020**, *56*, 6102–6105. | Kirsi Salorinne, Renee W. Y. Man, Paul A. Lummis, Maryam Sabooni Asre Hazer, Sami Malola, Jacky C.-H. Yim, Alex J. Veinot, Wenxia Zhou, Hannu Häkkinen, Masakazu Nambo and Cathleen M. Crudden |
| 4 | $[Au_{10}(bisNHC)_4Br_2](OOCCF_3)_2$ | *J. Am. Chem. Soc.* **2022**, *144*, 2056–2061. | Renee W. Y. Man, Hong Yi, Sami Malola, Shinjiro Takano, Tatsuya Tsukuda, Hannu Häkkinen, Masakazu Nambo, and Cathleen M. Crudden |
| 5 | $Cd_{32}Te_{14}(SR)_{36}(PR_3)_4$ | *J. Am. Chem. Soc.* **2025**, *147*, 7507–7512. | Saryvoudh A. Mech, Joshua O. Gibson, Fuyan Ma, Lukasz M. Dobrzycki, and Chenjie Zeng |
| 6 | $Pd_6(SC_6H_{11})_{12}$ | *Nano Lett.* **2024**, *24*, 13054–13060. | Chenhao Ruan, Changqing Meng, Kunpeng Wang, Yanshuang Li, Huixin Xiang, Hao Yan, Wen Wu Xu, Meng Zhou, and Chuanhao Yao |
| 7 | $Ag_2(dppm)_2(AcO)_2$ | *J. Am. Chem. Soc.* **2024**, *146*, 28572–28579. | Huixin Xiang, Yanze Wang, Xinqi Xu, Chenhao Ruan, Kunpeng Wang, Wanyu Cheng, Meng Zhou, Xiaowang Liu, and Chuanhao Yao |
| 8 | $PdAu_{24}(SR)_{18}$ | *J. Am. Chem. Soc.* **2015**, *137*, 10833–10840. | Kyuju Kwak, Qing Tang, Minseok Kim, De-en Jiang, and Dongil Lee |

| | | | |
|---|---|---|---|
| 9 | $[Ag_{78}(^{i}PrPhS)_{30}(dppm)_{10}Cl_{10}]4+$ | *Angew. Chem. Int. Ed.* **2021**, *60*, 4231. | W.-J. Zhang, Z. Liu, K.-P. Song, C. M. Aikens, S.-S. Zhang, Z. Wang, C.-H. Tung, D. Sun |
| 10 | $[Ag_{155}(CyS)_{40}(TC4A)_5Cl_2]$ | *Angew. Chem. Int. Ed.* **2022**, *61*, e202206742. | Zhi Wang, Fahri Alkan, Christine M. Aikens, Mohamedally Kurmoo, Zhen-Yi Zhang, Ke-Peng Song, Chen-Ho Tung, and Di Sun |
| 11 | $(DBA)_4Cu_4I_4$ | *Angew. Chem. Int. Ed.* **2023**, *62*, e202217784. | Qingsong Hu, Chengkai Zhang, Xian Wu, Guijie Liang, Lei Wang, Xiaowei Niu, Zhi Wang, Wei-Dan Si, Yibo Han, Ruiqin Huang, Jiawen Xiao, and Di Sun |
| 12 | $[Cu_{31}(4\text{-MeO-}PhCC)_{21}(dppe)_3](ClO_4)_2$ | *J. Am. Chem. Soc.* **2023**, *145*, 10355–10363. | Tao Jia, Zong-Jie Guan, Chengkai Zhang, Xiao-Zhao Zhu, Yun-Xin Chen, Qian Zhang, Yang Yang, and Di Sun |
| 13 | $[Au_{13}Ag_{12}(PPh_3)_{10}Cl_8]SbF_6$ | *Nat. Commun.* **2020**, *11*, 6019. | Zhaoxian Qin, Jiangwei Zhang, Chongqing Wan, Shuang Liu, Hadi Abroshan, Rongchao Jin and Gao Li |
| 14 | $Cu_{50}S_{12}(SC(CH_3)_3)_{20}(CF_3COO)_{12}$ | *J. Am. Chem. Soc.* **2023**, *145*, 25673–25685. | Cheng Xu, Yuhao Jin, Hao Fang, Huijuan Zheng, Jesse C. Carozza, Yanxiong Pan, Ping-Jie Wei, Zhenyi Zhang, Zheng Wei, Zheng Zhou, and Haixiang Han |
| 15 | $[Au_6(m\text{-}PhDP)_4](PF_6)_2(1\cdot(PF_6)_2)$ | *Nat. Commun.* **2017**, *8*, 576. | Md. Abu Bakar, Mizuho Sugiuchi, Mitsuhiro Iwasaki, Yukatsu Shichibu and Katsuaki Konishi |
| 16 | $Au_2Cu_6(PPh_2Py)_2(SC_{10}H_{15})_6$ | *Angew. Chem. Int. Ed.* **2016**, *55*, 3611. | X. Kang, S. Wang, Y. Song, S. Jin, G. Sun, H. Yu, M. Zhu |

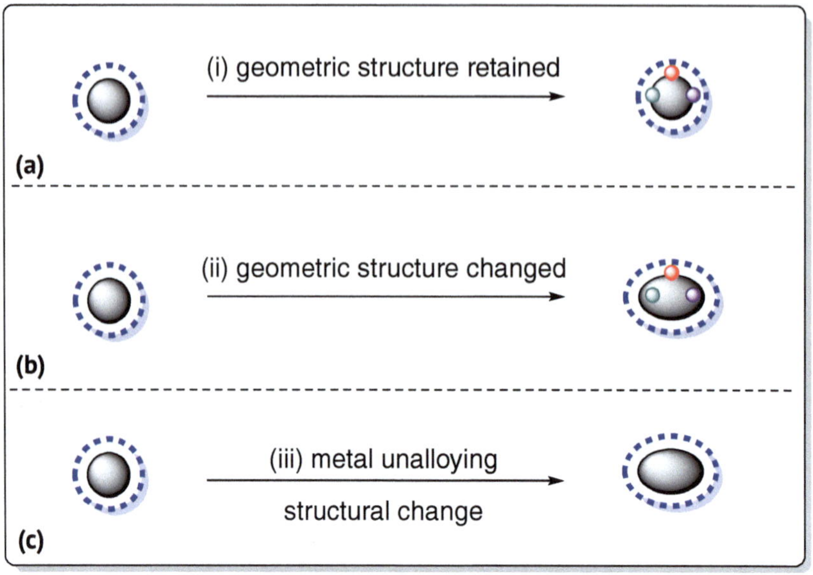


**Figure 4.3** Metal-exchange method for the synthesis of metal nanoclusters. (a) Replacement or addition of metal atoms while keeping the geometric structure of metal kernel basically unchanged. (b) Foreign metal alloying in metal kernel with significantly changed geometric structure. (c) Foreign metals-induced structural transformation without metal alloying.

Their atomically precise structures were successfully revealed, possessing the basically unchanged geometric structures with their parent metal nanoclusters. After that, this method has been widely applied by different research groups and diversified doped metal nanoclusters have been constructed with their parent nanoclusters' structures retained [13, 14]. For the second category, Wang and coworkers synthesized a series of $Au_{25-x}Ag_x$ nanoclusters by the reaction of $[Au_{11}(PPh_3)_8Cl_2]^+$ nanocluster with the $Ag(SCH_2CH_2Ph)$ chelate [15]. Notably, the $Au_{25-x}Ag_x$ nanoclusters are structurally different from the previously reported $[Au_{25-x}Ag_x(SCH_2CH_2Ph)_{18}]^-$. Their compositions were determined by mass spectroscopy to bc $[\Lambda u_{25-x}Ag_x(SCII_2CII_2Ph)_5(PPh_3)_5Cl_2]^{?+}$. Tlieir slructures are also distinct from the core-shelled $[Au_{25-x}Ag_x(SCH_2CH_2Ph)_{18}]^-$ and demonstrate a rod-like geometry. Another interesting finding of this work is that a 200-fold quantum yield boost in the photoluminescence of $[Au_{25-x}Ag_x(SCH_2CH_2Ph)_5(PPh_3)_5Cl_2]^{2+}$ was observed when $x > 12$. For the third category, Li and coworkers realized an Au nanocluster-to-Au nanocluster transformation in the presence of $Cu^{2+}$ ions [16]. Three stages are involved in this transformation, that is, the initial oxidation of $[Au_{25}(SCH_2CH_2Ph)_{18}]^-$ nanocluster by $Cu(NO_3)_2$ to give $[Au_{25}(SCH_2CH_2Ph)_{18}]^0$ nanocluster, the further oxidation of $[Au_{25}(SCH_2CH_2Ph)_{18}]^0$ nanocluster by $Cu(NO_3)_2$ to give $[Au_{25}(SCH_2CH_2Ph)_{18}]^+$ nanocluster and the last stage of decomposition of $[Au_{25}(SCH_2CH_2Ph)_{18}]^+$ nanocluster into $Au_{22}(SCH_2CH_2Ph)_{16}$ species and recombination of $Au_{22}(SCH_2CH_2Ph)_{16}$ into the final $Au_{44}(SCH_2CH_2Ph)_{32}$ nanocluster. During the whole transformation process, $Cu^{2+}$ ions are not doped into the metal kernel while acting as an oxidant to trigger the oxidation, decomposition, and recombination of the mother metal nanocluster $[Au_{25}(SCH_2CH_2Ph)_{18}]^-$.

For the synthetic method of antigalvanic reduction, Wu and coworkers systematically investigated the reactions between metal nanoclusters and metal ions, and found that the reduction of the metal cations such as $Ag^+$, $Cu^{2+}$, $Cd^{2+}$, and $Hg^{2+}$ by ultra-small Au nanoclusters is a common phenomenon. They proposed the concept of antigalvanic reduction [17] for the first time in 2012 and successfully synthesized a series of foreign metal-replaced $Au_{24}M(SR)_{18}$ (M = Cd, Hg, etc.) nanoclusters [18, 19] based on the concept. The antigalvanic reduction method [20] has been recognized as an efficient approach to the construction of doped metal nanoclusters. This method is distinct from the galvanic reduction method, while both methods are effective and have been employed for the synthesis of bimetallic nanoparticles and nanoclusters that were further applied in catalysis, sensing and materials science [21]. Typical examples on synthesis of metal nanoclusters by the metal-exchange method and antigalvanic reduction method are listed in Table 4.2.

The ligand-exchange method [22] is also based on a mother metal nanocluster, that is converted to another metal nanocluster or several metal nanoclusters in the presence of an organic ligand. This method focuses on the reconstruction of the organic surface on metal nanoclusters and can be classified into two categories (Figure 4.4): (i) replacement of organic ligand on the surface while keeping the metal kernel basically unchanged and (ii) organic ligand-induced structural transformation of metal nanocluster. The simple replacement of organic ligand on the surface of metal nanoclusters plays important roles in but not limited to the following two aspects. First, it helps in structural determination of metal nanoclusters. Negishi and coworkers synthesized two smallest water-soluble nanoclusters $Au_{15}(SG)_{13}$ and $Au_{18}(SG)_{14}$ (GSH = glutathione) as early as 2005 [23]. However, their crystal structures were not determined due to the flexibility of the glutathione ligand. This challenging problem has been resolved by Chen and coworkers 10 years later [24]. By ligand-exchange method, they successfully replaced the glutathione ligands on the surface of $Au_{18}(SG)_{14}$ by the rigid mercaptans ligand cyclohex anethiol, synchronously realizing the transfer of the water-soluble $Au_{18}(SG)_{14}$ nanocluster to an oil-soluble $Au_{18}(SC_6H_{11})_{14}$ nanocluster. The much easier crystal growth of $Au_{18}(SC_6H_{11})_{14}$ comparted to $Au_{18}(SG)_{14}$ and the further single-crystal X-ray diffraction (SCXRD) analysis successfully revealed the atomically precise structure of this kind of metal nanocluster. After three years, Yang and coworkers subsequently synthesized $Au_{16}(SAdm)_{12}$ by the ligand-exchange reaction of $Au_{15}(SG)_{13}$ and adamantane-1-thiolate [25]. The crystal structure of $Au_{16}(SAdm)_{12}$ nanocluster give important implications for the structure of $Au_{15}(SG)_{13}$ nanocluster. Second, it helps in surface functionalization of metal nanoclusters. Narouz and coworkers synthesized the first *N*-heterocyclic carbene (NHC)-functionalized metal nanoclusters $[Au_{11}(PPh_3)_7(NHC)Cl_2]^+$ and $[Au_{11}(PPh_3)_6(NHC)_2Cl_2]^+$. The successful preparation of this kind of functional metal nanocluster relies on the ligand-exchange reaction of $[Au_{11}(PPh_3)_8Cl_2]^+$ and the NHC ligand [26]. The NHC-functionalized Au nanoclusters demonstrate highly improved structural stability and catalytic activity toward the electrochemical reduction of $CO_2$.

**Table 4.2** Typical examples of metal nanoclusters prepared by metal-exchange and antigalvanic reduction methods.

| Entry | Nanocluster molecular formula | Publication | Authors |
|---|---|---|---|
| 1 | $[Ag_{13}Cu_{20}(PPh_3)_4(PET)_{24}](BF_4)$ | *Angew. Chem. Int. Ed.* **2024**, *63*, e202407214. | Y. Yao, W. Hao, J. Tang, K. Kirschbaum, C. G. Gianopoulos, A. Ren, L. Ma, L. Zheng, H. Li, Q. Li |
| 2 | $[Ag_{28}(AdmS)_{14}(S)_2(CF_3COO)_{10}(H_2O)_4]$ | *Angew. Chem. Int. Ed.* **2020**, *59*, 11898. | J.-S. Yang, Z. Han, X.-Y. Dong, P. Luo, H.-L. Mo, S.-Q. Zang |
| 3 | $Ag_{29-x}Au_x(BDT)_{12}(TPP)_4$, $x$=1–5 | *Angew. Chem. Int. Ed.* **2016**, *55*, 5749. | G. Soldan, M. A. Aljuhani, M. S. Bootharaju, L. G. AbdulHalim, M. R. Parida, A.-H. Emwas, O. F. Mohammed, O. M. Bakr |
| 4 | $[Pt_2Cu_{34}(PET)_{22}Cl_4](PPh_4)_2$ NCs | *J. Am. Chem. Soc.* **2021**, *143*, 12100–12107. | Sanghwa Lee, Megalamane S. Bootharaju, Guocheng Deng, Sami Malola, Hannu Häkkinen, Nanfeng Zheng, and Taeghwan Hyeon |
| 5 | $Au_{47}Cd_2(TBBT)_{31}$ | *Angew. Chem. Int. Ed.* **2020**, *59*, 3073. | S. Zhuang, D. Chen, L. Liao, Y. Zhao, N. Xia, W. Zhang, C. Wang, J. Yang, Z. Wu |
| 6 | $[(AuL)_6AgCu_6(C{\equiv}C)_3](BF_4)_7$ | *Angew. Chem. Int. Ed.* **2021**, *60*, 14381. | X.-L. Pei, Z.-J. Guan, Z.-A. Nan, Q.-M. Wang |
| 7 | $[Au_{21}(SR)_{12}(Ph_2PCH_2PPh_2)_2]^+$ | *Sci. Adv.* **2017**, *3*, e1603193. | Q. Li, T.-Y. Luo, M. G. Taylor, S.-X. Wang, X.-F. Zhu, Y.-B. Song, G. Mpourmpakis, N. L. Rosi, R.-C. Jin |
| 8 | $\{[Au_{13}Ag_{12}(PPh_3)_{10}Cl_8]\cdot[Au_{12}Ag_{13}(PPh_3)_{10}Cl_8]\}^{2+}$ | *Nano Res.* **2022**, *15*, 2971–2976. | Zhaoxian Qin, Song Hu, Wenhua Han, Zhiwen Li, Wen Wu Xu, Jingjing Zhang, and Gao Li |
| 9 | $MAg_{24}(SPhMe_2)_{18}$ (M = Ni, Pd, Pt) | *Chem. Mater.* **2020**, *32*, 10216–10226. | Minseok Kim, K. L. Dimuthu M. Weerawardene, Woojun Choi, Sang Myeong Han, Juwon Paik, Younghun Kim, Moon-Gun Choi, Christine M. Aikens, and Dongil Lee |
| 10 | $CdAu_{24}(SC2H4Ph)_{18}$ | *J. Am. Chem. Soc.* **2015**, *137*, 4018–4021. | Shuxin Wang, Yongbo Song, Shan Jin, Xia Liu, Jun Zhang, Yong Pei, Xiangming Meng, Man Chen, Peng Li, and Manzhou Zhu |
| 11 | $[Ag_{24}Au(SPhMe_2)_{18}]^-$ | *Angew. Chem. Int. Ed.* **2016**, *55*, 922. | M. S. Bootharaju, C. P. Joshi, M. R. Parida, O. F. Mohammed, O. M. Bakr |
| 12 | $Au_{24}Hg(PET)_{18}$ | *J. Am. Chem. Soc.* **2015**, *137*, 15350–15353. | Chuanhao Yao, Yue-jian Lin, Jinyun Yuan, Lingwen Liao, Min Zhu, Lin-hong Weng, Jinlong Yang, and Zhikun Wu |
| 13 | $Cd_{28}S_{17}I_{22}(PEt_3)_{12}$ | *J. Am. Chem. Soc.* **2025**, *147*, 17890–17901. | Cheng Xu, Zhenyi Zhang, Zheng Zhou, and Haixiang Han |

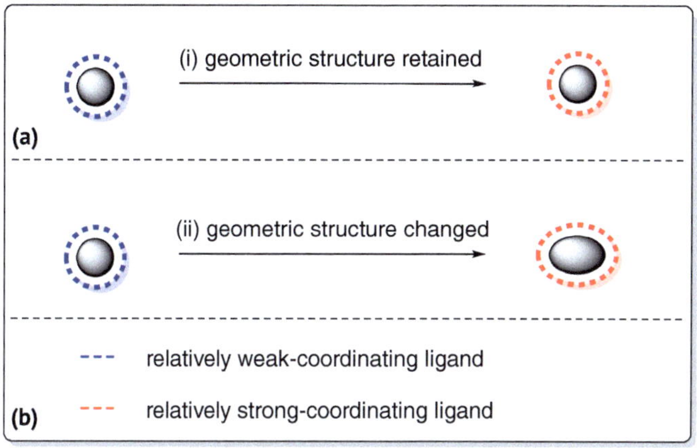


**Figure 4.4** Ligand-exchange method for the synthesis of metal nanoclusters. (a) Replacement of organic ligand on the surface while keeping the metal kernel basically unchanged. (b) Organic ligand-induced structural transformation of metal nanocluster.

Organic ligand-induced structural transformation of metal nanocluster contributes to the structural diversity to a large extent. As early as 2005, Shichibu and coworkers have employed ligand-exchange method for the large-scale synthesis of water-soluble $Au_{25}(SG)_{18}$ nanocluster from $[Au_{11}(PPh_3)_8Cl_2]^+$ [27]. After 20-year development, diversified nanocluster structures have been constructed by the ligand-exchange method. Typical examples are listed in Table 4.3.

It should be noted that ligand-exchange reaction generally occurs between weakly-coordinating ligand protected metal nanoclusters and strong-coordinating ligands. For example, the ligand-exchange between phosphine protected Au nanoclusters and thiols is very common because the Au–S interaction is much stronger than the Au–P interaction. The formation of ultra-stable Au–S bonds provides the driving force for the ligand-exchange process. However, the reverse process is also feasible. Li and coworkers investigated the reverse ligand-exchange reaction [28] between $[Au_{25}(SCH_2CH_2Ph)_{18}]^-$ nanocluster and $PPh_3$ (Figure 4.5). They found that this reaction could happen when excess amounts of $PPh_3$ are introduced into the system. In the presence of $PPh_3$, $[Au_{25}(SCH_2CH_2Ph)_{18}]^-$ nanocluster was sequentially transformed to $[Au_{13}(PPh_3)_6(SCH_2CH_2Ph)_3]^{2+}$, $[Au_{11}(PPh_3)_8(SCH_2CH_2Ph)]^+$, and finally $[Au_{25}(SCH_2CH_2Ph)_5(PPh_3)_5Cl_2]^{2+}$ nanoclusters. Too excessive $PPh_3$ (>100 equivalents per mole of the mother metal nanocluster) would lead to decomposition of the formed $[Au_{25}(SCH_2CH_2Ph)_5(PPh_3)_5Cl_2]^{2+}$ nanocluster. This case study indicates that the ligand-exchange processes between metal nanocluster and organic ligands are complex and cannot be simply predicted by comparing the bonding energy of the metal–ligand bonds. The reverse ligand-exchange process can occur, demonstrating that it is thermodynamically or kinetically favorable in certain reaction conditions. Reaction parameters such as the solvent, the temperature and the relative abundance of organic ligands all impact the reaction direction. This reverse ligand-exchange method was further applied by Zou, He, and other researchers for preparing metal nanoclusters with intriguing structures and catalytic performance [29–34].

**Table 4.3** Typical examples of metal nanoclusters prepared by ligand-exchange method via organic ligand-induced structural transformation.

| Entry | Nanocluster molecular formula | Publication | Authors |
|---|---|---|---|
| 1 | $Au_{25}(PET)_{18-x}(PA)_x$ | *Nanoscale*, **2020**, *12*, 6239–6242. | Christopher A. Hosier, Ian D. Anderson and Christopher J. Ackerson |
| 2 | $Au_4Cd_4(SR)_{12}$ | *J. Am. Chem. Soc.* **2022**, *144*, 14248–14257. | Huixin Xiang, Hao Yan, Jiaohu Liu, Ranran Cheng, Cong-Qiao Xu, Jun Li, and Chuanhao Yao |
| 3 | $Cd_1Au_{13}(DMBT)_6(DPPB)_3Cl$ | *J. Am. Chem. Soc.* **2023**, *145*, 4238–4245. | Yesen Tan, Ying Lv, Liyun Xu, Qinzhen Li, Jinsong Chai, Sha Yang, Haizhu Yu, and Manzhou Zhu |
| 4 | $[Au_{25}(SC_2Ph)_{17}(Por)_1]^-$ | *J. Am. Chem. Soc.* **2022**, *144*, 12310–12320. | Wataru Suzuki, Ryo Takahata, Yuki Chiga, Soichi Kikkawa, Seiji Yamazoe, Yoshiyuki Mizuhata, Norihiro Tokitoh, and Toshiharu Teranishi |
| 5 | $[PtAg_{24}(IPBT)_{18}](PPh_4)_2$ | *J. Am. Chem. Soc.* **2025**, *147*, 12546–12554. | Seungwoo Yoo, Dayeon Kim, Guocheng Deng, Yuping Chen, Kangjae Lee, Suhwan Yoo, Xiaolin Liu, Qing Tang, Yun Jeong Hwang, Taeghwan Hyeon, and Megalamane S. Bootharaju |
| 6 | $[Au_{25}(PPh_3)_{10}(SC_2H_4Ph)_5Cl_2]^{2+}$ | *J. Phys. Chem. C* **2009**, *113*, 17599–17603. | Huifeng Qian, Manzhou Zhu, Eric Lanni, Yan Zhu, Mark E. Bier, and Rongchao Jin |
| 7 | $[Au_{25}(PPh_3)_{10}(SePh)_5Cl_2]^q$ ($q = +1$ or $+2$) | *Chem. Mater.* **2016**, *28*, 2609–2617. | Yongbo Song, Shan Jin, Xi Kang, Ji Xiang, Huijuan Deng, Haizhu Yu, and Manzhou Zhu |
| 8 | $Au_{21}(S\text{-}Adm)_{15}$ | *J. Am. Chem. Soc.* **2016**, *138*, 10754–10757. | Shuang Chen, Lin Xiong, Shuxin Wang, Zhongyun Ma, Shan Jin, Hongting Sheng, Yong Pei, and Manzhou Zhu |
| 9 | $Au_{28}(S\text{-}c\text{-}C_6H_{11})_{20}$ | *J. Am. Chem. Soc.* **2016**, *138*, 1482–1485. | Yuxiang Chen, Chong Liu, Qing Tang, Chenjie Zeng, Tatsuya Higaki, Anindita Das, De-en Jiang, Nathaniel L. Rosi, and Rongchao Jin |
| 10 | $Au_{36}(SPh\text{-}Bu^t)_{24}$ | *Angew. Chem. Int. Ed.* **2012**, *51*, 13114–13118. | Zeng, C., Qian, H., Li, T., Li, G., Rosi, N.L., Yoon, B., Barnett, R.N., Whetten, R.L., Landman, U. and Jin, R. |
| 11 | $Ag_{25}(SPhMe_2)_{18}$ | *Chem. Mater.* **2016**, *28*, 3292–3297. | Megalamane S. Bootharaju, Chakra P. Joshi, Mohammad J. Alhilaly, and Osman M. Bakr |
| 12 | $Au_{24}(SCH_2Ph\text{-}^tBu)_{20}$ | *Nanoscale*, **2014**, *6*, 6458–6462. | Das, A.; Li, T.; Li, G.; Nobusada, K.; Zeng, C.; Rosi, N. L.; Jin, R. |
| 13 | $Au_{18}(SeC_6H_5)_{14}$ | *Nanoscale*, **2013**, *5*, 1176–1182. | Xu, Q.; Wang, S.; Liu, Z.; Xu, G.; Meng, X.; Zhu, M. |
| 14 | $Au_{30}(S\text{–}tBu)_{18}$ | *J. Phys. Chem. C* **2017**, *121*, 14929–14935. | Milan Rambukwella, Luca Sementa, Alessandro Fortunelli, and Amala Dass |

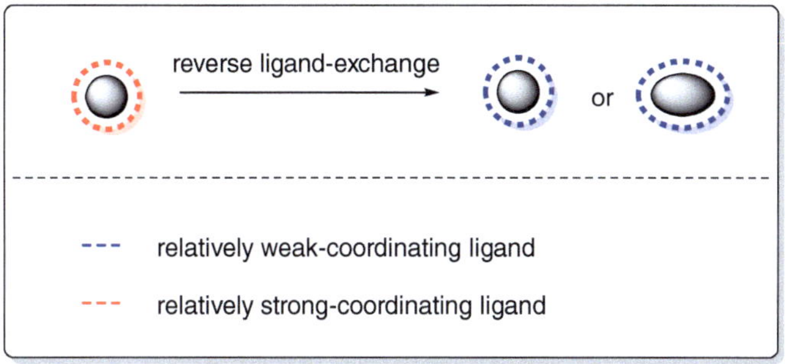


**Figure 4.5** Reverse ligand-exchange for the synthesis of metal nanoclusters.

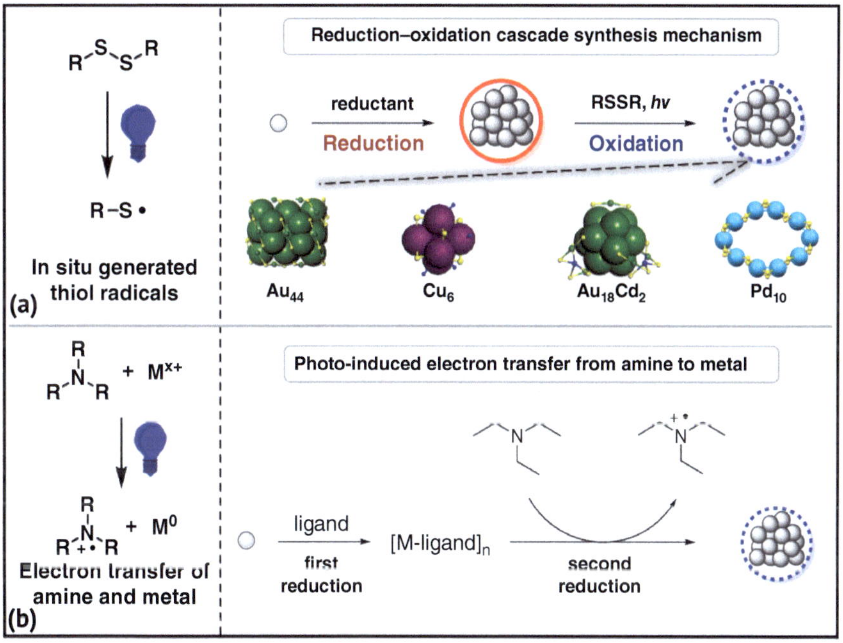


**Figure 4.6** Photo-induced synthesis of metal nanoclusters. (a) Photochemical route to thiolate-protected metal nanoclusters from disulfides. (b) Photochemical synthetic route to atomically precise Ag nanoclusters by using amine as the reductant.

The photo-induced synthetic method takes advantage of energy produced by light irradiation, thus significantly improve the compatibility of metal nanocluster synthesis for environmentally friendly organic ligands and reductants (Figure 4.6). For example, Zhu and coworkers developed a photochemical route to thiolate-protected metal nanoclusters from disulfides [35]. Thiolate-protected metal nanoclusters are usually prepared from thiols previously. This reagent is sensitive to oxidants such as atmospheric oxygen. Meanwhile, they are volatile and possess unpleasant smells, thus restricting the application of the method for acquiring metal nanoclusters with specific frameworks and surface ligands. Disulfides are low-cost, oxidant-stable and environmentally friendly sulfur reagents, while their employment in the synthesis of metal nanoclusters is quite unknown. The major challenge is their much weaker coordination ability and high oxidation potential compared to thiols. To overcome these issues, Man-Bo Li and coworkers introduce light irradiation into the synthesis. Metal precursors are initially reduced by $NaBH_4$ to give reduction state of metal intermediates, which are further oxidized by sulfur radicals generated in situ from disulfides under the photochemical reaction conditions (Figure 4.6a). The reduction–oxidation cascade route triggers the formation of metal nanoclusters with atomically precise structures. Various metal nanoclusters including $Au_{44}(SPh)_{28}$, $Cu_6$(2-S-pyridine)$_6$, $Au_{23}Pd_2(SPh)_5[P(4\text{-}FPh)_3]_{10}Br_2$, and $Au_{18}Cd_2(SPh)_{14}[P(4\text{-}FPh)_3]_2$, $Ni_{10}(4\text{-}MePhS)_{20}$, $Ni_{11}(PhS)_{22}$, $Pd_9(PhS)_{18}$, and $Pd_{10}(PhS)_{20}$ were synthesized in high yields under mild reaction conditions [36]. Very recently, the same research group synthesized an $Au_8Cu_6(SPh)_{14}[P(4\text{-}FPh)_3]_2$ nanocluster by the reduction–oxidation cascade method [37]. This metal nanocluster features high structural stability toward light irradiation because it is synthesized by the photochemical route. Meanwhile, it demonstrates dramatically high activity in photochemical alkyne and phosphonate activation. This photosynthesized metal nanocluster was then developed as an efficient catalyst for photocatalysis of alkyne functionalization by the phosphonate. Apart from the improvement of compatibility for organic ligands, photo-induced synthesis of metal nanoclusters enables usage of quite mild reductants, which are otherwise inaccessible in the conventional in situ synthetic method. Wang and coworkers developed a photochemical synthetic route to atomically precise Ag nanoclusters by using amine as the reductant [38]. The key stage of this protocol is the photo-induced single electron transfer from the amine (e.g. $Et_3N$) to the Ag-ligand intermediates (Figure 4.6b). This process directs the formation of atomically precise Ag nanoclusters in high yields with satisfactory selectivity. The as-obtained nanocluster structures should be ascribed to the thermodynamic along with kinetic control under the light irradiation reaction conditions. Their further investigations and the following work indicate the applicability of this route [39]. Amines including aliphatic and aromatic substituents are all available as the reductant for synthesizing metal nanoclusters bearing different metal kernels and surface ligands. This systematic study, combined with the photosynthetic method employing disulfides manifest the application of mild reagents and reductants in metal nanocluster synthesis under light irradiation.

Photo-induced synthesis not only contributes to the direct synthesis of metal nanoclusters but also be applied for the metal nanoclusters' structural transformation [40, 41]. Vivian Wing-Wah Yam and her research group systematically investigated this phenomenon by designing a unique ligand 1,2-bis(diphenylphosphino)ethene (dppee). This ligand contains two isomers, that is, *trans*-dppee and *cis*-dppee. *trans*-dppee is more stable than *cis*-dppee and their energy difference has been calculated to be 23 kJ/mol. Based on the *cis*-dppee, Vivian Wing-Wah Yam and coworkers synthesized an Au nanocluster $[Au_{10}(\textit{cis}\text{-dppee})_4S_4]^{2+}$, which can be selectively transformed to a new Au nanocluster $[Au_{18}(\textit{trans}\text{-dppee})_6S_8]^{2+}$ under light irradiation. In-depth studies demonstrate that the Au precursor *cis*-$[dppee(AuCl)_2]$ can also be transformed into *trans*-$[dppee(AuCl)_2]$ under the same reaction conditions, indicating that photo-induced isomerization of *cis*-dppee to *trans*-dppee triggers the transformation of $[Au_{10}(cis\text{-dppee})_4S_4]^{2+}$ to $[Au_{18}(trans\text{-dppee})_6S_8]^{2+}$ (Figure 4.7). This method is generalizable and the same research group realized the photo-induced synthesis of a $Au_{15}$ nanocluster from a $Au_{12}$ nanocluster (Figure 4.7) protected by the ligand of (2-(2-(diphenylphosphino)vinyl)-1,3-phenylene)*bis*(diphenylphosphine).

The intercluster reaction method is based on organic ligand-protected atomically precise mother nanoclusters, which interact with each other and produce new metal nanocluster(s) protected by organic ligands (Figure 4.8). Krishnadas and

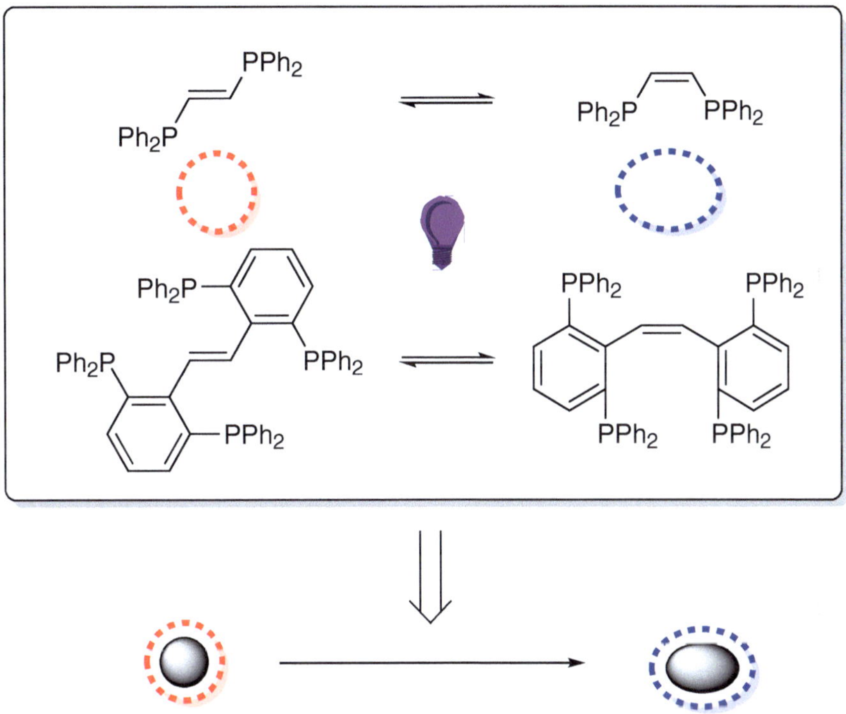


**Figure 4.7** Photo-induced transformation of metal nanoclusters.

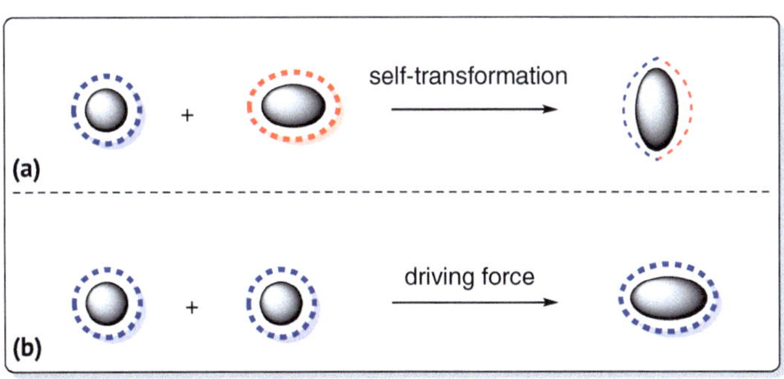


**Figure 4.8** Synthesis of metal nanoclusters by intercluster reaction. (a) Self-transformation based on interactions of two metal nanoclusters. (b) External force-driven transformation of metal nanoclusters by their self-reaction.

coworkers discovered this interesting phenomenon in 2016 [42] and then systematically investigated the reaction process [43–45]. The first report is based on the reaction of $Au_{25}(SR)_{18}$ and $Ag_{44}(SR)_{30}$ [42]. $Au_{25\text{-}x}Ag_x(SR)_{18}$ and $Au_xAg_{44\text{-}x}(SR)_{30}$ were selectively obtained by regulating the mole ratio of $Au_{25}(SR)_{18}$ and $Ag_{44}(SR)_{30}$. By employing different cluster precursors, they realized the synthesis of a series of novel metal nanoclusters that cannot be obtained by the other methods (Figure 4.8a). For example, $Au_{25}(SR)_{18}$ and $Ag_{25}(SR)_{18}$ with the similar geometric structures interact with each [43], generating $Au_mAg_n(SR)_{18}$ ($m = 1$–$24$) nanoclusters, merely by controlling the reactant compositions. They even captured one of the earliest events of the process, that is, the formation of the dianionic adduct of $[Au_{25}Ag_{25}(SR)_{36}]^{2-}$. The reaction of $Au_{25}(SR)_{18}$ and $Ir_9(SR)_6$ gives Ir-doped Au nanocluster $Au_{22}Ir_3(SR)_{18}$ as the final product with high selectivity [44]. The reaction of bimetallic nanocluster $MAg_{28}(BDT)_{12}(PPh_3)_4$ (BDT = 1,3-benzenedithiol and M = Ni, Pd, or Pt) and $Au_{25}(SR)_{18}$ affords a mixture of trimetallic $MAu_xAg_{28\text{-}x}(BDT)_{12}(PPh_3)_4$ ($x = 1$–$12$) and bimetallic $Ag_xAu_{25\text{-}x}(PET)_{18}$ ($x = 1$–$7$) nanoclusters as the final products [45]. There is a review paper written by Krishnadas and coworkers, describing the synthetic strategy of novel metal nanoclusters by intercluster reactions [46]. In addition to the in-cluster reactions between two different metal nanoclusters, the interaction between two molecules of the same metal nanocluster is also feasible for the synthesis of new nanoclusters. Suyama and coworkers revealed that the reaction between charge-natural $PtAu_{24}(SR)_{18}$ nanocluster and the equal amount of $[PtAu_{24}(SR)_{18}]^{2-}$ results in stoichiometric production of $[PtAu_{24}(SR)_{18}]^{-}$ [47]. This observation indicates that the spontaneous electron transfer between $[PtAu_{24}(SR)_{18}]^{2-}$ and $[PtAu_{24}(SR)_{18}]^{0}$ occurs. Gan and coworkers investigated the reaction between two molecules of the same $Au_{25}(SR)_{18}$ nanocluster under high pressure (Figure 4.8b) [48]. An unconventional $Au_{32}(SR)_{24}$ nanocluster bearing a bicuboid $Au_{14}$ kernel and three pairs of interlocked trimetric staples was obtained under the reaction conditions of 5G Pa pressure at room temperature for 30 minutes.

There are other methods not mentioned herein that are also contribute to the synthesis of metal nanoclusters and play important roles in regulating nanoclusters' structure and property. Meanwhile, the combination of these methods enables construction of structurally unique metal nanoclusters, which otherwise cannot be prepared by the single method. For example, the combination of metal-exchange and ligand-exchange methods produced catalytically active metal nanoclusters and highly photoluminescent metal nanoclusters with abundant metal kernel compositions and surface ligands. These nanoclusters are hardly prepared by the in situ synthesis or the sole usage of metal-exchange or ligand-exchange method. These methods are generally based on the transformation of metal nanoclusters triggered by driving forces, such as metal ions, organic ligands, acids, bases, temperatures, etc. [49]. We believe that more and more synthetic methods will be developed in the future, focusing on simpler, greener and large-scale synthesis of organic ligand-protected metal nanoclusters with diversified compositions and structures.

## 4.2 Characterization Techniques for Organic Ligand-protected Metal Nanoclusters

The most two important characterization techniques for organic ligand-protected metal nanoclusters are high-resolution mass spectroscopy (HRMS) and SCXRD, which determine the well-defined composition and atomically precise structure of the metal nanocluster. Apart from that, other characterization techniques such as ultraviolet and visible absorption spectroscopy (UV–vis), nuclear magnetic resonance spectroscopy, and circular dichroic spectroscopy are also commonly applied for identifying specific structures of metal nanoclusters. In this part, we introduce these important characterization techniques, mainly focusing on the HRMS, SCXRD, and UV–vis.

The characterization mechanism of HRMS is based on measuring the mass-to-charge ratio ($m/z$) of the ionized sample. It can be divided into four processes: (i) ionization of each component of the sample in an ion source, (ii) generation of charged ions with different mass-to-charge ratio, (iii) formation of ion beam by accelerating electric field, and (iv) obtaining mass spectrum in mass analyzer. Compared to other mass analyzers, the time-of-flight mass analyzer features simple construction, high sensitivity, and wide mass range, thus being advantageous in detection of metal nanoclusters because their molecular weights usually range from thousands to tens of thousands. Two kinds of ion sources are frequently equipped in HRMS for characterization of metal nanoclusters, that is, the matrix assisted laser desorption ionization (MALDI) and the electrospray ionization (ESI). MALDI-TOF-MS and ESI-TOF-MS (or ESI-MS) are both soft ionization mass spectroscopies, which are beneficial for acquirement of the molecular ion peak of the organic ligand-protected metal nanocluster.

As a case study, the ESI-MS spectrum of an Au–Cu alloyed $[Au_{13}Cu_2(PPh_3)_6(SC_2H_4Ph)_6]^+$ nanocluster protected by both phosphine and thiolate ligands is investigated herein (Figure 4.9) [50]. The horizontal axis represents $m/z$ value and the vertical axis represents the intensity of the species with specific $m/z$ value. There are two obvious peaks shown in the spectrum, bearing the $m/z$ value of 5084.61 and 4822.24 Da, respectively. The first one can be well assigned to the composition $Au_{13}Cu_2(PPh_3)_6(SC_2H_4Ph)_6$ of the metal nanocluster and the other one represents the composition $Au_{13}Cu_2(PPh_3)_5(SC_2H_4Ph)_6$. This kind of fragmentation with one missed $PPh_3$ in the mass spectrum is possible and probably due to the loss of this fragment during ionization process. In addition to determination of the composition, HRMS is also useful for conclusion of the charge states of metal nanoclusters. As shown in the inset picture of Figure 4.9, the isotopic patterns of the peak at 5084.61 Da is enlarged. The peak separation of 1 indicates the one positively changed status of this peak, suggesting that the charge state of the $Au_{13}Cu_2(PPh_3)_6(SC_2H_4Ph)_6$ nanocluster would be +1. Therefore, if the peak separation is 0.5 or 0.33, the species detected should be +2 or +3 charged and the charge state of as-tested metal nanocluster would be +2 or +3.

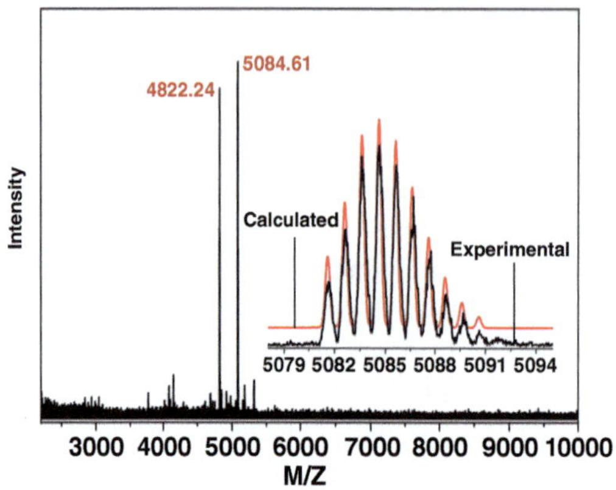


**Figure 4.9** ESI-MS spectrum of $[Au_{13}Cu_2(PPh_3)_6(SC_2H_4Ph)_6]^+$ nanocluster.

In addition to composition determination of organic ligand-protected metal nanoclusters, HRMS is helpful in structural elucidation as well. Chakraborty and coworkers investigated the intermediate fragments during stepwise dissociation of organic ligand-protected silver nanoclusters [51]. They applied trapped ion mobility mass spectrometry to determine collision cross-sections of the fragments and compared them with the structures simulated by density functional theory. Based on this technique, they resolved the geometric structures of the primary fragments of $[Ag_{29}L_{12}]^{3-}$ nanocluster (L = 1,3-benzene dithiolate). For example, the $[Ag_{24}L_9]^{2-}$, $[Ag_{19}L_6]^-$, and $[Ag_5L_3]^-$. The corresponding results demonstrate that $[Ag_{29}L_{12}]^{3-}$ nanocluster dissociates via two-step elimination of $[Ag_5L_3]^-$ to form $[Ag_{19}L_6]^-$, which further dissociates by $Ag_2$ loss generating species such as $[Ag_{17}L_6]^-$ and $[Ag_{15}L_6]^-$ and $[Ag_{13}L_6]^-$. This investigation reflects a competition between remaining the eight-electron structure of the metal kernel and increasing the steric strain of ligands and the surface staples.

Another important application of HRMS is understanding formation mechanism of metal nanoclusters. The metal nanocluster formation process can be effectively detected by MALDI-TOF-MS and ESI-MS. Early studies by Amala Dass and Rongchao Jin research groups have shown that MALDI-TOF-MS is able to capture the intermediates during the formation of metal nanoclusters. Based on this technique, they found that mixtures of different sized metal nanoclusters were formed initially and then gradually transformed to the thermodynamically stable metal nanocluster with a size-focusing process. Five years later, Luo research group investigated the formation of the star metal nanocluster $[Au_{25}(SR)_{18}]^-$ in details by ESI-MS [52]. To control the reaction rate during the synthesis, they employed carbon monoxide as the mild and efficient reducing agent, which significantly slows down the reaction and make the reaction quenchable for ESI-MS analysis. Based on 29 stable intermediates that they detected in ESI-MS spectra, they successfully drew a detailed map for the size-focusing process during the formation of $[Au_{25}(SR)_{18}]^-$ nanocluster. This process consists of the isoelectronic addition, disproportionation, and comproportionation stages that are shown in Figure 4.10. Inspired by this work, the same research group revealed the seed-growth mechanism from $[Au_{25}(SR)_{18}]^-$ to $[Au_{44}(SR)_{26}]^-$ nanocluster and the foreign

$$[Au_{23}(SR)_{16}]^- + 2\,[Au(SR)Cl]^- \xrightarrow{\text{isoelectronic addition}} [Au_{23}(SR)_{18}]^- + 2\,Cl^-$$

$$[Au_{20}(SR)_{15}]^- \xrightarrow{\text{disproportionation}} [Au_{15}(SR)_{12}]^- + [Au_{25}(SR)_{18}]^-$$

$$[Au_{21}(SR)_{16}]^- + [Au_{29}(SR)_{20}]^- \xrightarrow{\text{comproportionation}} 2\,[Au_{25}(SR)_{18}]^-$$

**Figure 4.10** Isoelectronic addition, disproportionation, and comproportionation stages described for the size-focusing process during formation of $[Au_{25}(SR)_{18}]^-$ nanocluster identified by ESI-MS spectra.

metal-alloying mechanism for the production of $[Au_{12}Ag_{32}(SR)_{30}]^{4-}$ nanocluster [53, 54]. The other research groups also utilized ESI-MS to systematically investigate the reaction mechanism of metal nanoclusters [55, 56]. Chen and coworkers wrote a review paper focusing on the electrospray ionization mass spectrometry [57]. They systematically described this powerful platform for analysis of noble metal nanoclusters. This review would be inspiring for researchers to acquire comprehensive information for this characterization technique.

The working principle of SCXRD is based on detection of diffraction points that are generated from the characteristic X-ray acting on the single crystals mounted on the goniometer. The diffraction lines in the spatial distribution of location and intensity is closely related to the single-crystal structures. All of the three-dimensional structural information including the bond length, bond angle and the packing mode are available by SCXRD analysis. Therefore, it has become the most powerful tool for the structural determination of organic ligand-protected metal nanoclusters. In fact, it is the elegant work in structural determination of $Au_{102}(SC_6H_4COOH)_{44}$ [1] and $Au_{25}(SC_2H_4Ph)_{25}$ [2] by SCXRD that start to attract extensive attention of chemists and material scientists. The technique of SCXRD highly promotes the development of metal nanocluster chemistry. For example, the structure and size transition from metal nanoclusters to nanocrystals is a central issue in nanoscience. Therefore, structural determination of metal nanoclusters in the transition size and structure range is of high significance for understanding the evolution from nanoclusters to nanocrystals. The gold nanocluster of $Au_{144}(SR)_{60}$ falls into this transition range. Although its synthesis [58], purifications [59, 60], and composition determination by ESI-MS [61] have been achieved nearly two decades ago by different research groups, the atomically precise structure of $Au_{144}(SR)_{60}$ remains unknown. Attempts on characterizations and computations for its structure all lead to unsatisfactory results [62–64]. This challenging issue was resolved by the research group of Yan in 2018 by SCXRD [65]. By using a special organic ligand $PhCH_2SH$ with appropriate rigidity between that of $PhCH_2CH_2SH$ and para-mercaptobenzoic acid, they successfully obtained single crystals of $Au_{144}(SCH_2Ph)_{60}$ that suitable for its structural determination by SCXRD. The atomically precise structure of $Au_{144}(SCH_2Ph)_{60}$ is shown in Figure 4.11. The intra- and intercluster weak interactions in the crystals

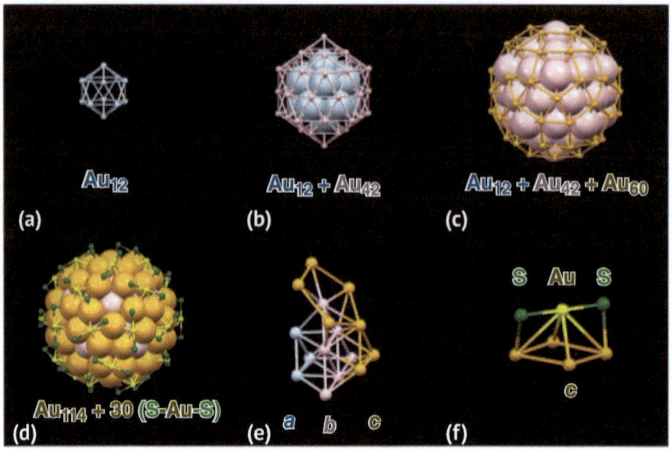


**Figure 4.11** Atomically precise structure of $Au_{144}(SCH_2Ph)_{60}$ revealed by SCXRD. *Source:* [65]/with permission of American Association for the Advancement of Science – AAAS.

play important roles in growth of high-quality single crystals of the metal nanocluster. Various Au–Au bonds with different lengths and angles are revealed in the kernel of $Au_{144}(SCH_2Ph)_{60}$, verifying that this nanocluster is a transition size in terms of Au–Au bonding character. The structure and its high thermodynamic stability further provide an atomic-level perspective that metals possess good thermal extensibility compared to other common nonmetal solid materials.

The atomically precise structure of organic ligand-protected metal nanoclusters revealed by SCXRD is particularly crucial for metal nanocluster catalysis. As we mentioned in the second chapter, the in situ generated metal nanoclusters demonstrate high activity in catalysis. However, the structure–catalytic property relationship is difficult to be acquired due to their ill-defined structure, which limits the structural optimization based on the relationship. SCXRD enables the structural determination of metal nanoclusters at atomic level, thus being beneficial for understanding the catalytic mechanism and the subsequential structural optimization. T. Liu and coworkers systematically investigated the correlation of metal nanoclusters' structure and catalytic property [66]. For example, they studied the methane catalytic oxidation performance of two structurally similar gold nanoclusters $Au_{24}(PPh_3)_{10}(SC_2H_4Ph)_5Cl_2$ and $Au_{25}(PPh_3)_{10}(SC_2H_4Ph)_5Cl_2$. From the molecular formula, only one gold atom difference is observed. More importantly, SCXRD reveals that both nanoclusters have the identical geometric structure, and one gold atom is lost in the central position of $Au_{24}(PPh_3)_{10}(SC_2H_4Ph)_5Cl_2$, compared to that of $Au_{25}(PPh_3)_{10}(SC_2H_4Ph)_5Cl_2$ (Figure 4.12). Their studies indicate that the loss of one central atom highly enhances the catalytic activity of the gold nanocluster in methane oxidation reaction. More efficient $CH_3OH$ production was observed by $Au_{24}(PPh_3)_{10}(SC_2H_4Ph)_5Cl_2$. Based on a reversible structural transformation between $Au_{24}(PPh_3)_{10}(SC_2H_4Ph)_5Cl_2$ and $Au_{25}(PPh_3)_{10}(SC_2H_4Ph)_5Cl_2$ enabled by $H_2O_2$ oxidation and $PPh_3$ reduction, they further realized the reversible switching of catalytic activity by shuttling an atom into and out of gold nanoclusters. Control

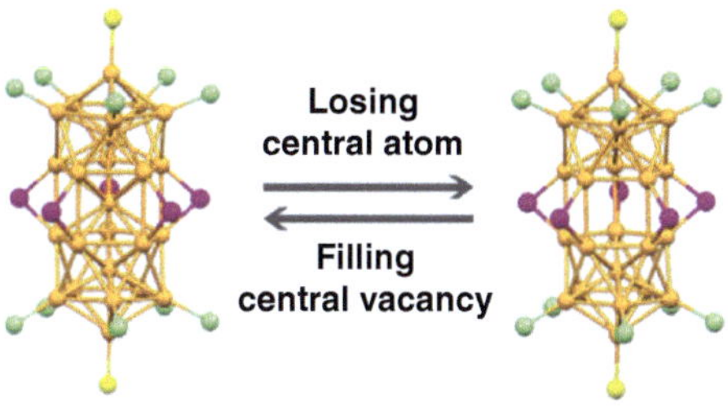


**Figure 4.12** $Au_{24}(PPh_3)_{10}(SC_2H_4Ph)_5Cl_2$ and $Au_{25}(PPh_3)_{10}(SC_2H_4Ph)_5Cl_2$ nanoclusters with one atom alteration. *Source:* [66]/with permission of John Wiley & Sons.

experiments and the calculations based on the SCXRD-revealed clusters' structures help to draw a full view of the catalytic process.

UV–vis is the most frequently used characterization technique for the initial identification of the as-obtained metal nanoclusters protected by organic ligand. UV–vis equipment is much smaller and cheaper than that of HRMS and SCXRD, thus being beneficial for set up and operation. The principle of UV–vis for identification of metal nanoclusters is based on their discrete and molecule-like energy levels, which are distinct from that of metal nanoparticles. Metal nanoparticles generally show a broad surface plasmon resonance (SPR) absorption peak in UV–vis. For example, the SPR peak of Au nanoparticles locates in about 520 nm. In sharp contrast, metal nanoclusters show step-like absorption profiles due to their molecule-like HOMO-LUMO transitions. The UV–vis spectra of metal nanoclusters suggest significant structure-dependent characteristic. One of the most important works in metal nanocluster chemistry, reported by Zhu and coworkers, highlights the correlation between crystal structure and UV–vis optical properties of $[Au_{25}(SC_2H_4Ph)_{18}]^-$ nanocluster [2]. As shown in Figure 4.13, owing to the strong quantum size effects, $[Au_{25}(SC_2H_4Ph)_{18}]^-$ shows three well-defined bands at about 400 nm (3.1 eV), 450 nm (2.75 eV), and 680 nm (1.8 eV) in UV–vis spectrum. Time-dependent density functional theory (TDDFT) calculations well explain the molecule-like optical absorption spectrum, which are reflected in the Kohn–Sham molecular orbitals. The absorption peak at 400 nm arises principally from an interband transition (sp←d). The absorption peak at 450 nm arises from mixed intraband (sp←sp) and interband (sp←d) transitions. The absorption peak at 400 nm corresponds to a LUMO←HOMO transition, which is essentially an intraband (sp←sp) transition.

Metal nanoclusters with distinct geometric and electronic structures possess different UV–vis spectra. Thus, it is convenient to preliminarily identify metal nanoclusters by UV–vis spectra. The UV–vis spectrum is not only important for reflecting the electronic structure and optical properties, but also useful in monitoring the formation process of metal nanoclusters. In addition, this technique has been widely applied by different research groups to investigate the structural and catalytic stability of metal nanoclusters in catalysis.

It should be pointed out that other characterization techniques such as the Fourier transform infrared spectrometer, nuclear magnetic resonance spectroscopy, circular dichroic spectroscopy, transient absorption spectroscopy, transmission electron microscope. Dynamic light scattering, thermogravimetric analysis, electron paramagnetic resonance, X-ray photoelectron spectroscopy, and X-ray absorption fine

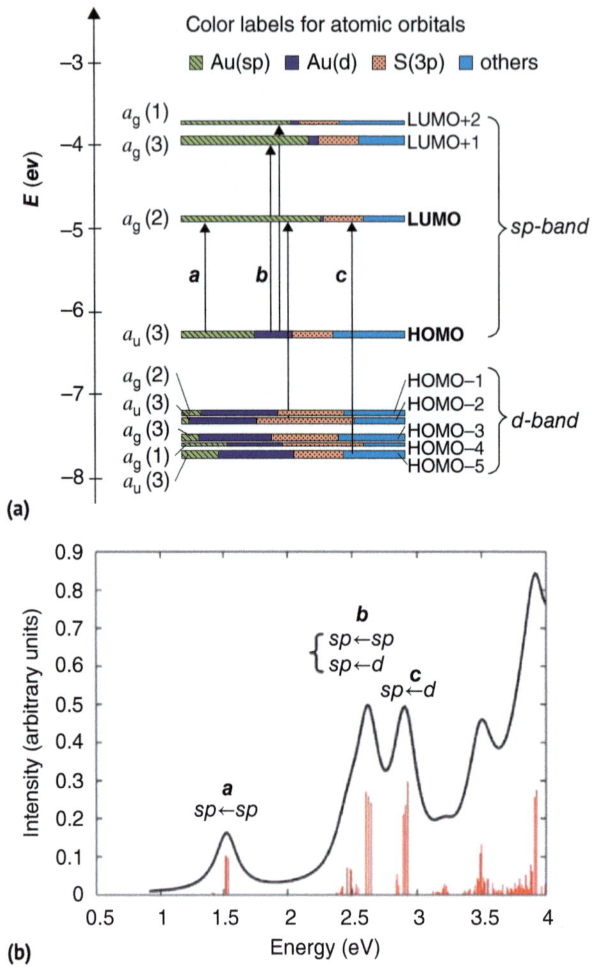


**Figure 4.13** Structure-optical property correlation of $[Au_{25}(SC_2H_4Ph)_{18}]^-$ nanocluster. *Source:* [2]/with permission of American Chemical Society.

structure are important as well to reveal the structural characteristics of metal nanoclusters. Combined usage of these methods gives more information of the metal kernels, surface ligands, and their synergistic effect in catalysis.

## 4.3 Atomically Precise Structures of Organic Ligand-protected Metal Nanoclusters

Atomically precise structures provide basis for investigating the structure-catalytic property correlation of organic ligands protected metal nanoclusters. The surface organic ligand, on the one hand, stabilizes the geometric and electronic structure of the metal kernel, and, on the other hand, cooperates with the metal activation sites to regulate the activity and selectivity of metal nanoclusters. In this part, we briefly introduce the atomically precise structures of organic ligands protected metal nanoclusters in the aspects of metal atoms arrangement in the kernel and the metal–ligand interactions on the surface.

Before talking about the structures of organic ligand-protected metal nanoclusters, we need to clarify that this kind of metal nanocluster includes a wide range of metal species [67, 68]. As described by Ulrich Schubert in 2011: "According to F. A. Cotton (Acc. Chem. Res., 1969, 2, 240), a metal cluster consists of a finite number of metal atoms which are connected through a major part by metal–metal bonds. The term "cage compound" was proposed for polynuclear compounds connected by ligand bridges without metal–metal bonding. In this article the term "cluster" is also for the latter compounds in line with current practice." Thus, compared to a special definition of the metal nanocluster that "a group of two or more metal atoms where direct and substantial metal bonding is present," a general definition of the metal nanocluster should be "a number of metal centers are grouped close together which can have direct metal bonding interactions or interactions through a bridging ligand." Herein, we mainly discuss the structures of metal nanoclusters with direct and substantial metal–metal bonds.

As we know, metal atoms adopt three kinds of the closest packing modes in bulk metals and most of metal nanocrystals (and metal nanoparticles). They are the face-centered cubic (FCC), body-centered cubic and hexagonal close-packed (HCP), respectively. The metal nanocrystal (and metal nanoparticle) differentiates from the bulk metal mainly by the long-range ordering or short-range ordering of the closest packing mode. Their metal atoms' packing modes should not be significantly different from each other. However, the packing modes of metal atoms in metal nanoclusters are much more diverse. Figure 4.14 shows some representative metal kernels of organic

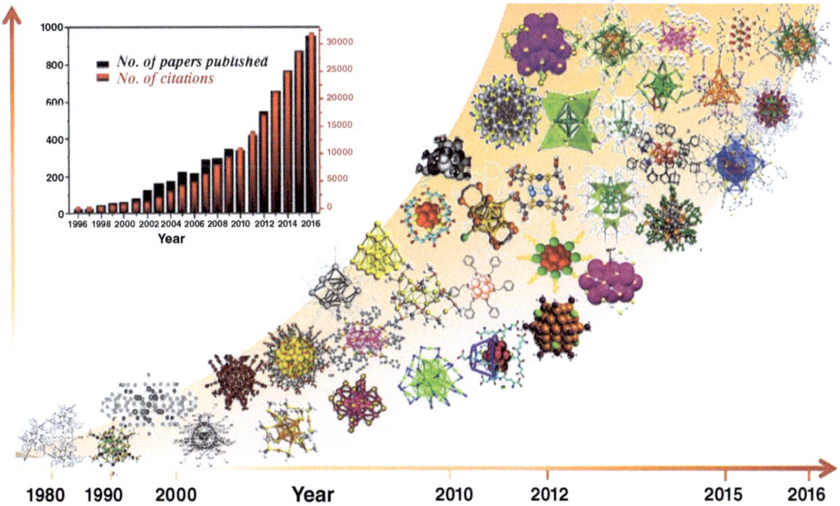


**Figure 4.14** Representative metal kernels of organic ligand-protected metal nanoclusters. *Source:* [69]/with permission of American Chemical Society.

**Figure 4.15** The icosahedral $Au_{13}$ kernel (a) and bi-icosahedral $Au_{25}$ kernel (b) of $[Au_{25}(SR)_{18}]^-$ and $[Au_{25}(SR)_5(PPh_3)_5Cl_2]^{2+}$ nanoclusters.

ligand-protected metal nanoclusters, in which the metal atoms adopt diversified packing modes compared to bulk metals and form abundant metal kernels geometry.

We take two famous metal kernels in Figure 4.14 as examples to describe the metal–metal bonding information in detail. As shown in Figure 4.15, both metal nanoclusters have the same number of gold atoms while their compositions are different. $[Au_{25}(SR)_{18}]^-$ was synthesized by several groups, while its structural determination was completed by Zhu and coworkers in 2008 [2]. $[Au_{25}(SR)_5(PPh_3)_5Cl_2]^{2+}$ was structurally determined by Shichibu and coworkers in 2007 for the first time [70], although metal nanoclusters with its similar structures have been investigated by Teo and his research group from 1984 to 1990 [71–73]. $[Au_{25}(SR)_{18}]^-$ nanocluster bears an $Au_{13}$ kernel, in which the gold atoms neither adopt the common FCC nor BCC/HCP packing modes. The thirteen gold atoms arrange to form an icosahedral geometry (Figure 4.15a). The other gold atoms and organic ligands surround the $Au_{13}$ kernel and constitute the whole $[Au_{25}(SR)_{18}]^-$ nanocluster. For the $[Au_{25}(SR)_5(PPh_3)_5Cl_2]^{2+}$ nanocluster, all of the 25 gold atoms consist of an $Au_{25}$ kernel, which can be viewed as a bi-icosahedron via vertex sharing of two icosahedral $Au_{13}$ (Figure 4.15b). The organic ligands surround on the surface of the vertex-sharing bi-icosahedral $Au_{25}$ kernel.

The one-dimensional growth [74–76] of the icosahedral $Au_{13}$ kernel can lead to metal nanoclusters with longer metal kernels and intriguing photothermal and near-infrared photoluminescence performance (Figure 4.16a). The diversified combination modes of icosahedral $M_{13}$ units further trigger the discovery of structurally diverse metal nanoclusters. This kind of one-dimensional growth of metal atoms' packing mode is not limited to the icosahedral $Au_{13}$ kernel. It has been observed in other metal nanoclusters such as $Au_{28}(SR)_{20}$, $Au_{36}(SR)_{24}$, $Au_{44}(SR)_{28}$, and $Au_{52}(SR)_{32}$, in which several $T_d$ $Au_4$ units grow along the axis and constitute the gold kernel of these nanoclusters. The one-dimensional growth of Au units can be reflected on the absorption spectrum and employed for structural prediction. For example, $Au_{76}(SR)_{44}$ shows an intense absorption band at 1350 nm. Its structure has been predicted to follow the growth pattern of a series of structurally similar metal nanoclusters, that is, $Au_{28}(SR)_{20}$, $Au_{36}(SR)_{24}$, $Au_{44}(SR)_{28}$, $Au_{52}(SR)_{32}$, $Au_{60}(SR)_{36}$, and $Au_{68}(SR)_{40}$ to be a double helical kernel made up of 16 $T_d$ $Au_4$ units (Figure 4.16b) [77].

Famous mathematical geometries are also observed in the metal kernels of organic ligand-protected metal nanoclusters [78]. Platonic and Archimedean

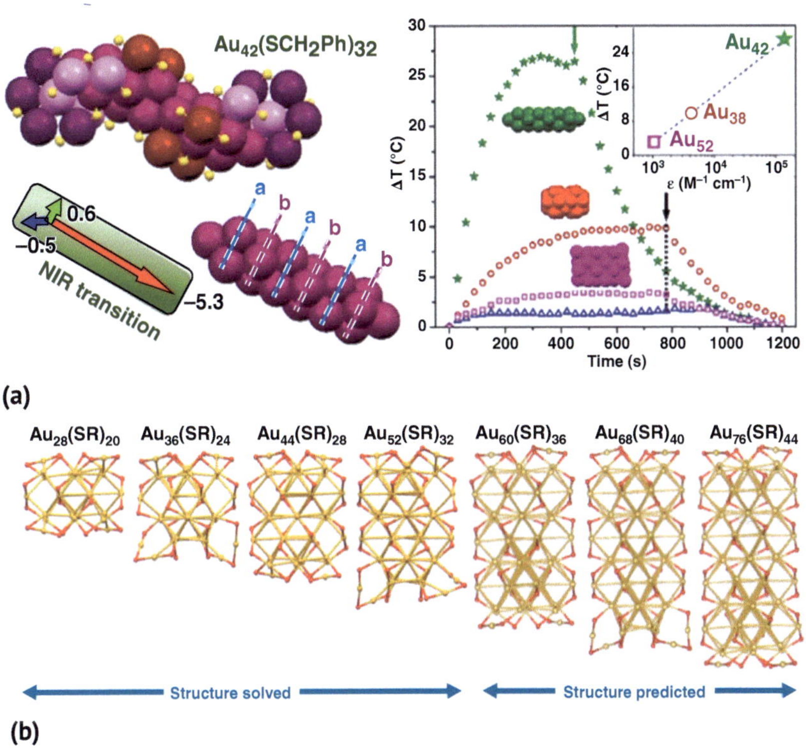


**Figure 4.16** (a) One-dimensional growth of the icosahedral $Au_{13}$ kernel and the intriguing photothermal and near-infrared photoluminescence performance of the as-obtained metal nanoclusters. (b) $T_d$ $Au_4$ kernel observed in metal nanoclusters protected by organic ligand.

solids are well-known polyhedra with high symmetry. Platonic solids contain five common polyhedra of tetrahedron, hexahedron, octahedron, dodecahedron, and icosahedron. All of them are constructed from a single type of regular polygon, that is, triangle, square or pentagon with equivalent vertices and equivalent edges. These five Platonic solids are regular polyhedra and belong to one of the highly symmetric point groups of tetrahedral ($T_d$), octahedral ($O_h$), and icosahedral ($I_h$). A lot of organic ligand-protected metal nanoclusters bear Platonic solid geometry inside. Among them, the simple tetrahedral $M_4$ and hexahedral $M_6$ nanoclusters are the most common ones. Archimedean solids are often called semiregular solids, featuring a single type of vertex, that is, equivalent vertices surrounded by regular polygons. In addition to chiral Archimedean solids, the other 13 centrosymmetric ones have also been found in organic ligand-protected atomically precise metal nanoclusters, mainly in truncated tetrahedron, cuboctahedron, truncated cube, truncated octahedron and rhombicuboctahedron. In Figure 4.17, we choose some typical metal nanocluster structures and show their polyhedral kernels with Platonic and Archimedean solids, which should be ascribed to their unique metal–metal bonding modes.

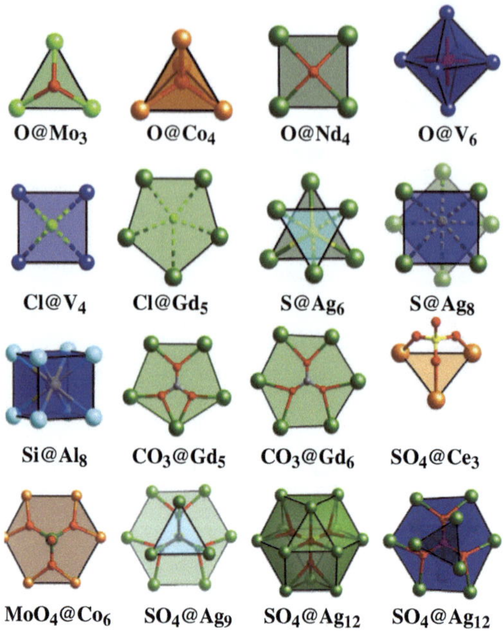


**Figure 4.17** Platonic and Archimedean solids in metal nanoclusters protected by organic ligand. *Source:* [78]/with permission of Royal Society of Chemistry.

In addition to the diversified metal kernel structures, the organic surface of organic ligand-protected atomically precise metal nanoclusters contains diverse staples and abundant metal–ligand interactions as well that vary from each other with the type of ligands [69, 79]. Figure 4.18 shows representative metal–ligand interactions on the surface of metal nanoclusters with gold as a representative metal. Phosphines have been utilized for stabilization of metal nanoclusters a long time ago. Typical metal kernels such as $Au_{11}$, $Au_6$, $Au_9$, and $Au_{13}$ were explored to be stabilized by phosphine ligands. They prefer to bind directly with the metal kernel and form a pure organic layer on the surface. Thus, all of metal atoms are involved in the metal kernel in this case. *N*-heterocyclic carbenes act similarly with phosphines and bind to the metal kernel directly. Another commonly used organic ligand, thiolates, display quite different binding mode compared to phosphines and *N*-heterocyclic carbenes. As shown in Figure 4.18, more than one thiolate and gold atoms connect with each other and form a staple that locates on the surface of the metal kernel. Thus, core–shell structures are generally observed in thiolate-protected metal nanoclusters and the [metal-thiolate]$_n$ staples constitute the shell protecting the metal kernel from aggregation or decomposition. Alkynes and the nitrogen-based ligands were recently developed by the research group of Quan-Ming Wang and show some interesting motif structures on the surface. Both types of organic ligands were considered to be difficult to stabilize the metal nanoclusters previously because the corresponding metal–ligand interactions were thought to be too weak. Quan-Ming Wang and coworkers discovered the potential of alkynes and nitrogen-based ligands in stabilizing metal nanoclusters with ultra-stable structures.

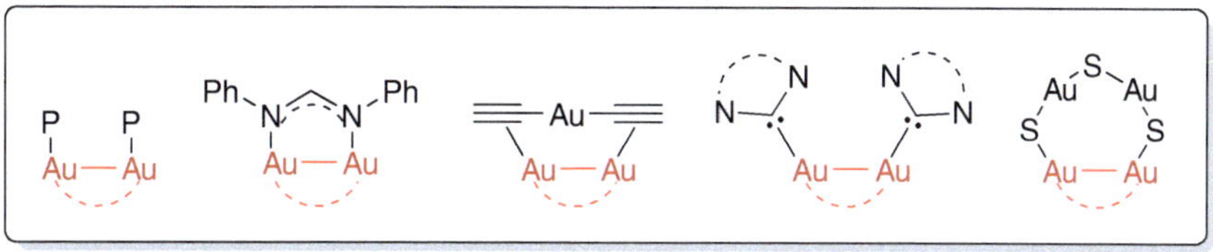


**Figure 4.18** Representative metal–ligand interactions on the surface of metal nanoclusters protected by organic ligand.

The surface structure of metal nanoclusters is particularly important for catalysis because it is the place where substrates' activation happens. Herein, we briefly described the metal–ligand interactions on the surface of metal nanoclusters. We will systematically introduce the surface structures in the following chapters, focusing on their important roles in catalysis. Of course, metal kernels also play crucial roles in regulating the activity and selectivity of metal nanocluster catalysis by electronic effect. Considering the important dual functionalities of organic ligands in stabilizing the metal kernel and creating activation sites, we divide the following chapters of organic ligand-protected metal nanocluster catalysis by the category of organic ligands: thiolate-protected metal nanocluster for catalysis; phosphine-protected metal nanocluster for catalysis; alkyne-protected metal nanocluster for catalysis; carbene-protected metal nanocluster for catalysis; nitrogen ligand-protected metal nanocluster for catalysis; pincer ligand-protected metal nanocluster for catalysis; hydride-doping in metal nanocluster for catalysis.

## References

**1** P. D. Jadzinsky, G. Calero, C. J. Ackerson, D. A. Bushnell, R. D. Kornberg, Structure of a Thiol Monolayer-Protected Gold Nanoparticle at 1.1 Å Resolution. *Science* **2007**, *318*, 430–433.

**2** M. Zhu, C. M. Aikens, F. J. Hollander, G. C. Schatz, R. Jin, Correlating the Crystal Structure of a Thiol-Protected $Au_{25}$ Cluster and Optical Properties. *J. Am. Chem. Soc.* **2008**, *130*, 5883–5885.

**3** M. Brust, M. Walker, D. Bethell, D. J. Schiffrin, R. Whyman, Synthesis of Thiol-derivatised Gold Nanoparticles in a Two-phase Liquid–Liquid System. *J. Chem. Soc., Chem. Commun.* 1994, *7*, 801–802.

**4** B. M. Barngrover, C. M. Aikens, The Golden Pathway to Thiolate-Stabilized Nanoparticles: Following the Formation of Gold(I) Thiolate from Gold(III) Chloride. *J. Am. Chem. Soc.* **2012**, *134*, 12590–12595.

**5** S. G. Booth, A. Uehara, S.-Y. Chang, C. La Fontaine, T. Fujii, Y. Okamoto, T. Imai, S. L. M. Schroeder, R. A. W. Dryfe, The Significance of Bromide in the Brust–Schiffrin Synthesis of Thiol Protected Gold Nanoparticles. *Chem. Sci.* **2017**, *8*, 7954–7962.

**6** P. J. G. Goulet, R. B. Lennox, New Insights into Brust–Schiffrin Metal Nanoparticle Synthesis. *J. Am. Chem. Soc.* **2010**, *132*, 9582–9584.

**7** Y. Li, O. Zaluzhna, B. Xu, Y. Gao, J. M. Modest, Y. J. Tong, Mechanistic Insights into the Brust–Schiffrin Two-Phase Synthesis of Organo-chalcogenate-Protected Metal Nanoparticles. *J. Am. Chem. Soc.* **2011**, *133*, 2092–2095.

**8** M. Zhu, E. Lanni, N. Garg, M. E. Bier, R. Jin, Kinetically Controlled, High-Yield Synthesis of $Au_{25}$ Clusters. *J. Am. Chem. Soc.* **2008**, *130*, 1138–1139.

**9** X. Kang, H. Chong, M. Zhu, $Au_{25}(SR)_{18}$: the Captain of The Great Nanocluster Ship. *Nanoscale* **2018**, *10*, 10758–10834.

**10** S. Wang, Q. Li, X. Kang, M. Zhu, Customizing the Structure, Composition, and Properties of Alloy Nanoclusters by Metal Exchange. *Acc. Chem. Res.* **2018**, *51*, 2784–2792.

**11** J.-P. Choi, C. A. Fields-Zinna, R. L. Stiles, R. Balasubramanian, A. D. Douglas, M. C. Crowe, R. W. Murray, Reactivity of $[Au_{25}(SCH_2CH_2Ph)_{18}]^{1-}$ Nanoparticles with Metal Ions. *J. Phys. Chem. C.* **2010**, *114*, 15890–15896.

**12** S. Wang, Y. Song, S. Jin, X. Liu, J. Zhang, Y. Pei, X. Meng, M. Chen, P. Li, M. Zhu, Metal Exchange Method Using $Au_{25}$ Nanoclusters as Templates for Alloy Nanoclusters with Atomic Precision. *J. Am. Chem. Soc.* **2015**, *137*, 4018–4021.

**13** A. Ghosh, O. F. Mohammed, O. M. Bakr, Atomic-Level Doping of Metal Clusters. *Acc. Chem. Res.* **2018**, *51*, 3094–3103.

**14** S. Hossain, Y. Niihori, L. V. Nair, B. Kumar, W. Kurashige, Y. Negishi, Alloy Clusters: Precise Synthesis and Mixing Effects. *Acc. Chem. Res.* **2018**, *51*, 3114–3124.

**15** S. Wang, X. Meng, A. Das, T. Li, Y. Song, T. Cao, X. Zhu, M. Zhu, R. Jin, A 200-fold Quantum Yield Boost in the Photoluminescence of Silver-Doped $Ag_xAu_{25-x}$ Nanoclusters: The 13th Silver Atom Matters. *Angew. Chem. Int. Ed.* **2014**, *53*, 2376–2380; *Angew. Chem.* **2014**, *126*, 2408–2412.

**16** M.-B. Li, S.-K. Tian, Z. Wu, R. Jin, $Cu^{2+}$ induced formation of $Au_{44}(SC_2H_4Ph)_{32}$ and Its High Catalytic Activity for the Reduction of 4-nitrophenol at Low Temperature. *Chem. Commun.* **2015**, *51*, 4433–4436.

**17** Z. Wu, Anti-galvanic Reduction of Thiolate-protected Gold and Silver Nanoparticles. *Angew. Chem. Int. Ed.* **2012**, *51*, 2934–2938; *Angew. Chem.* **2012**, *124*, 2988–2992.

**18** C. Yao, Y.-J. Lin, J. Yuan, L. Liao, M. Zhu, L.-H. Weng, J. Yang, Z. Wu, Mono-cadmium vs Mono-mercury Doping of $Au_{25}$ Nanoclusters. *J. Am. Chem. Soc.* **2015**, *137*, 15350–15353.

**19** L. Liao, S. Zhou, Y. Dai, L. Liu, C. Yao, C. Fu, J. Yang, Z. Wu, Mono-Mercury Doping of $Au_{25}$ and the HOMO/LUMO Energies Evaluation Employing Differential Pulse Voltammetry. *J. Am. Chem. Soc.* **2015**, *137*, 9511–9514.

**20** Z. Gan, N. Xia, Z. Wu, Discovery, Mechanism, and Application of Antigalvanic Reaction. *Acc. Chem. Res.* **2018**, *51*, 2774–2783.

**21** X. Liu, D. Astruc, From Galvanic to Anti-Galvanic Synthesis of Bimetallic Nanoparticles and Applications in Catalysis, Sensing, and Materials Science. *Adv. Mater.* **2017**, *29*, 1605305.

**22** X. Kang, M. Zhu, Transformation of Atomically Precise Nanoclusters by Ligand-Exchange. *Chem. Mater.* **2019**, *31*, 9939–9969.

**23** Y. Negishi, K. Nobusada, T. Tsukuda, Glutathione-Protected Gold Clusters Revisited: Bridging the Gap between Gold(I)–Thiolate Complexes and Thiolate-Protected Gold Nanocrystals. *J. Am. Chem. Soc.* **2005**, *127*, 5261–5270.

**24** S. Chen, S. Wang, J. Zhong, Y. Song, J. Zhang, H. Sheng, Y. Pei, M. Zhu, The Structure and Optical Properties of the [Au18(SR)14] Nanocluster. *Angew. Chem. Int. Ed.* **2015**, *54*, 3145–3149; *Angew. Chem.* **2015**, *127*, 3188–3192.

**25** S. Yang, S. Chen, L. Xiong, C. Liu, H. Yu, S. Wang, N. L. Rosi, Y. Pei, M. Zhu, Total Structure Determination of $Au_{16}(S\text{-}Adm)_{12}$ and $Cd_1Au_{14}(StBu)_{12}$ and Implications for the Structure of $Au_{15}(SR)_{13}$. *J. Am. Chem. Soc.* **2018**, *140*, 10988–10994.

**26** M. R. Narouz, K. M. Osten, P. J. Unsworth, R. W. Y. Man, K. Salorinne, S. Takano, R. Tomihara, S. Kaappa, S. Malola, C.-T. Dinh, J. D. Padmos, K. Ayoo, P. J. Garrett, M. Nambo, J. H. Horton, E. H. Sargent, H. Häkkinen, T. Tsukuda, C. M. Crudden, N-heterocyclic Carbene-functionalized Magic-number Gold Nanoclusters. *Nat. Chem.* **2019**, *11*, 419–425.

**27** Y. Shichibu, Y. Negishi, T. Tsukuda, T. Teranishi, Large-Scale Synthesis of Thiolated $Au_{25}$ Clusters via Ligand Exchange Reactions of Phosphine-stabilized $Au_{11}$ Clusters. *J. Am. Chem. Soc.* **2005**, *127*, 13464–13465.

**28** M.-B. Li, S.-K. Tian, Z. Wu, R. Jin, Peeling the Core–Shell $Au_{25}$ Nanocluster by Reverse Ligand-Exchange. *Chem. Mater.* **2016**, *28*, 1022–1025.

**29** J. Zou, W. Fei, Y. Qiao, Y. Yang, Z. He, L. Feng, M.-B. Li, Z. Wu, Combined Synthesis of Interconvertible $Au_{11}Cd$ and $Au_{26}Cd_5$ for Photocatalytic Oxidations Involving Singlet Oxygen. *Chin. Chem. Lett.* **2023**, *34*, 107660.

**30** Z. He, Y. Yang, J. Zou, Q. You, L. Feng, M.-B. Li, Z. Wu, Partial Phosphorization: a Strategy to Improve Some Performance(s) of Thiolated Metal Nanoclusters Without Notable Reduction of Stability. *Chem. Eur. J.* **2022**, *28*, e202200212.

**31** C. Liu, Y. Zhao, T.-S. Zhang, C.-B. Tao, W. Fei, S. Zhang, M.-B. Li, Asymmetric Transformation of Achiral Gold Nanoclusters with Negative Nonlinear Dependence Between Chiroptical Activity and Enantiomeric Excess. *Nat. Commun.* **2023**, *14*, 3730.

**32** W. Fei, Y. Tao, Y. Qiao, S.-Y. Tang, M.-B. Li, Structural Modification and Performance Regulation of Atomically Precise Metal Nanoclusters by Phosphine. *Polyoxometalates* **2023**, *2*, 9140043.

**33** C. Liu, Y. Li, Z. He, Y. Yang, C. Wu, W. Fan, W. W. Xu, M.-B. Li, Reduction–Oxidation Cascade Strategy for Reforming a $Au_{13}$-Kerneled Gold Thiolate Nanocluster. *J. Phys. Chem. Lett.* **2023**, *14*, 11558–11564.

**34** Q. Li, B. Huang, S. Yang, H. Zhang, J. Chai, Y. Pei, M. Zhu, Unraveling the Nucleation Process from a Au(I)-SR Complex to Transition-Size Nanoclusters. *J. Am. Chem. Soc.* **2021**, *143*, 15224–15232.

**35** Z.-M. Zhu, Y. Zhao, H. Zhao, C. Liu, Y. Zhang, W. Fei, H. Bi, M.-B. Li, Photochemical Route for Synthesizing Atomically Precise Metal Nanoclusters from Disulfide. *Nano Lett.* **2023**, *23*, 7508–7515.

**36** J.-Q. Fan, K. Cen, H. J. Xu, H. Y. Wang, Y. Yang, Z.-M. Zhu, H. Liu, D. Chen, W. Fan, M.-B. Li, Photochemical Synthesis of Group 10 Metal Nanoclusters for Electrocatalysis. *Nanoscale* **2023**, *15*, 19079–19084.

**37** Y. Zhao, Z.-M. Zhu, W. Fan, W. Zhu, J.-J. Yang, Y. Tao, W. Fei, H. Bi, S. Zhang, M.-B. Li, Photosynthesis of $Au_8Cu_6$ Nanocluster For Photocatalysis In Oxidative Functionalization of Alkynes. *Nat. Commun.* **2024**, *15*, 9632.

**38** Y.-X. Wang, J. Zhang, H.-F. Su, X. Cui, C.-Y. Wei, H. Li, X.-M. Zhang, Photochemical Synthesis of Atomically Precise Ag Nanoclusters. *ACS Nano* **2023**, *17*, 11607–11615.

**39** Y.-X. Wang, F.-Q. Zhang, Z. Qi, X. Zhao, N. Zhang, H. Li, H. Li, X.-M. Zhang, Photochemical Synthesis of $Ag_{12}Cu_7$ Nanocluster with Cuprophilicity-related Long-lived Phosphorescence. *Aggregate* **2024**, *6*, e675.

**40** A. Jana, M. Jash, A. K. Poonia, G. Paramasivam, M. R. Islam, P. Chakraborty, S. Antharjanam, J. Machacek, S. Ghosh, K. N. V. D. Adarsh, T. Base, T. Pradeep, Light-activated Intercluster Conversion of an Atomically Precise Silver Nanocluster. *ACS Nano* **2021**, *15*, 15781–15793.

**41** R. Saito, K. Isozaki, Y. Mizuhata, M. Nakamura, Synthesis of $N_2$-Type Superatomic Molecules. *J. Am. Chem. Soc.* **2024**, *146*, 20930–20936.

**42** K. R. Krishnadas, A. Ghosh, A. Baksi, I. Chakraborty, G. Natarajan, T. Pradeep, Intercluster Reactions Between $Au_{25}(SR)_{18}$ and $Ag_{44}(SR)_{30}$. *J. Am. Chem. Soc.* **2016**, *138*, 140–148.

**43** K. R. Krishnadas, A. Baksi, A. Ghosh, G. Natarajan, T. Pradeep, Structure-Conserving Spontaneous Transformations Between Nanoparticles. *Nat. Commun.* **2016**, *7*, 13447.

**44** S. Bhat, A. Baksi, S. K. Mudedla, G. Natarajan, V. Subramanian, T. Pradeep, $Au_{22}Ir_3(PET)_{18}$: An Unusual Alloy Cluster through Intercluster Reaction. *J. Phys. Chem. Lett.* **2017**, *8*, 2787–2793.

**45** E. Khatun, P. Chakraborty, B. R. Jacob, G. Paramasivam, M. Bodiuzzaman, W. A. Dar, T. Pradeep, Intercluster Reactions Resulting in Silver-Rich Trimetallic Nanoclusters. *Chem. Mater.* **2020**, *32*, 611–619.

**46** K. R. Krishnadas, A. Baksi, A. Ghosh, G. Natarajan, A. Som, T. Pradeep, Interparticle Reactions: An Emerging Direction in Nanomaterials Chemistry. *Acc. Chem. Res.* **2017**, *50*, 1988–1996.

**47** M. Suyama, S. Takano, T. Nakamura, T. Tsukuda, Stoichiometric Formation of Open-Shell $[PtAu_{24}(SC_2H_4Ph)_{18}]^-$ via Spontaneous Electron Proportionation between $[PtAu_{24}(SC_2H_4Ph)_{18}]^{2-}$ and $[PtAu_{24}(SC_2H_4Ph)_{18}]^0$. *J. Am. Chem. Soc.* **2019**, *141*, 14048–14051.

**48** Z. Gan, N. Xia, N. Yan, S. Zhuang, J. Dong, Y. Zhao, S. Jiang, Q. Tao, Z. Wu, Compression-Driven Internanocluster Reaction for Synthesis of Unconventional Gold Nanoclusters. *Angew. Chem. Int. Ed.* **2021**, *60*, 12253–12257; *Angew. Chem.* **2021**, *133*, 12361–12365.

**49** X. Zou, X. Kang, M. Zhu, Recent Developments in the Investigation of Driving Forces for Transforming Coinage Metal Nanoclusters. *Chem. Soc. Rev.* **2023**, *52*, 5892–5967.

**50** Y. Yang, C. Chen, G.-Y. Xu, J. Yuan, S.-F. Ye, L. Chen, Q.-L. Lv, G. Luo, J. Yang, M.-B. Li, Z. Wu, An Efficient Nanocluster Catalyst for Sonogashira Reaction. *J. Catal.* **2021**, *401*, 206–213.

51 P. Chakraborty, S. Malola, M. Neumaier, P. Weis, H. Häkkinen, M. M. Kappes, Elucidating the Structures of Intermediate Fragments during Stepwise Dissociation of Monolayer-Protected Silver Clusters. *Angew. Chem. Int. Ed.* **2023**, *62*, e202305836; *Angew. Chem.* **2023**, *135*, e202305836.

52 Z. Luo, V. Nachammai, B. Zhang, N. Yan, D. T. Leong, D.-E. Jiang, J. Xie, Toward Understanding the Growth Mechanism: Tracing All Stable Intermediate Species from Reduction of Au(I)–Thiolate Complexes to Evolution of $Au_{25}$ Nanoclusters. *J. Am. Chem. Soc.* **2014**, *136*, 10577–10580.

53 Q. Yao, Y. Feng, V. Fung, Y. Yu, D.-E. Jiang, J. Yang, J. Xie, Precise Control of Alloying Sites of Bimetallic Nanoclusters Via Surface Motif Exchange Reaction. *Nat. Commun.* **2017**, *8*, 1555.

54 Q. Yao, X. Yuan, V. Fung, Y. Yu, D. T. Leong, D.-E. Jiang, J. Xie, Understanding Seed-mediated Growth of Gold Nanoclusters at Molecular Level. *Nat. Commun.* **2017**, *8*, 927.

55 C. Zeng, C. Liu, Y. Pei, R. Jin, Thiol Ligand-induced Transformation of $Au_{38}(SC_2H_4Ph)_{24}$ to $Au_{36}(SPh\text{-}t\text{-}Bu)_{24}$. *ACS Nano* **2013**, *7*, 6138–6145.

56 M. S. Bootharaju, L. Sinatra, O. M. Bakr, Distinct Metal-exchange Pathways of Doped $Ag_{25}$ Nanoclusters. *Nanoscale* **2016**, *8*, 17333–17339.

57 T. Chen, Q. Yao, R. R. Nasaruddin, J. Xie, Electrospray Ionization Mass Spectrometry: A Powerful Platform for Noble-Metal Nanocluster Analysis. *Angew. Chem. Int. Ed.* **2019**, *58*, 11967–11977; *Angew. Chem.* **2019**, *131*, 12093–12103.

58 N. K. Chaki, Y. Negishi, H. Tsunoyama, Y. Shichibu, T. Tsukuda, Ubiquitous 8 and 29 kDa Gold: Alkanethiolate Cluster Compounds: Mass-Spectrometric Determination of Molecular Formulas and Structural Implications. *J. Am. Chem. Soc.* **2008**, *130*, 8608–8610.

59 H. Qian, R. Jin, Controlling Nanoparticles with Atomic Precision: The Case of $Au_{144}(SCH_2CH_2Ph)_{60}$. *Nano Lett.* **2009**, *9*, 4083–4087.

60 H. Qian, R. Jin, Ambient Synthesis of $Au_{144}(SR)_{60}$ Nanoclusters in Methanol. *Chem. Mater.* **2011**, *23*, 2209–2217.

61 C. A. Fields-Zinna, R. Sardar, C. A. Beasley, R. W. Murray, Electrospray Ionization Mass Spectrometry of Intrinsically Cationized Nanoparticles, $[Au_{144/146}(SC_{11}H_{22}N(CH_2CH_3)_3^+)_x(S(CH_2)_5CH_3)_y]^{x+}$. *J. Am. Chem. Soc.* **2009**, *131*, 16266–16271.

62 D. Bahena, N. Bhattarai, U. Santiago, A. Tlahuice, A. Ponce, S. B. H. Bach, B. Yoon, R. L. Whetten, U. Landman, M. Jose-Yacaman, STEM Electron Diffraction and High-Resolution Images Used in the Determination of the Crystal Structure of the $Au_{144}(SR)_{60}$ Cluster. *J. Phys. Chem. Lett.* **2013**, *4*, 975–981.

63 O. Lopez-Acevedo, J. Akola, R. L. Whetten, H. Grönbeck, H. Häkkinen, Structure and Bonding in the Ubiquitous Icosahedral Metallic Gold Cluster $Au_{144}(SR)_{60}$. *J. Phys. Chem. C.* **2009**, *113*, 5035–5038.

64 J. Koivisto, K. Salorinne, S. Mustalahti, T. Lahtinen, S. Malola, H. Häkkinen, M. Pettersson, Vibrational Perturbations and Ligand–Layer Coupling in a Single Crystal of $Au_{144}(SC_2H_4Ph)_{60}$ Nanocluster. *J. Phys. Chem. Lett.* **2014**, *5*, 387–392.

65 N. Yan, N. Xia, L. Liao, M. Zhu, F. Jin, R. Jin, Z. Wu, Unraveling the long-Pursued $Au_{144}$ structure By X-Ray Crystallography. *Sci. Adv.* **2018**, *4*, eaat7259.

**66** X. Liu, E. Wang, M. Zhou, Y. Wan, Y. Zhang, H. Liu, Y. Zhao, J. Li, Y. Gao, Y. Zhu, Asymmetrically Doping a Platinum Atom into a $Au_{38}$ Nanocluster for Changing the Electron Configuration and Reactivity in Electrocatalysis. *Angew. Chem. Int. Ed.* **2022**, *61*, e202207685; *Angew. Chem.* **2022**, *134*, e202207685.

**67** F. A. Cotton, Strong Homonuclear Metal-Metal Bonds. *Acc. Chem. Res.* **1969**, *2*, 240–247.

**68** U. Schubert, Cluster-Based Inorganic–Organic Hybrid Materials. *Chem. Soc. Rev.* **2011**, *40*, 575–582.

**69** I. Chakraborty, T. Pradeep, Atomically Precise Clusters of Noble Metals: Emerging Link Between Atoms and Nanoparticles. *Chem. Rev.* **2017**, *117*, 8208–8271.

**70** Y. Shichibu, Y. Negishi, T. Watanabe, N. K. Chaki, H. Kawaguchi, T. Tsukuda, Biicosahedral Gold Clusters $[Au_{25}(PPh_3)_{10}(SC_nH_{2n+1})_5Cl_2]^{2+}$ ($n = 2–18$): A Stepping Stone to Cluster-Assembled Materials. *J. Phys. Chem. C.* **2007**, *111*, 7845–7847.

**71** B. K. Teo, K. Keating, Novel Triicosahedral Structure of the Largest Metal Alloy Cluster: $[(Ph_3P)_{12}Au_{13}Ag_{12}C_{16}]^{m+}$. *J. Am. Chem. Soc.* **1984**, *106*, 2224–2226.

**72** B. K. Teo, M. C. Hong, H. Zhang, D. B. Huang, Cluster of Clusters: Structure of the 37-Atom Cluster $[(p\text{-}Tol_3P)_{12}Au_{18}Ag_{19}Br_{11}]^{2\oplus}$ and a Novel Series of Supraclusters Based on Vertex-Sharing Icosahedra. *Angew. Chem. Int. Ed.* **1987**, *26*, 897–900.

**73** B. K. Teo, H. Zhang, X. Shi, Cluster of Clusters: A Modular Approach to Large Metal Clusters. Structural Characterization of a 38-atom Cluster $[(p\text{-}Tol_3P)_{12}Au_{18}Ag_{20}Cl_{14}]$ Based on Vertex-Sharing Triicosahedra. *J. Am. Chem. Soc.* **1990**, *112*, 8552–8562.

**74** S. Jin, X. Zou, L. Xiong, W. Du, S. Wang, Y. Pei, M. Zhu, Bonding of Two 8-Electron Superatom Clusters. *Angew. Chem. Int. Ed.* **2018**, *57*, 16768–16772; *Angew. Chem.* **2018**, *130*, 17010–17014.

**75** L. Luo, Z. Liu, A. Mazumder, R. Jin, Raising Near-Infrared Photoluminescence Quantum Yield of $Au_{42}$ Quantum Rod to 50% in Solutions and 75% in Films. *J. Am. Chem. Soc.* **2024**, *146*, 27993–27997.

**76** Y. Li, Y. Song, X. Zhang, T. Liu, T. Xu, H. Wang, D.-E. Jiang, R. Jin, Atomically Precise $Au_{42}$ Nanorods with Longitudinal Excitons for an Intense Photothermal Effect. *J. Am. Chem. Soc.* **2022**, *144*, 12381–12389.

**77** Y. Li, M. Zhou, R. Jin, Programmable Metal Nanoclusters with Atomic Precision. *Adv. Mater.* **2021**, *33*, e2006591.

**78** X.-M. Luo, Y.-K. Li, X.-Y. Dong, S.-Q. Zang, Platonic and Archimedean Solids in Discrete Metal-Containing Clusters. *Chem. Soc. Rev.* **2023**, *52*, 383–444.

**79** K. Pyo, V. D. Thanthirige, K. Kwak, P. Pandurangan, G. Ramakrishna, D. Lee, Ultrabright Luminescence from Gold Nanoclusters: Rigidifying the Au(I)–Thiolate Shell. *J. Am. Chem. Soc.* **2015**, *137*, 8244–8250.

# 5

# S- and P-based Ligand-protected Metal Nanocluster Catalysis

## 5.1 Thiolate-protected Metal Nanocluster for Catalysis

### 5.1.1 Catalytic Oxidations and Reductions

Since the discovery of the core–shell structure of $Au_{102}(SC_6H_4COOH)_{44}$ and $Au_{25}(SC_2H_4Ph)_{18}$ by single-crystal X-ray crystallography, thiolate-protected metal nanoclusters bearing diversified metal kernels and abundant surface environment have attracted wide interest in catalysis. Inspired by the work from Masatake Haruta and coworkers that ultra-small gold nanoclusters (<2.0 nm) immobilized on titanosilicalite exhibited high catalytic activity toward the oxidation of propene, the catalytic potential of thiolate-protected gold nanoclusters in oxidation reactions have been discovered by different research groups. Oxidation reactions employing molecular oxygen is of major interest as a result of the green and mild reaction conditions. Thiolate-protected gold nanoclusters exhibit high reactivity toward $O_2$. It has been investigated by several group that oil-soluble $[Au_{25}(SC_2H_4Ph)_{18}]^-$ and water-soluble $[Au_{25}(SG)_{18}]^-$ nanoclusters can be easily oxidized by $O_2$ and produce the charge-neutral $[Au_{25}(SC_2H_4Ph)_{18}]^0$ and $[Au_{25}(SG)_{18}]^0$. Zhu and coworkers revealed that the gold kernel, instead of the surface thiolate ligands, first undergoes one electron loss. The oxidized $[Au_{25}(SC_2H_4Ph)_{18}]^0$ and $[Au_{25}(SG)_{18}]^0$ and be reduced to $[Au_{25}(SC_2H_4Ph)_{18}]^-$ and $[Au_{25}(SG)_{18}]^-$ readily by common reductants such as the $NaBH_4$ (Figure 5.1). More importantly, SCXRD analysis demonstrates that the oxidized $[Au_{25}(SC_2H_4Ph)_{18}]^0$ and $[Au_{25}(SG)_{18}]^0$ nanoclusters remain their geometric structures compared to the $[Au_{25}(SC_2H_4Ph)_{18}]^-$ and $[Au_{25}(SG)_{18}]^-$ [1]. The reversible charge-state conversion of $Au_{25}(SR)_{18}$ system indicates that the nanocluster can be utilized in catalytic oxidations and reductions.

The early work by Zhu and coworkers shows that the $Au_{25}(SR)_{18}$ nanocluster is able to catalyze the styrene oxidation in the presence of $O_2$ and selectively produce benzaldehyde as the major product (Figure 5.2) [2]. The oxidation reaction

*Metal Nanocluster Catalysis: From Fundamentals to Applications*, First Edition. Manzhou Zhu and Man-Bo Li. © 2026 Wiley-VCH GmbH. All rights reserved, including rights for text and data mining and training of artificial intelligence technologies or similar technologies. Published 2026 by Wiley-VCH GmbH.

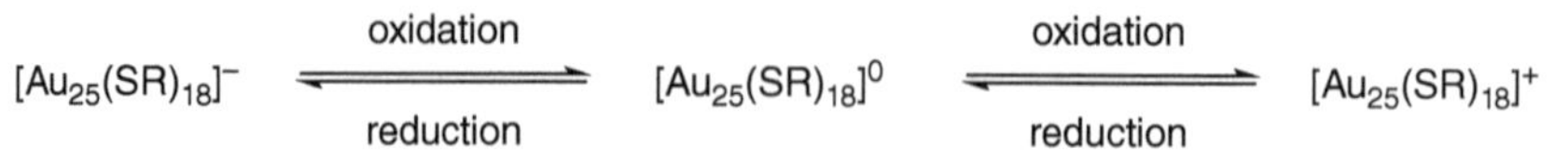


**Figure 5.1** Reversible charge-state conversion of $Au_{25}(SR)_{18}$ nanocluster.

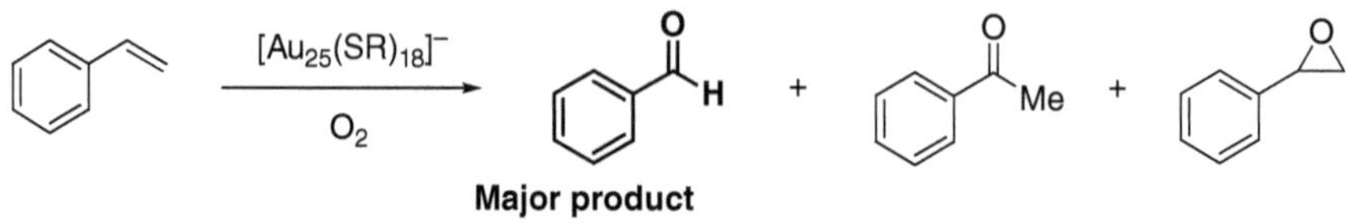


**Figure 5.2** $Au_{25}(SR)_{18}$ nanocluster-catalyzed styrene oxidation reactions.

was carried out at about 80 °C for 12–24 hours with toluene as the solvent. A 35% conversion of styrene was observed even with a part per million level of $Au_{25}(SR)_{18}$ nanocluster catalysts in the reaction solution. Styrene epoxide and acetophenone was obtained as the minor products accompanied with benzaldehyde. The catalytic selectivity can be switched by using the same $Au_{25}(SR)_{18}$ nanocluster catalyst while slightly modifying the reaction conditions. Liu and coworkers investigated hydroxyapatite (HAP) immobilized $Au_{25}(SG)_{18}$ nanocluster-catalyzed styrene oxidation reaction and found that styrene epoxide was obtained as the major product in the presence of anhydrous *tert*-butyl hydroperoxide (TBHP) as the oxidant [3]. The reaction solution (toluene) and reaction temperature (80 °C) are the same to that in the work reported by Zhu and coworkers [2]. Ninety-two percent yield of styrene epoxide was obtained in 12 hours, which is significantly higher than conventional nano-catalysts (ca. 50%–60% yields) with the same Au loading.

Under high pressure, the thiolate-protected gold nanoclusters are capable of catalyzing oxidation of cyclohexane using $O_2$ as the oxidant. Cyclohexanol and cyclohexanone were obtained as the major products with almost equal yields with total selectivity (cyclohexanol + cyclohexanone) up to 99%. Notably, the thiolate-protected gold nanoclusters demonstrate size-dependent catalytic activity in the cyclohexane oxidation (Figure 5.3a). Liu and coworkers prepared four thiolate-protected gold nanoclusters with the same surface ligand but different numbers of gold atoms in the kernel [4]. They are $Au_{10}(SG)_{10}$, $Au_{18}(SG)_{14}$, $Au_{25}(SG)_{18}$ and $Au_{39}(SG)_{24}$. The turnover frequency of the four nanoclusters catalyzed cyclohexane oxidation monotonically increased with an increase in the kernel size, reaching values as high as 18,500 $h^{-1}$ per Au atom for $Au_{39}(SG)_{24}$, and thereafter decreased with a further increase when the number of gold kernel reach to approximately 85. These experimental results demonstrate that although the surface activation sites are very important for the catalysis, the kernel and the electronic effect also play crucial roles in regulating the catalytic activity of thiolate-protected gold nanoclusters. The same research group investigated the thiolate-protected gold nanoclusters catalyzed alcohol aerobic oxidation reactions and found the metal kernel effect as well (Figure 5.3b). They prepared $Au_{25}(SC_{12}H_{25})_{18}$ nanocluster and central palladium-doped nanocluster $Pd_1Au_{24}(SC_{12}H_{25})_{18}$ [5]. Both nanoclusters have the same thiolate surface and the similar geometric structure. However, they found

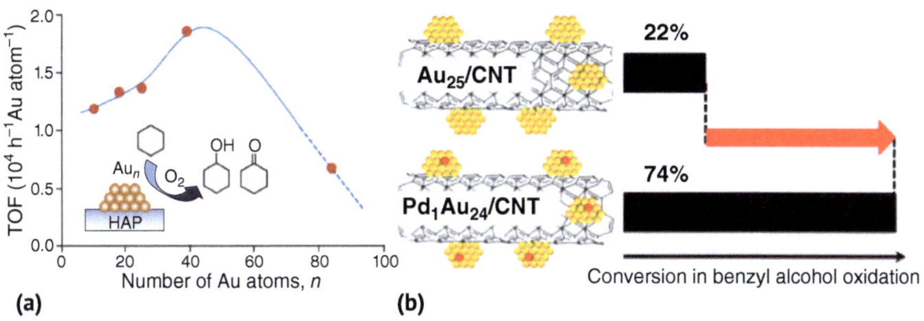


**Figure 5.3** Metal kernel impacts catalytic activity of thiolate-protected gold nanoclusters in cyclohexane and alcohol oxidation reactions.

that carbon nanotube (CNT)-immobilized $Au_{25}(SC_{12}H_{25})_{18}$ and $Pd_1Au_{24}(SC_{12}H_{25})_{18}$ demonstrated completely different activity toward oxidation of benzyl alcohol. $Pd_1Au_{24}(SC_{12}H_{25})_{18}$ catalyzed the oxidation reaction and showed much higher catalytic yields than $Au_{25}(SC_{12}H_{25})_{18}$ (74% vs. 22%). Benzaldehyde, benzoic acid, and benzoate were obtained as the oxidized products. Because the palladium is doped in the center of the nanocluster, it would not affect the surface of the nanocluster so much. Thus, the catalytic difference of $Au_{25}(SC_{12}H_{25})_{18}$ and $Pd_1Au_{24}(SC_{12}H_{25})_{18}$ should be ascribed to the modulation of the electronic structure of nanoclusters by intercluster electron transfer from the palladium atom to the gold atoms. In other word, the metal kernel plays roles in affecting the catalytic activity of thiolate-protected gold nanoclusters by electronic effect.

Apart from the oxidation reactions described above, the thiolate-protected gold nanoclusters have shown promising catalytic activity in many other oxidation reactions [6–8], such as the carbon monoxide oxidation to carbon dioxide, the selective oxidation of sulfide to sulfoxide, the oxidative coupling of amines to imines. All of these interesting organic transformations are based on the fundamental redox ability of the thiolate-protected gold nanoclusters, which retain their atomically precise structures during catalysis. Of course, this feature also endows the nanoclusters with catalytic capacity in reduction reactions.

A typical work from Zhu and coworkers indicates that the core–shell structure of $Au_{25}(SR)_{18}$ nanocluster is particularly beneficial for control the catalytic selectivity of hydrogenation reactions [9]. $\alpha,\beta$-Unsaturated ketones and aldehydes were used as substrates and $Au_{25}(SC_{12}H_{25})_{18}$ and $Au_{25}(SC_2H_4Ph)_{18}$ were used as the nanocluster catalysts. It was found that the C=O bond was selectively hydrogenated under a $H_2$ flow (under atmospheric pressure) and the C=C bond was untouched. The reaction conditions were quite mild and the catalytic reduction was efficient. Only 1.0 mg of $Au_{25}(SR)_{18}$ nanocluster triggered the hydrogenation of 0.1 mmol of benzalacetone at 0 °C for three hours, giving the unsaturated alcohol as the product with 22% conversion and 100% selectivity. They investigated the hydrogenation of $\alpha,\beta$-unsaturated ketones and aldehydes with different substituents and found that the catalytic selectivity are uniformly high for all the substrates. Control experiments combined the DFT calculations revealed the catalytic mechanism toward the high

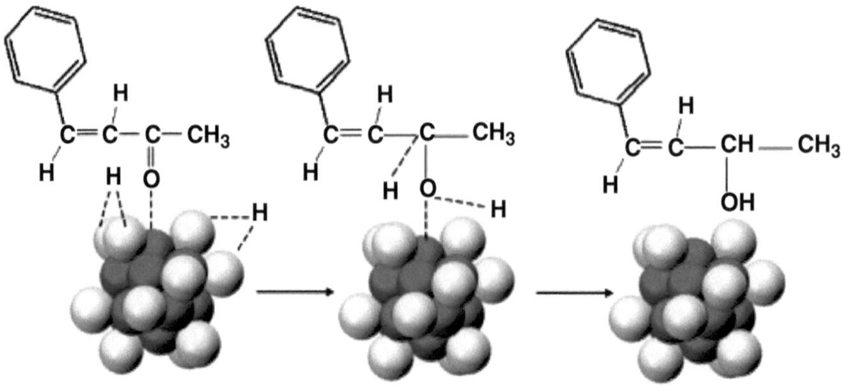


**Figure 5.4** $Au_{25}(SR)_{18}$ nanocluster-catalyzed selective hydrogenation of α, β-unsaturated ketones and aldehydes. *Source:* [9]/with permission of John Wiley & Sons.

selectivity (Figure 5.4). The $Au_{13}$ kernel of $Au_{25}(SR)_{18}$ nanocluster is electron-rich, which favors adsorption of electron-deficient C=O instead of C=C. Meanwhile, the electron-poor $Au_{12}(SR)_{18}$ surface of $Au_{25}(SR)_{18}$ nanocluster provides suitable sites for the activation of $H_2$. The synergistic activation of the C=O bond and $H_2$ by electron-rich kernel and electron-poor surface endows the catalytic activity and high selectivity of $Au_{25}(SR)_{18}$ nanocluster in the hydrogenation of α, β-unsaturated ketones and aldehydes.

Inspired by the highly selective hydrogenation of C=O bonds catalyzed by $Au_{25}(SR)_{18}$, the same research group further developed $Au_{25}(SR)_{18}$-catalyzed semi-hydrogenation of terminal alkynes. Almost 100% conversion of a range of terminal alkynes with >99% selectivity for alkene products was achieved using ethanol/$H_2O$ as the mixed solvent and pyridine as the base at 100 °C [10]. Interestingly, only terminal alkynes demonstrate reactivity. Internal alkynes cannot be reduced by $Au_{25}(SR)_{18}$ nanocluster catalysis. This unique selectivity should be attributed to the activation effect of the open triangular $Au_3$ facet on the surface of $Au_{25}(SR)_{18}$. The open triangular $Au_3$ facet activates the terminal alkyne via a R–C≡C–[$Au_{25}(SR)_{18}$] intermediate. This kind of terminal activation mode inhibits the catalytic transformation of internal alkynes under the same reaction conditions. The thiolate-protected gold nanocluster-catalyzed hydrogenation in the presence of $H_2$ is further applied for selective reduction of aldehydes. Li and coworkers synthesized a thermally robust $Au_{99}(SPh)_{42}$ nanocluster, which exhibited high activity toward the hydrogenation of 4-nitrobenzaldehyde [11]. Up to 93.4% conversion was observed. More importantly, this reaction gave 4-nitrobenzyl alcohol as the exclusive product, and the other byproducts such as 4-aminobenzaldehyde and 4-aminobenzyl alcohol were not observed. The robust $Au_{99}(SPh)_{42}$ nanocluster can be recovered and reused without loss of catalytic activity or selectivity.

The last example we are discussing in this part is distinct from the above reports. In this case, $Au_{25}(SR)_{18}$ nanocluster was utilized as a single-electron transfer catalyst, triggering a unique intramolecular cascade reaction of 2-nitrobenzonitrile [12].

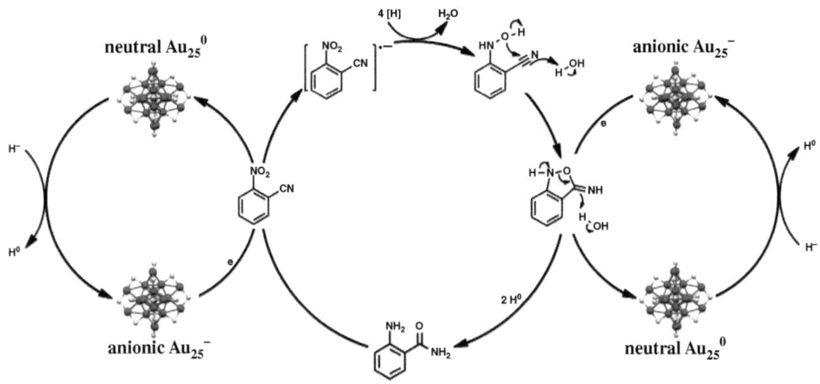


**Figure 5.5** $Au_{25}(SR)_{18}$ nanocluster-catalyzed reduction of 2-nitrobenzonitrile via single-electron transfer process. *Source:* [12]/with permission of Springer Nature.

The reaction was conducted in THF at room temperature, using 1 wt% of $Au_{25}(SR)_{18}$ nanocluster as the catalyst and $NaBH_4$ as the reductant. The 2-nitrobenzonitrile was transformed to 2-aminobenzamide with 87% conversion and 96% selectivity. During the catalysis, the single-electron redox process was observed by time-dependent UV–vis spectra. In addition, nitrogen radical was successfully captured by EPR, suggesting the single-electron transfer between $Au_{25}(SR)_{18}$ nanocluster and the substrate 2-nitrobenzonitrile. The proposed catalytic cycle well explains the formation process of the product 2-aminobenzamide (Figure 5.5). Specifically, $[Au_{25}(SR)_{18}]^-$ interacts with 2-nitrobenzonitrile, undergoing a single-electron transfer process and affording charge-neutral $[Au_{25}(SR)_{18}]^0$ and 2-nitrobenzonitrile radical anion. 2-(Hydroxyamino) benzonitrile is readily generated as an intermediate under the reduction conditions. This species undergoes an intramolecular cyclization and produces a five-membered heterocyclic intermediate, which is eventually transformed to the final product 2-aminobenzamide via another single-electron transfer process with $[Au_{25}(SR)_{18}]^-$. The oxidized charge-neutral $[Au_{25}(SR)_{18}]^0$ can be reduced by $NaBH_4$ to afford the initial negatively charged $[Au_{25}(SR)_{18}]^-$ catalyst, which is involved in the next catalytic cycle. This work suggests another catalytic pathway of the thiolate-protected metal nanocluster, which can be used as an electron-transfer catalyst to trigger unique organic transformation via radical generation. This catalytic strategy is distinct compared to the common oxidation or hydrogenation reactions catalyzed by thiolate-protected metal nanoclusters.

### 5.1.2 Catalytic Cross-coupling Reactions

Cross-coupling reactions enable direct combination of two molecules, constituting an efficient way to multifunctional compounds. We discussed this kind of important reaction in Section 2.4 "In situ generated metal nanocluster catalyzed cross-coupling reactions." The in situ generated metal nanoclusters exhibit high activity (TON up to $10^6$) toward cross-coupling reactions, while their sensitivity to the surrounding environment limits the further structural determination and catalytic mechanism studies. Employing organic ligands to protect the metal kernel provides the opportunity for the above investigation at atomic level [13]. The thiolate-protected $Au_{25}(SR)_{18}$ nanocluster is still the earlier one involved in catalytic cross-coupling reactions investigations. It was found in the year of 2012 that $[Au_{25}(SC_2H_4Ph)_{18}]^-$ was able to catalyze Ullmann coupling reaction without the addition of copper [14]. 1 mg of $[Au_{25}(SC_2H_4Ph)_{18}]^-$ catalyzed the homocoupling of 0.2 mmol of iodobenzene at 130 °C for two days, affording the desired product biphenyl in 99.8% yield. Detailed investigations revealed that the reaction only underwent smoothly by employing DMF as the solvent. The other solvents such as toluene gave poor yields. This observation should be ascribed to the fact that the oxidized $Au_{25}(SC_2H_4Ph)_{18}$ after catalysis can be reduced by DMF to regenerate the $[Au_{25}(SC_2H_4Ph)_{18}]^-$ catalyst. Mechanism studies combined with DFT calculations suggest that the Au...I interaction plays important roles in the activation of the C—I bond and the electron-transfer process between substrates and $[Au_{25}(SC_2H_4Ph)_{18}]^-$ catalyst would be the key step instead of the conventional organometallic elementary reactions. Specifically, the iodobenzene is first adsorbed on the surface of $[Au_{25}(SC_2H_4Ph)_{18}]^-$ via Au...I interactions to form the $[Au_{25}](PhI)_2$ intermediate. The C—I bond is then activated by the positively charged exterior $Au_{12}$ shell of the nanocluster. Electron transfer from the nanocluster to iodobenzene gives the $[Au_{25}Ph_2]I_2$ intermediate, which undergoes reductive elimination to afford the final cross-coupling product. The oxidized $Au_{25}(SC_2H_4Ph)_{18}$ nanocluster can be reduced by DMF to regenerate the catalyst.

Following work by several research groups revealed that the surface ligand and the metal kernel both affected the catalytic activity of the thiolate-protected metal nanoclusters. As shown in Figure 5.6, by employing the Ullmann hetero-coupling reaction of 4-methyl-iodobenzene and 4-nitroiodobenzene as the model reaction, Li and coworkers investigated the ligand effect on catalytic activity of $Au_{25}(SR)_{18}$ nanocluster [15]. Four $Au_{25}(SR)_{18}$ nanoclusters with different surface thiolate ligands (R = Nap, Ph, $CH_2CH_2Ph$, and $n$-$C_6H_{13}$) was prepared and their catalytic activity in Ullmann reaction between 4-methyl-iodobenzene and 4-nitroiodobenzene was compared. Results showed that $Au_{25}(SR)_{18}$ nanocluster protected by aromatic thiolate ligands gave the better conversion and selectivity than the ones protected by the carbon-chain thiolate ligands. It was believed that the aromatic thiolate ligands on the nanocluster surface significantly decreased the activation energy of the C—I

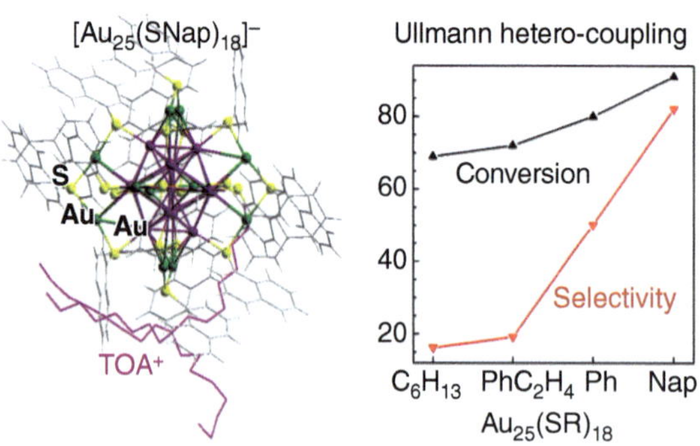


**Figure 5.6** Thiolate-protected metal nanoclusters-catalyzed Ullmann reaction. *Source:* [15]/with permission of American Chemical Society.

bond from −4.3 to −7.2 kcal/mol compared to the carbon-chain thiolate ligands on the surface. In addition to the surface ligand effect, the doping effect of the $Au_xM_{(25-x)}(SR)_{18}$ alloyed nanoclusters in Ullmann coupling reactions was explored as well [16]. It was found that only $Au_xCu_{(25-x)}(SR)_{18}$ demonstrated highly improved efficiency for catalyzing the Ullmann reaction of 4-iodoanisole among the investigated $Au_xM_{(25-x)}(SR)_{18}$ alloyed nanoclusters. Other alloyed nanoclusters including $Au_xAg_{(25-x)}(SR)_{18}$ and $Au_xPt_{(25-x)}(SR)_{18}$ showed similar catalytic performance to the undoped $Au_{25}(SR)_{18}$ nanocluster. Doping of Cu atoms in the thiolate-protected gold nanocluster would enhance its adsorption capacity for the 4-iodoanisole by stronger Cu...I interaction. Moreover, the Cu atoms' regulation on the electronic structure of the nanocluster would improve electron-transfer process between the activated substrates and the nanocluster catalyst, thus being favorable to the catalyzed Ullmann reactions. Phosphine-protected metal nanoclusters such as the $Au_{11}(PPh_3)_7I_3$ are also involved in catalyzing Ullmann reaction, which is beyond the topic discussed herein and will be described in the next section.

Using the aryl iodine (or other aryl halides) as the electrophile, by changing the nucleophiles, there are two cross-coupling reactions that are generally discussed by metal nanocluster catalysis. Owing to the excellent tolerance to functional groups and reaction conditions, Suzuki–Miyaura reaction (or Suzuki reaction) is the most widely used cross-coupling reaction in pharmaceutical industry. Abroshan and coworkers found that $[Au_{25}(SC_2H_4Ph)_{18}]^-$ catalyzed the Suzuki reaction of 4-iodoanisole and phenylboronic acid efficiently with ionic liquids as the solvent [17]. The ionic liquid, that is, 1-butyl-3-methylimidazolium cations help to remove the fragments of the thiolate-protected gold nanocluster by bonding to the surface thiolate ligands on the nanoclusters and thus resulting in exposure of the Au activation sites to catalyze the cross-coupling reaction. Sonogashira reaction with the alkyne as the nucleophile constructs C(sp)-C($sp^2$) bonds effectively. A palladium catalyst and a copper co-catalyst are generally required. The palladium catalyzes the transformation of electrophile via oxidative addition while the copper activates the nucleophile by forming a copper acetylide intermediate. Organic ligand protected metal nanoclusters possess a multinuclear metal kernel, providing an opportunity for designing the catalyst with composite activation sites on the surface. Yang and coworkers designed and successfully synthesized such nanocluster catalyst [18]. Its molecular formula was determined to be $Au_{13}Cu_2(PPh_3)_6(SC_2H_4Ph)_6$ and the atomically precise structure is shown in Figure 5.7. This nanocluster possesses an icosahedral $Au_{13}$ kernel that is similar to the $Au_{13}$ in $Au_{25}(SR)_{18}$ nanocluster. Two copper atoms locate at the contraposition of the $Au_{13}$ kernel by connecting to two triangular $Au_3(SC_2H_4Ph)_3$ motifs. DFT calculations revealed that the energy barriers for the oxidative addition of aryl bromide on the positions of $Au_1$, $Au_2$, $Au_3$, and Cu were 22.8, 35.0, 34.5, and 25.2 kcal/mol, demonstrating that the six equivalent gold atoms connected with the thiolate ligands are suitable for the activation of aryl bromides. Additionally, the copper sites would promote the generation of copper acetylide intermediates. The subsequential transmetalation between the [Ar–Au–Br] and [Cu–alkyne] intermediates on the surface of the $Au_{13}Cu_2(PPh_3)_6(SC_2H_4Ph)_6$ would afford the final product. Experimental results verified the envision and

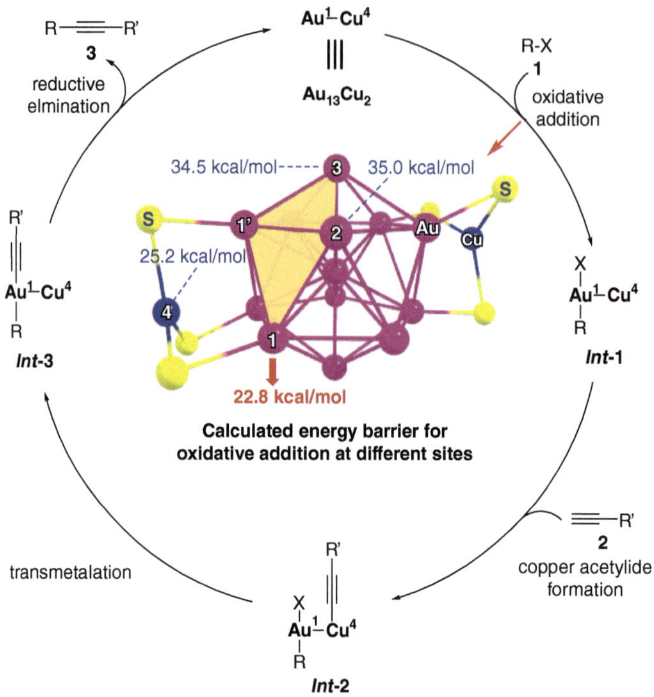


**Figure 5.7** $Au_{13}Cu_2(PPh_3)_6(SC_2H_4Ph)_6$ nanocluster-catalyzed Sonogashira reaction.

$Au_{13}Cu_2(PPh_3)_6(SC_2H_4Ph)_6$ nanocluster catalyzed the Sonogashira reaction in high yield even under atmosphere reaction conditions, which is not accessible for the Pd/Cu-catalyzed Sonogashira reaction because the oxygen in the air would lead to the homocoupling of alkynes and result in poor selectivity. This work manifests the advantages of metal nanocluster catalysis that several activation sites collaborate with each other efficiently on the surface of the nanocluster, triggering superior catalytic activity that otherwise cannot be achieved by single metal catalysts. It should be noted that the well-known thiolate-protected metal nanocluster $Au_{25}(SR)_{18}$ also exhibits activity toward the Sonogashira reaction after immobilization on $CeO_2$. Nevertheless, the catalytic yield is much lower than the copper alloyed nanocluster mentioned above. There are elegant reports by Sagadevan and coworkers that copper nanoclusters catalyze the Sonogashira reaction smoothly under light irradiation [19]. This unique transformation happens via photocatalysis pathway and will be described in detail in the next chapter "Metal nanocluster photo-catalysis."

The above reports show thiolate ligand-protected metal nanoclusters catalyzed cross-coupling reactions that are conducted under neutral (e.g., Sonogashira reaction) or reductive conditions (e.g., Ullmann reaction). In these cases, an electrophile is essential to couple with the nucleophile under neutral conditions or another electrophile in the presence of reducing reagents. The leaving group (e.g., I, Br, OAc) on the electrophile thus limits the atomic economy of the whole reaction because this part is not involved in the final product. Another important category of cross-coupling reactions is the oxidative cross-coupling, in which two nucleophiles couple with each other to give the coupling product. This different pathway avoids the usage of electrophiles with leaving groups, thus improving atomic economy of the cross-coupling reaction. Instead, an oxidant is needed in this case for regeneration of the reduced catalyst. The comparison of cross-coupling reaction, reductive cross-coupling reaction and oxidative cross-coupling reaction are shown in Figure 5.8 for readers' better understanding on the role of reductants and oxidants during the processes.

Metal nanocluster-catalyzed oxidative cross-coupling reactions feature low catalyst loading and high catalytic efficiency, because more activation sites are located on the surface of metal nanoclusters compared to the mono-metal complex catalyst. Moreover, green oxidants such as the oxygen molecule are generally used in metal nanocluster-catalyzed oxidative cross-coupling reactions. In sharp contrast, sophisticated and strong oxidants are essential in many metal complexes catalyzed oxidative cross-coupling. The affinity between metal nanoclusters' abundant surface and the oxygen molecule might be an important reason. As representatives, Yin and Sun respectively reported $Cu_4$ and $AuCu_{24}$ nanoclusters catalyzed oxidative cross-coupling reactions of alkynes [20, 21]. Results showed that the reaction underwent smoothly under mild conditions and the yields of the 1,3-diyne products were obtained in >90% yields. Moreover, Sun and coworkers [21] found that $AuCu_{24}$ nanocluster was able to catalyze the hetero-coupling of different substituent-functionalized alkynes under very mild reaction conditions (30 °C, $O_2$), producing the internal alkynes with high selectivity.

The last cross-coupling reaction we discuss herein is the tri-component coupling reaction of alkyne, amine, and aldehyde ($A^3$ coupling). This reaction was coined by Chao-Jun Li, who investigated this reaction for over 20 years. The activation of alkyne by metal catalysts is generally considered to be the rate-determining step (RDS). The as-obtained metal acetylide attacks imine, formed by condensation of amine and aldehyde, to give the final propargyl amine as the product. The activation of alkyne and condensation of amine and aldehyde both affect the efficiency of the reaction. Generally, this reaction requires high temperature and catalyst

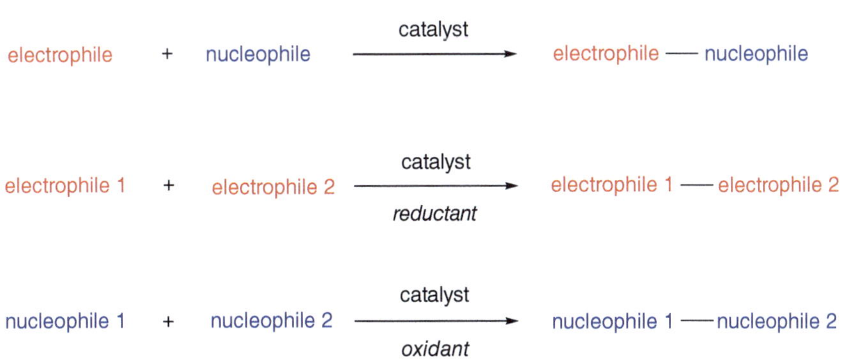


**Figure 5.8** Catalytic cycles of cross-coupling reaction, reductive cross-coupling reaction, and oxidative cross-coupling reaction.

loading. Owing to the high efficiency of gold nanoclusters in activating alkynes and facilitating nucleophilic addition process, the thiolate ligand-protected gold nanoclusters show intriguing performance in catalyzing $A^3$ coupling reactions. In 2016, Chen and Adachi reported $Au_{25}$-catalyzed $A^3$ coupling reaction, respectively [22, 23]. Interestingly, Yasushi Obora et al. used the core–shell structured $[Au_{25}(SC_2H_4Ph)_{18}]^-$ as the catalyst while Chen and coworkers [22] applied the rod-like $Au_{25}(PPh_3)_{10}(C{\equiv}CPh)_5X_2(X = Cl/Br)$ as the catalyst. Both nanoclusters catalyzed the tricomponent coupling reaction efficiently, giving propargyl amines in high yields with satisfactory selectivity (Figure 5.9a). In the next year, 2017, Li and coworkers designed a one-pot-two-step nanocluster synthetic method and successfully obtained a $Au_{13}Cd_2$ nanocluster catalyst that is protected by mixed ligands of thiolates and phosphines [24]. This nanocluster is generated from the reaction between $[Au_{25}(SC_2H_4Ph)_{18}]^-$ and $Cd(PPh_3)NO_3$. As we mentioned in Section 4.1, reverse ligand-exchange of thiolate ligand-protected $[Au_{25}(SC_2H_4Ph)_{18}]^-$ nanocluster with $PPh_3$ results in icosahedral $Au_{13}$ nanocluster as the product. Meanwhile, the Cd atom is easily doped in thiolate ligand-protected gold nanoclusters. The one-pot-two-step method combines both reverse ligand-exchange and Cd-doping steps, triggering the synthesis of $[Au_{13}Cd_2(PPh_3)_6(SC_2H_4Ph)_6(NO_3)_2]^{2+}$ nanocluster in one-pot with high efficiency. This nanocluster demonstrates high catalytic activity toward $A^3$ coupling reaction at room temperature with DCM as the solvent. Additionally, it can be recovered and recycled without notable loss of catalytic activity and selectivity. The exploration of the surface gold activation sites and location of Cd sites achieved by the one-pot-two-step method are considered to be crucial for its promising catalytic performance.

The metal nanocluster-catalyzed tricomponent coupling reactions can be expanded to other more inert substrates, such as the halomethane and ketone. Yuan and coworkers developed Cu nanoclusters-catalyzed coupling reaction of the alkyne, halomethane, and amine (AHA reaction) [25]. Three Cu nanoclusters,

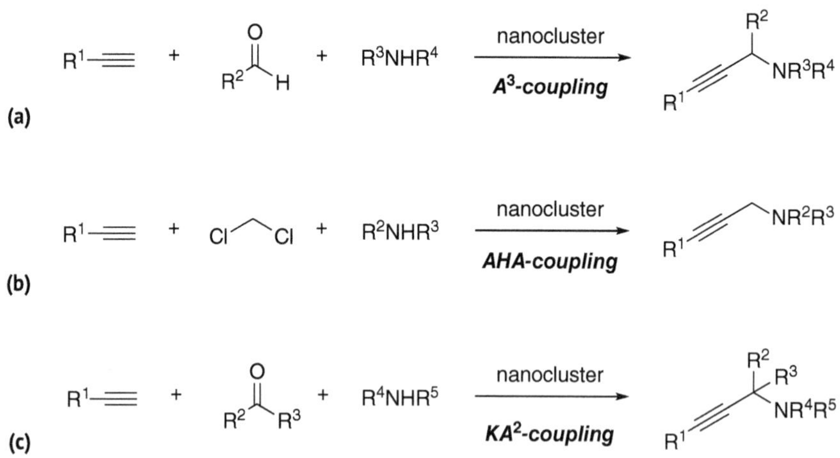


Figure 5.9 Metal nanoclusters-catalyzed $A^3$ coupling (a), AHA coupling (b), and $KA^2$ coupling (c) reactions.

namely $Cu_3I_2H(Dppm)_3$, $Cu_6(SC_7H_4NO)_6$, and $[Cu_{11}(TBBT)_9(PPh_3)_6](SbF_6)_2$ were synthesized, characterized, and immobilized on supports. It was envisioned that the surface copper atoms activated the alkyne, generating the copper acetylide intermediate. Meanwhile, methyleneammonium formed independently and was then adsorbed on the surface of the Cu nanocluster. The subsequential coupling between methyleneammonium and copper acetylide would give the desired product (Figure 5.9b). Experimental results showed that the three Cu nanoclusters did catalyze the AHA coupling reaction efficiently, affording propargyl amines with high TONs (>300). This value is significantly higher than that reported previously. Fan and coworkers developed a PNP pincer ligand-modified thiolate ligand-protected Au–Cd nanocluster $Au_{17}Cd_2(PNP)_2(SR)_{12}$ [26]. This nanocluster features an intriguing coordination mode, that is, the tridentate PNP pincer ligand takes an unconventional di-coordination mode (we will systematically introduce this kind of metal nanocluster in Chapter 6, Section 6.3.2), which results in "open nitrogen sites" on the surface of the nanocluster. These basic sites can be used to activate the slightly acidic alkynes. Meanwhile, the doped Cd atoms on the surface of the nanocluster would promote the condensation of ketone and amine. Based on the proposed activation mode, Fan and coworkers [26] developed highly efficient ketone, amine, alkyne coupling ($KA^2$ coupling) reaction catalyzed by $Au_{17}Cd_2(PNP)_2(SR)_{12}$ nanocluster (Figure 5.9c). As a result of the much lower reactivity of ketone compared to aldehyde, the $KA^2$ coupling reaction requires harsher reaction conditions than the $A^3$ coupling reaction. More than 5 mol% of metal catalyst and higher than 100 °C temperature are generally used for triggering the $KA^2$ coupling reaction. In sharp contrast, $Au_{17}Cd_2(PNP)_2(SR)_{12}$ nanocluster catalyzes the $KA^2$ coupling reaction at room temperature, furnishing the desired product with TON up to 9000. Mechanism studies revealed that the synergy between "open nitrogen sites" and the metallic Cd sites on the surface of the nanocluster contributes to the high efficiency of the nanocluster. The thiolate ligands also play important roles in activating amine/ketone and retaining the total structure of $Au_{17}Cd_2(PNP)_2(SR)_{12}$.

### 5.1.3 Catalytic $C_1$ Molecules Conversion

$C_1$ molecules represent small molecules with only one carbon atom. Typical examples include $CO$, $CO_2$, $CH_4$, $CH_3OH$, and HCOOH. They are primarily derived from the organic wastes, biomass, natural gas, etc. Their efficient conversion into high value-added chemicals and fuels is not only helpful for resolving the serious environmental problem but also beneficial for addressing the issue of the decreased resources of crude oil. Metal catalysts are promising candidates for the activation of $C_1$ molecules. Some well-known $C_1$ molecules conversion processes, such as the Fischer–Tropsch synthesis, water–gas shift, oxidative coupling of methane, and methanol to olefin transformation, are realized by homogeneous or heterogeneous metal catalysts. In spite of the significant progress, the current methods cannot meet the industrial demand. The applied homogeneous metal catalysts are generally sensitive to the harsh reaction conditions in industrials, while the robust heterogeneous metal catalysts generally do not have the atomically precise structures and restrict the efficient structural optimization. Thiolate ligand-protected metal nanoclusters bear both robust structures that can be well defined at atomic level and abundant surface activation sites that can be precisely regulated. Therefore, they have attracted wide research interest for catalytic $C_1$ molecules conversion [27]. In this part, we select CO conversion, $CO_2$ conversion, and $CH_4$ conversion as representatives to manifest the feature of thiolate ligand-protected metal nanocluster in catalytic $C_1$ molecule conversion.

CO conversion includes two important processes, that is, the CO oxidation and syngas Fischer–Tropsch synthesis. We discuss CO oxidation process in this part. The atomically precise structures of metal nanoclusters provide the basis for investigating catalytic mechanism at atomic level. Elegant work from Wu and coworkers reveals that there are three types of gold sites for $[Au_{25}(SC_2H_4Ph)_{18}]^-$ in catalyzing CO oxidation reaction [28], namely $Au^+$, $Au^{\delta+}(0 < \delta < 1)$, and $Au^{\delta-}(0 < \delta < 1)$. These gold activation sites can be identified in FTIR spectra and show signals at 2157, 2117, and 2076 $cm^{-1}$, respectively. The $Au^{\delta+}$ sites are active at relatively low temperature while the $Au^+$ and $Au^{\delta-}$ sites are active at higher temperatures. The low temperature reaction mechanism of CO oxidation has been confirmed as the Mars van Krevelen mechanism. Specifically, CO molecules are activated on the dethiolated gold sites and then the lattice oxygen atom participates in the oxidation. In contrast, the complete removal of the surface ligands on the thiolate ligand-protected gold nanoclusters results in Langmuir–Hinshelwood mechanism as the major reaction pathway, in which CO molecules as well as $O_2$ molecules are activated by the uncovered gold atoms. $H_2O$ has been confirmed to play important roles in thiolate ligand-protected gold nanocluster-catalyzed CO oxidation reaction. $[Au_{25}(SC_{12}H_{25})_{18}]^-$ showed almost no catalytic activity in the absence of $H_2O$, while it catalyzed the reaction in 96.2% yield with 2.3 vol% $H_2O$ in the catalytic system. In situ characterizations such as FTIR and XPS suggest that the thiolate ligands on the surface of $[Au_{25}(SC_{12}H_{25})_{18}]^-$ nanocluster are easily to be shed under $H_2O$ vapor, thus forming activation sites for CO oxidation. This proposal was then verified by DFT calculations.

With the highly efficient activation sites on the surface, thiolate ligand-protected metal nanoclusters have been investigated for CO oxidations. It was found that $[Au_{25}(SC_{12}H_{25})_{18}]$ nanocluster immobilized on $CeO_2$ rods was highly active toward adsorption of CO when the thiolate ligands were partially removed by calcination. The maximum activity of this nanocluster was achieved when it was treated at 200–240 °C, while it showed low activity when treated at 150 °C. Control experiments revealed that 200 °C might be a critical temperature for $[Au_{25}(SC_{12}H_{25})_{18}]^-$ nanocluster exhibiting catalytic activity, because the thiolate ligands are partially removed and the Au sites can be exposed for activating CO and $O_2$. Metal nanoclusters with different kernel and surface ligand demonstrate distinct catalytic activity toward CO oxidation reaction. Chen and coworkers found that $Au_{28}(SR)_{20}$ and $Au_{38}(SR)_{24}$ catalyzed the conversion from CO to $CO_2$ at lower temperature [29], compared to $Au_{25}(SR)_{18}$. Moreover, the ligand type on the surface significantly impacts the catalytic activity of nanoclusters with certain metal kernels.

For example, $Au_{28}(S\text{-}c\text{-}C6H11)_{20}$ (S-c-C6H11 = cyclohexanethiolate) nanocluster exhibited superior catalytic performance in CO oxidation reaction while $Au_{28}(SPh\text{-}tBu)_{20}$ (SPh-*t*Bu = p-*tert*-butylphenylthiophenol) nanocluster with the same gold kernel demonstrated much lower activity. It was rationalized that the staple gold atoms acted as the activation sites and less steric hindrance on the surface of $Au_{28}(S\text{-}c\text{-}C6H11)_{20}$ resulted in its higher activity. After removing the surface ligands, both $Au_{28}$ nanoclusters exhibited the similar catalytic performance. It is also the Rongchao Jin research group that investigated the surface effect of $Au_{38}$ and $Au_{36}$ nanoclusters on CO oxidation activity. The results showed that the catalytic activity of $Au_{38}$ nanoclusters shows the following order: $Au_{38}(SCH_2CH_2Ph)_{24} > Au_{38}(SPh)_{24} > Au_{38}(o\text{-}MBT)_{24}$ (o-MBT = o-methyl thiophenol), and the catalytic activity of $Au_{36}$ nanoclusters shows the following order instead: $Au_{36}(S\text{-}c\text{-}C_5H_9)_{24}$ ($S\text{-}c\text{-}C_5H_9$ = cyclopentanethiol) $> Au_{36}(SPh)_{24} \approx Au_{36}(SPh\text{-}tBu)_{24} > Au_{36}(2\text{-}SNap)_{24}$ (2-SNap = 2-thionaphthol). The corresponding results imply that the steric hindrance existed among the ligand is an important factor in the catalytic performance since it prevents the adsorption of CO onto the gold activation sites.

Apart from the surface ligand effect, the support effect, size effect and doping effect also play important roles in thiolate ligand-protected metal nanoclusters-catalyzed CO oxidation reaction. As shown in Figure 5.10a, $[Au_{25}(SC_2H_4Ph)_{18}]^-$ nanoclusters immobilized on supports of $TiO_2$, $CeO_2$ or $Fe_2O_3$ present different activity toward CO oxidation [30, 31]. Specifically, $[Au_{25}(SC_2H_4Ph)_{18}]^-$ immobilized on $TiO_2$ remains inactive below 200 °C, while $[Au_{25}(SC_2H_4Ph)_{18}]^-$ immobilized on $CeO_2$ or $Fe_2O_3$ show moderate activity with an onset temperature of 50 °C. For the size effect, it has been revealed that thiolate ligand-protected gold nanoclusters with larger size generally showed higher catalytic activity than that with smaller gold kernels [32, 33]. Rongchao Jin et al. classified the as-investigated nanoclusters into three types, that is, the metallic nanocluster with a size larger than 2.3 nm ($Au_{333}$), the transition nanocluster with a size of 2.3–1.7 nm ($Au_{144}$) and the nonmetallic nanocluster with a size smaller than 1.7 nm ($Au_{38}$, $Au_{25}$). The CO conversion over these thiolate ligand-protected gold nanoclusters increased with temperature, and the catalytic activity was determined to be $Au_{144} > Au_{333} > Au_{38} > Au_{25}$ (Figure 5.10b). Heteroatoms' doping also has influence in regulating catalytic activity of metal nanoclusters [34, 35]. An early work from Li research group demonstrates that doping Cu atoms into $Au_{25}(SC_2H_4Ph)_{18}$ nanocluster (i.e., $Cu_xAu_{25\text{-}x}(SC_2H_4Ph)_{18}$)

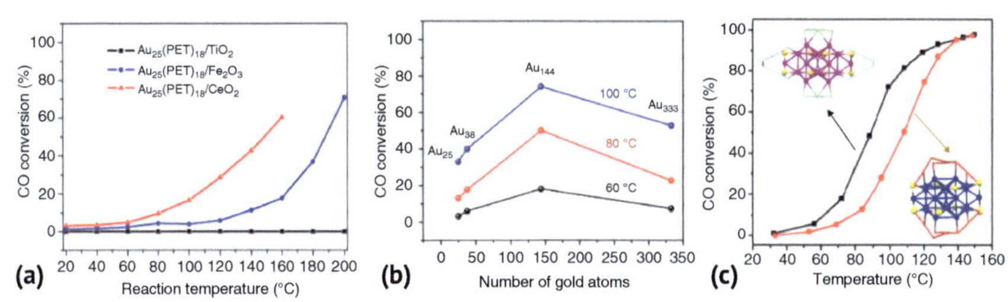


**Figure 5.10** Support effect (a), size effect (b), and doping effect (c) of thiolate ligand-protected metal nanoclusters-catalyzed CO oxidation reaction.

improves the catalytic activity toward CO oxidation while doping Ag atoms into the same nanocluster (i.e. $Ag_xAu_{25-x}(SC_2H_4Ph)_{18}$) results in a decreased catalytic activity instead [34]. An elegant work from Liu and coworkers indicates that the location of the doped foreign metals is very important for the catalytic performance. They synthesized a $Ag_2Au_{50}(SC_2H_4Ph)_{36}$ nanocluster, in which the two silver atoms act as linkers and connect two $Au_{25}(SC_2H_4Ph)_{18}$ moieties [35]. $Ag_2Au_{50}(SC_2H_4Ph)_{36}$ nanocluster is much more efficient than $Au_{25}(SC_2H_4Ph)_{18}$ in catalyzing CO oxidation (Figure 5.10c). In contrast, $Ag_xAu_{25-x}(SC_2H_4Ph)_{18}$ nanocluster with inner silver doping demonstrates unsatisfactory performance. In situ FTIR demonstrates that $Ag_2Au_{50}(SC_2H_4Ph)_{36}$ contains two kind of active sites including Ag sites and Au sites, while $Au_{25}(PET)_{18}$ and the inner shell-doped $Ag_xAu_{25-x}(SC_2H_4Ph)_{18}$ only consists of Au sites.

$CO_2$ conversion represents a green chemistry process because $CO_2$ is a nontoxic and abundant carbon resource. Nevertheless, the efficient conversion of $CO_2$ into liquid fuels and chemicals is quite challenging due to its high thermodynamic stability and kinetic inertness. Thiolate ligand-protected metal nanoclusters have been found to show promising catalytic performance toward $CO_2$ conversion under high temperature, electrolysis or photo irradiation conditions. Liu and coworkers made great progress in this research area [36]. They systematically investigated selective $CO_2$ hydrogenation catalyzed by metal nanoclusters. For example, they studied the impact of the periodic structure of thiolate ligand-protected gold nanocluster in their catalytic performance and found that larger size of nanoclusters is beneficial for acquiring higher catalytic activity under thermal activation conditions. More interestingly, $Au_{36}(TBBT)_{24}$ (TBBT = p-*tert*-butylphenylthiophenol) possessing the largest coordination number of Au and HOMO-LUMO energy gap showed a distinct catalytic property compared to $Au_{28}(TBBT)_{20}$, $Au_{44}(TBBT)_{28}$ and $Au_{52}(TBBT)_{32}$. HCOOH was found to the major product with $Au_{36}(TBBT)_{24}$ as the catalyst, while alcohols were obtained as the major products for $Au_{28}(TBBT)_{20}$, $Au_{44}(TBBT)_{28}$, and $Au_{52}(TBBT)_{32}$. Specifically, $Au_{28}(TBBT)_{20}$ catalyzed the reaction, giving $CH_3OH$ as the product. $Au_{44}(TBBT)_{28}$ and $Au_{52}(TBBT)_{32}$ catalyzed the reaction, affording $CH_3CH_2OH$ (Figure 5.11). In situ FTIR analysis rationalizes the observed results. It is revealed that $CH_3OH$ is generated from the hydrogenation of CO* toward HCO* and $CH_2O$* intermediates on $Au_{28}(TBBT)_{20}$ nanocluster and HCOOH was formed on $Au_{36}(TBBT)_{24}$ nanocluster via HCOO* intermediates. For $Au_{44}(TBBT)_{28}$ and $Au_{52}(TBBT)_{32}$ nanoclusters, OCO*, CO*, HCO*, $CH_2O$*, $CH_3CO$*, $CH_3CHO$*, and $CH_3CH_2O$* species are detected and imply that $CH_3CH_2OH$ is produced via C–C coupling of CO* and $CH_2$* species. DFT calculations illustrate that $Au_{36}(TBBT)_{24}$ has weak adsorption for CO* and HCOOH* intermediates because of its larger energy gap and coordination number. Particularly, $Au_{36}(TBBT)_{24}$ is prone to couple with HCOO* rather than COOH*. Meanwhile, it has weak adsorption for HCOOH*, which is helpful to produce HCOOH. $Au_{28}(TBBT)_{20}$, $Au_{44}(TBBT)_{28}$ and $Au_{52}(TBBT)_{32}$ nanoclusters have similar energy gaps with small coordination numbers, compared to $Au_{36}(TBBT)_{24}$ nanocluster. They are more likely to produce COOH* compared with HCOO* species, and also have relatively stronger adsorption with CO*, which prevents CO desorption and benefits further hydrogenation.

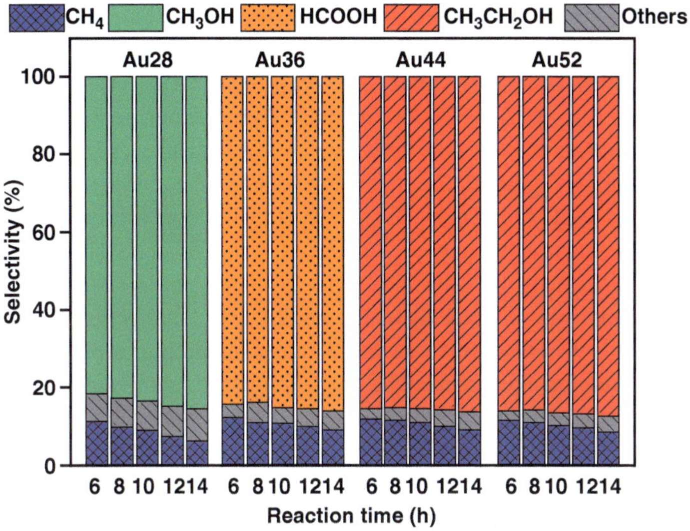


**Figure 5.11** Thiolate ligand-protected metal nanoclusters-catalyzed $CO_2$ hydrogenation reactions.

Additionally, $CH_2$* and CO* species are more inclined to co-adsorbing on the $Au_{44}(TBBT)_{28}$ and $Au_{52}(TBBT)_{32}$ nanoclusters, showing the feasibility of C–C coupling. Therefore, the reaction activity and selectivity of thiolate ligand-protected gold nanoclusters are sensitive to their geometrical and electronic structures.

Metal nanoclusters show interesting activity in catalyzing $CO_2$ carboxylation reactions. Liu and coworkers [36] found that doping foreign metals into silver nanocluster $Ag_{25}(DMBT)_{18}$ (DMBT = 2,4-dimethylbenzenethiol) significantly regulate their catalytic activity toward $CO_2$ carboxylation. The as-obtained $AuAg_{24}(DMBT)_{18}$, $PdAg_{24}(DMBT)_{18}$ and $PtAg_{24}(DMBT)_{18}$ nanoclusters all demonstrated higher structural stability and catalytic activity than undoped $Ag_{25}(DMBT)_{18}$. Among them, $AuAg_{24}(DMBT)_{18}$ exhibited the highest activity, which should be ascribed to the electron density-dominated regulation on catalytic performance. Manzhou Zhu and coworkers investigated the zeolitic imidazolate framework (ZIF) immobilized $Au_{25}(Cys)_{18}$ (Cys = cysteine) nanocluster catalysts in $CO_2$ carboxylation. It was found that the thickness of the ZIF shell plays an important role in affecting the catalytic activity of the composite catalyst. Sandwich-like structure of the composite catalyst with thick ZIF shell gave the best result, and ZIF-8@$Au_{25}(Cys)_{18}$@ZIF-67 showed superior activity. $CO_2$ temperature-programmed desorption (TPD) confirmed that the shell increased the adsorption and activation of $CO_2$ molecules.

$CO_2$ reduction reaction under electrolysis provides a relatively clean strategy for $CO_2$ conversion. Kauffman and coworkers investigated the electrochemical $CO_2$ reduction reaction catalyzed by metal nanoclusters in detail [37]. $[Au_{25}(SC_2H_4Ph)_{18}]^-$ showed excellent catalytic performance with an onset overpotential for CO formation from $CO_2$ reduction as low as 90 mV, which was lower than that of large Au nanoparticles (200–300 mV). Nearly 100% efficiency was reported on $[Au_{25}(SC_2H_4Ph)_{18}]^-$ nanocluster, which was 7–700 times than that of large Au nanoparticles. DFT calculations revealed that the superior activity was attributed to the slight redistribution of charge within the $[Au_{25}(SC_2H_4Ph)_{18}]^-$ nanocluster through the reversible $Au_{25}$–$CO_2$ weak interactions. They further explored that the charge state also impacts the catalytic activity of $Au_{25}(SC_2H_4Ph)_{18}$ nanocluster. $[Au_{25}(SC_2H_4Ph)_{18}]^-$ showed higher TOF, current density and CO Faraday efficiency than $[Au_{25}(SC_2H_4Ph)_{18}]^0$ and $[Au_{25}(SC_2H_4Ph)_{18}]^+$, which should be ascribed to the better affinity between negatively charged $Au_{25}(SC_2H_4Ph)_{18}$ nanocluster and the $CO_2$ molecule.

Doping foreign metals into thiolate ligand-protected metal nanoclusters also affects electrochemical $CO_2$ reduction reactions. Zhuang research groups, respectively, reported the Cd-doping effect in thiolate ligand-protected metal nanocluster-catalyzed electrochemical $CO_2$ reduction reaction [38]. Their results are consistent and suggest that the doped cadmium atoms greatly boost the electrochemical $CO_2$ reduction over gold nanoclusters. Specifically, Zhuang and coworkers [38] compared the catalytic activity of $Au_{47}Cd_2(TBBT)_{31}$ and $Au_{44}(TBBT)_{28}$ and found that $Au_{47}Cd_2(TBBT)_{31}$ showed significantly higher efficiency with a Faraday efficiency of 96% at −0.57 V. Control experiments suggest that the cadmium atoms have the capacity of stabilizing the intermediate COOH* and inhibiting the competitive hydrogen evolution reaction, thus resulting in better catalytic performance. Rongchao Jin et al. compared the catalytic activity of $Au_{19}Cd_2(SC_6H_{11})_{16}$ and the structurally similar $Au_{23}(SC_6H_{11})_{16}$. $Au_{19}Cd_2(SC_6H_{11})_{16}$ nanocluster was found to exhibit an enhanced CO Faraday efficiency of 90%–95% at −0.5 to −0.9 V, which was twice that of $Au_{23}(SC_6H_{11})_{16}$.

It has been previously confirmed that CO and $H_2$ are dominant product in copper nanoparticle-catalyzed $CO_2$ reduction reaction under electrolysis. In contrast, the results coming from Tang and coworkers demonstrate that Cu nanoclusters gave different selectivity [39]. Based on DFT calculations, they predicted that $Cu_{32}H_{20}L_{12}$ ($L = S_2P(O^iPr)_2$ and $S_2P(O^iPr)_2$ = diisopropanoloxy dithiophosphate) nanocluster would catalyze the electrochemical $CO_2$ reduction reaction and give HCOOH and $H_2$ as the preferred products at a lower and higher overpotentials, respectively. The production of HCOOH was proposed through a lattice-hydride mechanism, which cannot be achieved by Cu nanoparticles and Au nanoclusters. More importantly, their theoretical prediction was indeed proved by the experimental results that $Cu_{32}H_{20}L_{12}$ mainly produced HCOOH at a low overpotential while $H_2$ was formed at a higher overpotential. The formation of CO was basically not observed at all overpotentials.

Photocatalytic $CO_2$ conversion under metal nanocluster catalysis is charming area, while the corresponding reports are significantly rare compared to the thermal- and electro-activation modes. A typical example comes from Zhang and coworkers [40]. They developed the highly biocompatible hydro-soluble $Au_{22}(SG)_{18}$ nanocluster as the intercellular photosensitizer based on its high fluorescence properties. The as-obtained composite photocatalyst was able to efficiently catalyze the $CO_2$ conversion and produce acetic acid. The photoelectrons generated from thiolate ligand-protected metal nanocluster $Au_{22}(SG)_{18}$ under irradiation were applied by mediators and finally took part in the Wood–Ljungdahl pathway for $CO_2$ fixation to obtain the acetic acid. Notably, $Au_{22}(SG)_{18}$ nanocluster inhibits reactive species and maintains excellent survivability of bacterium to remain the high activity of the composite catalyst.

Compared to the activation of $CO_2$ and CO, $CH_4$ activation and conversion based on metal nanocluster catalysis are still full of longstanding challenges. The research group of Cai presented preliminary progress on this research field. They prepared two structurally similar nanoclusters, namely $Au_{25}(PPh_3)_{10}(SC_2H_4Ph)_5Br_2$ and $Au_{24}(PPh_3)_{10}(SC_2H_4Ph)_5Br_2$ [41]. Losing one central Au atom of $Au_{25}(PPh_3)_{10}(SC_2H_4Ph)_5Br_2$ leads to its transformation to $Au_{24}(PPh_3)_{10}(SC_2H_4Ph)_5Br_2$. It was found that the hollow $Au_{24}(PPh_3)_{10}(SC_2H_4Ph)_5Br_2$ displayed enhanced catalytic performance in thermal-driven $CH_4$ oxidation compared to the central Au-occupied $Au_{25}(PPh_3)_{10}(SC_2H_4Ph)_5Br_2$ nanocluster. EPR results demonstrate that hydroxy radicals are the active oxygen species and $CH_4$ reacts with $H_2O_2$ over $Au_{25}(PPh_3)_{10}(SC_2H_4Ph)_5Br_2$ or $Au_{24}(PPh_3)_{10}(SC_2H_4Ph)_5Br_2$ via the methoxy intermediate. The surface Au sites capped by the phosphine and thiolate ligands adsorb the hydroxy radicals efficiently, promoting $CH_4$ activation and adsorbing the as-obtained $CH_3$ intermediate. The higher activity of $Au_{24}(PPh_3)_{10}(SC_2H_4Ph)_5Br_2$ should be attributed to the fact that it enables two more pathways for the generation and subsequent transformation of hydroxy radicals. The same group further developed thiolate ligand-protected nickel nanoclusters as the catalysts for $CH_4$ activation under photo irradiation [27]. Based on the DFT calculations, they predicted that a series of Ni nanoclusters, that is, $Ni_4(SC_2H_4Ph)_8$, $Ni_5(SC_2H_4Ph)_{10}$ and $Ni_6(SC_2H_4Ph)_{12}$ would harvest solar energy in the visible range and trigger the light-driven $CH_4$ conversion. Experimental results verified that prediction and showed that the three nanoclusters catalyzed transformation of $CH_4$ and $O_2$ into liquid chemicals, such as the $CH_3OH$ and HCOOH. Photocurrent responses and EPR results revealed that $Ni_6(SC_2H_4Ph)_{12}$ prefer to activate $O_2$, producing superoxide as the active oxygen species which oxidizes the activated $CH_4$ that adsorbs on the surface of the thiolate ligand-protected nickel nanocluster.

### 5.1.4 Surface Modifying and Metal Doping Effect

In spite of the significant progress of thiolate ligand-protected metal nanoclusters in catalysis, the relatively strong metal-thiolate interactions inhibit the activation of substrates to some extent. Strategies such as the surface modification and foreign metal doping have been developed for creating specific activation sites for catalysis. In this part, we briefly discuss the surface modifying and metal doping effect of thiolate ligand-protected metal nanoclusters in catalysis. To break the inert Au—S bonds and create catalytic sites, Kawawaki and coworkers have studied the partial removal of surface ligand in thiolate ligand-protected metal nanoclusters that are loaded on metal oxide by calcination [42]. They prepared $PhCH_2CH_2S$- and *p*-mercaptobenzoic acid (*p*-MBA) co-protected core–shell $Au_{25}$ nanocluster $Au_{25}(PET, p\text{-}MBA)_{18}$ and immobilized the nanocluster on the surface of metal oxide. It was found that the S—C and Au—S bonds started to decompose at approximately 195 °C. The decomposed compounds were adsorbed on the surface of the metal oxide support and desorbed at 225 °C as $CO_2$ and styrene. As a result of the immobilization of metal oxide support, the structure of $Au_{25}$ nanoclusters retained even at 300 °C. When the calcination temperature reached to 500 °C, $Au_{25}$ nanoclusters started to aggregate and the decomposition species that adsorbed on the surface of metal oxide support cannot be removed completely. This study reveals that precisely controlling the calcination process allows partial removal of the thiolate ligand without changing the structures of thiolate ligand-protected metal nanoclusters. This method helps to modify the surface of catalytically inert metal nanoclusters, thereby enabling exposure of more metal sites that are able to participate in catalysis.

Sakamoto and coworkers also investigated the partial removal of thiolate ligands on the surface of the $Au_{25}$ nanocluster by calcination [43]. Being different from the work of Yuichi Negishi and coworkers, they immobilized the $Au_{25}(SC_2H_4Ph)_{18}$ nanocluster on thiolate (e.g., 1-dodecanethiolate)-modified carbon support. They proposed that partial removal of $PhCH_2CH_2S$ ligands on the surface of $Au_{25}(SC_2H_4Ph)_{18}$ would not affect the nanocluster's structure because the 1-dodecanethiolate grasped the $Au_{25}$ kernel and protected it from aggregation. Experimental results verified their proposal and the noncovalent ligand-support interactions played important roles in anchoring the $Au_{25}$ nanoclusters on carbon support. Of course, the calcination conditions are crucial for controlling the coordinated numbers of Au—Au and Au—S bonds, just like the conclusion made by Yuichi Negishi et al. The as-obtained active and robust catalysts with exposed Au activation sites showed superior catalytic ability in the aerobic oxidation reactions of alcohols. Kinetic experiments and DFT calculations suggest that partial removal of the surface thiolate ligands enriches the electron density of the Au nanocluster, thus enhancing oxygen adsorption. The adsorbed oxygen anion would extract hydrogen from the alkoxide and give the final product of benzoic acid. This mechanism studies reveal that the real activation sites of $Au_{25}(SC_2H_4Ph)_{18}$ nanocluster are not the Au sites coordinated to the thiolate ligand but the Au sites after the removal of thiolate ligands with rich electrons.

While partial removal of thiolate ligands constitutes an effective method for exposing the activation sites of the surface of thiolate ligand-protected metal nanoclusters, strategies such as the ligand exchange enable loading surface activation sites under milder conditions (e.g., room temperature). Suzuki and coworkers introduced the functional ligand porphyrin on the surface of thiolate ligand-protected metal nanoclusters by an exchange reaction of $Au_{25}(SC_2H_4Ph)_{18}$ with one or two porphyrin-thiolate (Por) ligands at room temperature (Figure 5.12a) [44]. Porphyrin can be introduced into the –[S–Au–S–Au–S]– staple of $Au_{25}(SC_2H_4Ph)_{18}$ nanocluster with controllable regioselectivity and quantity through ligand exchange reaction by regulating its molecular structure. The main factor controlling the reaction position of the ligand exchange is the steric hindrance of the incoming ligand rather than the electronic effect. The metal kernel of $Au_{25}(SC_2H_4Ph)_{18}$ becomes slightly distorted due to the spatial repulsion between the porphyrin–thiolate and $SC_2H_4Ph$.

In addition to the introduction of functional ligands, the ligand exchange method has been employed for introducing metal complexes on the surface of thiolate ligand-protected metal nanoclusters. Zhao and coworkers successfully replaced a

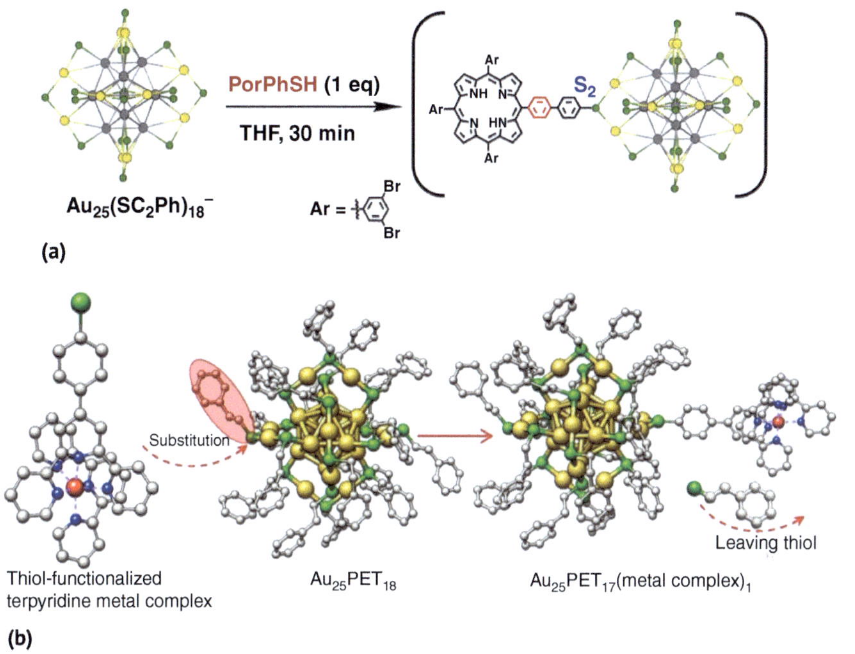


**Figure 5.12** Surface modification of $Au_{25}(SC_2H_4Ph)_{18}$ by ligand exchange reaction. *Source:* [44] and [45]/with permission of American Chemical Society and John Wiley & Sons.

single $SC_2H_4Ph$ ligand of $Au_{25}(SC_2H_4Ph)_{18}$ nanocluster by a thiolated metal complex and obtained $Au_{25}(SC_2H_4Ph)_{17}$(Fe complex)$_1$, $Au_{25}(SC_2H_4Ph)_{17}$(Co complex)$_1$, $Au_{25}(SC_2H_4Ph)_{17}$(Ni complex)$_1$, and $Au_{25}(SC_2H_4Ph)_{17}$(Ru complex)$_1$ nanoclusters (Figure 5.12b) [45]. The replacement is also verified to happen at the position of –[S–Au–S–Au–S]– staple. ESI-MS spectra of these newly formed metal nanoclusters showed molecular ion peaks with +2 valence states, demonstrating the successful replacement of a $SC_2H_4Ph$ ligand by a thiolated metal complex. Although their single-crystal structures were not obtained, emission spectra suggest their retained metal kernels compared the mother-cluster $Au_{25}(SC_2H_4Ph)_{18}$. The only difference is the metal complex grafted onto the surface. This kind of surface modification leads to significantly enhanced photoelectrochemical activity of the nanocluster in $CO_2$ reduction reaction. The enhanced catalytic activity has been proposed to be ascribed to the electron transition from the occupied orbital of $Au_{25}$ nanocluster to the unoccupied orbital under light irradiation. The photo-generated electrons are quickly transferred to a single metal site through the conjugated ligands to participate in the catalytic reaction. The applied extra electric field accelerates this process. The original $Au_{25}(SC_2H_4Ph)_{18}$ nanocluster cannot effectively transfer photo-generated electrons, thereby resulting in aggregation.

Metal doping constitutes another efficient method to construct activation sites on the surface of thiolate ligand-protected metal nanoclusters [46]. Yan Zhu and coworkers systematically investigated the catalytic synergism of atomically precise bimetallic nanoclusters. It was found that the doped foreign metals affected the catalytic performance of metal nanoclusters under various reaction conditions, including photocatalytic conditions, electrocatalytic conditions, heterogeneous conditions and also the homogeneous conditions. A typical example is the alloyed metal nanocluster-catalyzed $CO_2$ formylation reaction. Li and coworkers designed and synthesized three alloyed metal nanoclusters, namely $Cu_{12}Ag_{17}(SR)_{12}(PPh_3)_4$, $Cu_{12}Ag_{16}Au_1(SR)_{12}(PPh_3)_4$, and $Ag_{28}Au_1(SR)_{12}(PPh_3)_4$. They have the similar $MAg_{12}$ (M = Ag or Au) metal kernel but different surface structures [47]. The $Ag_{13}$ kernel of $Cu_{12}Ag_{17}(SR)_{12}(PPh_3)_4$ and $AuAg_{12}$ kernel of $Cu_{12}Ag_{16}Au_1(SR)_{12}(PPh_3)_4$ are both covered by $Cu_3(SR)_3$ and $Ag(PPh_3)$ motifs, while the $AuAg_{12}$ kernel of $Ag_{28}Au_1(SR)_{12}(PPh_3)_4$ is surrounded by $Ag_3(SR)_3$ and $Ag(PPh_3)$ motifs. The three nanoclusters exhibited distinct activity in $CO_2$ formylation reaction with the amine. Specifically, $Cu_{12}Ag_{17}(SR)_{12}(PPh_3)_4$ and $Cu_{12}Ag_{16}Au_1(SR)_{12}(PPh_3)_4$ demonstrated the similar activity and smoothly catalyzed the reaction, giving the desired product (Figure 5.13). In sharp contrast, $Ag_{28}Au_1(SR)_{12}(PPh_3)_4$ showed very low catalytic activity under the same reaction conditions. The in situ FTIR spectra of CO molecules adsorbed onto the three metal nanoclusters indicated that the Cu atoms on the surface $Cu_3(SR)_3$ motifs were potential activation sites rather than the Ag atoms from $Ag_3(SR)_3$ or $Ag(PPh_3)$ motifs, because none of CO molecules were absorbed on Ag sites. It was proposed that the reductant $PhSiH_3$ could adsorb at the Cu activation site of $Cu_3(SR)_3$ to form Cu—Si bonds and the hydrogen bridge between Cu and S atoms. Meanwhile, $CO_2$ would be absorbed at the activated Cu site via the Cu–C interaction. The hydrogen bridging Cu and S atoms would be transferred to the oxygen atom of the $CO_2$ molecule to form the key intermediate, which would be

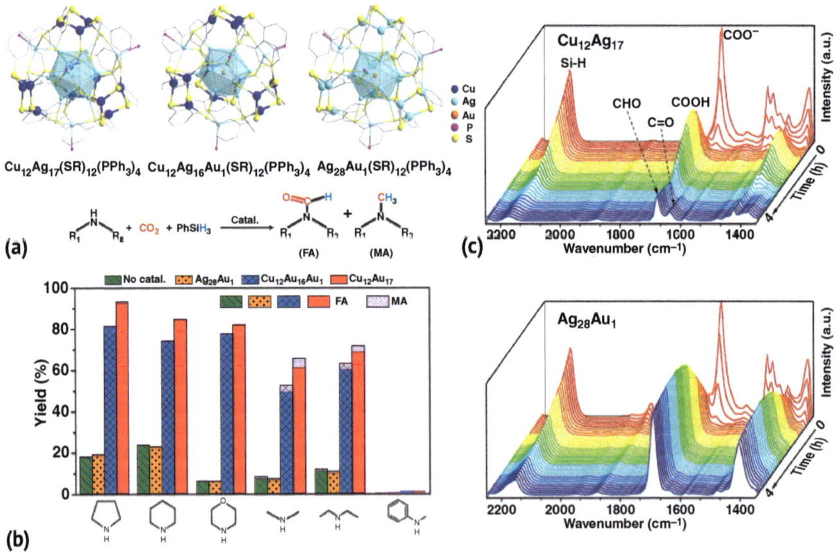


**Figure 5.13** Catalytic activity of structurally similar $Cu_{12}Ag_{17}(SR)_{12}(PPh_3)_4$, $Cu_{12}Ag_{16}Au_1(SR)_{12}(PPh_3)_4$, and $Ag_{28}Au_1(SR)_{12}(PPh_3)_4$ in $CO_2$ formylation reaction. *Source:* [36]/ with permission of American Chemical Society.

transformed to the final product smoothly. Compared to the high activity of the Cu sites for the transformations, the Ag sites on $Ag_3(SR)_3$ are catalytically inert for the $CO_2$ capture, hydrogen transfer, and successive recombination of the intermediate and substrates.

Another example is the thiolate ligand-protected metal nanoclusters catalyzed $CO_2$ reduction reaction under electrolysis. Experimental results from different research groups have verified that cadmium or nickel doped gold-thiolate nanoclusters demonstrated higher activity than the undoped mother gold nanoclusters in the electrochemical $CO_2$ reduction reaction. For example, the Cd-doped $Au_{19}Cd_2(SR)_{16}$ showed a twofold increase in activity compared to the undoped $Au_{23}(SR)_{16}$ nanocluster. Mechanism studies combined the DFT calculations indicate that the synergistic effect of Cd–S–Au motifs in $Au_{19}Cd_2(SR)_{16}$ facilitates the $CO_2$ reduction under electrolysis conditions. The similar effect was also observed in $Au_{24}Cd_1(SR)_{18}$ and $Au_4Ni_2(SR)_8$, compared to the undoped $Au_{25}(SR)_{18}$ and $Au_4(SR)_8$ nanoclusters (Figure 5.14) [48].

Apart from the surface heterometal atoms, the subsurface heteroatoms in bimetallic nanoclusters also affect their catalytic activities. Li and coworkers [49] synthesized a series of Au–Ag alloyed nanoclusters, that is, $Au_{19}Ag_4(SAdm)_{15}$, $Au_{20}Ag_1(SAdm)_{15}$, and $Au_{21}(SAdm)_{15}$ and investigated their catalytic performance in the cycloaddition reaction of $CO_2$ with epoxides toward cycle carbonates [49]. The three nanoclusters possess the similar core but different subsurface. $Au_{19}Ag_4(SAdm)_{15}$ bears a $Au_{10}Ag_3$ core protected by one $Ag(SAdm)_3$ motif and three trimeric $Au_3(SAdm)_4$ staples,

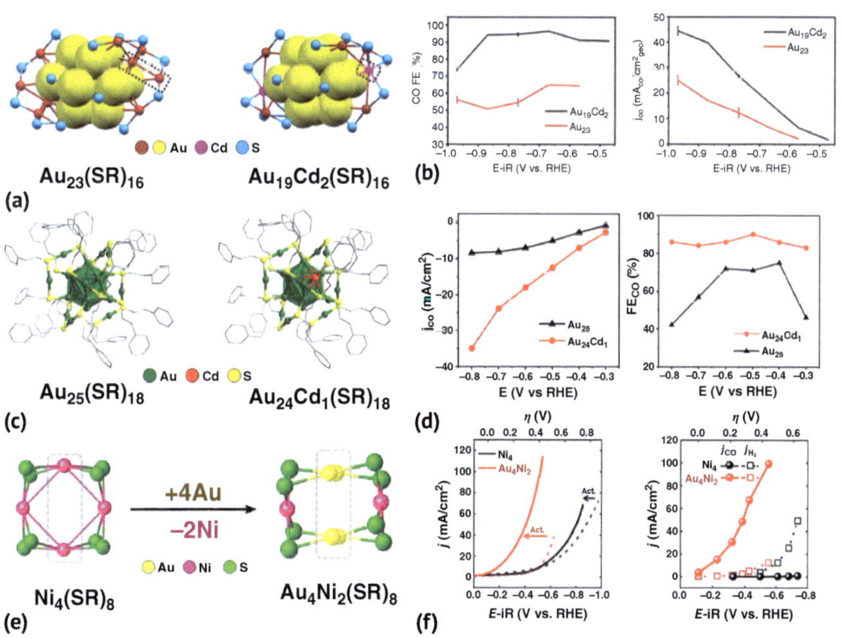


**Figure 5.14** Comparison of catalytic activity of thiolate ligand-protected metal nanoclusters and their alloyed analogs. *Source:* [36]/with permission of American Chemical Society.

where the subsurface $Au_3(SAdm)_4$ is on the shell of the $Au_{10}Ag_3$ core, and the surface $Ag(SAdm)_3$ is on the outer motif. $Au_{20}Ag_1(SAdm)_{15}$ possesses a $Au_{12}Ag_1$ core protected by a $Au_3(SAdm)_4$ and two $Au_2(SAdm)_3$, in which only one silver atom locates on the subsurface. $Au_{21}(SAdm)_{15}$ has a $Au_{12}$ core that is stabilized by an $Au(SAdm)_2$ and two $Au_3(SAdm)_4$ staples. It was found that the three nanocluster catalysts were presented in a descending sequence of $Au_{19}Ag_4(SAdm)_{15} > Au_{21}(SAdm)_{15} > Au_{20}Ag_1(SAdm)_{15}$ in the cycloaddition reaction of $CO_2$ with epoxides (Figure 5.15). In situ FTIR spectra demonstrate that the subsurface Ag atom in $Au_{20}Ag_1(SAdm)_{15}$ nanocluster is catalytically inert toward the reaction due to being partially capped by the surface motif, while the subsurface Ag atoms of $Au_{19}Ag_4(SAdm)_{15}$ are uncovered and catalytically active. DFT calculations also indicate that the activation barrier for the formation of the propylene carbonate complex on the subsurface silver sites of $Au_{19}Ag_4(SAdm)_{15}$ are less than that on the subsurface Ag site of $Au_{20}Ag_1(SAdm)_{15}$.

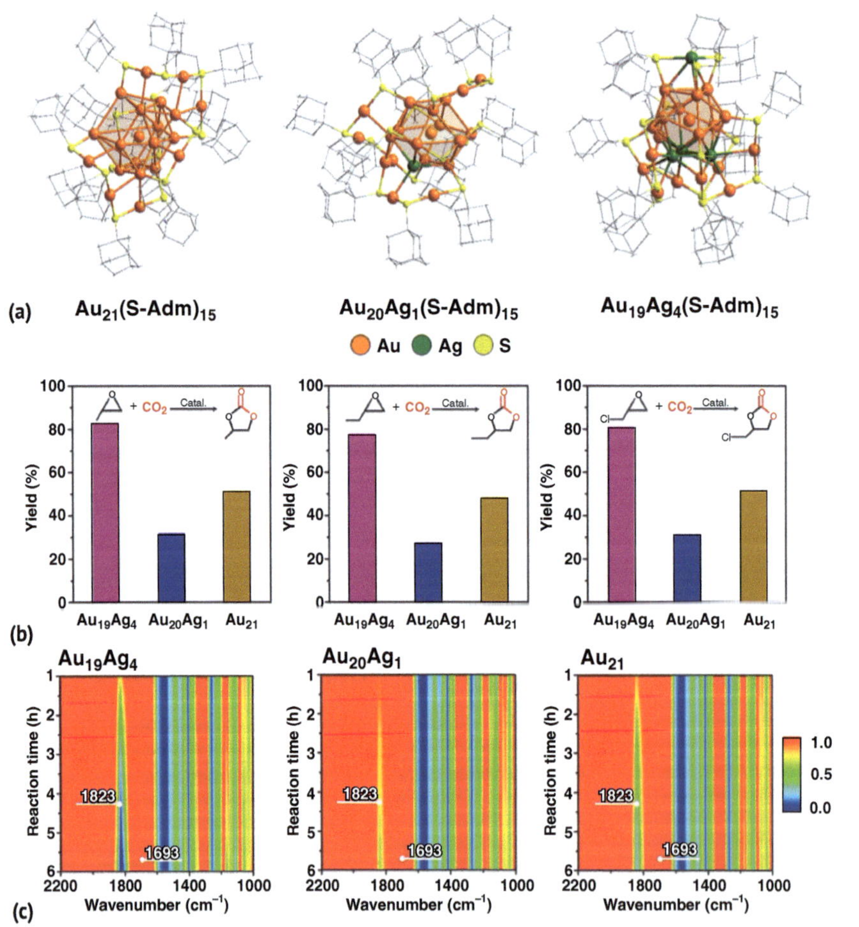


**Figure 5.15** Comparison of catalytic activity of thiolate ligand-protected metal nanoclusters and their alloyed analogs. *Source:* [36]/with permission of American Chemical Society.

Additionally, the overall formation energy for the transition states on the subsurface silver sites of $Au_{19}Ag_4(SAdm)_{15}$ is also less than that on the subsurface Ag site of $Au_{20}Ag_1(SAdm)_{15}$, further rationalizing the higher catalytic efficiency of Ag-doped $Au_{19}Ag_4(SAdm)_{15}$ nanocluster.

## 5.2 Phosphine-protected Metal Nanocluster for Catalysis

### 5.2.1 Structural Features

Compared to the [metal core + metal-thiolate surface] structure of thiolate ligand-protected metal nanoclusters, phosphine ligand-protected metal nanoclusters possess the structural feature of [metal core + phosphine surface] [50]. This difference is ascribed to the anionic and neutral characters of thiolate and phosphine ligands, respectively. Actually, the synthesis and structural characterizations of phosphine ligand-protected metal nanoclusters can trace back to 1960s, which is earlier than the discovery of thiolate ligand-protected metal nanoclusters. Malatesta and coworkers from Italy reported the $PPh_3$-protected $Au_5$ nanocluster, which was named as the triphenylphosphine-gold(0)gold(I) compounds on that time [51]. After that, D. M. P. Mingos and others started to synthesize and characterize various gold nanocluster composed of 5–13 gold atoms and surface phosphine ligands by XCXRD in the 1970s and 1980s. However, the related research did not attract much attention due to the instability of phosphine ligand-protected metal nanoclusters. With the development of catalysis science and technology, scientists start to realize the importance of metal nanoclusters, which are generally the initial active species and contain crucial information for understanding the catalytic mechanism. Additionally, advanced synthesis and separation technologies allow large-scale preparation of structurally stable and pure metal nanoclusters protected by phosphine ligands.

Great progress has been made in this research area and diversified structures have been revealed. As a result of the relatively weaker metal–phosphine interaction compared to the strong metal–thiolate interaction, the metal activation sites are more easily to be exposed on the surface of phosphine ligand-protected metal nanoclusters. Thus, their potential in catalysis attracts great attention of researchers. On the aspect of metal nanocluster compositions and structures, phosphine ligands also facilitate the discovery of unique nanoclusters that is otherwise difficult to be achieved by the thiolate ligands. For example, atomically precise palladium nanoclusters with reduced Pd atoms are highly interesting metal species because of their catalytic potential. As we mentioned in the first chapter, Pd has been widely applied in industrial catalysis. Meanwhile, Pd-catalyzed cross-coupling reaction has been awarded the 2010 Nobel Prize award. Further studies revealed that not only reduced Pd complexes but also Pd nanoparticles were catalytically active. However, the multidispersity nature of Pd nanoparticles restricted the in-depth understanding of nano-Pd-catalyzed reactions. Thus, the synthesis of atomically precise reduced Pd nanoclusters is crucial for revealing the structure-catalytic property correlation

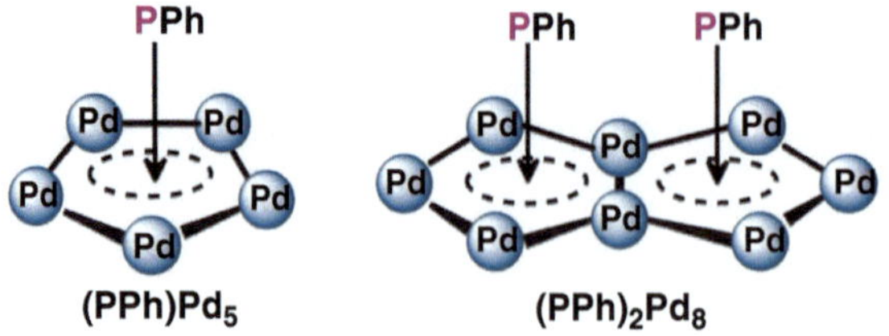


**Figure 5.16** Reduced Pd nanoclusters with phosphine as the crucial ligand. *Source:* [52]/with permission of John Wiley & Sons.

of nano palladium catalysts. However, several decades efforts by different research groups are failed. The as-obtained species are either unreduced $Pd^{II}$-thiolate nanoclusters or partially reduced Pd nanoclusters with polydisperse size and undefined structures. The breakthrough comes from the elegant work of Zhuang, You, and coworkers [52, 53]. By using the phosphine as a relatively weaker coordination reagent, which is mixed with the palladium precursors before the addition of thiolate ligands, they successfully obtained the first atomically precise Pd nanocluster $Pd_8B(PPh_2)_3(PPh_3)_2(SR)_6$ with reduced Pd atoms (Figure 5.16). With this unique synthesis method with phosphine as the crucial ligand, the same research group obtained a series of reduced Pd nanoclusters, such as the $Pd_8(PPh)_2(PPh_3)_2(SAdm)_6$ nanosheet.

From the compositions of both $Pd_8$ nanoclusters, the average valence state of Pd atoms is around +1. The $Pd_8$ kernel of $Pd_8B(PPh_2)_3(PPh_3)_2(SR)_6$ represents a rod-like bipyramid, in which a boron atom is located. The outer shell of the Pd nanocluster consists of three $PPh_2$ at the waist position, two $PPh_3$ at the top position and six thiolate ligands at the corner position. Direct Pd—Pd bonds are observed in the reduced Pd nanocluster, ranging from 2.67 to 2.91 Å. This nanocluster demonstrates interesting activity in catalyzing oxygen oxidation reaction. Compared to various Pd species, including the $Pd_2$ complex, Pd nanoparticles and the commercial Pt/C catalyst, the $Pd_8$ nanocluster shows the highest efficiency and the specific activity of the nanocluster catalyst at 0.9 V for the oxygen oxidation reaction was determined to be as high as $-2.72\ mACm^{-2}$. It was found that the reduced Pd activation sites play important roles in adsorbing and activating $O_2$ and facilitating the transformation of the O* intermediate. Compared to the rod-like $Pd_8B(PPh_2)_3(PPh_3)_2(SR)_6$, another $Pd_8$ nanocluster $Pd_8(PPh)_2(PPh_3)_2(SAdm)_6$ demonstrates a plane shape. Eight Pd atoms are placed on one planc, forming a fused bicycle. Two PPh locate in the two centers of the bicycle, completing the $Pd_8(PPh)_2$ kernel. Two $PPh_3$ ligands and six thiolate ligands surround and stabilize the $Pd_8(PPh)_2$ kernel. In each ring of the bicycle $Pd_8(PPh)_2$ kernel, one P-3s and two P-3p orbitals of the centroid phosphine atom in PPh interacts with the five Pd atoms to yield three six-centered two-electron bonds, obeying the Hückel $4n + 2$ rule for aromaticity with $n = 1$. The aromaticity endows the $Pd_8$ nanosheet with high structural stability, which has been applied for photothermal conversion with 73.3% efficiency.

Another unique cluster structure stabilized by phosphine ligands is the $Au^I$-S nanoclusters. Yan, Yao, and coworkers made great progress in this research area [54–63]. They designed a series of bidentate phosphine ligands and employed them to stabilize structurally diversified gold-sulfido nanoclusters (Figure 5.17). These phosphine-ligand-protected gold-sulfido nanoclusters are generally composed of a gold-sulfido

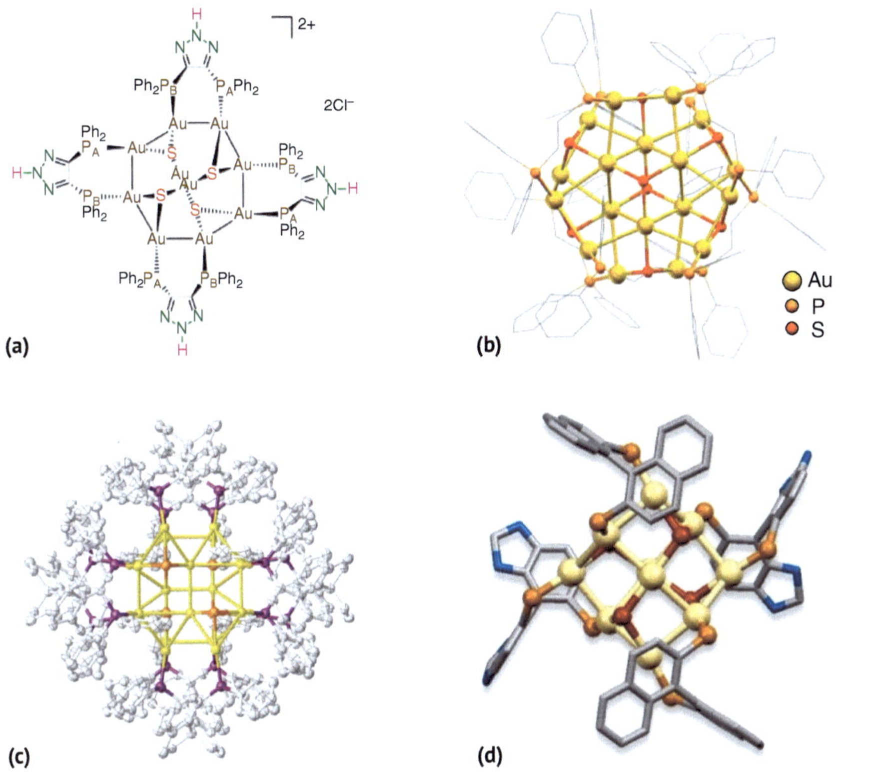


**Figure 5.17** Gold-sulfido nanoclusters stabilized by bidentate phosphine ligands.

kernel with positively charged gold atoms and multi-coordinated sulfur atoms (e.g., $\mu_3$-S), and surface phosphine ligands. They exhibit interesting stimulus-response properties. For example, eight [($\mu_3$-S)$Au_3$] units composed cubic $(C_{36}H_{28}P_2)_{12}Au_{24}S_8$ nanocluster was found to be converted into six [($\mu_3$-S)$Au_3$] units composed triangular prism $(C_{36}H_{28}P_2)_9Au_{18}S_6$ nanocluster under conditions of improved temperature [54]. In situ NMR spectra revealed that the structural transformation process from $(C_{36}H_{28}P_2)_{12}Au_{24}S_8$ to $(C_{36}H_{28}P_2)_9Au_{18}S_6$ occurred at 50 °C. The reversible process from $(C_{36}H_{28}P_2)_9Au_{18}S_6$ to $(C_{36}H_{28}P_2)_{12}Au_{24}S_8$ is also feasible and happens under the conditions of a reduced temperature. Apart from the temperature-induced thermo-reversible cage-to-cage transformation, the gold-sulfido nanoclusters demonstrate photoinduced structural transformation phenomenon. Yao and coworkers designed a bidentate phosphine ligand bis(diphenylphosphino)ethene (dppee) [55]. The *cis* isomer of this ligand would be transformed to the *trans* isomer under photoirradiation. By employing the *cis*- and *trans*- isomers of bis(diphenylphosphino) ethene, respectively, they synthesized two nanoclusters of $Au_{10}$ (*cis*-dppee)$_4$ ($\mu_3$-S)$_4$ and $Au_{18}$ (*trans*-dppee)$_6$ ($\mu_3$-S)$_8$. Interestingly, $Au_{10}$ (*cis*-dppee)$_4$ ($\mu_3$-S)$_4$ can be converted into $Au_{18}$ (*trans*-dppee)$_6$ ($\mu_3$-S)$_8$ nanocluster upon photoirradiation. Detailed experiments revealed that the driving force of the nanocluster transformation was the isomerization of the dppee ligands from the cis to the trans under the light irradiation conditions. The similar photo-induced nanocluster transformation process was observed in another tridentate phosphine ligand-protected gold nanocluster. Yan and coworkers synthesized a *trans*-tridentate phosphine ligand-protected $Au_{12}$ nanocluster, in which the *trans*-tridentate phosphine ligand is (2-(2-(diphenylphosphino)vinyl)-1,3-phenylene)*bis*(diphenylphosphine) [56]. This gold-sulfido nanocluster was found to be converted to another $Au_{15}$ nanocluster under UV light irradiation. Interestingly, being different from the transformation from $Au_{10}$(*cis*-dppee)$_4$($\mu_3$-S)$_4$ to $Au_{18}$(*trans*-dppee)$_6$($\mu_3$-S)$_8$, the *trans*-tridentate phosphine ligand on the $Au_{15}$ nanocluster remained *trans* conformation. Moreover, a reversible transformation from the $Au_{15}$ to $Au_{12}$ was also observed at improved temperature (~85 °C). Mechanism investigation revealed that the photo- and temperature-induced reversible structural transformation between the dodecanuclear and pentadecanuclear gold-sulfido nanoclusters was triggered by photoirradiation or high temperature-induced dynamic dissociation and reorganization processes, which probably involves the *trans*- and *cis*-isomerization of the tridentate phosphine ligands.

The phosphine ligand-protected gold-sulfido nanoclusters enable introduction of chirality with diversified structures. Various chiral phosphine ligands, such as 2,2′-bis-(diphenylphosphino)-1,1′-binaphthyl (BINAP), 7,7′-bis(diphenylphosphino)-1,1′-spirobiindane (SDP) and 5,5′-bis(diphenylphosphino)-4,4′-bi-1,3-benzodioxole (SEGPHOS) were successfully introduced on the surface of the gold-sulfido kernel by Yan, Yao, and coworkers. [57–60] These chiral phosphine ligands not only bring the organic molecular chirality but also regulate the asymmetric arrangement of metal atoms in the gold-sulfido kernel and the phosphine ligand on the surface, thus endowing phosphine ligand-protected gold-sulfido nanoclusters with hierarchically chiral structures. Interesting chiral phenomena such as self-discrimination and self-sorting were observed in these unique nanoclusters. The diversified chiral structures of phosphine ligand-protected gold-sulfido nanoclusters would be beneficial for developing metal nanocluster asymmetric catalysis strategy in the future.

The employment of phosphine ligand in stabilizing metal nanoclusters triggers the discovery of the smallest chiral gold nanocluster and its self-assembly into crystalline nanocubes. Shi and coworkers applied *R*- and *S*-BINAP to synthesize chiral gold nanoclusters. Trinuclear $Au_3$[(*R*) Tol-BINAP]$_3$Cl and $Au_3$[(*S*)-Tol-BINAP]$_3$Cl were obtained respectively, which were considered to be the smallest chiral gold nanoclusters with only three metal atoms [64]. An intriguing experimental phenomenon was observed for the chiral $Au_3$ nanocluster, that is, the UV–vis, CD, and CPL spectra of the dissolved nanocluster (in DCM) gradually changed with addition of the poor solvent (i.e., hexane). The photoluminescence intensity also increased with the addition of hexane. XRD and TEM gave the answer, which demonstrate that the poor solvent triggers the self-assembly of the nanocluster into cubic morphology with an average edge length of 366 nm. The self-assembly results in the change on UV–vis, CD, and CPL spectra of the metal nanocluster. This interesting finding represents a unique feature of chiral phosphine ligand-protected metal nanoclusters. By using the phosphine ligand, the same research

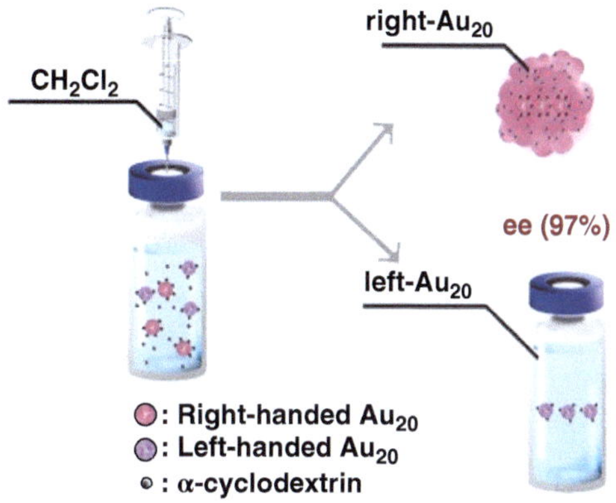


**Figure 5.18** Kinetic resolution of phosphine ligand-protected chiral $Au_{20}(L)_4Cl_4$ nanocluster by α-cyclodextrin. *Source:* [65]/with permission of John Wiley & Sons.

group synthesized a chiral $Au_{20}(L)_4Cl_4$ nanocluster, where the phosphine ligand L = tris(2-(diphenylphosphino)ethyl) phosphine [65]. Of course, due to the fact that the ligand is achiral, the obtained nanocluster is racemic. Zhiyong Tang and coworkers developed a kinetic resolution method and successfully obtained the enantiopure $Au_{20}(L)_4Cl_4$ nanocluster by the resolving reagent of α-cyclodextrin. The principle of the kinetic resolution is the selective combination and assembly of α-cyclodextrin with the right-handed $Au_{20}(L)_4Cl_4$ nanocluster (Figure 5.18). The O—H hydrogen bonds play important roles in forming the favorable configuration of $Au_{20}(L)_4Cl_4$-α-cyclodextrin assemblies.

The last structural feature of the phosphine ligand-protected metal nanoclusters that we present here is the construction of uncoordinated sites on the cluster surface. Li and coworkers identified a highly stable $Au_{20}$ gas-phase nanocluster, which possesses an energy gap of 1.77 eV [66]. This value is greater than that of $C_{60}$. IR, trapped-ion electron diffraction and HAADF-STEM confirmed the tetrahedral structure of the gas-phase $Au_{20}$ nanocluster. The structure is a fragment of the FCC lattice of bulk gold with a small structural relaxation. However, the large HOMO-LUMO gap suggests the molecular character of this nanocluster. Li and coworkers [66] proposed that $Au_{20}$ nanocluster would be a unique molecule with atomic packing similar to bulk gold while with distinct properties. For example, owing to its stable structure and highly exposed surface gold atoms, the $Au_{20}$ nanocluster should be a promising catalyst. Inspired by the observed stable $Au_{20}$ gas-phase nanocluster, Lai-Sheng Wang et al. tried to synthesize structurally similar $Au_{20}$ nanocluster in solution. To maintain the unique structural and electronic properties of the tetrahedral $Au_{20}$ nanocluster, phosphine ligands were selected because they only form dative bonds to the gold atoms and the tetrahedral $Au_{20}$ core and electronic structure would be retained to a large extent. Thiolate ligands have too strong chemical interactions with the gold atoms and would alter the geometric and electronic structure of the tetrahedral $Au_{20}$. Theoretical calculations revealed that the four apex sites-protected $Au_{20}(PPh_3)_4$ nanocluster would be chemically inert and experimentally separatable. Meanwhile, sixteen uncoordinated surface sites would make this nanocluster highly beneficial in catalysis (Figure 5.19a). They indeed observed the

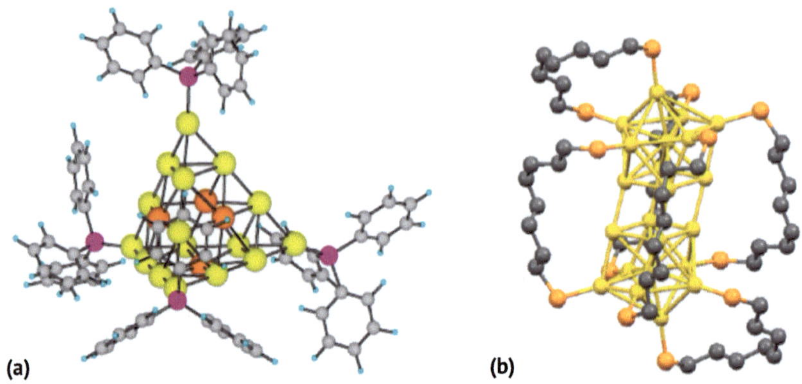


**Figure 5.19** Theoretically predicted $Au_{20}$ nanocluster with 16 uncoordinated surface sites (a) and the experimentally synthesized $Au_{22}$ nanocluster with eight uncoordinated surface sites (b) protected by phosphine ligands.

ESI-MS signal and TEM morphology of a similar tetrahedral $[Au_{20}(PPh_3)_8]^{2+}$ nanocluster from the sample prepared in solution. However, the synthesis was not selective and the yield the desired nanocluster is quite low. The synthetic issue restricts the structural determination and the subsequent property discovery [67].

For developing selective synthesis strategy to the desired $Au_{20}$ nanocluster, Lai-Sheng Wang et al. modified the structure of phosphine ligands. A series of di-phosphine ligands $Ph_2P(CH_2)_nPPh_2$ with different chain lengths were tried to synthesize the desired $Au_{20}$ nanocluster with high selectivity. Their efforts were verified to be effective and the in situ ESI-MS spectra demonstrated that highly selective gold nanoclusters were obtained with $n = 3 - 6$. However, the desired $Au_{20}$ nanocluster was still not available. Further increasing the chain length of the di-phosphine, $Ph_2P(CH_2)_8PPh_2$ gave inspiring results. It was found that a gold nanocluster $Au_{22}(Ph_2P(CH_2)_8PPh_2)_6$ was obtained in high yield with satisfactory selectivity [68]. SCXRD revealed the structure of this nanocluster and indicated eight uncoordinated gold sites on the surface (Figure 5.19b). Applying the similar synthetic strategy, Lai-Sheng Wang et al. synthesized a series of di-phosphine-protected gold nanoclusters. Bearing eight uncoordinated gold sites, the catalytic activity of $Au_{22}(Ph_2P(CH_2)_8PPh_2)_6$ was then investigated. Wu and coworkers found that $TiO_2$ immobilized $Au_{22}(Ph_2P(CH_2)_8PPh_2)_6$ nanocluster showed catalytic activity toward the CO oxidation even at room temperature, although the formation of $CO_2$ is not so efficient [69]. This observation is completely different from the $TiO_2$ immobilized $Au_{25}(SR)_{18}$ nanocluster catalyst, because none of CO oxidation was detected at room temperature with $Au_{25}(SR)_{18}$. The eight uncoordinated gold atoms would play important roles in the intriguing catalytic activity of $Au_{22}(Ph_2P(CH_2)_8PPh_2)_6$. Further investigation suggests that this nanocluster is able to adsorb hydrogen and activate hydrogen gas even better than platinum, manifesting the potential of $Au_{22}(Ph_2P(CH_2)_8PPh_2)_6$ in hydrogen evolution reaction and hydrogenation reactions.

### 5.2.2 Catalytic Property

As we mentioned in the last part, the weaker metal–phosphine interactions trigger the discovery of structurally diversified metal nanoclusters that are difficult to be constructed by the thiolate ligands. On the other hand, the dative bonding is advantageous for ligand exchange and reversible removal, thus releasing the activation sites on the clusters' surface. To maintain the structural stability of metal nanoclusters, the combined usage of thiolate ligands and phosphine ligands are adopted. A typical example is the rod-like $Au_{25}(PPh_3)_{10}(SR)_5Cl_2$ nanocluster, which possesses good stability as well as promising catalytic activity. For example, Chen and coworkers prepared $Au_{25}(PPh_3)_{10}Cl_2[(SC_3H_6Si(OC_2H_5)_3]_5$ nanocluster and immobilized the nanocluster on $TiO_2$ support by hydrolysis reaction [70]. The thiolate ligands well protect the geometric structure of $Au_{25}(PPh_3)_{10}Cl_2[(SC_3H_6Si(OC_2H_5)_3]_5$ from decomposition, while the dative Au—P bonds favor the exposure of the Au activation sites. Therefore, this nanocluster catalyst demonstrates both structural stability and catalytic activity toward the oxidative coupling of benzylamines. Mechanism studies reveal that the nanocluster is exited under 455 nm of LED irradiation and activates $O_2$, generating superoxide radical anion that induces the oxidative coupling of benzylamines. The TOF of the catalytic reaction was determined to be 1522 $h^{-1}$, which is considerably higher than that of conventional gold catalysts. Moreover, the nanocluster catalysis exhibited good recyclability and stability for the oxidative coupling of a wide range of amines. The reversible phosphine ligand's detachment and relocation have been considered to be the key for the high stability and catalytic activity of this phosphine ligand-protected gold nanocluster.

Further functionalization of the surface ligands endows the phosphine ligand-protected gold nanoclusters with more promising catalytic activities. Liu and coworkers synthesized functional phosphine ligand-protected $Au_{11}$ nanocluster $Au_{11}(PPh_2Py)_7Br_3$ ($PPh_2Py$ = diphenyl-2-pyridylphosphine) and compared its catalytic hydrogenation activity with the nonfunctional phosphine ligand-protected $Au_{11}$ nanocluster $Au_{11}(PPh_3)_7Cl_3$. It was found that $Au_{11}(PPh_2Py)_7Br_3$ catalyzed the hydrogenation of 4-nitrobenzaldehyde efficiently in the absence of base, giving the desired product 4-nitrobenzyl alcohol in high yield with high selectivity [71]. In sharp contrast, $Au_{11}(PPh_3)_7Cl_3$ only catalyzed the same reaction smoothly with the addition of base, and the absence of base resulted in the loss of catalytic activity. Based on the atomically precise structure of both nanoclusters, DFT calculations revealed the probable catalytic cycle of $Au_{11}(PPh_2Py)_7Br_3$-catalyzed aldehyde hydrogenation reaction (Figure 5.20). Being similar to the $Au_{25}(PPh_3)_{10}Cl_2$ $[(SC_3H_6Si(OC_2H_5)_3]_5$-catalyzed oxidative coupling of benzylamines mentioned above, the one phosphine ligand removal is also the initial step during catalysis. This crucial step facilitates exposure of the Au sites to substrates, providing an activation site for the catalytic hydrogenation. Subsequently, the H—H bond of the molecular hydrogen is activated with the aid of the pyridine sites on the $PPh_2Py$ ligand. Finally, the 4-nitrobenzaldehyde is hydrogenated in the presence of activated $H_2$ and affords the 4-nitrobenzyl alcohol. Experimental results including UV–vis and ESI-MS spectra all support the proposed mechanism shown in Figure 5.20.

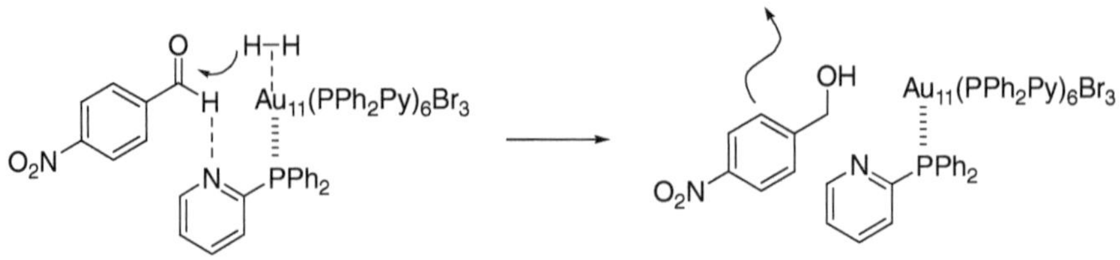


**Figure 5.20** Catalytic mode of $Au_{11}(PPh_2Py)_7Br_3$-catalyzed aldehyde hydrogenation reaction.

The above two examples demonstrate the advantage of phosphine ligand-protected metal nanoclusters in catalysis, that is, the dative metal–phosphine bonds enable reversible removal of the surface phosphine ligands and is therefore beneficial for releasing activation sites. This conclusion has been verified by the experimental results from different research groups [72–74], such as the $[Au_{11}(PPh_3)_8Cl_2]Cl$ nanocluster-catalyzed styrene oxidation developed by Zhenghua Tang, Shaowei Chen, et al., the $Au_6(Ph_2P\text{-o-tolyl})_6](NO_3)_2$ nanocluster-catalyzed alcohol oxidation developed by Gopinathan Sankar et al., the $Au_{101}(PPh_3)_{21}Cl_5$ nanocluster-catalyzed aerobic oxidation of cyclohexene developed by Vladimir B. Golovko and the $Au_{24}(PPh_3)_{10}(SR)_5Cl_2$ nanocluster-catalyzed $CO_2$ hydrogenation developed by Yan Zhu et al.

Apart from the above advantage on exposure of activation sites, the phosphine ligand-protected metal nanoclusters have demonstrated electron-transfer activity (Figure 1.2b) and been used for catalyzing photo-induced cyclization reactions. In this case, the whole nanocluster is used as a photosensor, thus triggering the single-electron transfer between organic substrates and the nanocluster under excited state. Herein, we use a typical example to illustrate this kind catalytic activity of phosphine ligand-protected metal nanoclusters. Hiraiand coworkers reported an interesting work on central metal-doped icosahedral $MAu_{12}$ nanoclusters. By using 1,2-bis(diphenylphosphino)ethane (dppe) as the phosphine ligand, they prepared five structurally similar $MAu_{12}(dppe)_5Cl_2$ nanoclusters, namely $RhAu_{12}(dppe)_5Cl_2$, $PdAu_{12}(dppe)_5Cl_2$, $IrAu_{12}(dppe)_5Cl_2$, $PtAu_{12}(dppe)_5Cl_2$ and $Au_{13}(dppe)_5Cl_2$ (Figure 5.21) [75]. It was found that the central metal atoms highly impact the HOMO LUMO gap, thus resulting in different absorption characters, redox properties and photoluminescence. Among these phosphine ligand-protected metal nanoclusters, $RhAu_{12}(dppe)_5Cl_2$, $IrAu_{12}(dppe)_5Cl_2$, and $PtAu_{12}(dppe)_5Cl_2$ show promising phosphorescence and long exited state lifetime. Based on the photophysics study, Hirai and coworkers [75] further investigated the catalytic activity of these structurally similar nanoclusters in photo-induced olefin cyclization reaction. $RhAu_{12}(dppe)_5Cl_2$, $IrAu_{12}(dppe)_5Cl_2$ and $PtAu_{12}(dppe)_5Cl_2$ that bear stable structure and long exited state lifetime exhibited much higher activity than $PdAu_{12}(dppe)_5Cl_2$ and $Au_{13}(dppe)_5Cl_2$. Additionally, compared to the phosphine ligand-protected metal nanoclusters, thiolate ligand-protected metal nanoclusters such as the $Au_{25}(SCH_2CH_2Ph)_{18}$ showed none of catalytic activity under the same reaction conditions (Figure 5.21).

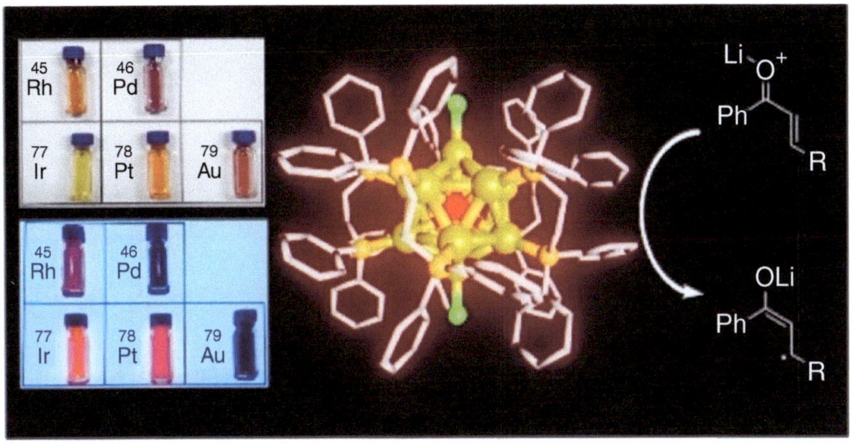


**Figure 5.21** Rh, Pd, Ir, Pt, and Au-kernelled $MAu_{12}(dppe)_5Cl_2$ nanoclusters and their catalytic activity in olefin cyclization reaction. *Source:* [75]/with permission of John Wiley & Sons.

### 5.2.3 Hydride-doping Effect

Compared to the thiolate ligand-protected metal nanoclusters, the hydride-doping phenomenon is much more common in the phosphine ligand-protected metal nanoclusters. The corresponding reports include but not limited to $[Au_{22}H_4(dppo)_6]^{2+}$ (dppo = 1,8-bis(diphenylphosphino)octane), $[Au_{22}H_3(dppee)_7]^{3+}$ (dppee = bis(2-diphenylphosphino)ethylether), $[Au_{11}HCl(dppee)_4]^+$, $[Au_9H(PPh_3)_8]^{2+}$, $[AuCu_{24}(dppp)_6H_{22}]^+$, $[Au_7H_5(PPh_3)_7]^{2+}$, and $[Cu_{25}H_{22}(PPh_3)_{12}]^+$ nanoclusters. The hydride-doping not only alters the electronic and geometric structures of nanoclusters but also regulates their catalytic activity and selectivity. In this part, we discuss the hydride-doping effect in phosphine ligand-protected metal nanoclusters and their catalytic performance.

The story of the hydride-doping in phosphine ligand-protected metal nanoclusters can trace back to the year of 2014, when Lai-Sheng Wang and coworkers successfully synthesized $Au_{22}(Ph_2P(CH_2)_8PPh_2)_6$ nanocluster with eight uncoordinated gold atoms. Inspired by the uncoordinated gold sites in this work, Hu and coworkers studied the hydrogen adsorption performance of $Au_{22}(Ph_2P(CH_2)_8PPh_2)_6$ nanocluster from the theoretical aspect. Theoretical calculations revealed that $Au_{22}(Ph_2P(CH_2)_8PPh_2)_6$ had the potential of adsorbing and activating $H_2$ [76]. They found that the nanocluster provided the waist adsorption position for maximum six hydrogen atoms. Gibbs energy for each hydrogen atom adsorption was calculated to be around zero at room temperature. This observation indicates that the hydrogen atom adsorption process is reversible, suggesting the catalytic hydrogenation activity of $Au_{22}(Ph_2P(CH_2)_8PPh_2)_6$. Additionally, the second hydrogen atom adsorption results in quite low Gibbs energy (−0.41 eV) of the total nanocluster structure, implying that the $Au_{22}(Ph_2P(CH_2)_8PPh_2)_6H_2$ nanocluster would be stable and isolable (Figure 5.22).

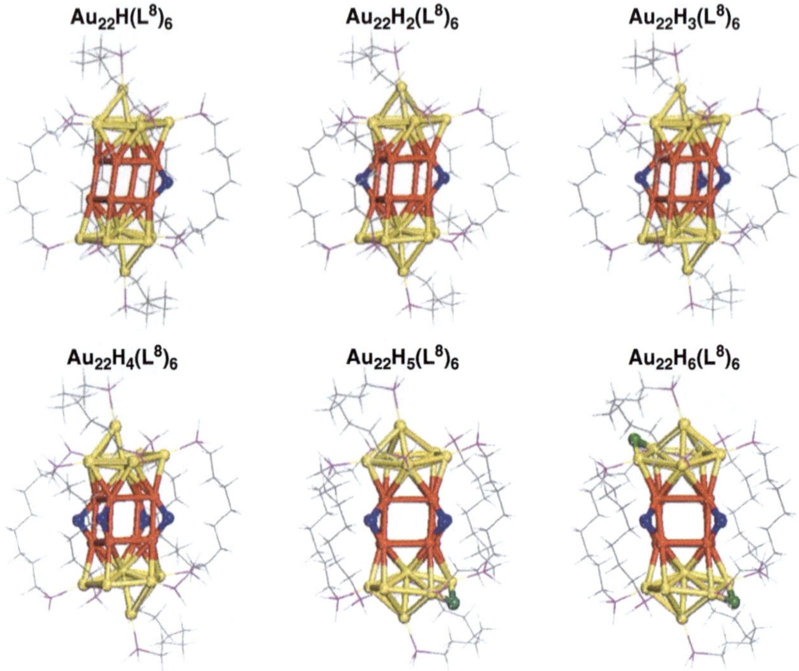


**Figure 5.22** Proposed $H_2$ adsorption and location in $Au_{22}(Ph_2P(CH_2)_8PPh_2)_6$ nanocluster. *Source:* [76]/with permission of Royal Society of Chemistry.

After two years, Dong and coworkers indeed synthesized this kind of hydride-doped gold nanocluster $Au_{22}(Ph_2P(CH_2)_8PPh_2)_6H_4$. The strong signal of the molecular ion peak of $[Au_{22}(Ph_2P(CH_2)_8PPh_2)_6H_4]^{2+}$ in ESI-MS spectrum verified the composition of the nanocluster [77]. Meanwhile, deuterated experiments by using $NaBD_4$ and EtOD instead of $NaBH_4$ and EtOH gave the exclusive ESI-MS peak of $[Au_{22}(Ph_2P(CH_2)_8PPh_2)_6D_4]^{2+}$, further confirming the existence of four hydrides in the phosphine-protected gold nanocluster (Figure 5.23). However, this nanocluster is not stable enough for its structural determination by SCXRD and slowly transformed to $Au_{22}(Ph_2P(CH_2)_8PPh_2)_6$ under the conditions of single-crystal growth. This result suggests that the $Au_{22}(Ph_2P(CH_2)_8PPh_2)_6$ nanocluster obtained in 2014 by Lai-Sheng Wang et al. might be formed via the same hydride loss process. Actually, the authors wrote in the paper that "During our attempt to observe $Au_{22}H_2(dppo)_6$, we found that $[Au_{22}H_x(dppo)_6]^{n+}$ species were present in some as-synthesized samples even without $H_2$ bubbling. In fact, $[Au_{22}H(dppo)_6]^{3+}$ was observed in the original mass spectrum along with $[Au_{22}M(dppo)_6]^{3+}$ (M = Na, K), and the protonated species was thought then to come from the electrospray process. Further experimentation led us to isolate a $[Au_{22}H_4(dppo)_6]^{2+}$ tetrahydride species, which was actually the initial product of the synthesis and the $Au_{22}(dppo)_6$ final crystal used for X-ray diffraction was formed through the loss of hydrogen during the slow crystallization process." This description indicates that the hydride-doping phenomenon in metal nanoclusters

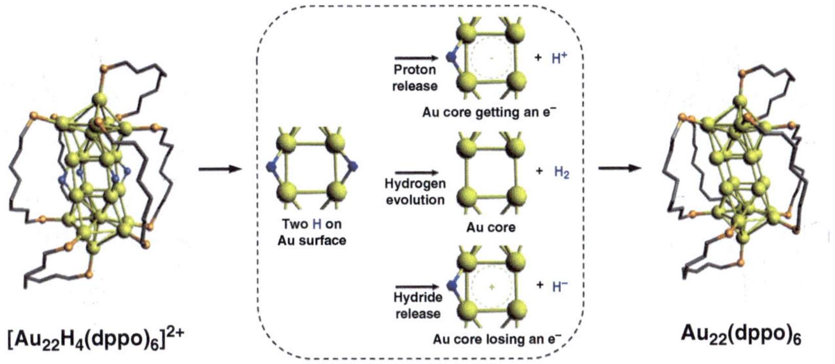


**Figure 5.23** Identification of the hydride-doped and phosphine-protected gold nanocluster $Au_{22}(Ph_2P(CH_2)_8PPh_2)_6H_4$. *Source:* [77]/with permission of John Wiley & Sons.

protected by phosphine ligands is common, but it is easy to be overlooked in the ESI-MS spectra. With the help of theoretical calculation and other characterization techniques such as NMR, Lai-Sheng Wang and coworkers further identified the possible positions of the four hydrides in the $Au_{22}(Ph_2P(CH_2)_8PPh_2)_6H_4$ nanocluster. The hydrogen loss mechanism was proposed, suggesting that the hydride-doped $Au_{22}(Ph_2P(CH_2)_8PPh_2)_6H_4$ nanocluster would be a versatile catalyst for various chemical transformations involving hydrogens, such as hydrogen evolution, hydrogenation, hydrogen atoms transfer, etc.

Inspired by the theoretical calculation-revealed catalytic potential of the hydride-doped gold nanocluster, the same research group synthesized a stable $[Au_{11}Cl_2(dppee)_4]^+$ nanocluster two years later [78], which can readily undergoes a $Cl/BH_4$ exchange, leading to the formation of $[Au_{11}(BH_4)Cl(dppee)_4]^+$ nanocluster and the subsequent hydride-doped nanocluster $[Au_{11}HCl(dppee)_4]^+$. More interestingly, this nanocluster undergoes a facile hydroauration reaction in the presence of alkynes, affording a stable Au-alkenyl intermediate that can be isolated and characterized. The further protonation of this intermediate results in the generation of Z-alkenes with high selectivity and regenerates the $[Au_{11}Cl_2(dppee)_4]^+$ nanocluster. The whole process has been described by Lai-Sheng Wang et al. and all of the intermediates have been captured and identified (Figure 5.24). The corresponding results suggest the catalytic activity of $[Au_{11}Cl_2(dppee)_4]^+$ toward the alkyne semi-hydrogenation reaction via the intermediate of hydride-doped nanocluster $[Au_{11}HCl(dppee)_4]^+$. Stoichiometric experimental results initially verify the efficient catalysis of this nanocluster, which triggers the formation of Z-alkene products with > 99% conversion and > 90 stereoselectivity.

Apart from the effort from Lai-Sheng Wang group, other research groups also made great contributions to the hydride-doped metal nanoclusters protected by phosphine ligands. Takano and coworkers experimentally observed the hydride-doped gold nanocluster for the first time in the year of 2018 [79]. They initially prepared a $[Au_9(PPh_3)_8]^{3+}$ nanocluster and then treated it with ethanol solution of $NaBH_4$ at room temperature. ESI-MS spectra were used to monitor the reaction process and

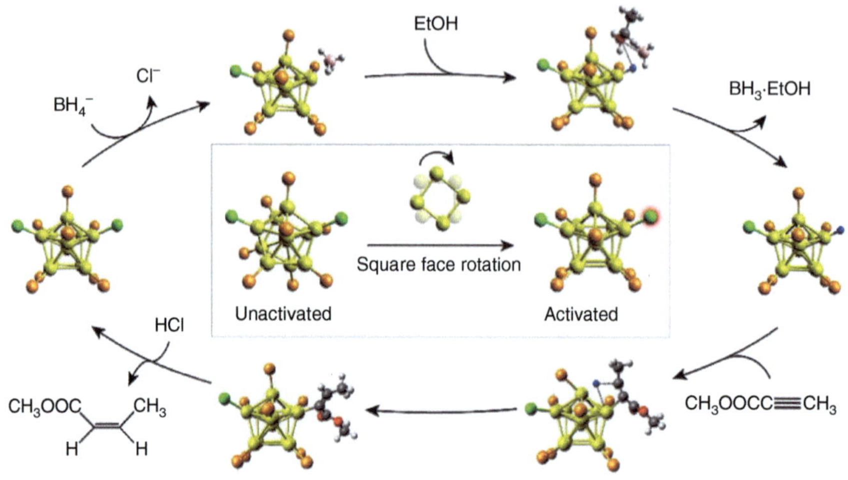


**Figure 5.24** Formation of hydride-doped gold nanocluster $[Au_{11}HCl(dppee)_4]^+$ and its potentially catalytic performance in alkyne semi-hydrogenation reaction. *Source:* [78]/with permission of American Chemical Society.

it was found that the characteristic peak of $[Au_9(PPh_3)_8]^{3+}$ disappeared completely with the addition of $NaBH_4$. Meanwhile, a new peak assigned to $[Au_9(PPh_3)_8H]^{2+}$ was observed a dominant peak in the ESI-MS spectrum. Deuterated and proton-exchange experiments further verified the addition of a hydride, because $NaBD_4$ results in 1 Da increment of the mass peak of the product and the hydride cannot be replaced by the surrounding protons in the medium. This is the first observation of the hydride-doping phenomenon in gold nanoclusters. However, as we mentioned above, the hydride-doped and phosphine-protected gold nanocluster is not as stable as its mother nanocluster and the structural determination by SCXRD has been confirmed to be failed. Actually, the peak of $[Au_9(PPh_3)_8H]^{2+}$ in ESI-MS spectrum gradually decreased with time, also suggesting the metastable nature of this nanocluster. Nevertheless, $[Au_9(PPh_3)_8H]^{2+}$ has a lifetime of several hours at room temperature and the hydride does not undergo exchange with protons in the solvent, which enables its structural characterization by NMR and other techniques. In the same year, Takano and coworkers prepared another hydride-doped metal nanocluster $[Au_8Pd(PPh_3)_8H]^+$ and observed its structural evolution to $[Au_9Pd(PPh_3)_8H]^{2+}$ and $[Au_{10}Pd(PPh_3)_8H]^{3+}$. The structural evolution process can be divided into two steps as follows [80]. (i) 6 Electron-$[Au_8Pd(PPh_3)_8]^{2+}$ is reduced to 8 electron-$[Au_8Pd(PPh_3)_8H]^+$ under reducing conditions. (ii) Sequential addition of two $Au^+$ units into the 8 electron-$[Au_8Pd(PPh_3)_8H]^+$ furnishes the final 8 electron-$[Au_{10}Pd(PPh_3)_8H]^{3+}$. The driving force of the sequential nanocluster transformation should be associated with the closure of the electronic shell in$[Au_8Pd(PPh_3)_8H]^+$ and the ligand shell in $[Au_{10}Pd(PPh_3)_8H]^{3+}$. Based on the corresponding results, Tatsuya Tsukuda and coworkers systematically introduce the hydride-doped and phosphine-protected metal nanoclusters in Acc. Chem. Res. With the title of "Hydride Doping of Chemically Modified Gold-Based Superatoms."

Tang and coworkers discovered a poly-hydride-doped gold nanocluster protected by $PPh_3$ [81]. The nanocluster $[Au_7(PPh_3)_7H_5]^{2+}$ was readily synthesized from $HAuCl_4$ and $PPh_3$ in the presence of $NaBH_4$. The hydrides on the surface of the gold nanocluster were verified by deuterated experiments and ESI-MS spectra. Once again, this hydride-doped and phosphine-protected gold nanocluster is quite instable and decomposed to give $[Au_8(PPh_3)_7]^{2+}$ upon light irradiation. Nevertheless, the geometric and electronic structures of the poly-hydride-doped gold nanocluster have been well characterized by NMR, ESI-MS, UV–vis, XPS, and DFT calculations as well. Being different from the electron-donating hydride in $[Au_{22}(Ph_2P(CH_2)_8PPh_2)_6H_4]^{2+}$ and $[Au_9(PPh_3)_8H]^{2+}$ nanoclusters, the hydride in $[Au_7(PPh_3)_7H_5]^{2+}$ possesses electron-withdrawing property, thus acting like a $H^-$ ligand instead of metal. That means the number of free electrons in $[Au_7(PPh_3)_7H_5]^{2+}$ is zero, and all of the gold atoms are positively charged. This kind of electronic feature endows $[Au_7(PPh_3)_7H_5]^{2+}$ nanocluster with high catalytic selectivity (98.2%) in the $CO_2$ reduction reaction, producing $H_2$, while $[Au_8(PPh_3)_7]^{2+}$ demonstrates significantly different catalytic selectivity and gives CO as the final product. DFT calculations revealed that the electron-withdrawing $H^-$ inhibited the $CO_2$ reduction reaction process compared to the electron-donating H.

In sharp contrast to the unstable hydride-doped gold nanoclusters, the hydride-doped copper nanoclusters display much improved stability and can be structurally determined by SCXRD. The same research group, Ma and coworkers successfully synthesized a $[AuCu_{24}(dppp)_6H_{22}]^+$ nanocluster, where dppp represents the bidentate phosphine ligand 1,3-bis(diphenyphosphino)propane [82]. This hydride-doped nanocluster presents high stability as well as exposed $Cu_3H_3$ units in specific surface cavities. In situ infrared spectra and theoretical calculations revealed that these $Cu_3H_3$ units effectively lowered the energy barrier for the transformation process from $CO_2$ to COCOH* intermediate under electrolysis. Blocking these $Cu_3H_3$ units by NaCl resulted in the switched selectivity of the electrocatalytic $CO_2$ reduction reaction (Figure 5.25). The hydride-doping phenomenon is very common in copper nanoclusters. Another typical example is the $[Cu_{25}(PPh_3)_{12}H_{22}]^+$ nanocluster reported by Nguyen and coworkers in 2015 [83]. This nanocluster also possesses a very stable geometric structure, which is beneficial for the performance investigations. Other reports on hydride-doped copper nanoclusters include $Cu_{36}(PPh_3)_6(SCH_2CH_2Ph)_{24}H_{10}Cl_2$, $[Cu_{36}(SPh)_{46}(^tBuNH_2)_{10}H_{32}]^{3+}$, etc. We will discuss these nanoclusters in Chapter 5, because they show very interesting photocatalytic activity.

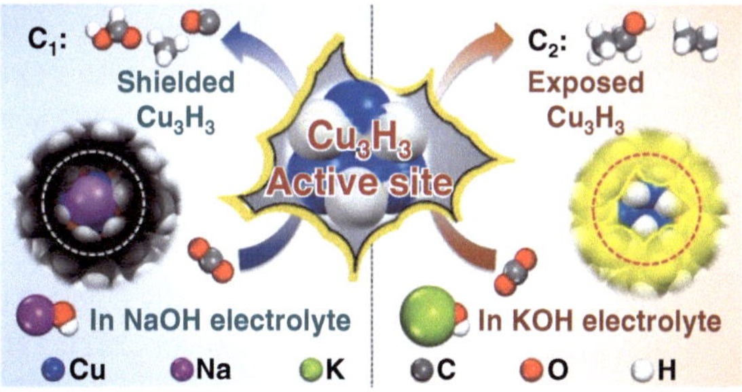


**Figure 5.25** Stable hydride-doped copper nanocluster $[AuCu_{24}(dppp)_6H_{22}]^+$ protected by phosphine ligand and its unique catalytic selectivity in electrocatalytic $CO_2$ reduction reaction. *Source:* [82]/with permission of John Wiley & Sons.

## References


1 S. Antonello, N. V. Perera, M. Ruzzi, J. A. Gascón, F. Maran, Interplay of Charge State, Lability, and Magnetism in the Molecule-like $Au_{25}(SR)_{18}$ Cluster. *J. Am. Chem. Soc.* **2013**, *135*, 15585–15594.

2 Y. Zhu, H. Qian, M. Zhu, R. Jin, Thiolate-Protected Aun Nanoclusters as Catalysts for Selective Oxidation and Hydrogenation Processes. *Adv. Mater.* **2010**, *22*, 1915–1920.

3 Y. Liu, H. Tsunoyama, T. Akita, T. Tsukuda, Efficient and Selective Epoxidation of Styrene with TBHP Catalyzed by $Au_{25}$ Clusters on Hydroxyapatite. *Chem. Commun.* **2010**, *46*, 550–552.

4 Y. Liu, H. Tsunoyama, T. Akita, S. Xie, T. Tsukuda, Aerobic Oxidation of Cyclohexane Catalyzed by Size-Controlled Au Clusters on Hydroxyapatite: Size Effect in the Sub-2 nm Regime. *ACS Catal.* **2011**, *1*, 2–6.

5 S. Xie, H. Tsunoyama, W. Kurashige, Y. Negishi, T. Tsukuda, Enhancement in Aerobic Alcohol Oxidation Catalysis of $Au_{25}$ Clusters by Single Pd Atom Doping. *ACS Catal.* **2012**, *2*, 1519–1523.

6 S. Yamazoe, K. Koyasu, T. Tsukuda, Nonscalable Oxidation Catalysis of Gold Clusters. *Acc. Chem. Res.* **2014**, *47*, 816–824.

7 G. Li, H. Qian, R. Jin, Gold Nanocluster-Catalyzed Selective Oxidation of Sulfide to Sulfoxide. *Nanoscale* **2012**, *4*, 6714–6717.

8 G. Li, R. Jin, Atomically Precise Gold Nanoclusters as New Model Catalysts. *Acc. Chem. Res.* **2013**, *46*, 1749–1758.

9 Y. Zhu, H. Qian, B. A. Drake, R. Jin, Atomically Precise $Au_{25}(SR)_{18}$ Nanoparticles as Catalysts for the Selective Hydrogenation of α,β-Unsaturated Ketones and Aldehydes. *Angew. Chem. Int. Ed.* **2010**, *49*, 1295–1298; *Angew. Chem.* **2010**, *122*, 1317–1320.

10 G. Li, R. Jin, Gold Nanocluster-Catalyzed Semihydrogenation: A Unique Activation Pathway for Terminal Alkynes. *J. Am. Chem. Soc.* **2014**, *136*, 11347–11354.

11 G. Li, C. Zeng, R. Jin, Thermally Robust $Au_{99}(SPh)_{42}$ Nanoclusters for Chemoselective Hydrogenation of Nitrobenzaldehyde Derivatives in Water. *J. Am. Chem. Soc.* **2014**, *136*, 3673–3679.

12 H. Chong, P. Li, S. Wang, F. Fu, J. Xiang, M. Zhu, Y. Li, $Au_{25}$ Clusters as Electron-Transfer Catalysts Induced the Intramolecular Cascade Reaction of 2-nitrobenzonitrile. *Sci. Rep.* **2013**, *3*, 3214.

13 J. Hu, Y.-M. Li, B. Zhang, X. Kang, M. Zhu, Atomically precise nanocluster-catalyzed coupling reactions. *Inorg. Chem. Front.* **2024**, *11*, 4974–5003.

14 G. Li, C. Liu, Y. Lei, R. Jin, $Au_{25}$ nanocluster-catalyzed Ullmann-type homocoupling reaction of aryl iodides. *Chem. Commun.* **2012**, *48*, 12005–12007.

15 G. Li, H. Abroshan, C. Liu, S. Zhuo, Z. Li, Y. Xie, H. J. Kim, N. L. Rosi, R. Jin, Tailoring the Electronic and Catalytic Properties of $Au_{25}$ Nanoclusters via Ligand Engineering. *ACS Nano* **2016**, *10*, 7998–8005.

16 R. Jin, G. Li, S. Sharma, Y. Li, X. Du, Toward Active-Site Tailoring in Heterogeneous Catalysis by Atomically Precise Metal Nanoclusters with Crystallographic Structures. *Chem. Rev.* **2021**, *121*, 567–648.

**17** H. Abroshan, G. Li, J. Lin, H. J. Kim, R. Jin, Molecular mechanism for the activation of $Au_{25}(SCH_2CH_2Ph)_{18}$ nanoclusters by imidazolium-based ionic liquids for catalysis. *J. Catal.* **2016**, *337*, 72–79.
**18** Y. Yang, C. Chen, G.-Y. Xu, J. Yuan, S.-F. Ye, L. Chen, Q.-L. Lv, G. Luo, J. Yang, M.-B. Li, Z. Wu, An efficient nanocluster catalyst for Sonogashira reaction. *J. Catal.* **2021**, *401*, 206–213.
**19** A. Sagadevan, K. Murugesan, O. M. Bakr, M. Rueping, Copper nanoclusters: emerging photoredox catalysts for organic bond formations. *Chem. Commun.* **2024**, *60*, 13858–13866.
**20** C. Yin, S. Liu, Z. Qin, Y. Zhang, G. Li, Z. Zhao, Butterfly-Like Tetranuclear Copper(I) Clusters for Efficient Alkyne Homocoupling Reactions. *Eur. J. Inorg. Chem.* **2021**, *2021*, 392–397.
**21** N. Sun, Y. Yun, K. Li, X. Ni, Y. Qin, X. Yang, H. Sheng, M. Zhu, Highly Efficient Conversion of Homocoupling and Heterocoupling of Terminal Alkynes Catalyzed by $AuCu_{24}$/AC-200. *Eur. J. Inorg. Chem.* **2023**, *26*, e202300161.
**22** Y. Chen, C. Liu, H. Abroshan, Z. Li, J. Wang, G. Li, M. Haruta, Phosphine/Phenylacetylide-ligated Au Clusters for Multicomponent Coupling Reactions. *J. Catal.* **2016**, *340*, 287–294.
**23** Y. Adachi, H. Kawasaki, T. Nagata, Y. Obora, Thiolate-protected Gold Nanoclusters $Au_{25}$(phenylethanethiol)$_{18}$: An Efficient Catalyst for the Synthesis of Propargylamines from Aldehydes, Amines, and Alkynes. *Chem. Lett.* **2016**, *45*, 1457–1459.
**24** M.-B. Li, S.-K. Tian, Z. Wu, Improving the Catalytic Activity of $Au_{25}$ Nanocluster by Peeling and Doping. *Chin. J. Chem.* **2017**, *35*, 567–571.
**25** J. Yuan, Y. Yun, Z. Tao, Y. N. Zhu, L. Li, H. Sheng, M. Zhu, Atomically Precise Cun($n$=3, 6 and 11) Nanocatalysts for Alkyne-Haloalkane-Amine (AHA) Coupling Reaction. *Eur. J. Inorg. Chem.* **2022**, *2022*, e202200159.
**26** J.-Q. Fan, Y. Yang, C.-B. Tao, M.-B. Li, Cadmium-Doped and Pincer Ligand-Modified Gold Nanocluster for Catalytic KA2 Reaction. *Angew. Chem. Int. Ed.* **2023**, *62*, e202215741; *Angew. Chem.* **2023**, *135*, e202215741.
**27** D. Yang, Y. Sun, X. Cai, W. Hu, Y. Dai, Y. Zhu, Y. Yang, Catalytic Conversion of C1 Molecules on Atomically Precise Metal Nanoclusters. *CCS Chem.* **2021**, *3*, 2771–2799.
**28** Z. Wu, D.-E. Jiang, A. K. P. Mann, D. R. Mullins, Z.-A. Qiao, L. F. Allard, C. Zeng, R. Jin, S. H. Overbury, Thiolate Ligands as a Double-Edged Sword for CO Oxidation on $CeO_2$ Supported $Au_{25}(SCH_2CH_2Ph)_{18}$ Nanoclusters. *J. Am. Chem. Soc.* **2014**, *136*, 6111–6122.
**29** Y. Chen, C. Liu, Q. Tang, C. Zeng, T. Higaki, A. Das, D.-E. Jiang, N. L. Rosi, R. Jin, Isomerism in $Au_{28}(SR)_{20}$ Nanocluster and Stable Structures. *J. Am. Chem. Soc.* **2016**, *138*, 1482–1485.
**30** X. Nie, H. Qian, Q. Ge, H. Xu, R. Jin, CO Oxidation Catalyzed by Oxide-Supported $Au_{25}(SR)_{18}$ Nanoclusters and Identification of Perimeter Sites as Active Centers. *ACS Nano* **2012**, *6*, 6014–6022.
**31** Y. Li, Y. Chen, S. D. House, S. Zhao, Z. Wahab, J. C. Yang, R. Jin, Interface Engineering of Gold Nanoclusters for CO Oxidation Catalysis. *ACS Appl. Mater. Interfaces.* **2018**, *10*, 29425–29434.

**32** M. Zhou, C. Zeng, Y. Chen, S. Zhao, M. Y. Sfeir, M. Zhu, R. Jin, Evolution from the Plasmon to Exciton State in Ligand-Protected Atomically Precise Gold Nanoparticles. *Nat. Commun.* **2016**, *7*, 13240.

**33** S. Li, X. Du, Z. Liu, Y. Li, Y. Shao, R. Jin, Size Effects of Atomically Precise Gold Nanoclusters in Catalysis. *Precis. Chem.* **2023**, *1*, 14–28.

**34** W. Li, C. Liu, H. Abroshan, Q. Ge, X. Yang, H. Xu, G. Li, Catalytic CO Oxidation Using Bimetallic $MxAu_{25-x}$ Clusters: A Combined Experimental and Computational Study on Doping Effects. *J. Phys. Chem. C.* **2016**, *120*, 10261–10267.

**35** X. Liu, G. Saranya, X. Huang, X. Cheng, R. Wang, M. Chen, C. Zhang, T. Li, Y. Zhu, $Ag_2Au_{50}(PET)_{36}$ Nanocluster: Dimeric Assembly of $Au_{25}(PET)_{18}$ Enabled by Silver Atoms. *Angew. Chem. Int. Ed.* **2020**, *59*, 13941–13946; *Angew. Chem.* **2020**, *132*, 14045–14050.

**36** X. Liu, X. Cai, Y. Zhu, Catalysis Synergism by Atomically Precise Bimetallic Nanoclusters Doped with Heteroatoms. *Acc. Chem. Res.* **2023**, *56*, 1528–1538.

**37** D. R. Kauffman, D. Alfonso, C. Matranga, H. Qian, R. Jin, Experimental and Computational Investigation of $Au_{25}$ Clusters and $CO_2$: A Unique Interaction and Enhanced Electrocatalytic Activity. *J. Am. Chem. Soc.* **2012**, *134*, 10237–10243.

**38** S. Zhuang, D. Chen, L. Liao, Y. Zhao, N. Xia, W. Zhang, C. Wang, J. Yang, Z. Wu, Hard-Sphere Random Close-Packed $Au_{47}Cd_2(TBBT)_{31}$ Nanoclusters with a Faradaic Efficiency of Up to 96% for Electrocatalytic $CO_2$ Reduction to CO. *Angew. Chem. Int. Ed.* **2020**, *59*, 3073–3077; *Angew. Chem.* **2020**, *132*, 3097–3101.

**39** Q. Tang, Y. Lee, D.-Y. Li, W. Choi, C. W. Liu, D. Lee, D.-E. Jiang, Lattice-Hydride Mechanism in Electrocatalytic $CO_2$ Reduction by Structurally Precise Copper-Hydride Nanoclusters. *J. Am. Chem. Soc.* **2017**, *139*, 9728–9736.

**40** B. Zhang, C. Chen, W. Chuang, S. Chen, P. Yang, Size Transformation of the $Au_{22}(SG)_{18}$ Nanocluster and Its Surface-Sensitive Kinetics. *J. Am. Chem. Soc.* **2020**, *142*, 11514–11520.

**41** X. Cai, G. Saranya, K. Shen, M. Chen, R. Si, W. Ding, Y. Zhu, Reversible Switching of Catalytic Activity by Shuttling an Atom into and out of Gold Nanoclusters. *Angew. Chem. Int. Ed.* **2019**, *58*, 9964–9968; *Angew. Chem.* **2019**, *131*, 10069–10073.

**42** T. Kawawaki, Y. Kataoka, M. Hirata, Y. Akinaga, R. Takahata, K. Wakamatsu, Y. Fujiki, M. Kataoka, S. Kikkawa, A. S. Alotabi, S. Hossain, D. J. Osborn, T. Teranishi, G. G. Andersson, G. F. Metha, S. Yamazoe, Y. Negishi, Creation of High-Performance Heterogeneous Photocatalysts by Controlling Ligand Desorption and Particle Size of Gold Nanocluster. *Angew. Chem. Int. Ed.* **2021**, *60*, 21340–21350; *Angew. Chem.* **2021**, *133*, 21510–21520.

**43** K. Sakamoto, S. Masuda, S. Takano, T. Tsukuda, Partially Thiolated $Au_{25}$ Cluster Anchored on Carbon Support via Noncovalent Ligand–Support Interactions: Active and Robust Catalyst for Aerobic Oxidation of Alcohols. *ACS Catal.* **2023**, *13*, 3263–3271.

**44** W. Suzuki, R. Takahata, Y. Chiga, S. Kikkawa, S. Yamazoe, Y. Mizuhata, N. Tokitoh, T. Teranishi, Control over Ligand-Exchange Positions of Thiolate-Protected Gold Nanoclusters Using Steric Repulsion of Protecting Ligands. *J. Am. Chem. Soc.* **2022**, *144*, 12310–12320.

45 J. Zhao, A. Ziarati, A. Rosspeintner, T. Bürgi, Anchoring of Metal Complexes on $Au_{25}$ Nanocluster for Enhanced Photocoupled Electrocatalytic $CO_2$ Reduction. *Angew. Chem. Int. Ed.* **2024**, *63*, e202316649; *Angew. Chem.* **2015**, *127*, 3188–3192.

46 Z. Liu, X.-M. Yang, Y. Wang, S. Wang, Atomically precise coinage metal nanoclusters with defects in catalysis. *Nano Res.* **2025**, *18*, 94907141.

47 G. Li, J. Hou, X. Lei, D. Li, E. Yu, W. Hu, X. Cai, X. Liu, M. Chen, Y. Zhu, Reactivity and Recyclability of Ligand-Protected Metal Cluster Catalysts for $CO_2$ Transformation. *Angew. Chem. Int. Ed.* **2023**, *62*, e202216735; *Angew. Chem.* **2023**, *135*, e202216735.

48 H. Seong, Y. Jo, V. Efremov, Y. Kim, S. Park, S. M. Han, K. Chang, J. Park, W. Choi, W. Kim, C. H. Choi, J. S. Yoo, D. Lee, Transplanting Gold Active Sites into Non-Precious-Metal Nanoclusters for Efficient $CO_2$-to-CO Electroreduction. *J. Am. Chem. Soc.* **2023**, *145*, 2152–2160.

49 G. Li, X. Sui, X. Cai, W. Hu, X. Liu, M. Chen, Y. Zhu, Precisely Constructed Silver Active Sites in Gold Nanoclusters for Chemical Fixation of $CO_2$. *Angew. Chem. Int. Ed.* **2021**, *60*, 10573–10576; *Angew. Chem.* **2021**, *133*, 10667–10670.

50 K. Konishi, M. Iwasaki, Y. Shichibu, Phosphine-Ligated Gold Clusters with Core+exo Geometries: Unique Properties and Interactions at the Ligand–Cluster Interface. *Acc. Chem. Res.* **2018**, *51*, 3125–3133.

51 L. Malatesta, L. Naldini, G. Simonetta, F. Cariati, Triphenylphosphine-gold(0)/gold(I) Compounds. *Coord. Chem. Rev.* **1966**, *1*, 255–262.

52 S. Zhuang, D. Chen, Q. You, W. Fan, J. Yang, Z. Wu, Thiolated, Reduced Palladium Nanoclusters with Resolved Structures for the Electrocatalytic Reduction of Oxygen. *Angew. Chem. Int. Ed.* **2022**, *61*, e202208751; *Angew. Chem.* **2022**, *134*, e202208751.

53 Q. You, X.-L. Jiang, W. Fan, Y.-S. Cui, Y. Zhao, S. Zhuang, W. Gu, L. Liao, C.-Q. Xu, J. Li, Z. Wu, Pd8 Nanocluster with Nonmetal-to-Metal- Ring Coordination and Promising Photothermal Conversion Efficiency. *Angew. Chem. Int. Ed.* **2024**, *63*, e202313491; *Angew. Chem.* **2024**, *136*, e202313491.

54 L.-L. Yan, L.-Y. Yao, M. Ng, V. W.-W. Yam, Stimuli-Responsive and Structure-Adaptive Three-Dimensional Gold(I) Cluster Cages Constructed via "De-aurophilic" Interaction Strategy. *J. Am. Chem. Soc.* **2021**, *143*, 19008–19017.

55 L.-Y. Yao, V. W.-W. Yam, Photoinduced Isomerization-Driven Structural Transformation Between Decanuclear and Octadecanuclear Gold(I) Sulfido Clusters. *J. Am. Chem. Soc.* **2015**, *137*, 3506–3509.

56 L.-L. Yan, V. Wing-Wah Yam, Photo- and Temperature-Induced Reversible Structural Transformation between Dodecanuclear and Pentadecanuclear Gold(I) Sulfido Complexes. *J. Am. Chem. Soc.* **2023**, *145*, 7454–7461.

57 L.-Y. Yao, T. K.-M. Lee, V. W.-W. Yam, Thermodynamic-Driven Self-Assembly: Heterochiral Self-Sorting and Structural Reconfiguration in Gold(I)–Sulfido Cluster System. *J. Am. Chem. Soc.* **2016**, *138*, 7260–7263.

58 L.-Y. Yao, L. Qin, Z. Chen, J. Lam, V. W.-W. Yam, Assembly of Luminescent Chiral Gold(I)-Sulfido Clusters via Chiral Self-Sorting. *Angew. Chem. Int. Ed.* **2024**, *63*, e202316200; *Angew. Chem.* **2024**, *136*, e202316200.

**59** L.-Y. Yao, K.-H. Low, V. W.-W. Yam, A Gold Quartet Framework with Reversible Anisotropic Structural Transformation Accompanied by Luminescence Response. *Chem* **2019**, *5*, 2418–2428.

**60** L.-L. Yan, Z. Chen, V. W.-W. Yam, Heterochiral Self-Discrimination Driven Dimerization of Polynuclear Gold(I)-Sulfido Complexes with Enhanced Phosphorescence. *Angew. Chem. Int. Ed.* **2025**, *64*, e202419062; *Angew. Chem.* **2025**, *137*, e202419062.

**61** L.-Y. Yao, V. W.-W. Yam, Diphosphine-Stabilized Small Gold Nanoclusters: From Crystal Structure Determination to Ligation-Driven Symmetry Breaking and Anion Exchange Properties. *J. Am. Chem. Soc.* **2016**, *138*, 15736–15742.

**62** L.-Y. Yao, V. W.-W. Yam, Dual Emissive Gold(I)–Sulfido Cluster Framework Capable of Benzene–Cyclohexane Separation in the Solid State Accompanied by Luminescence Color Changes. *J. Am. Chem. Soc.* **2021**, *143*, 2558–2566.

**63** L.-L. Yan, V. W.-W. Yam, Evolution of Polynuclear Gold(I) Sulfido Complexes from Clusters and Cages to Macrocycles. *J. Am. Chem. Soc.* **2024**, *146*, 609–616.

**64** L. Shi, L. Zhu, J. Guo, L. Zhang, Y. Shi, Y. Zhang, K. Hou, Y. Zheng, Y. Zhu, J. Lv, S. Liu, Z. Tang, Self-Assembly of Chiral Gold Clusters into Crystalline Nanocubes of Exceptional Optical Activity. *Angew. Chem. Int. Ed.* **2017**, *56*, 15397–15401; *Angew. Chem.* **2017**, *129*, 15599–15603.

**65** Y. Zhu, H. Wang, K. Wan, J. Guo, C. He, Y. Yu, L. Zhao, Y. Zhang, J. Lv, L. Shi, R. Jin, X. Zhang, X. Shi, Z. Tang, Enantioseparation of $Au_{20}(PP_3)_4Cl_4$ Clusters with Intrinsically Chiral Cores. *Angew. Chem. Int. Ed.* **2018**, *57*, 9059–9063; *Angew. Chem.* **2018**, *130*, 9197–9201.

**66** J. Li, X. Li, H.-J. Zhai, L.-S. Wang, $Au_{20}$: A Tetrahedral Cluster. *Science* **2003**, *299*, 864–867.

**67** Q.-F. Zhang, X. Chen, L.-S. Wang, Toward Solution Syntheses of the Tetrahedral $Au_{20}$ Pyramid and Atomically Precise Gold Nanoclusters with Uncoordinated Sites. *Acc. Chem. Res.* **2018**, *51*, 2159–2168.

**68** J. Chen, Q.-F. Zhang, T. A. Bonaccorso, P. G. Williard, L.-S. Wang, Controlling Gold Nanoclusters by Diphospine Ligands. *J. Am. Chem. Soc.* **2014**, *136*, 92–95.

**69** Z. Wu, G. Hu, D.-E. Jiang, D. R. Mullins, Q.-F. Zhang, L. F. Allard, Jr., L.-S. Wang, S. H. Overbury, Diphosphine-Protected $Au_{22}$ Nanoclusters on Oxide Supports Are Active for Gas-Phase Catalysis without Ligand Removal. *Nano Lett.* **2016**, *16*, 6560–6567.

**70** H. Chen, C. Liu, M. Wang, C. Zhang, N. Luo, Y. Wang, H. Abroshan, G. Li, F. Wang, Visible Light Gold Nanocluster Photocatalyst: Selective Aerobic Oxidation of Amines to Imines. *ACS Catal.* **2017**, *7*, 3632–3638.

**71** C. Liu, H. Abroshan, C. Yan, G. Li, M. Haruta, One-Pot Synthesis of $Au_{11}(PPh_2Py)_7Br_3$ for the Highly Chemoselective Hydrogenation of Nitrobenzaldehyde. *ACS Catal.* **2016**, *6*, 92–99.

**72** R. H. Adnan, J. M. L. Madridejos, A. S. Alotabi, G. F. Metha, G. G. Andersson, A Review of State of the Art in Phosphine Ligated Gold Clusters and Application in Catalysis. *Adv. Sci.* **2022**, *9*, e2105692.

**73** J. Kilmartin, R. Sarip, R. Grau-Crespo, D. Di Tommaso, G. Hogarth, C. Prestipino, G. Sankar, Following the Creation of Active Gold Nanocatalysts from Phosphine-Stabilized Molecular Clusters. *ACS Catal.* **2012**, *2*, 957–963.

**74** B. G. Donoeva, D. S. Ovoshchnikov, V. B. Golovko, Establishing a Au Nanoparticle Size Effect in the Oxidation of Cyclohexene Using Gradually Changing Au Catalysts. *ACS Catal.* **2013**, *3*, 2986–2991.

**75** H. Hirai, S. Takano, T. Nakashima, T. Iwasa, T. Taketsugu, T. Tsukuda, Doping-Mediated Energy-Level Engineering of $M@Au_{12}$ Superatoms (M=Pd, Pt, Rh, Ir) for Efficient Photoluminescence and Photocatalysis. *Angew. Chem. Int. Ed.* **2022**, *61*, e202207290; *Angew. Chem.* **2022**, *134*, e202207290.

**76** G. Hu, Z. Wu, D.-E. Jiang, Stronger-than-Pt hydrogen adsorption in a $Au_{22}$ nanocluster for the hydrogen evolution reaction. *J. Mater. Chem. A.* **2018**, *6*, 7532–7537.

**77** J. Dong, Z.-H. Gao, Q.-F. Zhang, L.-S. Wang, The Synthesis, Bonding, and Transformation of a Ligand-Protected Gold Nanohydride Cluster. *Angew. Chem. Int. Ed.* **2021**, *60*, 2424–2430; *Angew. Chem.* **2021**, *133*, 2454–2460.

**78** J. Dong, J. R. Robinson, Z.-H. Gao, L.-S. Wang, Selective Semihydrogenation of Polarized Alkynes by a Gold Hydride Nanocluster. *J. Am. Chem. Soc.* **2022**, *144*, 12501–12509.

**79** S. Takano, H. Hirai, S. Muramatsu, T. Tsukuda, Hydride-Doped Gold Superatom $(Au_9H)^{2+}$: Synthesis, Structure, and Transformation. *J. Am. Chem. Soc.* **2018**, *140*, 8380–8383.

**80** S. Takano, S. Hasegawa, M. Suyama, T. Tsukuda, Hydride Doping of Chemically Modified Gold-Based Superatoms. *Acc. Chem. Res.* **2018**, *51*, 3074–3083.

**81** L. Tang, Y. Luo, X. Ma, B. Wang, M. Ding, R. Wang, P. Wang, Y. Pei, S. Wang, Poly-Hydride $[AuI_7(PPh_3)_7H_5](SbF_6)_2$ cluster complex: Structure, Transformation, and Electrocatalytic $CO_2$ Reduction Properties. *Angew. Chem. Int. Ed.* **2023**, *62*, e202300553; *Angew. Chem.* **2023**, *135*, e202300553.

**82** X. Ma, C. Fang, M. Ding, Y. Zuo, X. Sun, S. Wang, Atomic-Level Elucidation of Lattice-Hydrogens in Copper Catalysts for Selective $CO_2$ Electrochemical Conversion toward $C_2$ Products. *Angew. Chem. Int. Ed.* **2025**, *64*, e202500191; *Angew. Chem.* **2025**, *137*, e202500191.

**83** T.-A. D. Nguyen, Z. R. Jones, B. R. Goldsmith, W. R. Buratto, G. Wu, S. L. Scott, T. W. Hayton, A $Cu_{25}$ Nanocluster with Partial Cu(0) Character. *J. Am. Chem. Soc.* **2015**, *137*, 13319–13324.

# 6

# C- and N-based Ligand-protected Metal Nanocluster Catalysis

## 6.1 Alkyne-protected Metal Nanocluster for Catalysis

The reports of alkyne adsorption on the surface of bulk metals or metal nanoparticles can date back to half a century ago. Soon after the determination of crystal structures of $Au_{25}(SR)_{25}$ and $Au_{102}(pMBA)_{44}$ nanoclusters, protection of metal nanoclusters by alkyne ligands have been investigated. An early work comes from Maity and coworkers, who demonstrated that ligand exchange between polyvinylpyrrolidone (PVP)-protected gold precursors and alkynes was capable of preparing alkyne-protected gold nanoclusters [1–3]. The compositions of these nanoclusters were determined by ESI-MS, FTIR, NMR, and other characterizations to be $Au_{34}(C{\equiv}CPh)_{16}$, $Au_{43}(C{\equiv}CPh)_{22}$, $Au_{54}(C{\equiv}CPh)_{26}$, etc. Other early contributions of alkyne-protected metal nanoclusters include the $\left[Au_{25}(C{\equiv}CPh)_5(PPh_3)_{10}X_2\right]^{2+}$ and $Au_{20}(C{\equiv}CPh)_x$ nanoclusters, which were reported by Li and Tang [4, 5]. However, only ESI-MS or calculated results were given in these works and the crystal structures were still not obtained, probably due to the instability of these nanoclusters. In 2013, Kobayashi and coworkers determined the first atomically precise structure of the alkyne-protected metal nanocluster [6] namely $\left[Au_8(dppp)_4(C{\equiv}CPh)_2\right]^{2+}$. This nanocluster was prepared by ligand-exchange method of $[Au_8(dppp)_4]^{2+}$ with the alkyne PhC≡CH under alkaline reaction conditions (Figure 6.1). By replacing PhC≡CH with other alkynes, such as $CH_3(CH_2)_3C{\equiv}CH$ and pyridylethynyl compounds, they synthesized the similar $\left[Au_8(dppp)_4(C{\equiv}CR)_2\right]^{2+}$ (R = functional group) nanoclusters and systematically investigated the resonance-coupled electronic perturbation of metal nanoclusters through π-conjugated Groups.

The first alkyne-protected metal nanocluster synthesized by the conventionally direct reduction method was discovered by Wan and coworkers in the year of 2015 (Figure 6.2) [7]. The composition of the nanocluster was determined by ESI-MS analysis to be $\left[Au_{19}(Hdppa)_4(C{\equiv}CPh)_2\right]^{2+}$ (Hdppa = *N*,*N*-bis-(diphenylphosphino)amine).

*Metal Nanocluster Catalysis: From Fundamentals to Applications*, First Edition. Manzhou Zhu and Man-Bo Li. © 2026 Wiley-VCH GmbH. All rights reserved, including rights for text and data mining and training of artificial intelligence technologies or similar technologies. Published 2026 by Wiley-VCH GmbH.

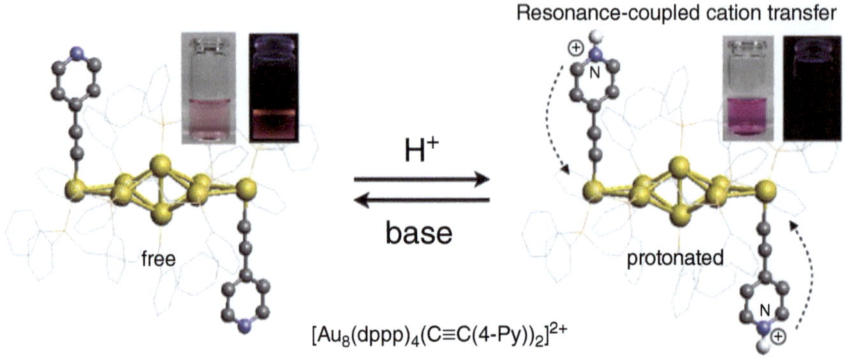


**Figure 6.1** The first crystal structure of alkyne-protected metal nanocluster $[Au_8(dppp)_4(C{\equiv}CPh)_2]^{2+}$. *Source:* [6]/with permission of American Chemical Society.

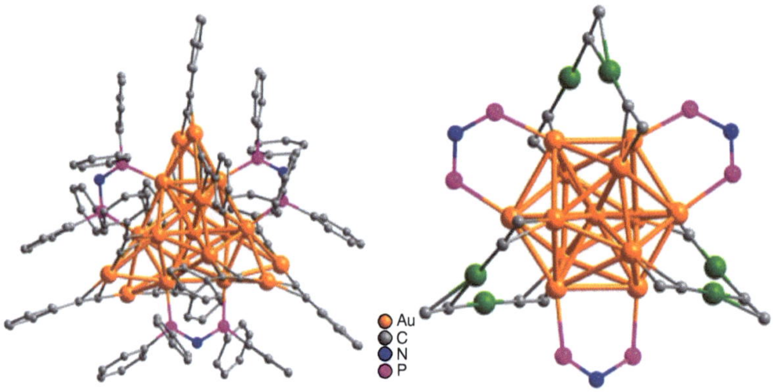


**Figure 6.2** The first alkyne-protected metal nanocluster $[Au_{19}(Hdppa)_4(C{\equiv}CPh)_2]^{2+}$ that is prepared by the direct reduction method. *Source:* [7]/with permission of American Chemical Society.

Further crystal structure revealed the unique [C≡C–Au–C≡C–Au–C≡C] motifs at the interface of the metal kernel that have not been discovered in the previous reports on alkyne-protected metal nanoclusters. This kind of motif is similar but still different from the [S–Au–S–Au–S] staple motif in thiolate-protected metal nanoclusters. This kind of interfacial bonding mode suggest that there is still a large room for discovering alkyne-protected metal nanoclusters with diversified structures, given the large number of the thiolate-protected metal nanoclusters already reported. Actually, this paper triggers the discovery of various metal nanoclusters that are protected by the alkyne ligands. Up to now, there are more than one hundred crystal structures of alkyne-protected metal nanoclusters that have been reported. The intriguing performance such as the high catalytic activity of this kind of nanoclusters has also been explored from then on.

Diverse metal-alkyne coordination modes have been found in alkyne-protected metal nanoclusters. Taking the alkyne ligand-protected gold nanocluster as an

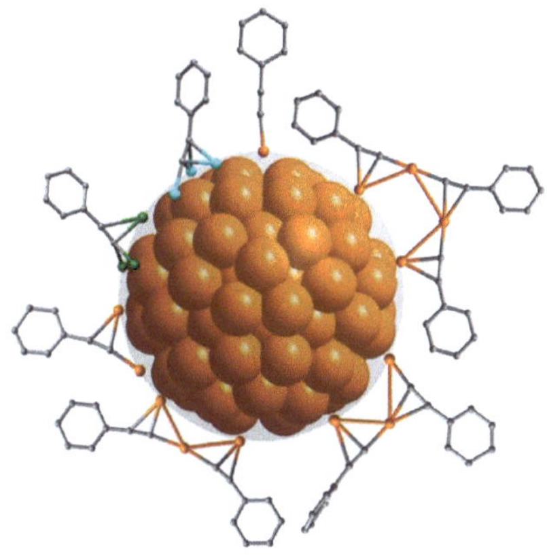

**Figure 6.3** The diverse coordination modes in alkyne ligand-protected gold nanoclusters. *Source:* [8]/with permission of American Chemical Society.

example, it demonstrates even more diversified surface motifs than the thiolate ligand-protected gold nanocluster. Figure 6.3 shows some typical coordination modes in alkyne ligand-protected gold nanoclusters, including the gold-alkyne $\sigma$ interactions such as the $\mu_1$, $\mu_2$, and $\mu_3$ coordination modes and the different gold-alkyne $\pi$ interactions. Other hybrid coordination modes involve both $\sigma$ and $\pi$ interactions are also observed. These diverse metal-alkyne coordination modes trigger the formation of alkyne ligand-protected metal nanoclusters with diversified geometric structures. Of course, combination of the metal doping method and employment of the mixture ligands of alkynes and others such as thiolates and phosphines further contribute the structural diversity of alkyne ligand-protected metal nanoclusters [8].

The alkyne-protected metal nanocluster constitutes an ideal platform for the investigation of structure-property correlation, because the surface alkynes well protect the metal kernel from decomposition or aggregation and meanwhile are able to significantly impact the electronic property of the whole nanocluster. Before we focus on the catalytic performance of alkyne ligand protected metal nanoclusters, we talk about the synthesis, structure, and photoluminescence features of this kind of nanocluster. Quan-Ming Wang research group made great effort in this research area. By solely using alkynes or combining alkynes with other ligands, they prepared a lot of alkyne ligand-protected metal nanoclusters. Meanwhile, they developed several efficient synthesis and isolation methods toward this kind of nanocluster. Intriguing ultrafast excited-state dynamics, thermochromic luminescence and photoluminescence were discovered. In the following, we select some typical examples for detailed discussion.

They developed the efficient synthesis strategy and isolation method toward alkyne ligand-protected metal nanoclusters. For cluster synthesis, Hu and coworkers developed $PhSiH_2$ as the reducing agent and realized the efficient synthesis of largest structurally characterized cluster of clusters $[Ag_{93}(PPh_3)_6(C{\equiv}CR)_{50}]^{3+}$ (R=4-$CH_3OC_6H_4$) [9]. This silver nanocluster features the structure of (i) a four concentric core–shell $Ag_{15}@Ag_{54}@Ag_{18}@Ag_6$ kernel of $C_{2v}$ symmetry, (ii) a novel bicapped hexagonal prism $Ag_{15}$, (iii) a central silver atom which has the highest coordination number of 14, and (iv) a Ino decahedron $M_{13}$ that acts as the building block for constructing the cluster of clusters $Ag_{93}$. For cluster isolation, Li and coworkers developed the gel column chromatography method to purify the alkyne ligand-protected gold nanoclusters with different sizes. A $Au_{25}$ nanocluster $[Au_{25}L_{18}]^{-}$, a $Au_{67}$ nanocluster $[Au_{67}L_{32}Cl_4]^{3-}$ and a $Au_{106}$ nanocluster $[Au_{106}L_{40}Cl_{12}]^{4-}$ (L = 3,5-bis(trifluoromethyl) phenylacetylide) were synthesized in one pot and successfully isolated by the gel column chromatography [10]. Specifically, $Au_{25}$ nanocluster with the smallest size bears the lowest polarity and the biggest $Au_{106}$ nanocluster has the heist polarity. Thus $Au_{25}$, $Au_{67}$ and $Au_{102}$ were sequentially eluted out. Size-dependent properties such as the HOMO–LUMO gap were revealed based on these alkyne ligand-protected metal nanoclusters.

The most luminous metal nanocluster was protected by the alkyne ligand. Shi and coworkers observed the near-unity near-infrared (NIR) phosphorescent quantum yield of a 4-(*tert*-butyl)phenylacetylene ($^t$BuPhC≡C) protected Au–Cu nanocluster [11]. Two isostructural nanoclusters, that is, $Au_{22}(^tBuPhC{\equiv}C)_{18}$ and $Au_{16}Cu_6(^tBuPhC{\equiv}C)_{18}$, were synthesized based on the in situ synthetic method. Specifically, $Au_{22}(^tBuPhC{\equiv}C)_{18}$ was synthesized by using a methanol solution of sodium hydroxide as the reductant, while $Au_{16}Cu_6(^tBuPhC{\equiv}C)_{18}$ was obtained by using the stronger *tert*-butylamine or sodium borohydride as the reducing agent as a result of the lower reduction potential of $Cu^{2+}$ precursor compared to $Au^{3+}$. Atomically precise structures revealed by SCXRD demonstrate that both nanoclusters have the same kernel geometry and surface ligand array (Figure 6.4a). The only difference comes from the replacement of six gold atoms by the copper atoms in the position of $Au_3Cu_3$ ring and $Au_2Cu$ staples. Compared to the undoped $Au_{22}(^tBuPhC{\equiv}C)_{18}$ nanocluster, the copper doping results in the structural contraction in $Au_{16}Cu_6(^tBuPhC{\equiv}C)_{18}$. More interestingly, compared to the 10% photoluminescence quantum yield (PLQY) of $Au_{22}(^tBuPhC{\equiv}C)_{18}$, $Au_{16}Cu_6(^tBuPhC{\equiv}C)_{18}$ exhibited a PLQY of >99% at room temperature in the solution. This kind of near-unity PLQY was observed for the first time in metal nanoclusters. Further investigation on the ultrafast excited-state dynamics of $Au_{22}(^tBuPhC{\equiv}C)_{18}$ and $Au_{16}Cu_6(^tBuPhC{\equiv}C)_{18}$ indicates that the copper atoms doping effectively suppresses the nonradiative decay and promotes the intersystem crossing rate, thus leading to the highly improved PLQY (Figure 6.4b). This work gives an inspiring discussion on the doping method for preparing luminous metal nanoclusters.

Notably, $Au_{22}(^tBuPhC{\equiv}C)_{18}$ demonstrates interesting thermochromic luminescence phenomenon. As mentioned above, the photoluminescence of this nanocluster is significantly lower than the copper-doped $Au_{16}Cu_6(^tBuPhC{\equiv}C)_{18}$. However, Han and coworkers found that $Au_{22}(^tBuPhC{\equiv}C)_{18}$ displayed an unusual temperature-dependent photoluminescence change [12]. That is, this nanocluster shows an emission peak at about 655 nm at 80 K and the emission intensity decreases with the increasement of the temperature. This experimental phenomenon is usual and similar to the previous reports. When the temperature reaches to 185 K, a new emission peak at 580 is observed, which increases with the increasement of the

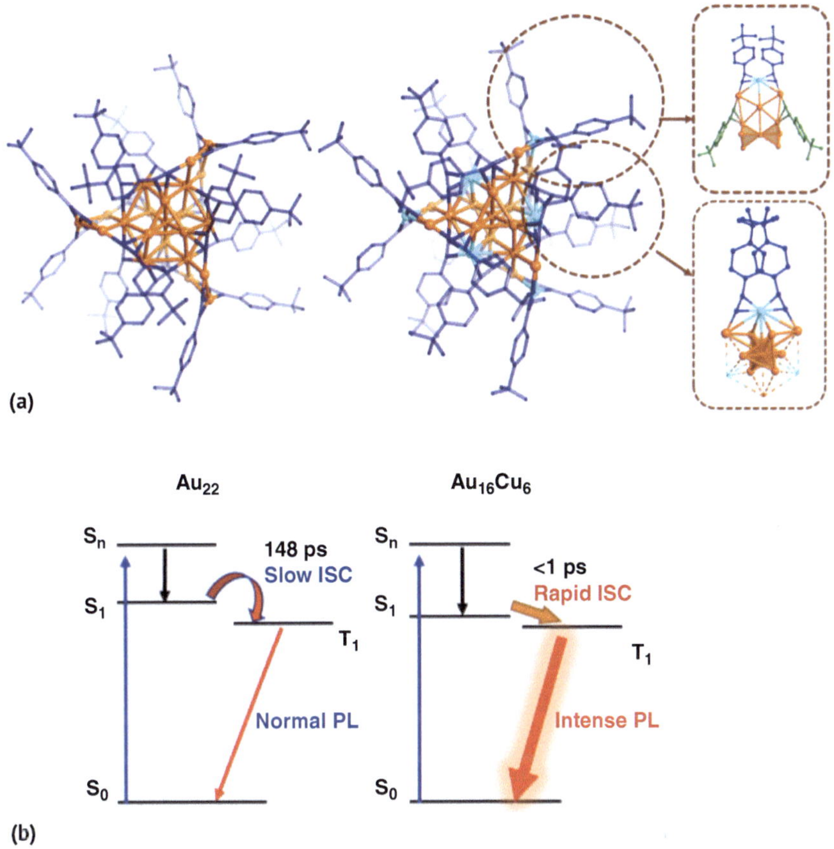


**Figure 6.4** (a) Structures of $Au_{22}$($^t$BuPhC≡C)$_{18}$ and $Au_{16}Cu_6$($^t$BuPhC≡C)$_{18}$ nanoclusters. (b) Origin of the difference on photoluminescence of $Au_{22}$($^t$BuPhC≡C)$_{18}$ and $Au_{16}Cu_6$($^t$BuPhC≡C)$_{18}$. *Source:* [11]/with permission of American Association for the Advancement of Science – AAAS.

temperature to 290 K. Meanwhile, a long lifetime of 1.28 μs is observed. The red-shift and increasement of the emission peak with increased temperature suggest the thermally activated delayed fluorescence (TADF) mechanism of this alkyne ligand-protected metal nanocluster. Further experimental and theoretical results confirmed this TADF phenomenon.

Alkyne-protection also triggers the formation of large-sized metal nanoclusters. For example, Wang, Hu, and coworkers [13, 14] synthesized the largest alkynyl-protected FCC gold nanocluster $[Au_{110}(4\text{-}CF_3C_6H_4C{\equiv}C)_{48}]^{2-}$ and the largest all-alkynyl-protected gold nanocluster $[Au_{156}(4\text{-}CF_3C_6H_4C{\equiv}C)_{60}]^{6-}$, and investigated their ultrafast excited-state dynamics behaviors [13, 14]. Based on the structural analysis and ultrafast excited-state dynamics results, two interesting conclusions were made. (i) $[Au_{110}(4\text{-}CF_3C_6H_4C{\equiv}C)_{48}]^{2-}$ exhibits nonmetallic behavior and can be recognized as the largest face-centered cubic (fcc)-structured nonmetallic gold nanocluster. (ii) $[Au_{156}(4\text{-}CF_3C_6H_4C{\equiv}C)_{60}]^{6-}$ possesses both metallic and molecular behaviors and can be recognized as the smallest gold nanocluster showing metallic electron dynamics. These studies are helpful to find the critical size in terms of nonmetallic to metallic transition of gold nanoclusters, which is a long-pursued but quite challenging task in the research area of metal nanocluster.

By using the $Et_3SiC{\equiv}C\text{–}H$ as the reagent to react with a precursor nanocluster $[O(AuL)_3Ag](BF_4)_2$, where L= phenylbis(2-pyridyl)phosphine (PPhpy2), Pei and coworkers synthesized a rarely reported trimetallic nanocluster $[(AuL)_6AgCu_6(C{\equiv}C)_3](BF_4)_7$ [15]. The copper doping significantly regulate the clusters' photoluminescence. Compared to its structural isomer $[(AuL)_6Ag_7(C{\equiv}C)_3](BF_4)_7$ that shows decreased emission intensity with increased temperature, $[(AuL)_6AgCu_6(C{\equiv}C)_3](BF_4)_7$ exhibits a thermochromic luminescence. This phenomenon should be originated to the fact that Cu-doping significantly increases the energy level of ($C_2$-Cu)-involved occupied orbital. Therefore, the emissive state from $C_2$-Cu to metal core, apart from the other one from $PPhpy_2$ ligands to metal core, produces a dual-emission at low temperature.

As we mentioned above, alkyne ligands enable the synthesis of metal nanocluster in the presence of very weak reducing reagents, such as the NaOH/$CH_3OH$. Thus, compared to other ligands (e.g. thiolates), alkyne ligands protection strategy is beneficial for the introduction of reduction-sensitive functional groups. Guan and coworkers reported the gram-scale synthesis of silver nanoclusters $[Ag_{21}(Ph_2PO_2)_{10}(4\text{-}CHOPhC{\equiv}C)_6]SbF_6$ that bears multiple surface aldehyde groups. These functional aldehyde groups would allow late-stage functionalization of metal nanoclusters [16]. By using chiral (R/S)-(±)-1-phenylethylamines as the reaction partners, they successfully realized the clusters' surface functionalization by a aldimine condensation reaction and obtained a series of homochiral metal nanoclusters. Very recently, Zheng and coworkers applied the functional alkynes as the protection ligand, realizing the construction of a cage-like $Cu_8$ nanocluster with both circular polarized luminescence and long persistent luminescence [17]. The key to success is the stepwise ligand engineering design strategy on the surface of $Cu_8$ nanocluster. Specifically, achiral N-benzylprop-2-yn-1-amine and $PPh_3$ were initially introduced on the surface of $Cu_8$ to enable the efficient photoluminescence of this nanocluster. Second, the similar alkyne ligand, namely chiral (*R*/*S*)-2,3-dihydro-N-methyl-N-2-propynyl-1H-inden-1-amine was used to replace achiral N-benzylprop-2-yn-1-amine. The replacement further introduced chirality and triggered the circular polarized luminescence performance of the $Cu_8$. Finally, replacement of $PPh_3$ by tris(4-chlorophenyl)phosphine further endowed this $Cu_8$ nanocluster with long persistent luminescence. This sequential strategy well regulates the surface of the alkyne ligand-protected metal nanocluster and lead to the nanocluster with multifunctional properties.

Apart from the work from Quan-Ming Wang group, the other research groups also contribute a lot to the field of alkyne ligand-protected metal nanoclusters. Jia and coworkers synthesized a phosphine/alkyne mixed ligand protected copper nanocluster $[Cu_{31}(4\text{-MeO-PhC}\equiv C)_{21}(dppe)_3](ClO_4)_2$ (dppe = 1,2-bis(diphenylphosphino)ethane) and revealed the inherent chiral metal core of this nanocluster [18]. The $Cu_{31}$ nanocluster bears a distorted icosahedral $Cu_{13}$ kernel, with its top four trigonal faces being capped by a $Cu_6$ tile and surrounded by two sets of three $Cu_2$ units. As a result of the impact from the surface alkyne and phosphine ligands, the four parts of the copper kernel arranged asymmetrically, thus endowing the copper kernel with chirality. Besides, an intriguing NIR-II emission phenomenon at about 1300 nm was observed for the $Cu_{31}$ nanocluster crystal. Tao and coworkers utilized 4-(tert-butyl)phenylacetylene as the protecting ligand and obtained a silver nanocluster $[Ag_{24}(C\equiv C^tBu)_{20}]^{4+}$. An $[^{99}TcO_4]^-$ anion was found to be entrapped within the cavity of the $Ag_{24}$ nanocluster, constituting a $[^{99}TcO_4]^-$ core–$[Ag_{24}(C\equiv C^tBu)_{20}]^{4+}$ shell structure [19]. This unique cluster structure results in an unprecedented photo-induced charge transfer phenomenon from the cluster shell to the anion core.

The diversified coordination modes of alkynes on the surface of metal kernels also lead to structural isomerism in the alkyne ligand-protected metal nanoclusters. Deng and coworkers synthesized and investigated the electronic properties of two $Ag_{20}Cu_{12}(C\equiv CR)_{24}$ nanoclusters bearing the completely same composition but different geometric and electronic structures, where R = 3,5-bis(trifluoromethyl) phenyl group [20]. Notably, although the metal nanocluster structural isomerism has been found previously, this is the first time that this phenomenon is discovered in alloyed metal nanoclusters. One of the $Ag_{20}Cu_{12}(C\equiv CR)_{24}$ nanocluster is not stable and can be transformed to the other one under the dissolved conditions. Apart from the structural isomerism, the appearance of alkyne ligands also contributes to the hierarchical chirality of metal nanoclusters. Deng and coworkers synthesized a $[Ag_{22}Cu_7(C\equiv CR)_{16}(PPh_3)_5Cl_6]^-$ nanocluster, where R = 3,5-bis(trifluoromethyl)phenyl group [21]. Owing to the asymmetric arrangement of the metal atoms and surface ligands, the metal kernel and the whole nanocluster demonstrate chirality, although they did not introduce any chiral ligand during nanocluster synthesis. The most interesting thing is that the cluster enantiomers can be well isolated based on a chiral enrichment process. R- or S-(3,5-dioxa-4-phosphacyclohepta[2,1-a;3,4-a']dinaphthalen-4-yl)dimethylamine) was used as optical enrichment agents, and one of the $[Ag_{22}Cu_7(C\equiv CR)_{16}(PPh_3)_5Cl_6]^-$ nanocluster enantiomers was slowly converted into corresponding enantiomer. The non-covalent interactions, such as C–H...F, and $\pi\ldots\pi$, are anticipated to drive the optical enrichment.

As the relatively weak coordination between metals and alkynes, the alkyne ligand-protected metal nanoclusters demonstrate promising catalytic activity. Qin and coworkers prepared a $Au_{13}Ag_{16}$ nanocluster, where 24 alkyne 2-(prop-2-ynyloxy)-benzonitrile ligands are located on the surface (Figure 6.5) [22]. This nanocluster was able to catalyze the $A^3$ coupling reaction of aldehyde, amine, and alkyne, giving the corresponding propargylamine in high yield (89%). Interestingly, Comparison of the catalytic performance of different metal nanoclusters showed that the

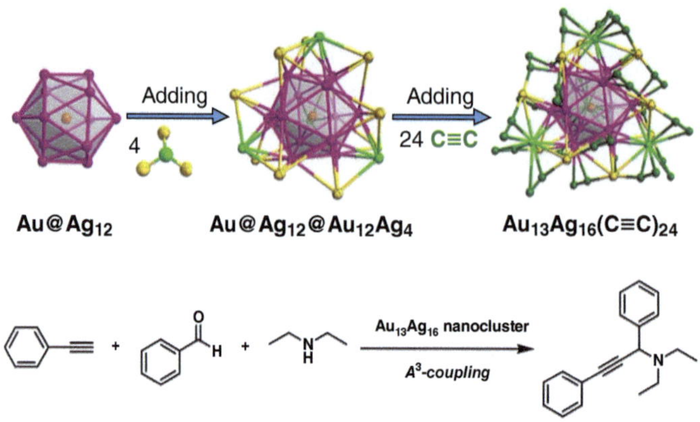


**Figure 6.5** Alkyne ligand-protected $Au_{13}Ag_{16}$ nanocluster and its high catalytic activity toward $A^3$ coupling reaction. *Source:* [22]/with permission of John Wiley & Sons.

thiolate ligand-protected metal nanoclusters such as $Au_{25}(SR)_{18}$, $Au_{38}(SR)_{24}$, and $Au_{25\text{-}x}Ag_x(PPh_3)_{10}(SPh)_5Cl_2$ were catalytically inert and only gave the product in quite low yields. In contrast, the alkyne ligand-protected metal nanoclusters such as $Au_{25}(PPh_3)_{10}(C{\equiv}CPh)_5Cl_2$ and the $Au_{13}Ag_{16}$ nanocluster investigated herein exhibited equally high catalytic activity in the $A^3$ coupling reaction. Thus, the alkyne ligand is beneficial to produce structurally stable and catalytically active metal nanoclusters.

By using the 3,5-bis(trifluoromethyl)phenylacetylene as the protecting ligand, which is the most commonly used alkyne ligand by Taeghwan Hyeon group, Deng and coworkers synthesized a AgCu alloyed nanocluster $[Ag_{15}Cu_6(C{\equiv}CR)_{18}(dppe)_2]^-$. This nanocluster possesses a unique body-centered cubic (bcc) $Ag_{11}Cu_4$ kernel, which is surrounded by another two copper atoms, two ($Ag_2$dppe) and the surface alkyne ligands [23]. Although a series of characterizations demonstrate that this nanocluster is stable under ambient conditions, it is predicted that one alkyne ligand is easily removed during catalyzing the $CO_2$ electroreduction reaction. The as-exposed metal atom would act as the activation site for the adsorption and activation of $CO_2$. Notably, the stripping of an alkyne ligand is reversible and this nanocluster shows high structural stability. Its exceptional catalytic performance toward the electrochemical transformation from $CO_2$ to CO (faradaic efficiency of 91.3%) can be retained in a prolonged test of 145 hours.

Wan, Yuan, He, and coworkers conducted systematic investigations on the catalytic performance of alkyne ligand-protected metal nanoclusters [24–26]. For example, they synthesized a alkyne/phosphine ligands co-protected gold nanocluster $[Au_{34}(PhC{\equiv}C)_{14}(Ph_3P)_6](SO_3CF_3)_2$. This nanocluster possesses an 18-electron superatomic structure, while presents significantly different catalytic performance compared to the similar 18-electron superatom nanoclusters protected by other ligands [24]. The surface coordination structure plays a critical role in improving the catalytic performance of $[Au_{34}(PhC{\equiv}C)_{14}(Ph_3P)_6](SO_3CF_3)_2$ toward the chemoselective hydrogenation of 4-nitrostyrene with $H_2$ as the reducing agent. The Au(I) sites derived from the PhC≡C–Au-C≡CPh monomeric staple motif and the steric hindrance of PhC≡C and $Ph_3P$ on the surface of the nanocluster have been identified to be the activation sites. The turnover frequency (TOF) and the turnover number (TON) of this alkyne-protected gold nanocluster reach to 1322.5 $h^{-1}$ and 23,500, respectively, representing the highest values observed among the gold nanocluster catalysts toward the same reaction. The same "alkyne effect" has also been observed in silver nanoclusters. Quan-Ming Wang et al. prepared two structurally similar $Ag_{21}$ nanoclusters, that is, $[Ag_{21}L_3(O_2PPh_2)_6]SbF_6$ and $[Ag_{21}(C{\equiv}CC_6H_3\text{-}3,5\text{-}OMe)_6(O_2PPh_2)_{10}]SbF_6$, where L is a calixarene (4-*tert*-butylthiacalix-[4]arene). Both $Ag_{21}$ nanoclusters bear an identical icosahedral kernel that is surrounded by eight peripheral silver atoms. Due to the well-protection of the $Ag_{21}$ nanocluster ligands, $Ag_{21}L_3(O_2PPh_2)_6]SbF_6$ is highly robust even at 90 °C under oxidative conditions. However, it was revealed that the catalytic performance of this nanocluster was significantly lower than that of $[Ag_{21}(C{\equiv}CC_6H_3\text{-}3,5\text{-}OMe)_6(O_2PPh_2)_{10}]SbF_6$. Specifically, $[Ag_{21}(C{\equiv}CC_6H_3\text{-}3,5\text{-}OMe)_6(O_2PPh_2)_{10}]SbF_6$ catalyzed the reduction of 4-nitrophenol in 16 minutes and gave the final product 4-aminophenol in quantitative yield, whereas the reaction was not completed even after 1.5 hours when $Ag_{21}L_3(O_2PPh_2)_6]SbF_6$ was used as a catalyst under the same reaction conditions.

In addition to the high catalytic activity of the alkyne ligand-protected metal nanoclusters, they are also ideal platforms for the investigation of doping-dependent catalytic performance. Yuan and coworkers prepared two similar nanoclusters $Ag_{25}Cu_4Cl_6(dppb)_6(PhC{\equiv}C)_{12}(SO_3CF_3)_3$ and $Ag_{25}Cu_4Cl_6H_8(dppb)_6(PhC{\equiv}C)_{12}(SO_3CF_3)_3$ (dppb is 1,4-bis(diphenylphosphine)butane) [25]. Both nanoclusters possess almost the same composition and geometric structure. The only difference between them is that eight hydrides are doped in $Ag_{25}Cu_4Cl_6H_8(dppb)_6(PhC{\equiv}C)_{12}(SO_3CF_3)_3$ (Figure 6.6). As a result of this slight difference, $Ag_{25}Cu_4Cl_6H_8(dppb)_6(PhC{\equiv}C)_{12}(SO_3CF_3)_3$ demonstrated much higher catalytic reduction activity than $Ag_{25}Cu_4Cl_6(dppb)_6(PhC{\equiv}C)_{12}(SO_3CF_3)_3$. The former nanocluster catalyzed the reduction of 4-nitrophenol in the presence of $NaBH_4$, giving the desired product in quantitative yield within 100 minutes, while the latter one catalyzed the reduction under the same reaction conditions and gave only 24% yield of the product. Moreover, $Ag_{25}Cu_4Cl_6H_8(dppb)_6(PhC{\equiv}C)_{12}(SO_3CF_3)_3$ is robust during catalysis and can be recovered and recycled after centrifugation. Even after twenty cycles, it is still catalytically active and catalyzes the reduction with high TON of 3538. Control experiments including the deuterated exchange indicate that the hydrides in the nanocluster are not involved in the catalysis, and the activation of hydrides of $NaBH_4$ occurs on the surface of metal nanoclusters. However, the hydrides in $Ag_{25}Cu_4Cl_6H_8(dppb)_6(PhC{\equiv}C)_{12}(SO_3CF_3)_3$ still play pivotal roles in holding the nanocluster together by forming $\mu_3$-H-$Ag_3$ and $\mu_4$-H-$Ag_4$ bonds and stabilizing the positively charged states of Ag and Cu atoms on the surface. Therefore, the significantly higher catalytic activity of $Ag_{25}Cu_4Cl_6H_8(dppb)_6(PhC{\equiv}C)_{12}(SO_3CF_3)_3$ should be attributed to the

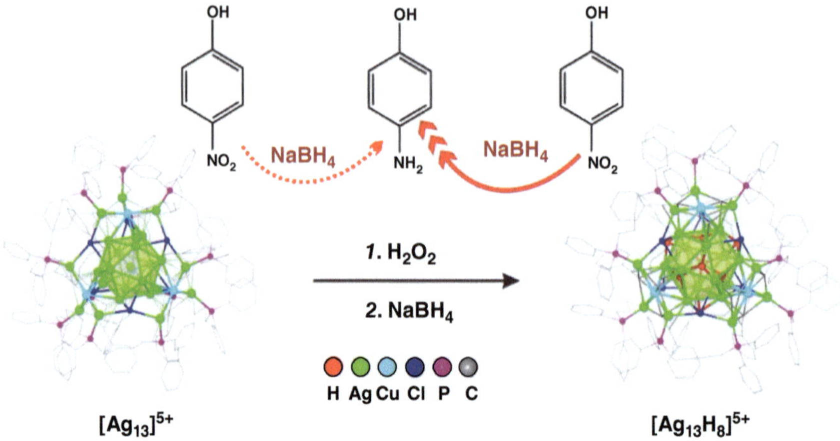


**Figure 6.6** Hydride-doping effect in catalysis of $Ag_{25}Cu_4Cl_6(dppb)_6(PhC{\equiv}C)_{12}(SO_3CF_3)_3$ and $Ag_{25}Cu_4Cl_6H_8(dppb)_6(PhC{\equiv}C)_{12}(SO_3CF_3)_3$ nanoclusters. *Source:* [25]/with permission of American Chemical Society.

electron-deficient Ag(I) and Cu(I) sites on the nanocluster's surface, which facilitate the activation of hydrogen of $NaBH_4$. In sharp contrast, $Ag_{25}Cu_4Cl_6(dppb)_6(PhC{\equiv}C)_{12}(SO_3CF_3)_3$ is an 8-electron superatom nanocluster and bears more electron-rich Ag and Cu atoms on its surface, which is not favored for the activation of hydrogen.

## 6.2 Carbene-protected Metal Nanocluster for Catalysis

A carbene is a divalent neutral carbon species, which was found one hundred years ago and generally considered as highly reactive (Figure 6.7a). It was hard to tame this kind of species until the Fischer carbene (Figure 6.7b) was prepared by Fischer and coworkers in 1964 [27]. The Fischer carbene is relatively stable because the divalent neutral carbon species is bonding to a transition metal. As a result of this contribution, Fischer was awarded the 1973 Nobel Prize. In the same year, VanDerveer and coworkers discovered the first example of a transition metal alkylidene, namely the Schrock carbene (Figure 6.7c) [28]. The application of these relatively stable carbene triggered the rapid development of olefin metathesis reaction, which was also awarded the Nobel Prize in 2005. Nevertheless, the crystalline carbene has not been obtained until the year of 1991 [29], when Anthony J. Arduengo III et al. reported the first completely stable and crystalline carbene (Figure 6.7d), that is the *N*-heterocyclic carbene (NHC). Nowadays, we can control the NHC to be short-lived or crystalline by regulating the substituents on the five-membered ring.

Owing to the proper interaction of NHC toward metals, the resulting metal-complexes show thermal and oxidative stability as well as catalytic activity. Therefore, NHCs have been widely applied as a ligand to regulate the stability

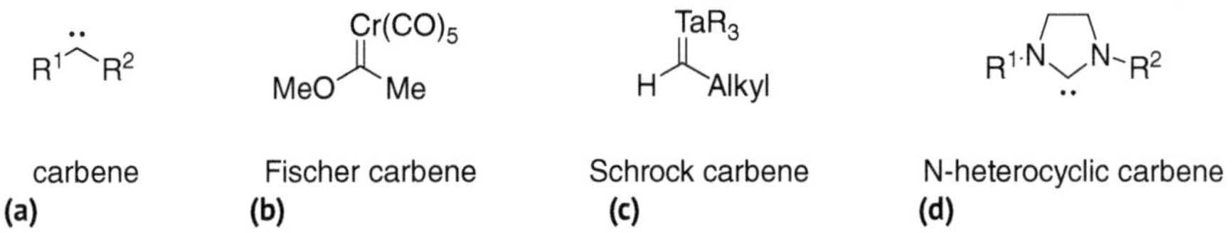


**Figure 6.7** Representatives of carbenes.

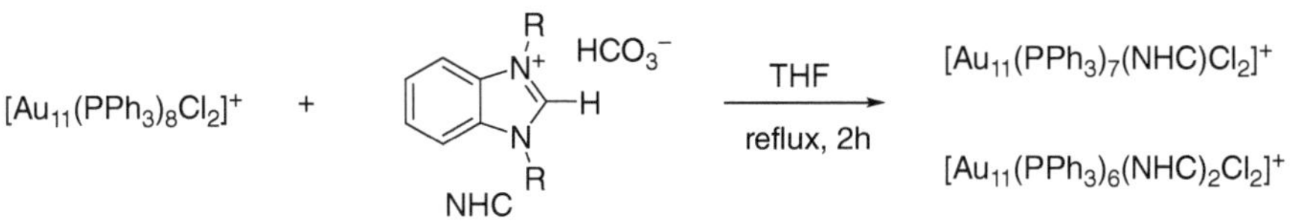


**Figure 6.8** The first preparation of carbene-protected metal nanoclusters by ligand-exchange method.

and catalytic activity of metals. The practical application of NHCs in material science can date back to 2009, when Vignolle and coworkers employed the NHC for preparation of gold nanoparticles [30]. The initial attempt of carbenes in stabilizing atomically precise metal nanoclusters was reported by Narouz and coworkers in 2019 [31]. Three SCXRD structures of NHC-modified $Au_{11}$ nanoclusters were obtained by ligand-exchange method. Specifically, $[Au_{11}(PPh_3)_8Cl_2]^+$ was used as the mother nanocluster and reacted with a series of NHC ligands, including di-isopropyl benzimidazolium hydrogen carbonate ($NHC^{i\text{-pr}}$), di-ethyl benzimidazolium hydrogen carbonate ($NHC^{Et}$), etc. The ligand-exchange reaction resulted in $[Au_{11}(PPh_3)_7(NHC^{i\text{-pr}})Cl_2]^+$, $[Au_{11}(PPh_3)_7(NHC^{Et})Cl_2]^+$, and $[Au_{11}(PPh_3)_6(NHC^{Et})_2Cl_2]^+$ in high yields (Figure 6.8). These products were successfully characterized by ESI-MS and SCXRD. Toward the stabilization of gold nanoclusters, although NHC ligands demonstrate the similar coordination mode with phosphine ligands, the replacement of several phosphine ligands by NHC ligands on the surface of $Au_{11}$ nanocluster leads to significant improvement of structural stability of the nanocluster. This phenomenon should be partially ascribed to the nature of NHC ligands, which possess much higher thermal stability than phosphine ligands. More importantly, the catalytic activity of the NHC-modified $Au_{11}$ nanocluster was highly improved simultaneously. It was found that $[Au_{11}(PPh_3)_7(NHC^{Me})Cl_2]^+$ catalyzed the electrochemical reduction of $CO_2$ under mild conditions, giving the desired product CO with more than 80% CO Faradaic efficiency, current density of up to 9 $mA/cm^2$ and CO mass activity of almost 30 A/g in the voltage of −1.0 V. In sharp contrast, the CO Faradaic efficiency, current density and CO mass activity were all much lower when $[Au_{11}(PPh_3)_8Cl_2]^+$ was employed as the catalyst.

After this initial result, Narouz, Man, DeJesus, Salorinne, Albright, and coworkers systematically investigated the NHC-protected metal nanoclusters [32–36]. They obtained $[Au_{13}(NHC)_9Cl_3]^{2+}$, $[Au_6(NHC)_4]^{2+}$, $[Au_{10}(NHC)_4Br_2]^{2+}$, $[Au_9(NHC)_4Br]^{2+}$, $[Au_{25}(NHC)_{10}Br_7]^{2+}$, and $[Au_8Pt(NHC)_8]^{2+}$ successively, in which the NHC ligands vary each other with different substituents and geometric structures. The as-obtained nanoclusters exhibit intriguing photoluminescence, chirality, and cluster formation process. For example, they synthesized the first full NHC ligand-protected gold nanocluster $[Au_{13}(NHC)_9Cl_3]^{2+}$ by the in situ synthetic method using Au(NHC)Cl as the precursor [32]. This nanocluster bears a wide HOMO–LUMO energy gap of 2.0 eV, thus demonstrating much higher stability than the phosphine ligand-protected gold nanocluster with the similar geometric structure. Meanwhile, this nanocluster is luminescent, with a photoluminescent quantum yield of 16.0%. This value is much higher than that of the most $Au_{13}$ nanoclusters protected by other kinds of ligands. The high thermal stability should be attributed to the natural stability of the NHC ligand, and the exceptional photoluminescence is ascribed to the multiple noncovalent interactions such as the C–H···π interaction, the π···π interaction, etc. Interestingly, Wang and coworkers prepared a bidentate NHC ligand-protected $Au_{13}$ nanocluster [37], whose composition has been identified as $[Au_{13}(bis\text{-}NHC)_5Cl_2]^{3+}$. Compared to the monodentate NHC ligand-protected $Au_{13}$ nanocluster $[Au_{13}(NHC)_9Cl_3]^{2+}$, $[Au_{13}(bis\text{-}NHC)_5Cl_2]^{3+}$ exhibited an even higher luminescent quantum yield (56%) although they are both 8-electron superatom with similar compositions. Unfortunately, the atomically precise structure of $[Au_{13}(bis\text{-}NHC)_5Cl_2]^{3+}$ has not been revealed, restricting the further structure-property correlation investigations.

By introducing chiral elements in NHC ligands, chiral metal nanoclusters protected by NHCs are formed. Man and coworkers prepared a chiral binuclear gold precursor, namely 2,2′-bis(methyl)-(S/R)binaphthyl-linked dinuclear Au(I) bis-*N*-heterocyclic carbene complexes ($Au_2(bisNHC)Br_2$) by the reaction of commercially available $Me_2SAuCl$ with bis-benzimidazolium salts in the presence of a base $K_2CO_3$ at 60 °C [33]. After a further reduction in the presence of $NaBH_4$, a chiral gold nanocluster $[Au_{10}(bisNHC)_4Br_2]^{2+}$ was obtained efficiently. It should be noted that the $Au_{10}$ kernel constitutes a chiral geometry, demonstrating that the chirality has been transferred from the outer NHC ligand to the inner $Au_{10}$ kernel. An intriguing intercluster transformation from $[Au_9(bisNHC)_4Br]^{2+}$ to $[Au_{10}(bisNHC)_4Br_2]^{2+}$ was also identified during the formation of the nanoclusters. Apart from the introduction of chirality, the use of NHC ligands also enables the introduction of heterometals into gold nanoclusters. By employing platinum as the alloying metal, two bimetallic nanoclusters, $[Au_8Pt(NHC)_8]^{2+}$ and $[Au_6Pt(NHC)_7]^{2+}$, were successfully obtained as the major and minor products, respectively. The substituents on nitrogen atoms of the NHC ligand are crucial for high formation of the bimetallic nanoclusters. Meanwhile, these functional groups also play important roles in regulate the redox property of nanoclusters, thus impacting their catalytic activity in electrochemical reduction of $CO_2$ to CO. Bi-elements on the NHC ligands also help to control the structure and property of as-obtained NHC ligand-protected metal nanoclusters. For example, introducing thiolate ligands on NHC constitute bidentate mixed carbene-thiolate ligands, which have a strong effect on the generation of final metal nanoclusters. The presence of the additional thiolate sites on NHC gave a $[Au_6(NHC^{SR})_4]^{2+}$ nanocluster that is difficult to be obtained by conventional NHC ligands.

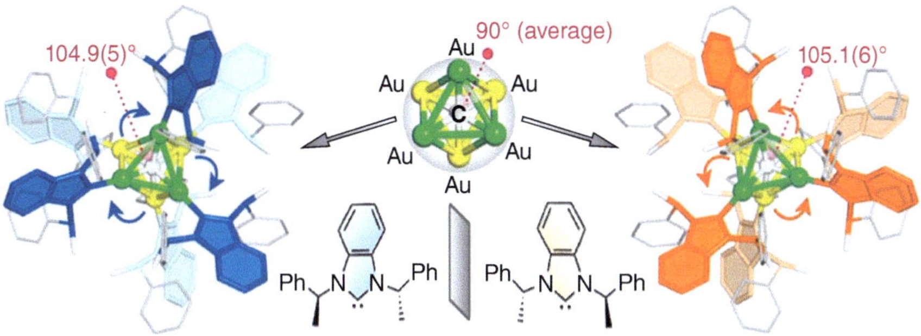


**Figure 6.9** Comparison of structures of the achiral NHC ligand-protected $[(C)Au_6(NHC)_6]^{2+}$ nanocluster and the chiral NHC ligand-protected $[(C)Au_6(NHC)_6]^{2+}$ nanocluster. *Source:* [39]/ with permission of American Chemical Society.

In addition to the contributions from Hannu Häkkinen, Tatsuya Tsukuda, and Cathleen M. Crudden [38], the other representative research groups that are involved in NHC ligand-protected metal nanoclusters are Mitsuhiko Shionoya, Shaung-Quan Zang, and Nanfeng Zheng. Pei and coworkers synthesized a carbon-centered chiral $Au_6$ nanocluster by using a NHC ligand bearing two stereocenters (Figure 6.9) [39]. The most interesting thing of this carbene ligand-protected gold nanocluster is that the chiral NHC ligand on the surface highly affects the geometric structure of the gold kernel. Specifically, Mitsuhiko Shionoya et al. synthesized an achiral NHC ligand-protected carbon-centered chiral $Au_6$ nanocluster $[(C)Au_6(NHC^{achiral})_6]^{2+}$ previously, in which all of the gold atoms are +1 charged. In contrast, by replacing one of the methyl groups by the phenyl group on the NHC ligand, the achiral ligand is converted to a chiral ligand bearing two stereocenters. Under the same reaction conditions, this chiral NHC ligand triggers the formation of a chiral $[(C)Au_6(NHC^{chiral})_6]^{2+}$ nanocluster, in which the charge state of gold, the composition and the coordination mode are all the same to that in the achiral $[(C)Au_6(NHC^{achiral})_6]^{2+}$ nanocluster. However, the $Au_6$ kernel in $[(C)Au_6(NHC^{chiral})_6]^{2+}$ is much more twisted than that in $[(C)Au_6(NHC^{achiral})_6]^{2+}$ (Figure 6.9). That is, the chirality of $[(C)Au_6(NHC^{chiral})_6]^{2+}$ originates from the central chirality of the surface ligands as well as the asymmetric arrangement of the gold atoms in the gold kernel. This kind of structural distortion in $[(C)Au_6(NHC^{chiral})_6]^{2+}$ is retained even this nanocluster is dissolved in solvent.

Luo and coworkers also designed a chiral NHC ligand, in which the skeleton is the same to that designed by Mitsuhiko Shionoya et al., mentioned above. Their difference comes from the substituents on the nitrogen sites of the NHC ligand. Two different substituents, namely the phenylethyl and 2-pyridylmethyl are located at the two different nitrogen atoms of the NHC ligand (Figure 6.10a). The chiral nanocluster $[Au_{13}(NHC)_8Br_4]^+$ showed high luminescence with the emission peak at about 700 nm, and fluorescence quantum yield of 61% [40]. Meanwhile, it demonstrated obvious circularly polarized luminescence (CPL) signal, which has potential utility in sensing applications in biology and medicine. Subsequently, the same research group designed another chiral NHC ligand (Figure 6.10b), in which the

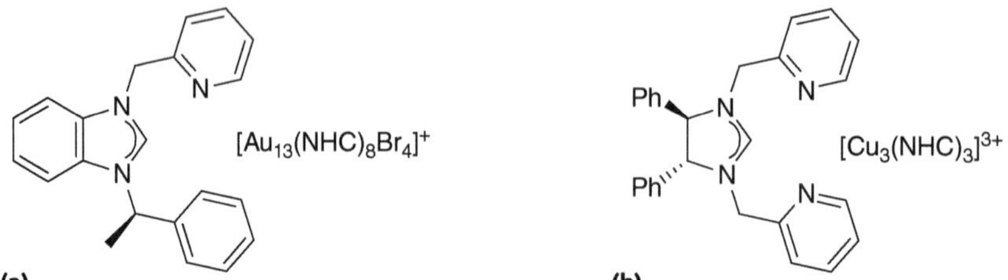


**Figure 6.10** Diversified structures of chiral NHC ligand utilized for the synthesis of chiral metal nanoclusters.

chirality comes from different positions of the scaffold of NHC ligands [41], compared to the aforementioned one. A $[Cu_3(NHC)_3]^{3+}$ nanocluster was obtained by utilizing this chiral NHC as the protecting ligand. The structure of this nanocluster is very simple, while it demonstrates very "unsimple" photoluminescent properties. $[Cu_3(NHC)_3]^{3+}$ exhibited an unprecedented photoluminescence quantum yield of 70%. Meanwhile its solution showed very bright circularly polarized phosphorescence with a dissymmetry factor of $2.1 \times 10^{-3}$. These properties, combined with the excellent solubility and photostability endow the chiral nanocluster with promising application as a pure blue circularly polarized phosphorescence ink for three-dimensional printing.

Nanfeng Zheng et al. [42–44] systematically investigated the catalytic performance of NHC ligand-protected metal nanoclusters, revealing their unique catalytic activity. First, they found that joint usage of the NHC ligand with other types of ligands such as the alkyne and halide would endows the as-obtained nanocluster with both high stability and catalytic activity. Second, they found that the NHC ligand-protected metal nanocluster has the potential of integrating advantages of metal nanoparticle catalysis and metal complex catalysis. We select two typical examples for detailed discussion.

Employing the mixed ligands of NHC (1,3-diisopropylbenzimidazolin-2-ylidene), alkyne (phenylacetylide) and halide (Br), a gold nanocluster $Au_{44}(NHC)_9(alkyne)_6Br_8$ was obtained efficiently (Figure 6.11a) [42]. Owing to the mixed ligand stabilization strategy, the Au–C bonds between NHC ligand and the gold kernel are slightly stronger than that discovered in $[Au_{25}(NHC)_{10}Br_7]^{2+}$, $[Au_{13}(NHC)_6Br_6]^-$, and $[Au_{11}(PPh_3)_7(NHC)Cl_2]^-$, which have been reported previously. The structural features one the one hand endow the nanocluster with high stability, and on the other hand create abundant activation sites on its surface. Specifically, this nanocluster retains its composition and structure even upon heating at 65 °C in the solution for 12 hours. Meanwhile, the halides are easily exchanged by thiolate ligands, indicating that these sites can be employed as activation sites. The catalytic results verified the assumption, and the $Au_{44}(NHC)_9(alkyne)_6Br_8$ nanocluster efficiently catalyzed the alkyne hydration reaction, giving the desired product ketone in up to 99% yield (Figure 6.11b). A unique orderly reactivity was observed for different kinds of alkyne substrates, indicating that the ligand exchange between the surface ligands on the $Au_{44}$ nanocluster and the substrate alkyne is crucial for the activation of

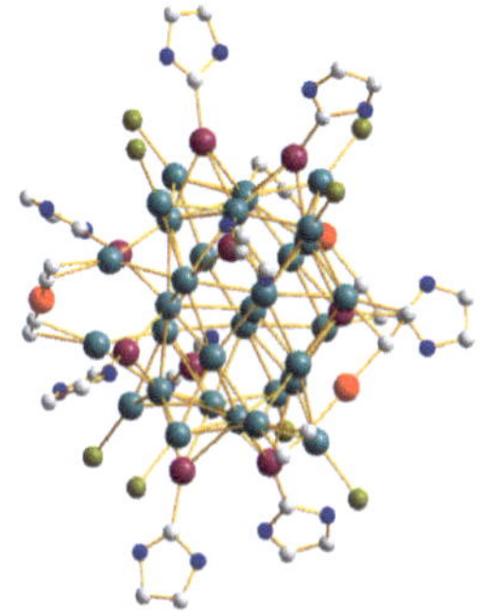

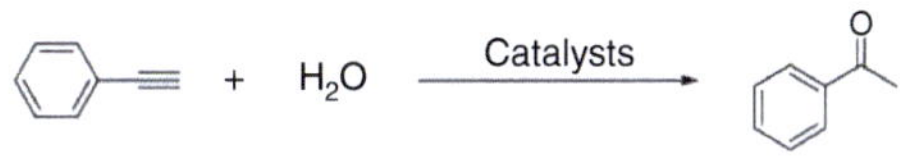


| Entry | Catalyst | Yield (%) |
|---|---|---|
| 1 | $Au_{44}(NHC)_9(alkyne)_6Br_8$ | >99 |
| 2 | Au(NHC)Cl | 22 |
| 3 | Au(NHC)Br | 26 |
| 4 | $Au_{25}(SR)_{18}$ | 26 |
| 5 | $Au_{23}(alkyne)_9(PPh_3)_6$ | 29 |
| 6 | $Au_{13}(NHC)_5Br$ | 28 |
| 7 | Au nanoparticle | 32 |

**(a)** **(b)**

**Figure 6.11** (a) NHC, alkyne and halide co-protected $Au_{44}(NHC)_9(alkyne)_6Br_8$ nanocluster. The gold atoms bonded with NHC and alkyne are marked with purple and red, respectively. (b) $Au_{44}(NHC)_9(alkyne)_6Br_8$ catalyzed alkyne hydration reaction.

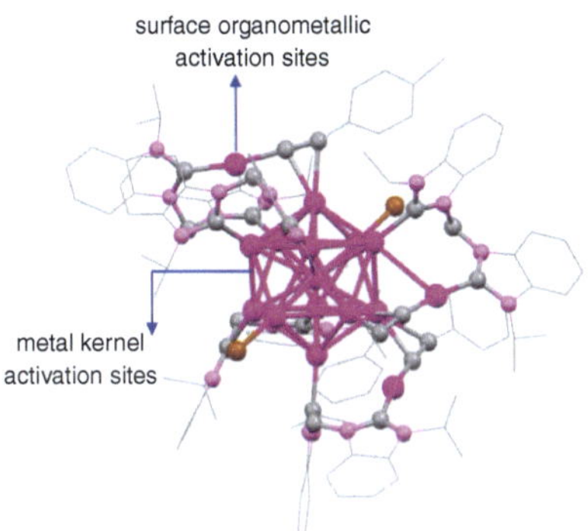


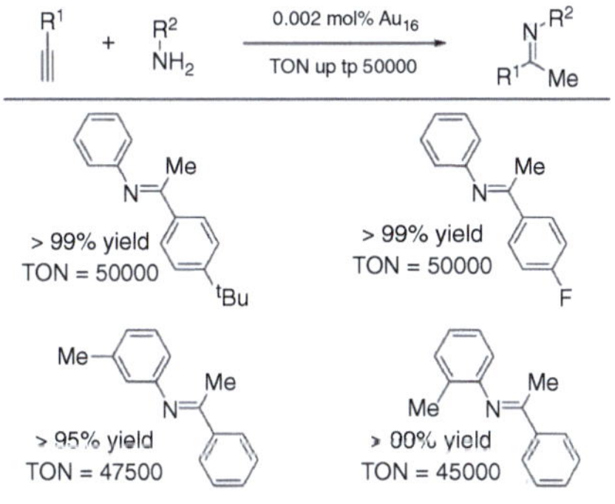


**(a)** **(b)**

**Figure 6.12** (a) Surface organometallic motifs anchored on the metallic kernel of $[Au_{16}(NHC)_5(alkyne)_3Br_2]^{3+}$ nanocluster. (b) $[Au_{16}(NHC)_5(alkyne)_3Br_2]^{3+}$ catalyzed alkyne hydroamination reaction.

alkynes. Meanwhile, the bulky carbene ligands also benefit the creation of uncoordinated surface metal activation sites for binding of water.

A subsequent work on the NHC ligand-protected gold nanocluster further manifests the catalytic features of this kind of nanocluster. Shen and coworkers synthesized a NHC, alkyne and halide co-protected nanocluster $[Au_{16}(NHC)_5(alkyne)_3Br_2]^{3+}$, which is robust as a result of the mixed ligand protection strategy mentioned above [43]. Meanwhile, this nanocluster features a structure that surface organometallic motifs are anchored on the metallic kernel (Figure 6.12a). This interesting structure enables combination of homogeneous molecular catalysis and heterogeneous nanoparticle catalysis in the single platform of metal nanocluster, thus triggering high activity of the $[Au_{16}(NHC)_5(alkyne)_3Br_2]^{3+}$ nanocluster. It catalyzed the alkyne hydroamination reaction with high efficiency, giving TON of 50,000 (Figure 6.12b). Mechanism studies revealed that the gold kernel and the surface organometallic motifs activated different substrates synergistically and produced the final imines in high yield with high selectivity. The principle that introducing both nano-catalytic activation sites and molecular organometallic activation sites in the nanocluster catalyst can be extended for the design of catalytically active metal nanoclusters. Based on this role, they further synthesized a gold nanocluster $[Au_{17}(NHC)_4(alkyne)_4Br_4]^+$, in which more surface organometallic motifs are anchored on the gold kernel. Indeed, this nanocluster showed even higher catalytic activity than $[Au_{16}(NHC)_5(alkyne)_3Br_2]^{3+}$ in the alkyne hydroamination reactions.

## 6.3 Nitrogen-based Ligand-protected Metal Nanocluster for Catalysis

As nanostructured materials with unique physicochemical properties, metal nanoclusters exhibit significant potential in catalysis, optoelectronics, and sensing applications. Among them, coin metal (gold, silver, copper) nanoclusters have garnered substantial research attention due to their intrinsic stability and facile synthetic accessibility. Organic ligands play pivotal roles in stabilizing metal cores while modulating the electronic structures and steric properties of nanoclusters. Nitrogen-containing ligands are traditionally regarded as less effective in stabilizing coin metal nanoclusters due to their relatively weak interactions with these soft-acid metals, posing challenges in constructing nanoclusters with specific geometric configurations. However, recent breakthroughs in designing nitrogen-containing multidentate ligands and fine-tuning nitrogen electronegativity have enabled the successful synthesis of a new class of highly stable coin metal nanoclusters stabilized by such ligands. These nanoclusters not only exhibit unique properties but also expand the structural diversity of metal nanoclusters, revealing unprecedented metal–ligand interaction mechanisms, and facilitating their applications in luminescence and catalysis. This section summarizes the latest advances in stabilizing coin metal nanoclusters using nitrogen-containing multidentate ligands, highlighting their distinctive structural features and intriguing properties from two perspectives: (i) direct coordination of multiple nitrogen atoms to stabilize the metal core and (ii) incorporation of nitrogen atoms within multidentate ligands without direct metal coordination.

Metal nanoclusters are multinuclear aggregates composed of multiple metal atoms with well-defined atomic compositions and structures. As the nascent state of condensed matter, they exhibit high reactivity and unique properties that are distinct from individual metal atoms and nanocrystals. Moreover, their precisely defined atomic structures provide a fundamental basis for investigating structure-property relationships, making them an emerging class of nanomaterials that have attracted significant attention [44–47]. Metal nanoclusters typically consist of a metal core and a surface ligand shell. Coin metals (gold, silver, copper) have garnered particular interest in the study of nanocluster structures and properties due to their inherent stability [48–50]. Surface ligands play crucial roles in stabilizing the geometric structures of metal nanoclusters through interactions with the metal core, while simultaneously modulating their electronic structures and properties [51–54]. Common ligands used for stabilizing coin metal nanoclusters include thiolates [55–57], phosphines [58, 59], carbenes [31, 42], and alkynes [8], which typically stabilize the metal core through monodentate or bidentate coordination. Multidentate ligands, widely applied in metal complexes, also exhibit unique roles in stabilizing metal nanoclusters. These ligands possess three or more coordinating atoms capable of forming multiple coordination bonds with metal centers. This multimodal coordination significantly enhances the binding strength between ligands and metal cores while providing more versatile strategies for modulating the structures and properties of metal nanoclusters. For instance, calixarenes are macrocyclic compounds composed of phenolic units bridged by methylene groups, featuring adjustable cavities and multiple coordination sites. Their conformational flexibility allows them to adapt to guest molecules through subtle structural adjustments, enabling specific recognition and binding [60–63]. Several research groups have reported calixarene-stabilized metal nanoclusters. For example, Guan and coworkers reported $Ag_{35}$ stabilized by calixarene and alkyne [64], as well as $Ag_{34}$, $AuAg_{33}$ [65], and fully calixarene-encapsulated $Ag_{44}$ [66] coin metal nanoclusters. In these systems, they observed interactions between thia-calix[4]arene ligands and metals, along with unique optical properties and enhanced stability. Wang and coworkers synthesized the $C_4$-symmetric $Ag_{88}$ nanocluster using *p-tert*-butylthiacalix[4]arene [67], which consists of eight $Ag_{11}$ secondary structural units surrounding a chromate ion $(CrO_4{}^{2-})$. Furthermore, they developed an even larger $Ag_{155}$ nanocluster [68], comprising 155 silver atoms organized into four nested structural units: $Ag_{13}$@$Ag_{42}$@$Ag_{30}$@$Ag_{70}$. These studies demonstrate the unique capabilities of multidentate ligands in stabilizing coin metal nanoclusters and expanding their structural diversity and functional properties.

Nitrogen atoms, with their small atomic radii, are typically designed as hard basic ligands that bind with relatively small 3d metals such as Fe, Co, and Ni to form metal complexes, which play important roles in catalysis and other fields. However, research on nitrogen-containing ligands in stabilizing metal nanoclusters has not received adequate attention, primarily due to their relatively weak interactions with soft-acid coin metals like gold and silver. Consequently, they were long considered incapable of stabilizing coin metal nanoclusters with specific geometric configurations. Recently, through the rational design of nitrogen-containing multidentate ligands, along with precise tuning of their electronegativity and coordination modes with other atoms, researchers have successfully synthesized a series of highly stable coin metal nanoclusters with unique properties. These advancements have enriched the family of metal nanoclusters and provided more versatile solutions for structural design and property modulation. Nitrogen-containing multidentate ligands stabilize coin metal cores and modulate their properties primarily through two distinct modes: the first mode involves multiple nitrogen atoms directly coordinating with the metal center to stabilize the metal core. This approach is significant for expanding the structural diversity of metal nanoclusters, understanding novel metal–ligand interaction mechanisms, and has led to the development of coin metal nanoclusters with unique luminescent and catalytic properties. Representative examples include multidentate amine ligand-stabilized gold, silver, and copper nanoclusters. The second mode features nitrogen atoms embedded within multidentate ligands without direct coordination to the metal. This design strategy creates available nitrogen sites for surface functionalization and catalytic applications of coin metal nanoclusters, opening new opportunities in these areas. A representative example is phosphine–nitrogen–phosphine (PNP) pincer ligand-stabilized gold nanoclusters. In the following sections, we will elaborate on the latest advancements in nitrogen-containing multidentate ligand-stabilized coin metal nanoclusters from these two perspectives.

### 6.3.1 Metal Nanoclusters Stabilized by Nitrogen-containing Multidentate Amine Ligands

As previously discussed, conventional monodentate, bidentate and primary, secondary, tertiary amine ligands preferentially bind to "hard acids" such as Fe, Co, and Ni but exhibit limited coordination ability with "soft acids" like Au and Ag, making them less effective in stabilizing multinuclear coin metal nanoclusters. However, this limitation can be overcome by introducing multiple nitrogen-based functional groups within a single ligand framework and enhancing the electron density at nitrogen centers through deprotonation or other strategies, thereby enabling robust anchoring of the metal core and yielding structurally stable coin metal nanoclusters (Figure 6.13a).

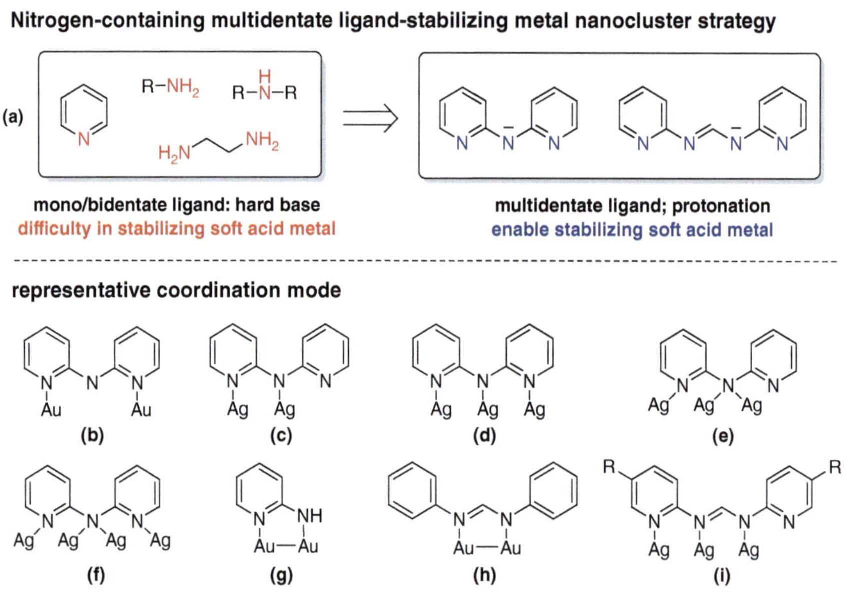


Figure 6.13 Multidentate amine ligand-stabilized coin metal nanoclusters. (a) Deprotonated and multidentate amine ligand stabilizing strategies. (b–i) Abundant coordination modes of multidentate amine ligand on the surface of metal nanoclusters.

Further structural modulation can be achieved by incorporating substituents (Figure 6.13i), which adjust the spatial configuration and coordination modes of nitrogen-containing multidentate amine ligands, tailoring their adaptability to stabilize diverse metal centers. The negatively charged amidinate ligand depicted in Figure 6.13 represents a class of nitrogen-containing multidentate amine ligands that effectively stabilize metal nanoclusters. Its four-electron three-center (NCN) moiety exhibits strong binding affinity with metals. Additionally, amidinate ligands possess multiple potential coordination sites, such as –NH and ≡N–, enabling bridging between multiple metal centers to form multinuclear cluster architectures. For instance, formamidinate ligands can bridge two gold atoms via covalent Au–N bonds while simultaneously enhancing Au–Au interactions, ultimately yielding structurally robust metal nanoclusters. This multidentate coordination strategy not only stabilizes the metal core but also provides precise control over the coordination environment, offering expanded opportunities for geometric and electronic structure modulation (Figure 6.13b–i) illustrates the diverse coordination modes observed in nitrogen-containing multidentate amine ligand-stabilized coin metal nanoclusters, including several unprecedented metal–ligand interaction patterns. These ligands can function either independently to stabilize the metal core or in synergy with other ligands to further optimize the stability and functionality of coin metal nanoclusters [69].

Yuan and coworkers [70] pioneered the use of amine ligands for stabilizing metal nanoclusters. They proposed a surface geometric mismatch strategy based on bulky ligands, employing triphenylphosphine ($PPh_3$) and bis(pyridyl)amine (Hdpa) as ligands to synthesize the well-defined gold nanocluster $Au_{32}(PPh_3)_8(dpa)_6$ (Figure 6.14a). This nanocluster consists of the cationic $[Au_{32}(PPh_3)_8(dpa)_6]^{2+}$ core and two $SbF_6^-$ counteranions. This work represents the first successful application of amine ligands in stabilizing coin metal nanoclusters [71]. During synthesis, deprotonation of Hdpa significantly enhances the electron density at the nitrogen center, enabling the dpa ligand to bridge two gold atoms via its pyridyl N-donors (Figure 6.14b). The resulting $Au_{32}$ nanocluster exhibits exceptional stability [72], retaining its structure for over one year under solid-state conditions and for more than 40 days in solution. Its stability was confirmed by UV–vis–NIR absorption spectroscopy (Figure 6.14c). Building on this success, the same group further utilized bulky $PPh_3$ and Hdpa to synthesize the precisely structured gold nanocluster $[Au_{23}(PPh_3)_{10}(dpa)_2Cl]^{2+}$ [73]. The surface geometric mismatch between the large $PPh_3$ ligands and Hdpa creates eight uncoordinated gold atoms on the $Au_{23}$ surface, providing accessible active sites for catalysis. Detailed studies revealed that $Au_{23}$ efficiently catalyzes the oxidation of benzyl alcohol, achieving 69% conversion with 100% selectivity for benzaldehyde under mild conditions, while maintaining excellent stability throughout the catalytic process (Figure 6.14d) [74, 75]. Beyond gold nanoclusters, nitrogen-containing multidentate amine ligands have also been successfully applied to stabilize silver nanoclusters [76]. Deprotonated Hdpa stabilizes the silver nanoclusters $[Ag_{21}(dpa)_{12}]^+$ and $[Ag_{22}(dpa)_{12}]^{2+}$, which exhibit remarkable stability. The different solubilities of $Ag_{21}$ and $Ag_{22}$, combined with the flexible arrangement of N-donors in the dpa ligands (Figure 6.13c–f), enable reversible interconversion between these two species [77]. In a distinct synthetic approach,

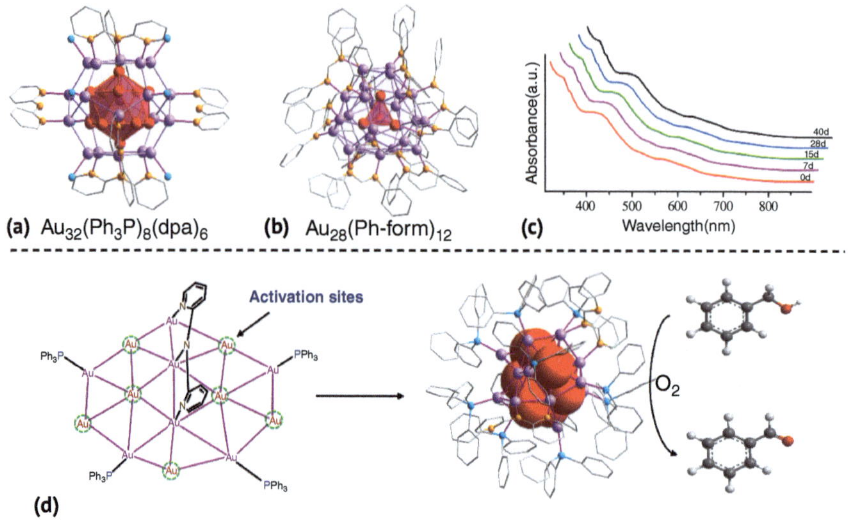


**Figure 6.14** Structures, stability, and catalytic activity of multidentate amine ligand-stabilized gold nanoclusters. (a) Atomically precise structure of typical nanocluster $Au_{32}(Ph_3P)_8(dpa)_6$ [70]. (b) Atomically precise structure of typical nanocluster $Au_{28}$(Ph-form)$_{12}$ [80]. (c) Structural stability of typical nanocluster $Au_{32}(Ph_3P)_8(dpa)_6$ [70]. (d) Catalytic performance of typical nanocluster $[Au_{23}(Ph_3P)_{10}(dpa)_2Cl]^{2+}$ [73]. Color code: Au = purple (the core gold atom is red); P = blue; N = orange; C = gray.

employing a larger excess of sodium borohydride ($NaBH_4$) as reducing agent facilitates the formation of a larger cluster, $Ag_{61}$. This represents the first reported one-dimensional rod-like silver nanocluster featuring vertex-sharing $Ag_{13}$ icosahedral units, formulated as $Ag_{61}(dpa)_{27}(SbF_6)_4$. Its strong electronic coupling effects and unique structure are anticipated to have significant applications in luminescence and related fields [78]. Nitrogen-containing multidentate amine ligand-stabilized metal nanoclusters exhibit unprecedented stability compared to other types of coin metal nanoclusters. For example, the recently reported gold fullerene nanocluster $[Au_{42}(PPh_3)_{10}(5\text{-}CF_3\text{-}dpa)_8]^{2+}$ features a characteristic core–shell structure (Au@$Au_{15}$@$Au_{26}$). Its extraordinary stability arises from the superatomic nature of the cluster and the strong metal–ligand interactions between gold and the nitrogen-containing multidentate amine ligands (Figure 6.13g) [79]. These studies have significantly advanced our understanding of metal–amine interfaces and nanocluster transformation mechanisms, while also promoting further research on nitrogen-containing multidentate amine ligand-stabilized metal nanoclusters and related nanostructured materials.

Beyond dpa ligands, Yuan and coworkers have also developed another class of amidinate nitrogen-containing multidentate ligands and reported their stabilized coin metal nanoclusters. One representative example is the gold nanocluster $[Au_{28}(\text{Ph-form})_{12}]^{2+}$ (Figure 6.14b), where the formamidinate moiety bridges two gold atoms to form covalent Au–N bonds (Figure 6.13h). This nanocluster exhibits excellent stability, retaining its structure in toluene solution at 80 °C for at least six days [80]. Amidinate nitrogen-containing multidentate ligands also effectively stabilize copper and silver nanoclusters. For instance, the tetradentate *N,N'*-bis(5-trifluoromethyl-2-pyridyl)formamidinate (Tf-dpf) stabilizes atomically precise copper nanoclusters $Cu_{11}H_3(Tf\text{-}dpf)_6(OAc)_2$ [81] and $[Cu_{12}H_3(Tf\text{-}dpf)_6(OAc)_2]^+$ [82]. The multidentate coordination feature (four N-donors) and high electron density on nitrogen atoms in Tf-dpf enable strong metal–ligand interactions, significantly enhancing the stability of copper nanoclusters [83]. Interestingly, metal–ligand interactions in amidinate-stabilized copper nanoclusters exhibit distinct behaviors. $Cu_{11}H_3(Tf\text{-}dpf)_6(OAc)_2$ contains three interfacial $\mu_5$-H atoms, while $[Cu_{12}H_3(Tf\text{-}dpf)_6(OAc)_2]^+$ possesses three internal $\mu_6$-H atoms. These structural differences lead to contrasting catalytic performances in nitro reduction reactions: $Cu_{11}$ (with $\mu_5$-H) exhibits high activity, whereas $Cu_{12}$ (with $\mu_6$-H) shows negligible activity. Deuterium labeling experiments confirmed that the $\mu_5$-H in $Cu_{11}$ participates in the catalytic cycle, while the $\mu_6$-H in $Cu_{12}$ does not. This work provides valuable atomic-level insights into the hydrogenation mechanism of copper catalysts [83]. Using tetradentate formamidinate ligands (Figure 6.13i), researchers have also synthesized fcc silver nanoclusters $[Ag_{52}(5\text{-F-dpf})_{16}Cl_4]^{2+}$ and $[Ag_{53}(5\text{-Me-dpf})_{18}]^{5+}$ [84]. The linear and parallel arrangement of formamidinate ligands facilitates the formation of fcc structures. The ligands selectively passivate {111} facets while stabilizing the cluster structure through π–π interactions and C-F…π interactions. This specific ligand arrangement prevents the formation of spike-like structures, enabling the construction of fcc silver nanoclusters. This strategy offers a novel approach for synthesizing fcc-structured metal nanoclusters [85]. These studies collectively demonstrate the immense potential of nitrogen-containing multidentate amine ligands in the synthesis and functional modulation of coin metal nanoclusters. Their unique coordination capabilities not only enhance cluster stability but also effectively regulate surface electronics and steric properties. The resulting families of amidinate-stabilized gold, silver, and copper nanoclusters exhibit remarkable performance, including high activity, selectivity, and stability in catalytic reactions.

### 6.3.2 PNP Pincer Ligand-stabilized Metal Nanoclusters

PNP pincer ligands represent a highly versatile family, typically composed of a pyridine unit and two ortho-substituted methyl diphenylphosphine groups. Conventionally, these pincer ligands exhibit a tridentate coordination mode (P, N, P) with a single metal center, forming stable metal–organic complexes. By strategically modifying the substituents at different positions of the ligand, the electronic properties and steric effects of the resulting metal centers can be finely tuned, enabling precise control over catalytic activity and selectivity [86–93]. Zhang, Fan, Zhang, and coworkers introduced PNP pincer ligands into the field of metal nanoclusters, synthesizing stable metal–pincer clusters [95–99]. They discovered a unique coordination behavior of these ligands on cluster surfaces: instead of adopting the canonical tridentate mode (binding to a single metal atom via P, N, P), the ligand's terminal phosphorus atoms coordinate to distinct metal

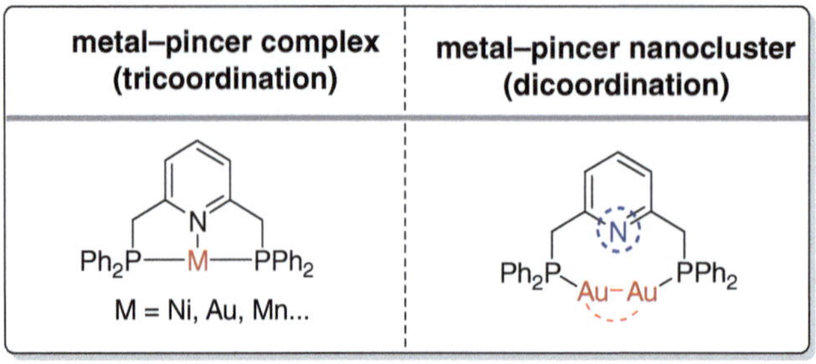


**Figure 6.15** The coordination pattern of the PNP pincer ligand.

atoms across the multicenter metal core. This bimetallic bidentate coordination mode liberates the rigid ligand's central nitrogen atom, creating "open organic nitrogen sites" on the cluster surface (Figure 6.15). These sites transform the pincer ligand from a mere "protective shell" for the metal core into a dynamic "bridge" connecting the cluster to external environments. This design not only stabilizes the metal core but also establishes organic active sites on the surface, enabling synergistic catalysis between the metal sites and the ligand-derived positions.

Based on the strategy of stabilizing metal nanoclusters with PNP pincer ligands, Man-Bo Li group has successively designed and synthesized metal–pincer clusters such as $[Au_8(PNP)_4]^+$ [95], $Au_{17}Cd_2(PNP)_2(SR)_{12}$ [96], and $Au_{35}Cd_2(PNP)(SR)_{18}Cl$ [97]. The $[Au_8(PNP)_4]^+$ cluster consists of eight gold atoms and four PNP pincer ligands. Although $[Au_8(PNP)_4]^+$ has a seven-electron configuration, it exhibits remarkable stability. This stability can be elucidated by comparing it with the structurally analogous $Au_8(DPPF)_4$, which has been previously reported. Both $[Au_8(PNP)_4]^+$ and $Au_8(DPPF)_4$ exhibit similar reduction potentials at −0.86 V. However, the oxidation potential of $[Au_8(PNP)_4]^+$ (0.09 V) is significantly lower than that of $Au_8(DPPF)_4$ (0.34 V), making it more prone to oxidation by losing an electron to form the stable seven-electron configuration. Notably, $[Au_8(PNP)_4]^+$ remains structurally unchanged in solution for up to one week [95]. As previously mentioned, PNP-stabilized metal clusters possess a unique bimetallic bidentate coordination mode, which endows them with "open organic nitrogen sites" on their surface. By leveraging these "open organic nitrogen sites" on the $Au_8$ cluster surface to activate the C–H bond of nitromethane, while simultaneously utilizing the gold sites to activate unsaturated aldehydes and ketones, the cluster demonstrates significantly higher catalytic activity in Michael addition reactions compared to organic Lewis bases or gold catalysts alone (Figure 6.16). Based on the atomically precise structure of the nanocluster, Zhang and coworkers have conducted in-depth mechanistic studies, confirming the synergistic catalytic mode of the $Au_8$ cluster. This provides a theoretical foundation for further tuning the catalytic performance of metal clusters. Furthermore, the rigid pincer ligands on the surface of $Au_8$, along with the distorted arrangement of the gold core, endow the metal cluster with chirality. By utilizing the interaction between the "open organic nitrogen sites" on the $Au_8$ surface and chiral acids, they achieved efficient separation of the enantiomers of chiral

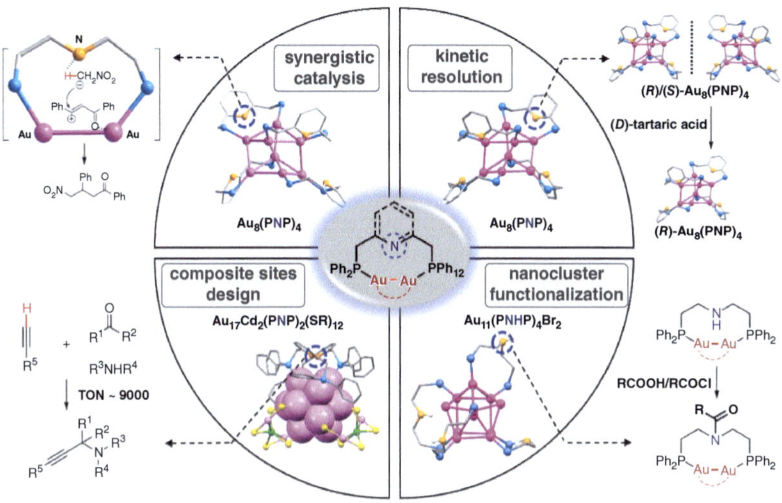


**Figure 6.16** PNP pincer ligand-stabilized gold nanoclusters and their diversified properties Color code: Au = purple; P = blue; N = orange; C = gray; S = yellow; Cd = green.

$Au_8$ clusters using the inexpensive and readily available tartaric acid. This approach enabled the chiral kinetic resolution of the cluster on a gram scale (Figure 6.16).

Based on the $Au_8(PNP)_4$ nanocluster stabilized by PNP ligands, Fan and coworkers further doped cadmium atoms to synthesize the $Au_{17}Cd_2(PNP)_2(SR)_{12}$ nanocluster [96]. This nanocluster features both cadmium sites and organic nitrogen sites on its surface. It exhibits exceptional catalytic activity in tandem reactions involving ketones, amines, and alkynes, enabling room-temperature three-component catalytic conversion to efficiently construct the important organic intermediate propargylamine. Cluster catalysis significantly lowers the reaction temperature (from 100 °C to ambient conditions) and catalyst loading compared to conventional copper catalysts like CuBr, achieving a turnover number (TON) of 9200 which 200 times higher than CuBr. Mechanistic studies reveal that the high catalytic performance of $Au_{17}Cd_2(PNP)_2(SR)_{12}$ arises from the synergistic catalysis of the $Au_{13}$ core, pyridine nitrogen activation sites in PNP, and cadmium sites in the $Au_2Cd(SR)_6$ motif (Figure 6.16). Furthermore, they synthesized a gold nanocluster with surface Lewis acid–base pairs, $Au_{35}Cd_2(PNP)(SR)_{18}Cl$ [97], which simultaneously hosts Lewis acidic CdCl and Lewis basic PNP catalytic centers. This structurally stable cluster demonstrates high activity and selectivity in the four-component Ugi reaction involving aldehydes, amines, carboxylic acids, and isocyanides.

In addition to high catalytic efficiency, Man-Bo Li et al. [98, 99] leveraged the reactivity of the "open organic nitrogen sites" on the surface of metal–pincer clusters to achieve surface functionalization and structural transformation of the clusters. Using a phosphine–nitrogen–hydrogen–phosphine (PNHP) pincer ligand to stabilize the gold core, they synthesized the $Au_{11}(PNHP)_4Br_2$ nanocluster. The NH site in PNHP remained uncoordinated, forming an open nitrogen site on the cluster surface. This not only further validated the unique coordination mode of metal–pincer nanoclusters proposed previously but also provided a predefined platform for surface functionalization. By employing acyl chlorides or carboxylic acids as amidation reagents, they achieved high-yield, selective amidation on the surface of $Au_{11}(PNHP)_4Br_2$ via simple organic amidation reactions (Figure 6.16). Structural analysis revealed that amidation preserved the number of gold atoms in the $Au_{11}$ core and their bonding mode with surface ligands but slightly altered the gold atom arrangement to a more symmetric configuration. This precise functionalization strategy was applied to: (i) introduce functional groups (e.g. halogens, trifluoromethyl, cyano, ester, natural products) onto cluster surfaces; (ii) create chiral environments at the cluster surface; and (iii) enhance the oxidation resistance and photostability of metal clusters. This method enables gram-scale cluster amidation with simplicity and efficiency [98]. Utilizing the unique "open organic nitrogen sites" on metal–pincer clusters, Man-Bo Li and coworkers further achieved structural transformation through directed interactions between nitrogen sites and inducers (e.g. nitromethane ($MeNO_2$), $Ag^+$). Experiments demonstrated that $MeNO_2$ and $AgNO_3$ induced the conversion of $[Au_8(PNP)_4]^+$ into novel $Au_{13}$ and $Au_9Ag_4$ clusters [99]. After structural tuning, femtosecond and nanosecond transient absorption spectroscopy revealed that the excited-state lifetimes of PNPP-stabilized $Au_{13}$ and $Au_9Ag_4$ were significantly longer than those of $Au_8$, with enhanced photoluminescence. This improvement arises from the localized excited-state characteristics of $Au_{13}/Au_9Ag_4$, contrasting with $Au_8$'s charge-transfer behavior, which shortens its excited-state lifetime. Crystalline $Au_9Ag_4$ exhibits a photoluminescence quantum yield of 27%, highlighting its potential in optoelectronic applications. The bimetallic bidentate coordination mode of PNP pincer ligands in metal clusters differs from traditional monometallic tridentate coordination. This mode not only stabilizes the metal core but also liberates nitrogen sites to create active surface sites, enabling applications in synergistic catalysis, chiral resolution, composite catalytic site design, and cluster surface functionalization.

In recent years, nitrogen-containing polydentate ligands have demonstrated unique advantages in the synthesis and functional tuning of coinage metal (gold, silver, copper) clusters. Through ligand design (e.g. polydentate amine ligands, PNP pincer ligands), researchers have overcome the limitations of traditional nitrogen containing ligands with weak interactions toward soft acidic coinage metals, achieving atomic-level precise control over cluster structures. On the one hand, by leveraging synergistic interactions between multiple nitrogen atoms and metals, along with enhanced electron density at nitrogen centers, a series of highly stable polydentate amine ligand-protected metal clusters have been synthesized. These studies have elucidated metal–amine interfacial mechanisms and enabled efficient catalytic performance. On the other hand, PNP pincer ligands employ a bimetallic bidentate coordination mode to form "open organic nitrogen sites" on cluster surfaces, providing new strategies for tuning catalytic activity and photoluminescence properties. These advancements have not only enriched the structural chemistry of metal clusters but also accelerated their practical applications in catalysis, energy conversion, and optoelectronic materials.

## References


1 P. Maity, H. Tsunoyama, M. Yamauchi, S. Xie, T. Tsukuda, Organogold Clusters Protected by Phenylacetylene. *J. Am. Chem. Soc.* **2011**, *133*, 20123–20125.

2 P. Maity, T. Wakabayashi, N. Ichikuni, H. Tsunoyama, S. Xie, M. Yamauchi, T. Tsukuda, Selective Synthesis of Organogold Magic Clusters $Au_{54}(C{\equiv}CPh)_{26}$. *Chem. Commun.* **2012**, *48*, 6085–6087.

3 P. Maity, S. Takano, S. Yamazoe, T. Wakabayashi, T. Tsukuda, Binding Motif of Terminal Alkynes on Gold Clusters. *J. Am. Chem. Soc.* **2013**, *135*, 9450–9457.

4 G. Li, R. Jin, Gold Nanocluster-Catalyzed Semihydrogenation: a Unique Activation Pathway for Terminal Alkynes. *J. Am. Chem. Soc.* **2014**, *136*, 11347–11354.

5 Q. Tang, D.-E. Jiang, Insights into the PhC≡C/Au Interface. *J. Phys. Chem. C.* **2015**, *119*, 10804–10810.

6 N. Kobayashi, Y. Kamei, Y. Shichibu, K. Konishi, Protonation-Induced Chromism of Pyridylethynyl-Appended [core+exo]-Type $Au_8$ Clusters. Resonance-Coupled Electronic Perturbation through π-Conjugated Group. *J. Am. Chem. Soc.* **2013**, *135*, 16078–16081.

7 X.-K. Wan, Q. Tang, S.-F. Yuan, D.-E. Jiang, Q.-M. Wang, Au19 Nanocluster Featuring a V-Shaped Alkynyl–Gold Motif. *J. Am. Chem. Soc.* **2015**, *137*, 652–655.

8 Z. Lei, X.-K. Wan, S.-F. Yuan, Z.-J. Guan, Q.-M. Wang, Alkynyl Approach toward the Protection of Metal Nanoclusters. *Acc. Chem. Res.* **2018**, *51*, 2465–2474.

9 F. Hu, R.-L. He, Z.-J. Guan, C.-Y. Liu, Q.-M. Wang, Disc-Like Silver Nanocluster $Ag_{93}$ Built with Bicapped Hexagonal Prismatic $Ag_{15}$ and Ino Decahedral $Ag_{13}$ Units. *Angew. Chem. Int. Ed.* **2023**, *62*, e202304134; *Angew. Chem.* **2023**, *135*, e202304134.

10 J.-J. Li, Z.-J. Guan, S.-F. Yuan, F. Hu, Q.-M. Wang, Enriching Structural Diversity of Alkynyl-Protected Gold Nanoclusters with Chlorides. *Angew. Chem. Int. Ed.* **2021**, *60*, 6699–6703; *Angew. Chem.* **2021**, *133*, 6773–6777.

11 W.-Q. Shi, L. Zeng, R.-L. He, X.-S. Han, Z.-J. Guan, M. Zhou, Q.-M. Wang, Near-Unity NIR Phosphorescent Quantum Yield from a Room-Temperature Solvated Metal Nanocluster. *Science* **2024**, *383*, 326–330.

12 X.-S. Han, X. Luan, H.-F. Su, J.-J. Li, S.-F. Yuan, Z. Lei, Y. Pei, Q.-M. Wang, Structure Determination of Alkynyl-Protected Gold Nanocluster $Au_{22}(tBuC{\equiv}C)_{18}$ and Its Thermochromic Luminescence. *Angew. Chem. Int. Ed.* **2020**, *59*, 2309–2312; *Angew. Chem.* **2020**, *132*, 2329–2332.

13 J.-Q. Wang, S. Shi, R.-L. He, S.-F. Yuan, G.-Y. Yang, G.-J. Liang, Q.-M. Wang, Total Structure Determination of the Largest Alkynyl-Protected fcc Gold Nanocluster $Au_{110}$ and the Study on Its Ultrafast Excited-State Dynamics. *J. Am. Chem. Soc.* **2020**, *142*, 18086–18092.

14 F. Hu, Z.-J. Guan, G. Yang, J.-Q. Wang, J.-J. Li, S.-F. Yuan, G.-J. Liang, Q.-M. Wang, Molecular Gold Nanocluster $Au_{156}$ Showing Metallic Electron Dynamics. *J. Am. Chem. Soc.* **2021**, *143*, 17059–17067.

**15** X.-L. Pei, Z.-J. Guan, Z.-A. Nan, Q.-M. Wang, Heterometallic Coinage Metal Acetylenediide Clusters Showing Tailored Thermochromic Luminescence. *Angew. Chem. Int. Ed.* **2021**, *60*, 14381–14384; *Angew. Chem.* **2021**, *133*, 14502–14505.

**16** Z.-J. Guan, R.-L. He, S.-F. Yuan, J.-J. Li, F. Hu, C.-Y. Liu, Q.-M. Wang, Ligand Engineering toward the Trade-Off between Stability and Activity in Cluster Catalysis. *Angew. Chem. Int. Ed.* **2022**, *61*, e202116965; *Angew. Chem.* **2022**, *134*, e202116965.

**17** H. Zheng, Y. Li, W.-W. Zhan, J. Zhou, G. Wu, C. Zhang, D. Sun, Y. Yang, Integrating Circular Polarized Luminescence and Long Persistent Luminescence in $Cu_8$ Cage-Like Clusters via Ligand Engineering. *Angew. Chem. Int. Ed.* **2025**, *64*, e202423787; *Angew. Chem.* **2025**, *137*, e202423787.

**18** T. Jia, Z.-J. Guan, C. Zhang, X.-Z. Zhu, Y.-X. Chen, Q. Zhang, Y. Yang, D. Sun, Eight-Electron Superatomic $Cu_{31}$ Nanocluster with Chiral Kernel and NIR-II Emission. *J. Am. Chem. Soc.* **2023**, *145*, 10355–10363.

**19** Y. Tao, N. Luan, C. Yang, J. Sun, K. Li, X. Dai, H. Zhang, Z. Chai, S. Wang, Y. Wang, Incorporation of the $^{99}TcO_4$– Anion within the $Ag_{24}(C{\equiv}C^tBu)_{20}{}^{4+}$ Cluster Unveiling the Unique Shell-to-Core Charge Transfer. *J. Am. Chem. Soc.* **2024**, *146*, 11018–11024.

**20** G. Deng, S. Malola, T. Ki, X. Liu, S. Yoo, K. Lee, M. S. Bootharaju, H. Häkkinen, T. Hyeon, Structural Isomerism in Bimetallic $Ag_{20}Cu_{12}$ Nanoclusters. *J. Am. Chem. Soc.* **2024**, *146*, 26751–26758.

**21** G. Deng, K. Lee, H. Deng, S. Malola, M. S. Bootharaju, H. Häkkinen, N. Zheng, T. Hyeon, Alkynyl-Protected Chiral Bimetallic $Ag_{22}Cu_7$ Superatom with Multiple Chirality Origins. *Angew. Chem. Int. Ed.* **2023**, *62*, e202217483; *Angew. Chem.* **2023**, *135*, e202217483.

**22** Z. Qin, S. Sharma, C. q. Wan, S. Malola, W. w. Xu, H. Häkkinen, G. Li, A Homoleptic Alkynyl-Ligated $[Au_{13}Ag_{16}L_{24}]^{3-}$ Cluster as a Catalytically Active Eight-Electron Superatom. *Angew. Chem. Int. Ed.* **2021**, *60*, 970–975; *Angew. Chem.* **2021**, *133*, 983–988.

**23** G. Deng, J. Kim, M. S. Bootharaju, F. Sun, K. Lee, Q. Tang, Y. J. Hwang, T. Hyeon, Body-Centered-Cubic-Kernelled $Ag_{15}Cu_6$ Nanocluster with Alkynyl Protection: Synthesis, Total Structure, and $CO_2$ Electroreduction. *J. Am. Chem. Soc.* **2023**, *145*, 3401–3407.

**24** X.-K. Wan, T.-T. Liu, N.-L. Li, Q. Dai, J. Wei, Q.-M. Wang, Tailoring Atomically Precise Gold Nanoclusters for Boosting Selective Hydrogenation of Nitrostyrene with $H_2$. *ACS Nano* **2025**, *19*, 11371–11380.

**25** S.-F. Yuan, Z.-J. Guan, Q.-M. Wang, Identification of the Active Species in Bimetallic Cluster Catalyzed Hydrogenation. *J. Am. Chem. Soc.* **2022**, *144*, 11405–11412.

**26** R.-L. He, F. Hu, Z.-J. Guan, Q.-M. Wang, Large Scale Synthesis of a Stable Prefunctionalized Silver Nanocluster. *Angew. Chem. Int. Ed.* **2024**, *63*, e202410827; *Angew. Chem.* **2024**, *136*, e202410827.

**27** E. O. Fischer, A. Maasböl, On the Existence of a Tungsten Carbonyl Carbene Complex. *Angew. Chem. Int. Ed.* **1964**, *3*, 580–581.

**28** D. G. VanDerveer, R. Eisenberg, Rhodium(I) dithiolene complexes. Synthesis, structure, and dynamic behavior. *J. Am. Chem. Soc.* **1974**, *96*, 4994–4996.

**29** A. J. Arduengo, III, R. L. Harlow, M. Kline, A Stable Crystalline Carbene. *J. Am. Chem. Soc.* **1991**, *113*, 361–363.

**30** J. Vignolle, T. D. Tilley, N-Heterocyclic Carbene-Stabilized Gold Nanoparticles and Their Assembly into 3D Superlattices. *Chem. Commun.* **2009**, *14*, 7230–7232.

**31** M. R. Narouz, K. M. Osten, P. J. Unsworth, R. W. Y. Man, K. Salorinne, S. Takano, R. Tomihara, S. Kaappa, S. Malola, C.-T. Dinh, J. D. Padmos, K. Ayoo, P. J. Garrett, M. Nambo, J. H. Horton, E. H. Sargent, H. Häkkinen, T. Tsukuda, C. M. Crudden, N-heterocyclic Carbene-Functionalized Magic-Number Gold Nanoclusters. *Nat. Chem.* **2019**, *11*, 419–425.

**32** M. R. Narouz, S. Takano, P. A. Lummis, T. I. Levchenko, A. Nazemi, S. Kaappa, S. Malola, G. Yousefalizadeh, L. A. Calhoun, K. G. Stamplecoskie, H. Häkkinen, T. Tsukuda, C. M. Crudden, Robust, Highly Luminescent Au13 Superatoms Protected by N-Heterocyclic Carbenes. *J. Am. Chem. Soc.* **2019**, *141*, 14997–15002.

**33** R. W. Y. Man, H. Yi, S. Malola, S. Takano, T. Tsukuda, H. Häkkinen, M. Nambo, C. M. Crudden, Synthesis and Characterization of Enantiopure Chiral Bis NHC-Stabilized Edge-Shared $Au_{10}$ Nanocluster with Unique Prolate Shape. *J. Am. Chem. Soc.* **2022**, *144*, 2056–2061.

**34** J. F. DeJesus, S. I. Jacob, Q. M. Phung, K. Mimura, Y. Aramaki, T. Ooi, M. Nambo, C. M. Crudden, If the Crown Fits: Sterically Demanding N-Heterocyclic Carbene Promotes the Formation of $Au_8Pt$ Nanoclusters. *J. Am. Chem. Soc.* **2024**, *146*, 23806–23813.

**35** K. Salorinne, R. W. Y. Man, P. A. Lummis, M. S. A. Hazer, S. Malola, J. C.-H. Yim, A. J. Veinot, W. Zhou, H. Häkkinen, M. Nambo, C. M. Crudden, Synthesis and Properties of an $Au_6$ Cluster Supported by a Mixed N-heterocyclic Carbene–Thiolate Ligand. *Chem. Commun.* **2020**, *56*, 6102–6105.

**36** E. L. Albright, T. I. Levchenko, V. K. Kulkarni, A. I. Sullivan, J. F. DeJesus, S. Malola, S. Takano, M. Nambo, K. Stamplecoskie, H. Häkkinen, T. Tsukuda, C. M. Crudden, N-Heterocyclic Carbene-Stabilized Atomically Precise Metal Nanoclusters. *J. Am. Chem. Soc.* **2024**, *146*, 5759–5780.

**37** X. Wang, R. Liu, L. Tian, J. Bao, C. Zhao, F. Niu, D. Cheng, Z. Lu, K. Hu, Highly Luminescent NHC-Stabilized $Au_{13}$ Clusters as Efficient Excited-State Electron Donors. *J. Phys. Chem. C.* **2022**, *126*, 18374–18382.

**38** C. A. Smith, M. R. Narouz, P. A. Lummis, I. Singh, A. Nazemi, C.-H. Li, C. M. Crudden, N-Heterocyclic Carbenes in Materials Chemistry. *Chem. Rev.* **2019**, *119*, 4986–5056.

**39** X.-L. Pei, P. Zhao, H. Ube, Z. Lei, K. Nagata, M. Ehara, M. Shionoya, Asymmetric Twisting of C-Centered Octahedral Gold(I) Clusters by Chiral N-Heterocyclic Carbene Ligation. *J. Am. Chem. Soc.* **2022**, *144*, 2156–2163.

**40** P. Luo, X.-J. Zhai, S. Bai, Y.-B. Si, X.-Y. Dong, Y.-F. Han, S.-Q. Zang, Highly Efficient Circularly Polarized Luminescence from Chiral $Au_{13}$ Clusters Stabilized by Enantiopure Monodentate NHC Ligands. *Angew. Chem. Int. Ed.* **2023**, *62*, e202219017; *Angew. Chem.* **2023**, *135*, e202219017.

**41** X.-H. Ma, Y. Si, J.-H. Hu, X.-Y. Dong, G. Xie, F. Pan, Y.-L. Wei, S.-Q. Zang, Y. Zhao, High-Efficiency Pure Blue Circularly Polarized Phosphorescence from Chiral N-Heterocyclic-Carbene-Stabilized Copper(I) Clusters. *J. Am. Chem. Soc.* **2023**, *145*, 25874–25886.

**42** H. Shen, Z. Xu, M. S. A. Hazer, Q. Wu, J. Peng, R. Qin, S. Malola, B. K. Teo, H. Häkkinen, N. Zheng, Surface Coordination of Multiple Ligands Endows N-Heterocyclic Carbene-Stabilized Gold Nanoclusters with High Robustness and Surface Reactivity. *Angew. Chem. Int. Ed.* **2021**, *60*, 3752–3758; *Angew. Chem.* **2021**, *133*, 3796–3802.

**43** H. Shen, Q. Wu, S. Malola, Y.-Z. Han, Z. Xu, R. Qin, X. Tang, Y.-B. Chen, B. K. Teo, H. Häkkinen, N. Zheng, N-Heterocyclic Carbene-Stabilized Gold Nanoclusters with Organometallic Motifs for Promoting Catalysis. *J. Am. Chem. Soc.* **2022**, *144*, 10844–10853.

**44** J. Yan, B. K. Teo, N. Zheng, Surface Chemistry of Atomically Precise Coinage–Metal Nanoclusters: From Structural Control to Surface Reactivity and Catalysis. *Acc. Chem. Res.* **2018**, *51*, 3084–3093.

**45** R. Jin, C. Zeng, M. Zhou, Y. Chen, Atomically Precise Colloidal Metal Nanoclusters and Nanoparticles: Fundamentals and Opportunities. *Chem. Rev.* **2016**, *116*, 10346–10413.

**46** J. Li, Gold Nanocluster Based Electrochemiluminescence Probe with 66% Quantum Yield and Its Mechanism. *Chin. Sci. Bull.* **2019**, *64*, 2661–2663.

**47** X. Kang, M. Zhu, The Precise Regulation of Nonlinear Optical Properties of Fluorescent Nanoclusters. *Chin. Sci. Bull.* **2022**, *67*, 4029–4030.

**48** J. Fang, Z. Liu, Y.-P. Xie, X. Lu, Synthesis, Crystal Structures and Photoelectric Properties of Copper(I) Alkynyl Nanoclusters. *Chin. Sci. Bull.* **2022**, *67*, 1119–1136.

**49** X. Cai, G. Saranya, K. Shen, M. Chen, R. Si, W. Ding, Y. Zhu, Reversible Switching of Catalytic Activity by Shuttling an Atom into and out of Gold Nanoclusters. *Angew. Chem. Int. Ed.* **2019**, *58*, 9964–9968; *Angew. Chem.* **2019**, *131*, 10069–10073.

**50** C. Zeng, Y. Chen, K. Kirschbaum, K. J. Lambright, R. Jin, Emergence of Hierarchical Structural Complexities in Nanoparticles and Their Assembly. *Science* **2016**, *354*, 1580–1584.

**51** T. Kawawaki, Y. Kataoka, M. Hirata, Y. Akinaga, R. Takahata, K. Wakamatsu, Y. Fujiki, M. Kataoka, S. Kikkawa, A. S. Alotabi, S. Hossain, D. J. Osborn, T. Teranishi, G. G. Andersson, G. F. Metha, S. Yamazoe, Y. Negishi, Creation of High-Performance Heterogeneous Photocatalysts by Controlling Ligand Desorption and Particle Size of Gold Nanocluster. *Angew. Chem. Int. Ed.* **2021**, *60*, 21340–21350; *Angew. Chem.* **2021**, *133*, 21510–21520.

**52** S. Wang, L. Tang, B. Cai, Z. Yin, Y. Li, L. Xiong, X. Kang, J. Xuan, Y. Pei, M. Zhu, Ligand Modification of $Au_{25}$ Nanoclusters for Near-Infrared Photocatalytic Oxidative Functionalization. *J. Am. Chem. Soc.* **2022**, *144*, 3787–3792.

**53** X.-K. Wan, J.-Q. Wang, Z.-A. Nan, Q.-M. Wang, Ligand effects in catalysis by atomically precise gold nanoclusters. *Sci. Adv.* **2017**, *3*, e1701823.

**54** S. Li, X.-Y. Dong, K.-S. Qi, S.-Q. Zang, T. C. W. Mak, Full-Color Tunable Circularly Polarized Luminescence Induced by the Crystal Defect from the Co-assembly of Chiral Silver(I) Clusters and Dyes. *J. Am. Chem. Soc.* **2021**, *143*, 20574–20578.

**55** B. Nieto-Ortega, T. Bürgi, Vibrational Properties of Thiolate-Protected Gold Nanoclusters. *Acc. Chem. Res.* **2018**, *51*, 2811–2819.

**56** N. A. Sakthivel, A. Dass, Aromatic Thiolate-Protected Series of Gold Nanomolecules and a Contrary Structural Trend in Size Evolution. *Acc. Chem. Res.* **2018**, *51*, 1774–1783.

**57** M. Zhu, Q. Yao, Z. Liu, B. Zhang, Y. Lin, J. Liu, M. Long, J. Xie, Surface Engineering Assisted Size and Structure Modulation of Gold Nanoclusters by Ionic Liquid Cations. *Angew. Chem. Int. Ed.* **2022**, *61*, e202115647; *Angew. Chem.* **2022**, *134*, e202115647.

**58** Z.-H. Gao, K. Wei, T. Wu, J. Dong, D.-E. Jiang, S. Sun, L.-S. Wang, A Heteroleptic Gold Hydride Nanocluster for Efficient and Selective Electrocatalytic Reduction of $CO_2$ to CO. *J. Am. Chem. Soc.* **2022**, *144*, 5258–5262.

**59** L.-L. Yan, L.-Y. Yao, M. Ng, V. W.-W. Yam, Stimuli-Responsive and Structure-Adaptive Three-Dimensional Gold(I) Cluster Cages Constructed via "De-aurophilic" Interaction Strategy. *J. Am. Chem. Soc.* **2021**, *143*, 19008–19017.

**60** Y. Bi, S. Du, W. Liao, Thiacalixarene-based nanoscale polyhedral coordination cages. *Coord. Chem. Rev.* **2014**, *276*, 61–72.

**61** S. B. Nimse, T. Kim, Biological applications of functionalized calixarenes. *Chem. Soc. Rev.* **2013**, *42*, 366–386.

**62** T. R. Tshikhudo, D. Demuru, Z. Wang, M. Brust, A. Secchi, A. Arduini, A. Pochini, Molecular Recognition by Calix[4]arene-Modified Gold Nanoparticles in Aqueous Solution. *Angew. Chem. Int. Ed.* **2005**, *44*, 2913–2916; *Angew. Chem.* **2005**, *117*, 2973–2976.

**63** N. Zhang, M. Wang, Y. Chen, C. Tang, Ligand Engineering of Coinage Metal Nanoclusters: The Synthesis and Properties of Calixarene Stabilized Nanoclusters. *Chin. Sci. Bull.* **2023**, *68*, 4318–4327.

**64** Z.-J. Guan, J.-L. Zeng, Z.-A. Nan, X.-K. Wan, Y.-M. Lin, Q.-M. Wang, Thiacalix[4]arene: New Protection for Metal Nanoclusters. *Sci. Adv.* **2016**, *2*, e1600323.

**65** Z.-J. Guan, F. Hu, S.-F. Yuan, Z.-A. Nan, Y.-M. Lin, Q.-M. Wang, The Stability Enhancement Factor Beyond Eight-Electron Shell Closure in Thiacalix[4]Arene-Protected Silver Clusters. *Chem. Sci.* **2019**, *10*, 3360–3365.

**66** L.-M. Zheng, W.-Q. Shi, F. Hu, Z.-J. Guan, Q.-M. Wang, All-Calixarene-Protected Silver Nanocluster with All Silver Atoms in a Face-Centered Cubic Arrangement. *J. Am. Chem. Soc.* **2024**, *146*, 25101–25107.

**67** Z. Wang, H.-F. Su, Y.-W. Gong, Q.-P. Qu, Y.-F. Bi, C.-H. Tung, D. Sun, L.-S. Zheng, A Hierarchically Assembled 88-Nuclei Silver-Thiacalix[4]arene Nanocluster. *Nat. Commun.* **2020**, *11*, 308.

**68** Z. Wang, F. Alkan, C. M. Aikens, M. Kurmoo, Z.-Y. Zhang, K.-P. Song, C.-H. Tung, D. Sun, An Ultrastable 155-Nuclei Silver Nanocluster Protected by Thiacalix[4]arene and Cyclohexanethiol for Photothermal Conversion. *Angew. Chem. Int. Ed.* **2022**, *61*, e202206742; *Angew. Chem.* **2022**, *134*, e202206742.

**69** S.-F. Yuan, W.-D. Liu, C.-Y. Liu, Z.-J. Guan, Q.-M. Wang, Nitrogen Donor Protection for Atomically Precise Metal Nanoclusters. *Chem. Eur. J.* **2022**, *28*, e202104445.

**70** S.-F. Yuan, C.-Q. Xu, J. Li, Q.-M. Wang, A Ligand-Protected Golden Fullerene: The Dipyridylamido $Au_{32}^{8+}$ Nanocluster. *Angew. Chem. Int. Ed.* **2019**, *58*, 5906–5909; *Angew. Chem.* **2019**, *131*, 5967–5970.

**71** J. Li, X. Li, H.-J. Zhai, L.-S. Wang, $Au_{20}$: A Tetrahedral Cluster. *Science* **2003**, *299*, 864–867.

**72** M. P. Johansson, D. Sundholm, J. Vaara, $Au_{32}$: A 24-Carat Golden Fullerene. *Angew. Chem. Int. Ed.* **2004**, *43*, 2678–2681; *Angew. Chem.* **2004**, *116*, 2732–2735.

**73** S.-F. Yuan, Z. Lei, Z.-J. Guan, Q.-M. Wang, Atomically Precise Preorganization of Open Metal Sites on Gold Nanoclusters with High Catalytic Performance. *Angew. Chem. Int. Ed.* **2021**, *60*, 5225–5229; *Angew. Chem.* **2021**, *133*, 5285–5289.

**74** N. de Silva, J.-M. Ha, A. Solovyov, M. M. Nigra, I. Ogino, S. W. Yeh, K. A. Durkin, A. Katz, A Bioinspired Approach for Controlling Accessibility in Calix[4]Arene-Bound Metal Cluster Catalysts. *Nat. Chem.* **2010**, *2*, 1062–1068.

**75** J. Chen, Q.-F. Zhang, T. A. Bonaccorso, P. G. Williard, L.-S. Wang, Controlling Gold Nanoclusters by Diphospine Ligands. *J. Am. Chem. Soc.* **2014**, *136*, 92–95.

**76** S.-F. Yuan, Z.-J. Guan, W.-D. Liu, Q.-M. Wang, Solvent-Triggered Reversible Interconversion of All-Nitrogen-Donor-Protected Silver Nanoclusters and Their Responsive Optical Properties. *Nat. Commun.* **2019**, *10*, 4032.

**77** X.-K. Wan, Z.-J. Guan, Q.-M. Wang, Homoleptic Alkynyl-Protected Gold Nanoclusters: $Au_{44}(PhC{\equiv}C)_{28}$ and $Au_{36}(PhC{\equiv}C)_{24}$. *Angew. Chem. Int. Ed.* **2017**, *56*, 11494–11497; *Angew. Chem.* **2017**, *129*, 11652–11655.

**78** S.-F. Yuan, C.-Q. Xu, W.-D. Liu, J.-X. Zhang, J. Li, Q.-M. Wang, Rod-Shaped Silver Supercluster Unveiling Strong Electron Coupling Between Substituent Icosahedral Units. *J. Am. Chem. Soc.* **2021**, *143*, 12261–12267.

**79** H.-W. Luyang, C.-Q. Xu, S.-F. Yuan, J. Li, Q.-M. Wang, Ligand-Protected Golden Fullerene $Au_{42}$ with Unprecedented Metal Coordination Number of 15. *Angew. Chem. Int. Ed.* **2025**, *64*, e202417904; *Angew. Chem.* **2025**, *137*, e202417904.

**80** S.-F. Yuan, R.-L. He, X.-S. Han, J.-Q. Wang, Z.-J. Guan, Q.-M. Wang, Robust Gold Nanocluster Protected with Amidinates for Electrocatalytic $CO_2$ Reduction. *Angew. Chem. Int. Ed.* **2021**, *60*, 14345–14349; *Angew. Chem.* **2021**, *133*, 14466–14470.

**81** C.-Y. Liu, S.-F. Yuan, S. Wang, Z.-J. Guan, D.-e. Jiang, Q.-M. Wang, Structural Transformation and Catalytic Hydrogenation Activity of Amidinate-Protected Copper Hydride Clusters. *Nat. Commun.* **2022**, *13*, 2082.

**82** X. Liu, D. Astruc, Atomically Precise Copper Nanoclusters and Their Applications. *Coord. Chem. Rev.* **2018**, *359*, 112–126.

**83** S.-F. Yuan, H.-W. Luyang, Z. Lei, X.-K. Wan, J.-J. Li, Q.-M. Wang, A Stable Well-Defined Copper Hydride Cluster Consolidated with Hemilabile Phosphines. *Chem. Commun.* **2021**, *57*, 4315–4318.

**84** F. Hu, H.-W. Luyang, R.-L. He, Z. J. Guan, S.-F. Yuan, Q.-M. Wang, Face-Centered Cubic Silver Nanoclusters Consolidated with Tetradentate Formamidinate Ligands. *J. Am. Chem. Soc.* **2022**, *144*, 19365–19371.

**85** S. Zhuang, L. Liao, J. Yuan, N. Xia, Y. Zhao, C. Wang, Z. Gan, N. Yan, L. He, J. Li, H. Deng, Z. Guan, J. Yang, Z. Wu, Fcc versus Non-fcc Structural Isomerism of Gold Nanoparticles with Kernel Atom Packing Dependent Photoluminescence. *Angew. Chem. Int. Ed.* **2019**, *58*, 4510–4514; *Angew. Chem.* **2019**, *131*, 4558–4562.

**86** M. Garbe, K. Junge, S. Walker, Z. Wei, H. Jiao, A. Spannenberg, S. Bachmann, M. Scalone, M. Beller, Manganese(I)-Catalyzed Enantioselective Hydrogenation of Ketones Using a Defined Chiral PNP Pincer Ligand. *Angew. Chem. Int. Ed.* **2017**, *56*, 11237–11241; *Angew. Chem.* **2017**, *129*, 11389–11393.

**87** K. Arashiba, Y. Miyake, Y. Nishibayashi, A Molybdenum Complex Bearing PNP-Type Pincer Ligands Leads to the Catalytic Reduction of Dinitrogen into Ammonia. *Nat. Chem.* **2011**, *3*, 120–125.
**88** D. Benito-Garagorri, K. Kirchner, Modularly Designed Transition Metal PNP and PCP Pincer Complexes based on Aminophosphines: Synthesis and Catalytic Applications. *Acc. Chem. Res.* **2008**, *41*, 201–213.
**89** S. Gu, R. J. Nielsen, K. H. Taylor, G. C. Fortman, J. Chen, D. A. Dickie, W. A. Goddard, T. B. Gunnoe, Use of Ligand Steric Properties to Control the Thermodynamics and Kinetics of Oxidative Addition and Reductive Elimination with Pincer-Ligated Rh Complexes. *Organometallics* **2020**, *39*, 1917–1933.
**90** M. Rauch, S. Kar, A. Kumar, L. Avram, L. J. W. Shimon, D. Milstein, Metal–Ligand Cooperation Facilitates Bond Activation and Catalytic Hydrogenation with Zinc Pincer Complexes. *J. Am. Chem. Soc.* **2020**, *142*, 14513–14521.
**91** L. Maser, C. Schneider, L. Vondung, L. Alig, R. Langer, Quantifying the Donor Strength of Ligand-Stabilized Main Group Fragments. *J. Am. Chem. Soc.* **2019**, *141*, 7596–7604.
**92** J. C. Borghs, L. M. Azofra, T. Biberger, O. Linnenberg, L. Cavallo, M. Rueping, O. El-Sepelgy, Manganese-Catalyzed Multicomponent Synthesis of Pyrroles through Acceptorless Dehydrogenation Hydrogen Autotransfer Catalysis: Experiment and Computation. *ChemSusChem* **2019**, *12*, 3083–3088.
**93** S. M. Chapp, N. D. Schley, Reversible C(sp3)-Si Oxidative Addition of Unsupported Organosilanes: Effects of Silicon Substituents on Kinetics and Thermodynamics. *J. Am. Chem. Soc.* **2021**, *143*, 5534–5539.
**94** Y. Park, S. P. Semproni, H. Zhong, P. J. Chirik, Synthesis, Electronic Structure, and Reactivity of a Planar Four-Coordinate, Cobalt–Imido Complex. *Angew. Chem. Int. Ed.* **2021**, *60*, 14376–14380; *Angew. Chem.* **2021**, *133*, 14497–14501.
**95** T.-S. Zhang, W. Fei, N. Li, Y. Zhang, C. Xu, Q. Luo, M.-B. Li, Open Nitrogen Site-Induced Kinetic Resolution and Catalysis of a Gold Nanocluster. *Nano Lett.* **2023**, *23*, 235–242.
**96** J.-Q. Fan, Y. Yang, C.-B. Tao, M.-B. Li, Cadmium Doped and Pincer Ligand-Modified Gold Nanocluster for Catalytic KA2 Reaction. *Angew. Chem. Int. Ed.* **2023**, *62*, e202215741; *Angew. Chem.* **2023**, *135*, e202215741.
**97** J.-Q. Fan, Y. Li, W. Wu Xu, M.-B. Li, Loading Lewis Acid/Base Pair on Metal Nanocluster for Catalytic Ugi Reaction. *Angew. Chem. Int. Ed.* **2025**, *64*, e202413861; *Angew. Chem.* **2025**, *137*, e202413861.
**98** Y. Zhang, S.-R. He, Y. Yang, T.-S. Zhang, Z.-M. Zhu, W. Fei, M.-B. Li, Preorganized Nitrogen Sites for $Au_{11}$ Amidation: A Generalizable Strategy toward Precision Functionalization of Metal Nanoclusters. *J. Am. Chem. Soc.* **2023**, *145*, 12164–12172.
**99** Y. Zhang, W. Zhang, T.-S. Zhang, C. Ge, Y. Tao, W. Fei, W. Fan, M. Zhou, M.-B. Li, Site-Recognition-Induced Structural and Photoluminescent Evolution of the Gold–Pincer Nanocluster. *J. Am. Chem. Soc.* **2024**, *146*, 9631–9639.

# 7

# Isomeric Metal Nanocluster Catalysis and Nanocluster Asymmetric Catalysis

## 7.1 Isomeric Metal Nanocluster Catalysis

Minor changes on surface or even kernel structures result in major difference on catalytic performance of metal nanoclusters. Representatives include the central atom switching, surface ligand exchange and one or two metal atoms lacking. For example, Wan and coworkers revealed that $Au_{38}$ nanoclusters bearing the same metal kernel while different surface ligands demonstrate completely different activity in hydrogenation reaction. Specifically, $[Au_{38}(PhC{\equiv}C)_{20}(PPh_3)_4]^{2+}$ and $[Au_{38}(SR)_{20}(PPh_3)_4]^{2+}$ nanoclusters have the same gold kernel, surface phosphine ligand and charge state [1]. The only difference comes from another organic ligand on the surface, one of which is alkyne and the other one is thiolate. $[Au_{38}(PhC{\equiv}C)_{20}(PPh_3)_4]^{2+}$ exhibits high activity and selectivity in catalyzing alkyne semihydrogenation, giving the corresponding alkene products in up to >99% yield and >99% selectivity. On the contrary, $[Au_{38}(SR)_{20}(PPh_3)_4]^{2+}$ nanocluster is catalytically inert toward this kind of reaction and results in recovery of the alkyne substrates. The same research group revealed the similar phenomenon in $Ag_{25}Cu_4Cl_6(dppb)_6(PhC{\equiv}C)_{12}(SO_3CF_3)_3$/$Ag_{25}Cu_4Cl_6H_8(dppb)_6(PhC{\equiv}C)_{12}(SO_3CF_3)_3$ nanoclusters and $[Ag_{21}(H_2BTCA)_3(O_2PPh_2)_6]^+$/$[Ag_{21}(RC{\equiv}C)_6(O_2PPh_2)_{10}]^+$ ($H_4$BTCA=*p-tert*-butylthiacalix-[4]arene) nanoclusters catalytic 4-nitrophenol reduction reactions [2, 3].

Research works from other groups also present the distinct catalytic difference of structurally similar metal nanoclusters. For example, Ruan and coworkers [4] prepared two $Pd_6(SR)_{12}$ nanoclusters. One of them is protected by cyclohexanethiol group ($SC_6H_{11}$) and the other one is protected by 2-phenylethanethiol group ($SC_2H_4Ph$). Interestingly, $Pd_6(SC_6H_{11})_{12}$ possesses an octahedral geometry and $Pd_6(SC_2H_4Ph)_{12}$ bears a tiara-like geometry. The structural discrepancy results in their completely different catalytic activities. For Sonogashira cross-coupling reaction, although $Pd_6(SC_6H_{11})_{12}$ and $Pd_6(SC_2H_4Ph)_{12}$ are both stable under the reaction conditions,

*Metal Nanocluster Catalysis: From Fundamentals to Applications*, First Edition. Manzhou Zhu and Man-Bo Li. © 2026 Wiley-VCH GmbH. All rights reserved, including rights for text and data mining and training of artificial intelligence technologies or similar technologies. Published 2026 by Wiley-VCH GmbH.

$Pd_6(SC_2H_4Ph)_{12}$ displays much higher catalytic activity compared to $Pd_6(SC_6H_{11})_{12}$ [4]. The reaction conversion and selectivity catalyzed by $Pd_6(SC_2H_4Ph)_{12}$ can be achieved in 90% and 100%, respectively. Experimental and theoretical results reveal that it is the difficulty on the initial stage (i.e. iodobenzene adsorption and oxidative addition) that results in the low activity of $Pd_6(SC_6H_{11})_{12}$ because of its octahedral geometry. Dong and coworkers also found that $[Cu_{58}H_{20}(SC_2H_4Ph)_{36}(PPh_3)_4]^{2+}$ and $[Cu_{57}H_{20}(SC_2H_4Ph)_{36}(PPh_3)_4]^{2+}$ nanoclusters with only one copper atom difference displayed different activity in azide-alkyne cycloaddition reaction (click reaction) [5]. $[Cu_{57}H_{20}(SC_2H_4Ph)_{36}(PPh_3)_4]^{2+}$ nanocluster with one copper lacking efficiently catalyzes the click reaction, while $[Cu_{58}H_{20}(SC_2H_4Ph)_{36}(PPh_3)_4]^{2+}$ nanocluster shows low activity. This kind of catalytic difference originates from the fact that loss of the single copper atom creates active catalytic sites for the activation of organic substrates.

The above examples show interesting catalytic phenomenon of structurally similar metal nanoclusters. However, the real isomeric metal nanocluster catalysis is rarely known. The major reason is the lack of strictly isomeric metal nanoclusters with the completely same molecular formula. One of the examples was reported by Tian and coworkers in 2016. They synthesized an $Au_{38}(SC_2H_4Ph)_{24}$ nanocluster by a modified in situ synthesis method and purified the nanocluster by preparative thin-layer chromatography [6]. ESI-MS gave intense peaks of $[Au_{38}(SC_2H_4Ph)_{24} + Cs]^+$ and $[Au_{38}(SC_2H_4Ph)_{24} + 2Cs]^{2+}$, indicating that this nanocluster is charged neutral. Therefore, this nanocluster has the completely same composition with the previously reported $Au_{38}(SC_2H_4Ph)_{24}$ nanocluster by Rongchao Jin et al. [7]. However, SCXRD revealed that both nanoclusters possessed different structures. Moreover, other properties such as their UV–vis absorptions and polarity were also different with each other. Both $Au_{38}(SC_2H_4Ph)_{24}$ nanoclusters have $Au_{23}$ kernel, while the atomic arrangement in the $Au_{23}$ kernel is quite different. The $Au_{23}$ kernel of the previously reported $Au_{38}(SC_2H_4Ph)_{24}$ nanocluster can be viewed as a face-fused bi-icosahedral, while this $Au_{38}(SC_2H_4Ph)_{24}$ nanocluster has a $Au_{23}$ kernel consisting of an icosahedral $Au_{13}$ and a $Au_{10}$ unit. Apart from the difference of their $Au_{23}$ kernels, the surface motifs are also quite different, as shown in Figure 7.1.

The structural difference results in distinct catalytic activity of the two $Au_{38}(SC_2H_4Ph)_{24}$ nanocluster isomers. Specifically, although both nanoclusters can efficiently catalyze 4-nitrophenol reduction at room temperature, their catalytic performance at low temperature is different. The $Au_{38}(SC_2H_4Ph)_{24}$ nanocluster reported by Zhikun Wu catalyzed the reduction of 4-nitrophenol with $NaBH_4$ as the reductant at 0 °C, giving the desired product 4-aminophenol in 44% yield. In sharp contrast, the $Au_{38}(SC_2H_4Ph)_{24}$ nanocluster reported by Rongchao Jin et al. showed none of catalytic activity under the same reaction conditions. Notably, the $Au_{38}(SC_2H_4Ph)_{24}$ nanocluster reported by Zhikun Wu can be slowly transformed to the previously reported $Au_{38}(SC_2H_4Ph)_{24}$ nanocluster isomer, indicating the more thermodynamically stable structure of the previously reported $Au_{38}(SC_2H_4Ph)_{24}$ isomer. This experimental phenomenon also explains the similar catalytic activity of both nanoclusters at room temperature, because one of which might be transformed to the other one during the catalysis at the temperature. While at lower

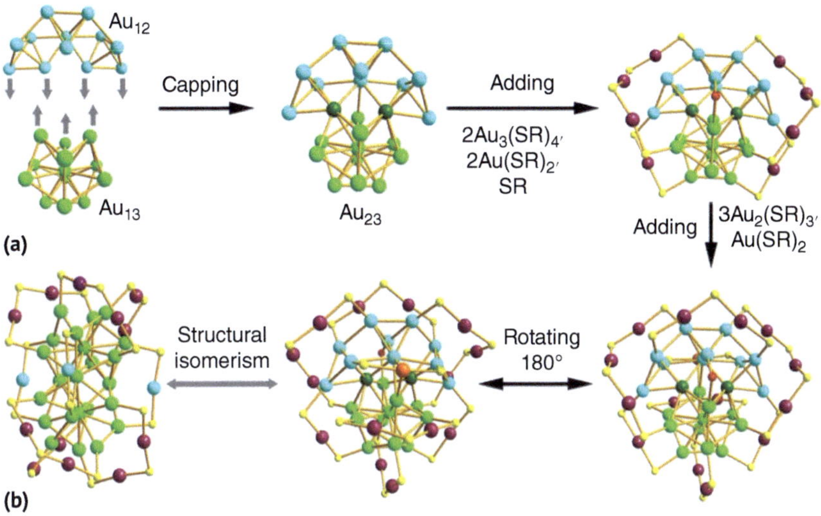


**Figure 7.1** (a) Structural comparison of two $Au_{38}(SC_2H_4Ph)_{24}$ nanocluster isomers. (b) Catalytic difference of two $Au_{38}(SC_2H_4Ph)_{24}$ nanocluster isomers. *Source:* [6]/with permission of Springer Nature.

temperature, the $Au_{38}(SC_2H_4Ph)_{24}$ nanocluster reported by Zhikun Wu presents its natural catalytic activity, which should be higher than the previously reported one.

Up to now, the reported metal nanocluster isomers are still quite limited, in a strict sense. The corresponding examples include [8–13]: two $Au_{38}(SC_2H_4Ph)_{24}$ nanocluster structural isomers; two $Au_{23}(C{\equiv}CBu^t)_{15}$ nanocluster structural isomers; two $[Au_{13}Ag_{12}(PPh_3)_{10}Cl_8]^+(SbF_6)^-$ nanocluster structural isomers; two $Au_{28}(CHT)_{20}$ (CHT = cyclohexanethiolate) nanocluster structural isomers; two $Au_{36}(DMBT)_{24}$ nanoclusters (DMBT = 3,5-dimethylbenzenethiol) nanocluster structural isomers; two $Au_{42}(TBBT)_{26}$ (TBBT = 4-*tert*-butylbenzenethiolate) nanocluster structural isomers; and two $Ag_{20}Cu_{12}(C{\equiv}CR)_{24}$ (HC≡CR = 3,5-bis(trifluoromethyl)phenylacetylene) nanocluster structural isomers. Although their catalytic performance has not been systematically investigated, we believe that more interesting phenomenon will be discovered concerning isomeric metal nanocluster catalysis in the future.

## 7.2 Metal Nanocluster Asymmetric Catalysis

Although there are a large number of reports on asymmetric catalysis by metal nanoparticles/nanoclusters, these catalysts generally do not have atomically precise structures, thus hindering the in-depth understanding on the catalytic mechanism. Organic ligand-protected, atomically precise metal nanoclusters provide structural basis for the corresponding research. On the other hand, it has been found that many catalytic centers of metalloenzymes have the structures of metal nanoclusters.

Therefore, chiral metal nanoclusters are expected to open up new avenues for developing dominating chiral catalysts toward challenging asymmetric catalysis [14].

An early work by Moberg and coworkers in 2012 have shown that [$H_4Ru_4(CO)_{10}$(chiral-phosphine)] and [$H_4Ru_4(CO)_{11}$(chiral-phosphine)] (chiral-phosphine is the Josiphos or Walphos ligands employed) nanoclusters have the potential to catalyze asymmetric hydrogenation reactions of alkenes, giving excellent conversion (99%–100%) and relatively good enantioselectivity (up to 90% enantio excess [ee] in certain cases) [15]. For a long time after that, no significant progress has been achieved concerning organic ligand-protected metal nanocluster asymmetric catalysis. The major challenge would be the construction of atomically precise metal nanoclusters bearing structural stability, catalytic activity, and chiral structures.

After 12 years of this work, Wang and coworkers showed an interesting work on metal nanocluster asymmetric catalysis [16]. They prepared a pair of enantiomeric metal nanoclusters, namely [$Au_7Ag_8$(dppf)$_3$(L-proline)$_6$]($BF_4$)$_2$ and [$Au_7Ag_8$(dppf)$_3$(D-proline)$_6$]($BF_4$)$_2$ (dppf = 1,1′-bis(diphenylphosphino)ferrocene). SCXRD revealed structure demonstrate that dppf molecules are generally located on the surface of gold atoms and the chiral proline molecules are generally connected with the silver atoms (Figure 7.2), thus creating chiral environment for activation of substrates. This nanocluster is highly stable and can catalyze the reaction between acetone and 4-nitrobenzaldehyde at room temperature. Optimized reaction conditions gave almost 100% conversion and 70% ee of the Aldol product. The TON for each activation site of [$Au_7Ag_8$(dppf)$_3$(L-proline)$_6$]($BF_4$)$_2$ or [$Au_7Ag_8$(dppf)$_3$(D-proline)$_6$]($BF_4$)$_2$ nanoclusters is calculated to be 11, which is three times higher than that using the proline as the catalyst. In their proposed mechanism, the gold

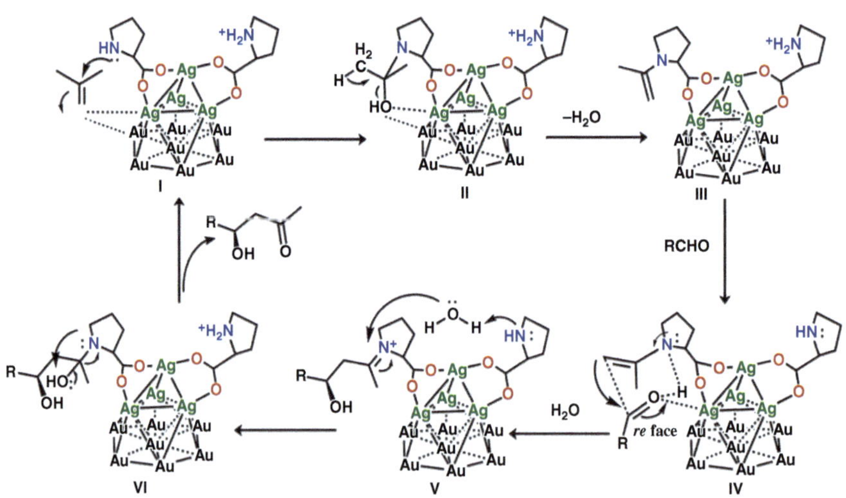


**Figure 7.2** Structure of chiral nanocluster [$Au_7Ag_8$(dppf)$_3$(proline)$_6$]($BF_4$)$_2$ and catalytic mechanism of this chiral nanocluster in asymmetric Aldol reaction. *Source:* [16]/with permission of American Chemical Society.

part of the metal kernel is beneficial for maintaining the structural stability of the chiral nanocluster, while the silver sites and proline sites cooperatively activate the acetone and aldehyde, endowing the high catalytic activity of this metal nanocluster (Figure 7.2).

Considering the abundant chirality and surface activation sites of metal nanoclusters, compared to chiral organic molecules and chiral metal complexes, it is believed that further investigation on organic ligand-protected metal nanocluster asymmetric catalysis would trigger the discovery of more effective activation mode. This would provide opportunity for challenging asymmetric reactions that are difficult to be resolved by conventional chiral catalysts.

## References


1 X.-K. Wan, J.-Q. Wang, Z.-A. Nan, Q.-M. Wang, Ligand Effects in Catalysis by Atomically Precise Gold Nanoclusters. *Sci. Adv.* **2017**, *3*, e1701823.

2 Z.-J. Guan, R.-L. He, S.-F. Yuan, J.-J. Li, F. Hu, C.-Y. Liu, Q.-M. Wang, Ligand Engineering Toward the Trade-off Between Stability and Activity in Cluster Catalysis. *Angew. Chem. Int. Ed.* **2022**, *61*, e202116965; *Angew. Chem.* **2022**, *134*, e202116965.

3 S.-F. Yuan, Z.-J. Guan, Q.-M. Wang, Identification of the Active Species in Bimetallic Cluster Catalyzed Hydrogenation. *J. Am. Chem. Soc.* **2022**, *144*, 11405–11412.

4 C. Ruan, C. Meng, K. Wang, Y. Li, H. Xiang, H. Yan, W. W. Xu, M. Zhou, C. Yao, Octahedral vs Tiara-like $Pd_6(SR)_{12}$ Clusters. *Nano Lett.* **2024**, *24*, 13054–13060.

5 C. Dong, R.-W. Huang, A. Sagadevan, P. Yuan, L. Gutiérrez-Arzaluz, A. Ghosh, S. Nematulloev, B. Alamer, O. F. Mohammed, I. Hussain, M. Rueping, O. M. Bakr, Isostructural Nanocluster Manipulation Reveals Pivotal Role of One Surface Atom in Click Chemistry. *Angew. Chem. Int. Ed.* **2023**, *62*, e202307140; *Angew. Chem.* **2023**, *135*, e202307140.

6 S. Tian, Y.-Z. Li, M.-B. Li, J. Yuan, J. Yang, Z. Wu, R. Jin, Structural Isomerism in Gold Nanoparticles Revealed by X-ray Crystallography. *Nat. Commun.* **2015**, *6*, 8667.

7 H. Qian, W. T. Eckenhoff, Y. Zhu, T. Pintauer, R. Jin, Total Structure Determination of Thiolate-protected Au38 Nanoparticles. *J. Am. Chem. Soc.* **2010**, 132, 8280.

8 Z. Qin, J. Zhang, C. Wan, S. Liu, H. Abroshan, R. Jin, G. Li, Atomically Precise Nanoclusters with Reversible Isomeric Transformation for Rotary Nanomotors. *Nat. Commun.* **2020**, *11*, 6019.

9 Z.-J. Guan, F. Hu, J.-J. Li, Z.-R. Wen, Y.-M. Lin, Q.-M. Wang, Isomerization in Alkynyl-Protected Gold Nanoclusters. *J. Am. Chem. Soc.* **2020**, *142*, 2995–3001.

10 G. Deng, S. Malola, T. Ki, X. Liu, S. Yoo, K. Lee, M. S. Bootharaju, H. Häkkinen, T. Hyeon, Structural Isomerism in Bimetallic $Ag_{20}Cu_{12}$ Nanoclusters. *J. Am. Chem. Soc.* **2024**, *146*, 26751–26758.

**11** X. Liu, W. W. Xu, X. Huang, E. Wang, X. Cai, Y. Zhao, J. Li, M. Xiao, C. Zhang, Y. Gao, W. Ding, Y. Zhu, *De novo* Design of $Au_{36}(SR)_{24}$ Nanoclusters. *Nat. Commun.* **2020**, *11*, 3349.

**12** S. Zhuang, L. Liao, J. Yuan, N. Xia, Y. Zhao, C. Wang, Z. Gan, N. Yan, L. He, J. Li, H. Deng, Z. Guan, J. Yang, Z. Wu, Fcc versus Non-fcc Structural Isomerism of Gold Nanoparticles with Kernel Atom Packing Dependent Photoluminescence. *Angew. Chem. Int. Ed.* **2019**, *58*, 4510–4514; *Angew. Chem.* **2019**, *131*, 4558–4562.

**13** N. Xia, J. Yuan, L. Liao, W. Zhang, J. Li, H. Deng, J. Yang, Z. Wu, Structural Oscillation Revealed in Gold Nanoparticles. *J. Am. Chem. Soc.* **2020**, *142*, 12140–12145.

**14** Q.-Y. Zhang, L. Zhao, Chiral Metal Cluster and Nanocluster Complexes and Their Application in Asymmetric Catalysis. *Tetrahedron Lett.* **2018**, *59*, 310–316.

**15** V. Moberg, R. Duquesne, S. Contaldi, O. Röhrs, J. Nachtigall, L. Damoense, A. T. Hutton, M. Green, M. Monari, D. Santelia, M. Haukka, E. Nordlander, Efficient Cluster-Based Catalysts for Asymmetric Hydrogenation of α-Unsaturated Carboxylic Acids. *Chem. Eur. J.* **2012**, *18*, 12458–12478.

**16** J.-Q. Wang, R.-L. He, W.-D. Liu, Q.-Y. Feng, Y.-E. Zhang, C.-Y. Liu, J.-X. Ge, Q.-M. Wang, Integration of Metal Catalysis and Organocatalysis in a Metal Nanocluster with Anchored Proline. *J. Am. Chem. Soc.* **2023**, *145*, 12255–12263.

# 8

# Metal Nanocluster Photocatalysis

## 8.1 Metal Nanocluster Photophysical Property

Metal nanoclusters with sizes ranging from about ten to a few hundred atoms constitute a special class of nanomaterials [1–4]. These nanoclusters are unique in that they can be represented with definite formulas (as opposed to an average size with size distribution), resembling well-defined organic molecules or organometallic compounds. Thus, metal nanoclusters are essentially inorganic–organic hybrid molecular compounds. Their ultra-small size (from subnanometer to 2–3 nm in diameter of the metal core) places them in between molecules (e.g. metal–ligand complexes) and plasmonic metal nanoparticles. Strong quantum size effects in nanoclusters are manifested in their photophysical properties, such as the discrete energy levels of electrons, multiple absorption bands, etc. [5]. These features endow metal nanoclusters with distinct catalytic activities from metal–ligand complexes and plasmonic metal nanoparticles. We introduce photophysical properties of metal nanoclusters from the following aspects: optical absorption, optical emission, and excited-state dynamics.

### 8.1.1 Absorption Properties

Conventional gold and silver nanoparticles exhibit surface plasmon resonances (SPRs), which account for their absorption, scattering, surface-enhanced Raman scattering (SERS), and many other properties. The SPR peak width of metallic nanoparticles is inherently broad because of ultrafast electron dephasing in the metallic state (on the order of femtoseconds). However, when metal nanoparticles enter the ultra-small size regime (e.g. <3 nm), strong quantum size effects come into play, and these ultra-small nanoparticles indeed become very sensitive to particle size, even a single atom difference [6]. Metal nanoclusters typically have discrete absorption peaks. Owing to the quantum confinement effect, the energy levels of metal

*Metal Nanocluster Catalysis: From Fundamentals to Applications*, First Edition. Manzhou Zhu and Man-Bo Li. © 2026 Wiley-VCH GmbH. All rights reserved, including rights for text and data mining and training of artificial intelligence technologies or similar technologies. Published 2026 by Wiley-VCH GmbH.

nanoclusters are discrete, which is very different from the continuous energy band of bulk metals. This leads to the appearance of discrete absorption peaks in the ultraviolet–visible–near-infrared region, rather than the broad absorption band like that of metal nanoparticles.

In 2008, Zhu and coworkers determined the crystal structure of the $Au_{25}(SR)_{18}$ nanocluster and collected its absorption spectrum. After the crystal structure of $[Au_{25}(SR)_{18}]^-$ $TOA^+$ (where TOA stands for tetraoctylammonium) being obtained (Figure 8.1a,b), Aikens and coworkers carried out time-dependent density functional theory (TD-DFT) calculations and reproduced the experimental spectrum [7]. Both the sp and d bands were found to be quantized. The experimental peak at 670 nm (peak a in Figure 8.1d) was revealed to be the highest occupied molecular orbital (HOMO)-to-lowest unoccupied molecular orbital (LUMO) transition, which is essentially an intraband (sp←sp) transition. The HOMO set is essentially of s character. Thus, transitions arising from the other occupied HOMO-n orbitals (belonging to the d band) are interband (sp←d) transitions, such as peak c, whereas peak bis of mixed type. In terms of atomic contributions, the HOMO and LUMO are composed, almost exclusively, of atomic orbital contributions from the 13 Au atoms in the icosahedral core rather than the

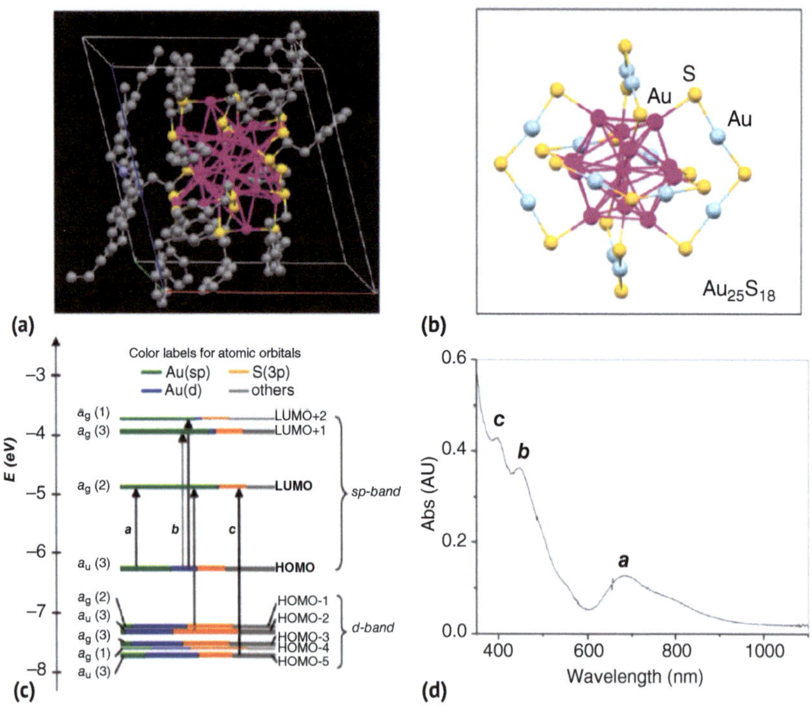


**Figure 8.1** (a) Crystal structure of $[Au_{25}(SC_2H_4Ph)_{18}]^-$ $TOA^+$. (b) Atomic arrangement of the $Au_{25}S_{18}$ framework. (c) Kohn–Sham orbital level diagram for $[Au_{25}(SR)_{18}]^-$. (d) Peak assignment of the absorption spectrum of $[Au_{25}(SR)_{18}]^-$. *Source:* Adapted from [7].

12 exterior Au atoms (Figure 8.1c). Thus, the first peak at 670 nm in the absorption spectrum can be viewed as a transition that is due entirely to the electronic and geometric structure of the $Au_{13}$ core.

The optical spectra and electronic structures of the nanocluster are the inherent characterization and vary with the size and geometric structure of metal nanoclusters. In 2015, the structure of $Au_{18}(SC_6H_{11})_{14}$ was reported independently by two different research groups [8, 9] (Figure 8.2a). Experimental and theoretical results showed that the optical spectrum of $Au_{18}(SC_6H_{11})_{14}$ match well. The electronic transitions associated with all absorption peaks were assigned to the d→sp interband transitions (Figure 8.2b,c). After that, Chen and coworkers [10] determined the structures of $Au_{21}$(S-Adm)$_{15}$ (Figure 8.2d) and $Au_{16}$(S-Adm)$_{12}$. DFT simulation showed that the HOMO–LUMO transition of $Au_{21}$(S-Adm)$_{15}$ is assigned to the Au sp←sp excitation (Figure 8.2e,f). For the $Au_{16}$(S-Adm)$_{12}$, the experimental absorption spectrum showed a weak shoulder peak at the 605 nm and theoretical

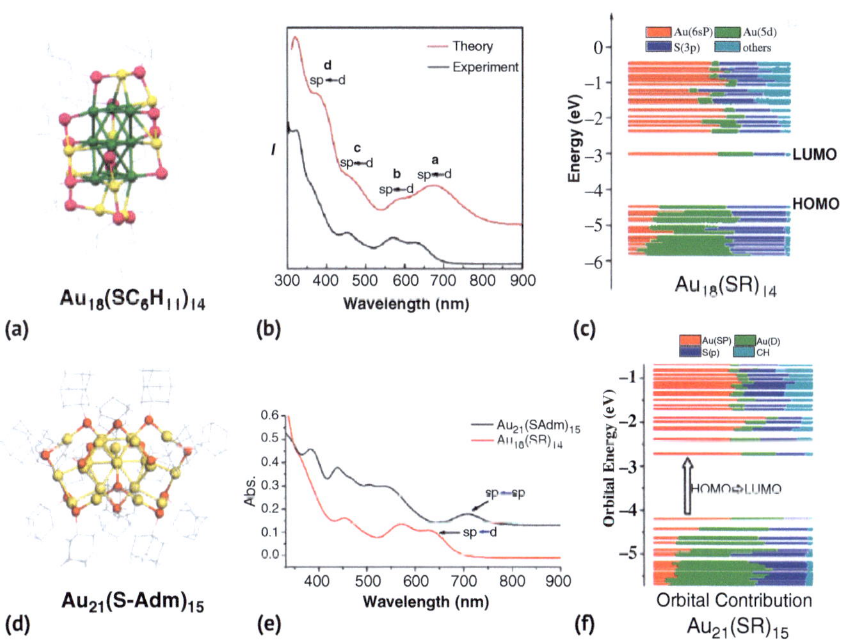


**Figure 8.2** (a) X-ray structure of $Au_{18}(SC_6H_{11})_{14}$, all hydrogen atoms are omitted for clarity. (b) Simulated UV–vis adsorption spectrum of $Au_{18}(SR)_{14}$. (c) Kohn–Sham orbital energy level diagram for $Au_{18}(SR)_{14}$. The energies are in eV. Each Kohn–Sham orbital is drawn to indicate the relative contributions (line length with color labels) of the atomic orbitals of Au(6sp) red, Au(5d) green, S(3p) blue, and other orbital contributions from C and H atoms are in cyan. (d) Total structure of the $Au_{21}$(S-Adm)$_{15}$ nanocluster. (e) UV–vis spectra of $Au_{21}(SR)_{15}$ nanoclusters. (f) Kohn–Sham orbitals of $Au_{21}(SR)_{15}$ nanoclusters. The orbital is drawn to indicate the relative contributions (line length with color labels) of the atomic orbitals of Au(6sp) red, Au(5d) green, and S(3p) blue, and other orbital contributions from C and H atoms cyan. *Source:* Adapted with permission from [8, 10].

simulation resulted in shoulder peak at and 630 nm, which is resulted from the HOMO→LUMO transitions, but the LUMO is largely contributed to the bare gold atom [11].

Li, Das, and coworkers reported atom-by-atom manipulation on metal nanoclusters based on the $Au_{21}$, $Au_{22}$, $Au_{22}Cd_1$, and $Au_{24}$ nanoclusters. The optical spectra changed with the sizes and structures for these nanoclusters. $Au_{22}Cd_1(SAdm)_{16}$ and $Au_{23}(SR)_{16}^-$ possess the same number of metal atoms. $Au_{22}Cd_1(SAdm)_{16}$ showed a red shift in absorption spectrum compared with that of $Au_{23}(SR)_{16}^-$ (Figure 8.3a–d) [12, 13]. Similarly, doping heterogeneous metal into Ag nanoclusters can affect the absorption spectra. Yan and coworkers successfully prepared Pd or Pt atoms-doped, thiolate ligands-protected Ag-rich 25-metal-atom nanoclusters. With a composition of $[PdAg_{24}(SR)_{18}]^{2-}$ (Figure 8.3e) or $[PtAg_{24}(SR)_{18}]^{2-}$, the obtained 25-metal-atom nanoclusters have a metal framework structure similar to that of widely investigated $Au_{25}(SR)_{18}^-$. $[PdAg_{24}(SR)_{18}]^{2-}$ displayed two absorption peaks centered at 483 and 628 nm. The optical spectra of $[PtAg_{24}(SR)_{18}]^{2-}$ showed similar profile, and the peaks were blue-shifted to 453 and 564 nm (Figure 8.3f) [14].

Das and coworkers obtained a hollow $[Au_{24}(PPh_3)_{10}(SC_2H_4Ph)_5X_2]^+$ nanocluster by treatment of the $[Au_{25}(PPh_3)_{10}(SC_2H_4Ph)_5X_2]^{2+}$ rod nanocluster with excess phosphine. The optical spectra of metal nanoclusters are related to their structures. $Au_{24}$ and $Au_{25}$ possessed the similar frame structure (Figure 8.4a).

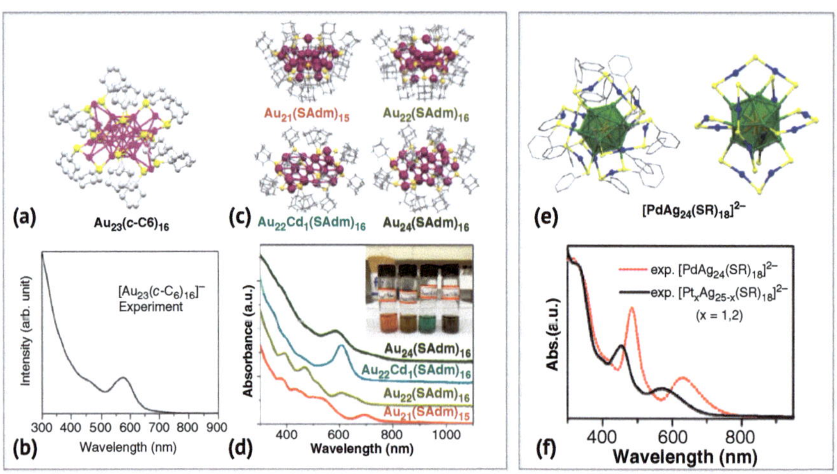


**Figure 8.3** (a) Crystal structure of a cyclohexanethiolate-protected $[Au_{23}(c\text{-}C_6)_{16}]^-$ nanocluster ($TOA^+$ counterion omitted). Color labels: magenta = Au, yellow = S, gray = C; all H atoms not shown. (b) Experimental UV–vis spectrum of $[Au_{23}(c\text{-}C_6)_{16}]^-$. (c) Crystal structure and (d) UV–vis spectra of $Au_{21}(SAdm)_{15}$, $Au_{22}(SAdm)_{16}$, $Au_{22}Cd_1(SAdm)_{16}$, and $Au_{24}(SAdm)_{16}$ NCs; inset: photographs of the dichloromethane solutions of the four NCs, from left to right are $Au_{21}(SAdm)_{15}$, $Au_{22}(SAdm)_{16}$, $Au_{22}Cd_1(SAdm)_{16}$, and $Au_{24}(SAdm)_{16}$, respectively. (e) Structure of the $[PdAg_{24}(SR)_{18}]^{2-}$ cluster revealed by X-ray single-crystal analysis. (f) Experimental UV–vis spectra of $[PdAg_{24}(SR)_{18}]^{2-}$ and $[PtAg_{24}(SR)_{18}]^{2-}$ nanoclusters. *Source:* [12]/American Chemical Society.

The main difference is a single Au atom in the center of rod core. The absorption spectrum of rod-like $Au_{24}$ showed prominent bands at 383 and 560 nm, whereas rod-like $Au_{25}$ displayed absorption peaks centered at 415 and 670 nm (Figure 8.4c,d) [15]. Besides, Wang and coworkers achieved shuttling a single Ag or Cu atom into centrally hollow, rod-shaped $Au_{24}$ nanoclusters, forming $Ag_1Au_{24}$ and $Cu_1Au_{24}$ nanoclusters (Figure 8.4b). It was found that the absorption peak of $Cu_1Au_{24}$ in the high-energy region had a slight blue shift, and the absorption peak of $Ag_1Au_{24}$ in the low-energy region had a slight red shift compared with that of $Au_{25}$ (Figure 8.4e) [16].

### 8.1.2 Optical Emission

Optical absorption is a prerequisite for photocatalysis, and photocatalysis is a process of converting light energy into chemical energy. Therefore, optical emission reflects the ability of the photocatalyst to convert the absorbed energy to a certain extent. Optical emission refers to the process of releasing energy in the

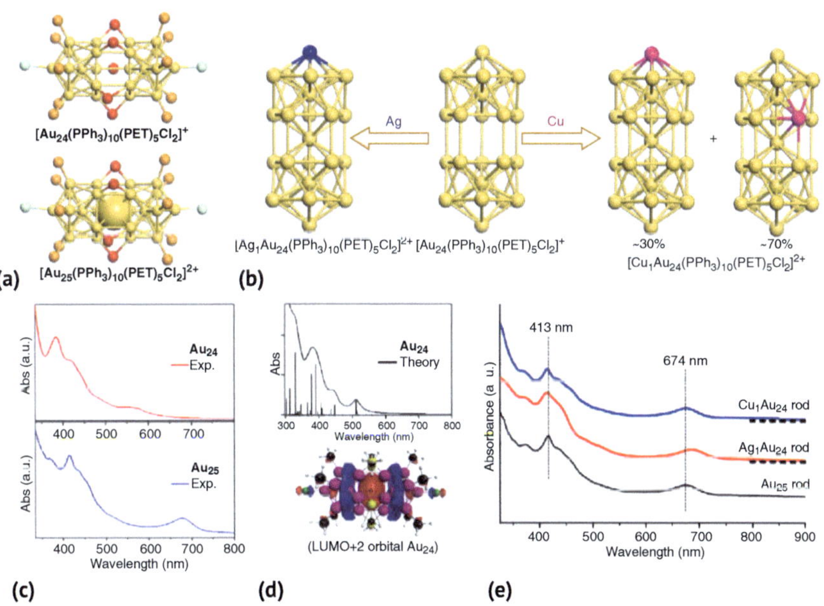


**Figure 8.4** (a) X-ray structures of $[Au_{24}(PPh_3)_{10}(PET)_5Cl_2]^+$ and $[Au_{25}(PPh_3)_{10}(PET)_5Cl_2]^{2+}$ nanoclusters. (b) Shuttling one Ag or Cu atom into the 24-atom hollow gold nanoparticle: pathways of single Ag/Cu entering the hollow $Au_{24}$ nanoparticle. Color code: Au, yellow; P, orange; S, red; Cl, green. Ag, blue; Cu, magenta. C and H atoms are not shown for clarity. (c) Experimental optical spectrum of $[Au_{24}(PPh_3)_{10}(PET)_5Cl_2]^+$ (in dichloromethane) and $[Au_{25}(PPh_3)_{10}(PET)_5Cl_2]^{2+}$ (in dichloromethane). (d) DFT simulated optical spectrum of $[Au_{24}(PH_3)_{10}(SCH_3)_5X_2]^+$ model cluster, and the LUMO+2 molecular orbital of $[Au_{24}(PH_3)_{10}(SCH_3)_5X_2]^+$. (e) UV–vis spectra of homo-gold $[Au_{25}(PPh_3)_{10}(PET)_5Cl_2]^{2+}$ (black line), single Ag doped $[Ag_1Au_{24}(PPh_3)_{10}(PET)_5Cl_2]^{2+}$ (red line), and single Cu-doped $[Cu_1Au_{24}(PPh_3)_{10}(PET)_5Cl_2]^{2+}$ (blue line). *Source:* Adapted with permission from Refs. [15, 16].

form of photons from the excited state to the ground state after absorbing photon energy. When an external light source illuminates a substance, photons carry a specific energy that impacts the atoms or molecules in the substance. If the energy of the photon matches the difference in the energy level of the electron inside the matter, according to the law of conservation of energy, the electron can absorb the photon and transition from the ground state (the lowest energy state) to the excited state (the higher energy state), which is the excitation process. After the electron transition to the excited state is not stable, it will quickly relax its energy by means of lattice vibration or intramolecular rotation, etc. Then, the electron returns to the ground state in the form of radiation transition, releasing photons to complete optical emission. Optical emission can also be divided into fluorescence and phosphorescence according to the difference in luminescence types. Fluorescence is the return of electrons from the excited singlet state back to the ground state. The speed is fast and the life is short, usually in the nanosecond level. The light source stops, the fluorescence is out. Phosphorescence is the return of electrons from the excited triplet state back to the ground state. As a result of spin prohibition, this kind of process is slow and the life is long, up to microseconds or even milliseconds. The light source can continue to glow for a while after being removed.

Compared to semiconductor quantum dots (QDs), metal nanoclusters generally exhibit superior photo-stability and precise compositions and structures, providing important support for the study of the optical emission properties of the nanoclusters [17] and therefore hold great promise for use in photocatalysis [18–24]. There are many factors that affect the optical emission characteristics of metal nanoclusters. For example, the core size, the rigidity of the structure, and the influence of peripheral ligands. In general, alloying and doping with heterogeneous metal can significantly improve the optical emission property of metal nanoclusters, which can lead to tens of times or even hundreds of times of enhancement. In addition, the surrounding environment also has a strong effect on the luminous intensity of metal nanoclusters, such as the temperature, solvent, pH, etc. Generally, the low temperature state is beneficial to the optical emission. Under normal circumstances, aggregation induction has a certain effect on the optical emission intensity of metal nanoclusters. By exploring these factors, the optical emission properties of metal nanoclusters can be effectively regulated, and the related applications, such as photocatalysis can be explored [17]. More commonly, researchers study the optical emission of metal nanoclusters that have been prepared. Some of these nanoclusters have inherent and intense fluorescence emission due to their unique geometry and electronic structure. Herein, we discuss typical metal nanoclusters with inherent optical emission, such as $Au_{25}$, $Au_{24}$, $Ag_{62}$, etc. It should be noted that optical emission properties of metal nanoclusters trigger many intriguing applications [25–42], such as optical waveguide [43], etc. Herein, we only briefly introduce the optical emission, which has relationship with the photocatalytic ability of metal nanoclusters.

Wu and coworkers explored how the optical emission of a series of $Au_{25}(SR)_{18}$ nanoclusters depended on the identity of the protecting ligands [44]. They showed that the optical emission increased with increasing charge donating capability of ligands, that is, $Au_{25}(SC_2H_4Ph)_{18}$ > $Au_{25}(SC_{12}H_{25})_{18}$ > $Au_{25}(SC_6H_{13})_{18}$ (all three of these $Au_{25}$ nanoclusters are in the "1⁻" charge state) (Figure 8.5b). In regard to the three $Au_{25}$ species, the fluorescence quantum yield seems to be parallel with the ligand's capability of donating electron density to the metal core via the S–Au bond (i.e. charge transfer capability of the ligand). The dodecyl group ($C_{12}H_{25}$–, long-chain) is longer than the hexyl group ($C_6H_{13}$–) and is more capable of pushing electron density to the sulfur atom (i.e. $S^{\delta-}$) and hence affects the $Au_{25}$ core through the S–Au bond. The $PhC_2H_4$– group has an even stronger electron donating capability than long-chain alkyl groups. Therefore, the order of charge donating capability follows $PhCH_2CH_2$ > $C_{12}H_{25}$ > $C_6H_{13}$. The different fluorescence intensities of the three $[Au_{25}(SR)_{18}]^-$ species indicate that the fluorescence is largely enhanced by charge transfer from the surface ligands to the metal core via S–Au bonds. The PL intensity of $Au_{25}(SR)_{18}$ was therefore enhanced by two methods (Figure 8.5a): (i) charge transfer from the ligands to the metal nanoparticle core (i.e. LMNCT) through the Au–S bonds, and (ii) direct donation of delocalized electrons of electron-rich atoms or groups of the ligands to the metal core.

As a result of the fact that metal nanoclusters consist of metal core and peripheral motifs, charge transfer between metal and ligand plays an important role in their optical emission. These processes include ligand to metal charge transfer (LMCT), ligand to metal–metal charge transfer (LMMCT), etc. The modulation on ligand is effective to regulate the optical emission of metal nanoclusters [45–55]. Li and coworkers prepared $[Ag_{62}S_{13}(S^tBu)_{32}](BF_4)_4$ with a core–shell structure. This nanocluster emits intense red light in both solution and solid state at room temperature with a max emission at 621 nm for solid state and 613 nm for sample

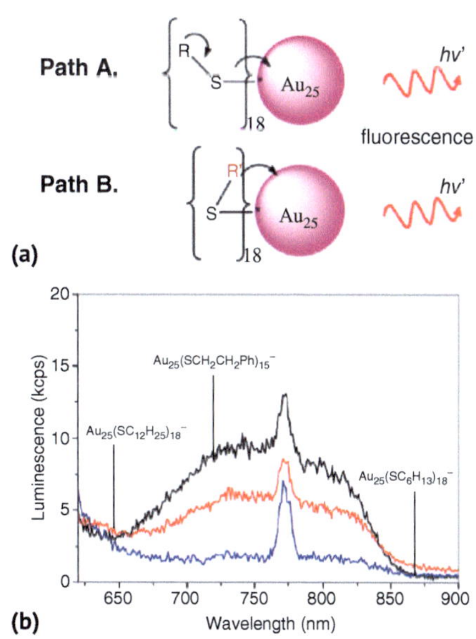


**Figure 8.5** (a) Schematic illustrations of the ligand engineering strategies for influencing the LMNCT process of $Au_{25}(SR)_{18}$ nanoclusters. (b) Weak fluorescence of $[Au_{25}(SR)_{18}]^-$ with different R groups ( $C_2H_4Ph$, $-C_{12}H_{25}$, and $-C_6H_{13}$). Note that the spike at ~770 nm is an artifact from the quartz cell; solvent, $CH_2Cl_2$; excitation, 514 nm (from a xenon arc source). *Source:* [44]/with permission of American Chemical Society.

solution (Figure 8.6a,c). The charge transfer from S(3p) to Ag(5s) is responsible for the bright emission [56]. Recently, Jin and coworkers determined the structure of $[Ag_{62}S_{12}(S^tBu)_{32}]^{2+}$ (NC-I) that composed of a face centered cubic $Ag_{14}$ core, a $Ag_{48}(S^tBu)_{32}$ shell, and 12 sulfur atoms. The nanocluster of $[Ag_{62}S_{12}(S^tBu)_{32}]^{2+}$ exhibits extremely weak fluorescence emission, which is ~1800 times weaker than $[Ag_{62}S_{13}(S^tBu)_{32}]^{4+}$ (NC-II) (Figure 8.6c, note the 900× scaling up) [57]. The femtosecond transient absorption spectra suggest that the quenching of photoluminescence is caused by the free valence electrons in the NC-I nanocluster (Figure 8.6d). The appearance of 4e in ${Ag_{14}}^{10+}$ may account for the disappearance of photoluminescence and the optical absorption difference in comparison to NC-II. This is due to the delocalization behavior of the 4e valence electrons in NC-I in comparison to zero valence electron in NC-II, which prevents the charge transfer from the ligands to the metal core. Thus, the formation of the triplet LMCT state is inhibited during the relaxation (Figure 8.6).

Apart from the ligand, the optical emission of metal nanoclusters is also significantly regulated by metal doping [58–61]. A typical example comes from the rod-like $[Ag_xAu_{25-x}(PPh_3)_{10}(SR)_5Cl_2]^{2+}$ nanoclusters, which were

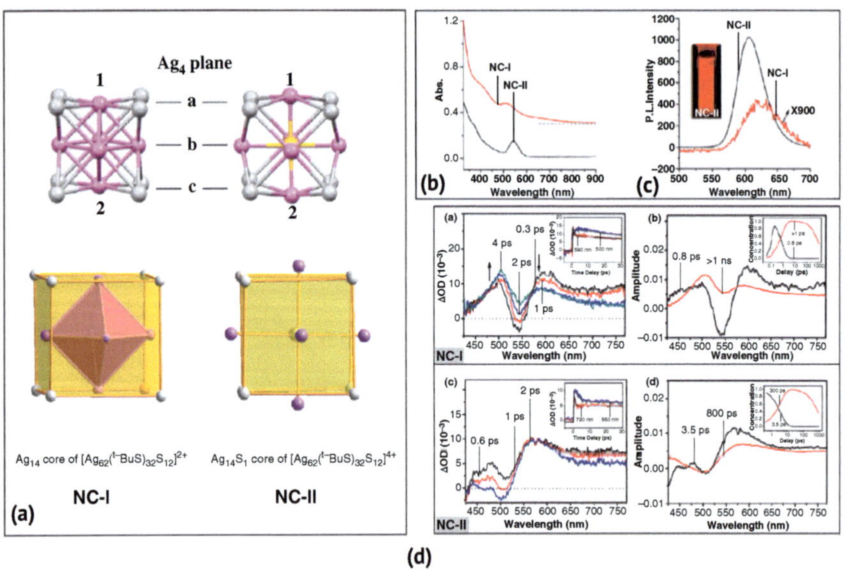


**Figure 8.6** (a) Structures of $Ag_{14}$ kernels in the NC-I and NC-II nanoclusters. (b) UV–vis absorption spectra of NC-I and NC-II. (c) Photoluminescence spectra of NC-I and NC-II (inset: photos of emissions from NC-II in MeOH at room temperature under ambient light and 365 nm excitation). (d) Evolution-associated transient absorption spectra at different time delays (a, c) and species associated difference spectra (SADS) obtained from singular value decomposition (SVD) combined with global fitting (b, d) for solutions of the NCs NC-II (a, b) and NC-I (c, d). The insets in (a, c) show the fitted kinetics at specific wavelengths, and the insets in (b, d) show the population dynamics of each component obtained from global fitting results, respectively. *Source:* [57]/with permission of American Chemical Society.

synthesized by "bottom up" and "top down" metal exchange method with different x ranges (Figure 8.7a). The "top down" method results in less-silver-doped $[Ag_xAu_{25-x}(PPh_3)_{10}(SR)_5Cl_2]^{2+}$ nanocluster ($x \leq 12$). On the contrary, a unique $Ag_{13}Au_{12}$ species can be found in the "bottom up" products, which shows ultrabright optical emission [62]. According to the X-ray structure analysis of Ag-doped $M_{25}$ (M = Au/Ag) rod-like nanoclusters, the doped Ag atoms first substitute the two Au atoms at the vertex site, and subsequently the Au atoms at the waist as well as in the center. When 13 Au atoms are exchanged by silver, strong distortion to the electronic structure occurs. Because of the strong sp hybridization of Au atoms due to the relativistic effect, the LUMO, which contributed by Au 6s and 6p in $Au_{25}$ nanocluster, will be significantly altered (Figure 8.7b,c). Such substitution gives rise to a larger energy gap. The ultrafast relaxation dynamics together with the DFT calculations indicates the boost in quantum yield is due to the high oscillator strength of these HOMO→LUMO transitions. The prominently longer excited lifetime with less nonradiative relaxation and larger energy gap in $Au_{12}Ag_{13}$ lead to much stronger and blue-shifted fluorescence compared with $Au_{25}$ rod. The excited-state relaxation scheme of $Au_{25}$ and $Au_{12}Ag_{13}$ rod nanoclusters is summarized in Figure 8.7d [63].

Bootharaju and coworkers also found the similar phenomenon. They used a direct synthesis involving the co reduction of Au and Ag salts in the presence

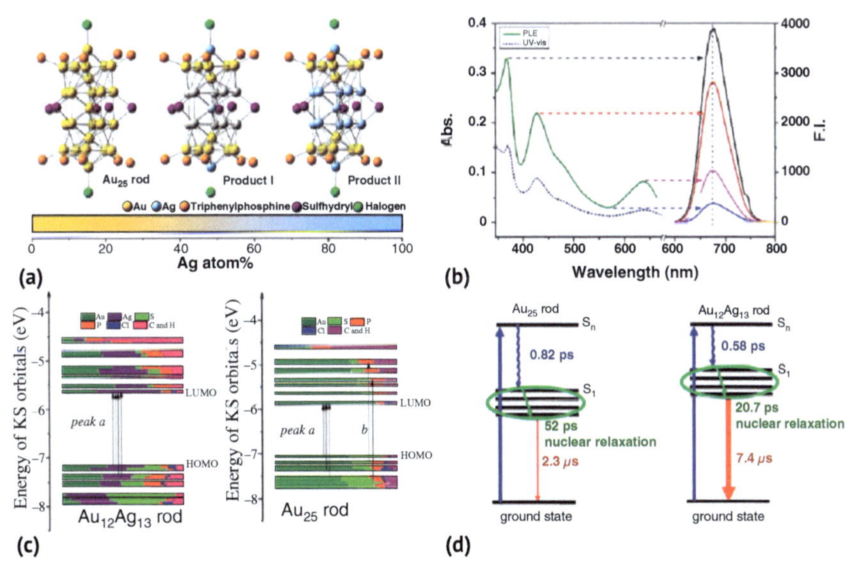


**Figure 8.7** (a) Crystal structure of homogold $[Au_{25}(PPh_3)_{10}(SR)_5Cl_2]^{2+}$ and $[Ag_xAu_{25-x}(PPh_3)_{10}(SR)_5Cl_2]^{2+}$ with different x range. (b) UV–vis absorption, photoexcitation (PLE), and photoluminescence properties of $[Ag_xAu_{25-x}(PPh_3)_{10}(SR)_5Cl_2]^{2+}$ ($x \leq 13$). (c) Population analysis of KS orbitals of $Au_{12}Ag_{13}$ rod and $Au_{25}$ rod. The Gauss broadening with a width at half-maximum of 0.1 eV is used to fit the optical curve based on the computed excitation energies by TD-DFT. (d) Relaxation Pathway for $Au_{25}$ Rod Cluster and Ag Alloying $Au_{12}Ag_{13}$ Cluster. *Source:* Adapted with permission from Refs. [62, 63].

of ligands. As the content of Au increases, the proportion of Au in the clusters increases. When the Au precursor concentration is 40 mmol%, the fluorescence QY of the cluster is as high as 24% (Figure 8.8). A red shift of PL emission was observed as the increase in Au concentration. In addition, by Au doping, a significant enhancement in the ambient stability of the $Ag_{29}$ cluster is achieved. Single-crystal analysis coupled with mass spectrometry, transient absorption spectroscopy, and nuclear magnetic resonance (NMR) provides insights into the optical emission mechanism and probable locations of Au atoms in the doped clusters. Femto-nanosecond transient absorption (fs-ns TA) and time-correlated single-photon counting (TCSPC) experiments showed that the heteroatoms doping perturbed the electronic relaxation and dynamics, and further led to a long lifetime and thus a high PL intensity (Figure 8.8d,e). The NMR results indicated that the possible locations of doped Au atoms are at the Ag atoms connected to the TPP ligands (Figure 8.8f) [64].

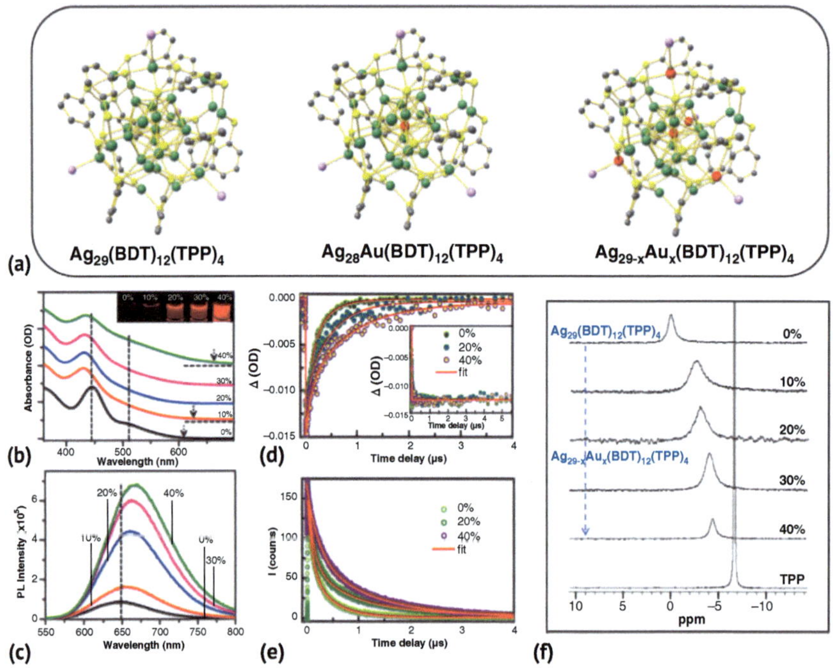


**Figure 8.8** (a) X-ray crystal structures of $Ag_{29}(BDT)_{12}(TPP)_4$, $Ag_{28}Au(BDT)_{12}(TPP)_4$, and possible locations of Au atoms in $Ag_{29-x}Au_x(BDT)_{12}(TPP)_4$ clusters, $x = 1–5$. Color spheres: green, Ag; red, Au; yellow, S; pink, P; gray, C. For simplicity, H atoms are omitted. (b) UV–vis and (c) PL spectra of $Ag_{29}$ and Au-doped $Ag_{29}$ clusters synthesized using different amounts (mmol%) of Au. Inset: a photograph of $Ag_{29}$ and Au-doped $Ag_{29}$ clusters under a UV lamp (using 365 nm light). (d) Transient kinetic traces corresponding to the GSB from the TA spectra of $Ag_{29}$ and its Au-doped NCs. Inset: kinetics in 0.0–5.5 ns window. (e) Kinetics of fluorescence excited-state decay of $Ag_{29}$ and Au-doped clusters at the emission maxima measured by TCSPC. All the solid red lines are fits of the kinetic traces. (f) P NMR spectra of pure TPP, $Ag_{29}(BDT)_{12}(TPP)_4$, and $Ag_{29-x}Au_x(BDT)_{12}(TPP)_4$ clusters. *Source:* [64]/John Wiley & Sons.

### 8.1.3 Excited-state Dynamics

The excited-state dynamics of metal nanoclusters refers to the disciplinary field that studies the electron transition from the ground state to the excited state of metal nanoclusters after being excited by light and other effects, as well as a series of dynamic processes such as relaxation, transfer, and energy transfer of the excited-state electrons subsequently. These processes involve interactions at the electronic, atomic and molecular levels, which are crucial for understanding the photophysical properties of metal nanoclusters and their applications in fields such as photocatalysis and photodetection. Among them, ultrafast electron dynamics spectroscopy is a powerful experimental technique often used to study the dynamics of excited states of substances. It uses femtosecond-level laser pulses to excite the sample, and then measures the absorption changes of the sample to the detection light at different time delays, thereby obtaining information such as the lifetime, relaxation process, and intermediate states of the excited state.

Understanding the correlation between the atomic structure and optical properties of gold nanoclusters is essential for exploration of their functionalities and applications involving light harvesting and electron transfer. Zhou and coworkers [65] report the femto-nanosecond excited-state dynamics of a periodic series of face-centered cubic (FCC) gold nanoclusters (including $Au_{28}$, $Au_{36}$, $Au_{44}$, and $Au_{52}$), which exhibit a set of unique features compared with other similar sized clusters. (Figure 8.9a) Molecular-like ultrafast $S_n{\rightarrow}S_1$ internal conversions

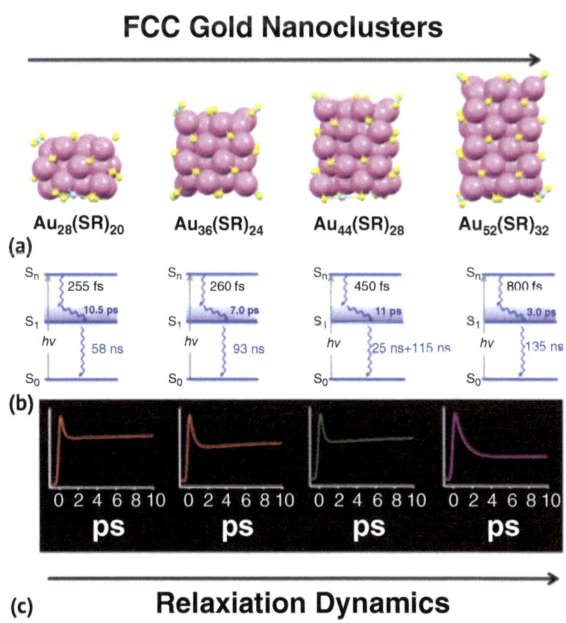


**Figure 8.9** (a) Crystal structures of FCC gold nanoclusters viewed by core and shell gold atoms (yellow = S; R groups are not shown; other colors = Au). (b) Schematic relaxation model of four FCC gold nanoclusters. (c) Nanosecond TA decay traces of four FCC gold nanoclusters. *Source:* [65]/with permission of American Chemical Society.

(i.e. radiationless electronic transitions) are observed in the relaxation dynamics of FCC periodic series (Figure 8.9b). Excited-state dynamics with near-HOMO–LUMO gap excitation lacks ultrafast decay component, and only the structural relaxation dominates in the dynamical process, which proves the absence of core–shell relaxation. Interestingly, both the relaxation of the hot carriers and the band-edge carrier recombination become slower as the size increases (Figure 8.9c). The evolution in excited-state properties of this FCC series offers new insight into the structure-dependent properties of metal nanoclusters, which will benefit their optical energy harvesting and photocatalytic applications.

Trapped in the synthesis and characterization of atomically precise gold clusters with particle sizes exceeding 2 nm, researchers are wondering exactly where the size boundary of gold nanoclusters lies from non-metals to metals. In 2021, Quan-Ming Wang research group successfully synthesized the largest pure acetylene ligand-protected, structurally clear gold nanocluster $(Ph_4P)_6\left[Au_{156}(C{\equiv}CR)_{60}\right]$ (R = 4-$CF_3C_6H_4$–) (abbreviated as $Au_{156}$) to date. $Au_{156}$ has a rod-shaped structure, with the long and short axes of its metal core being 2.38 and 2.04 nm, respectively. This nanocluster not only exhibits the molecular characteristics of energy level separation but also has metallic electronic dynamic behavior. It is currently the smallest metallic gold nanocluster in size.

Time-resolved TA spectroscopy allows people to track the excited-state dynamics. Owing to strong quantum confinement effects, molecular-state nanoclusters have power-independent electron relaxation dynamics with long excited-state lifetimes. On the contrary, metallic nanoclusters show strong plasmonic bleaches and power-dependent dynamics. This study focuses on the excited-state dynamics of the gold nanocluster $Au_{156}$ and finds that its electronic dynamic behavior significantly depends on the excitation wavelength [66]. Based on the transient absorption (TA) test, under the excitation of 400 and 500 nm, $Au_{156}$ exhibits metallic electron dynamics characteristics, manifested as a pump power-dependent electron relaxation process, which conforms to the dual-temperature model, that is, the electron temperature increases due to electron–electron scattering caused by femtosecond laser excitation. However, when the excitation wavelengths are 600 and 700 nm, $Au_{156}$ exhibits molecular electronic dynamics, and its electronic relaxation process is independent of the pump power (Figure 8.10). The fast (1.15 ps) and slow (23.3 ps) attenuation processes are determined through double-exponential fit, corresponding to the internal transition of the excited state and the transition from the lowest excited state to the ground state, respectively. This excitation wavelength sensitivity reveals that $Au_{156}$ has both molecular discrete energy levels and metallic band structures. When the excitation energy is high, electrons transition to the band region and exhibit metallic properties; when the energy is low, they are excited to discrete energy levels and show molecular properties. The unique rod-like structure of $Au_{156}$ (with a long axis exceeding 2.2 nm) enables it to exhibit metallic properties even with a relatively small number of atoms, providing an important model for the electronic structure characteristics that connect small molecular clusters with large-sized nanoparticles.

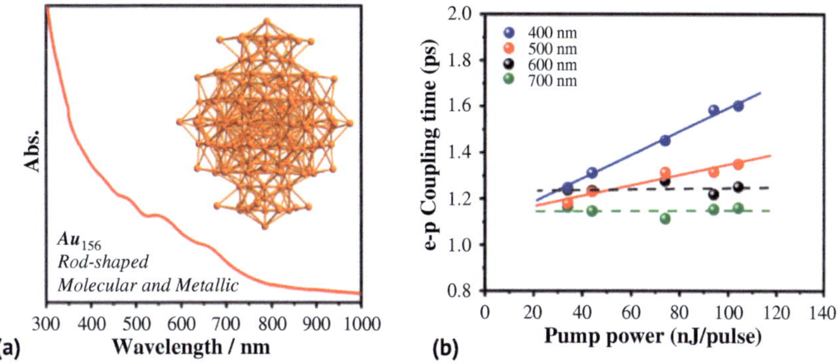


**Figure 8.10** (a) The core structure, steady-state absorption spectrum and (b) ultrafast kinetics of $Au_{156}$. *Source:* [66]/with permission of American Chemical Society.

Gold nanoclusters have potential biological applications as near-infrared (NIR) emitting materials, but they typically exhibit a low photoluminescence quantum yield (PLQY) in solutions at room temperature. Shi and coworkers [67] discovered that by forming the $Au_{16}Cu_6$ cluster through copper substitution, the PLQY can be increased to more than 60% in an oxygen-containing solution at room temperature. The presence of copper leads to an ultrafast intersystem crossing to a long-lived triplet state and suppresses non-radiative decay. Specifically, the excited-state dynamics of $Au_{22}$ and $Au_{16}Cu_6$ were probed using femtosecond and nanosecond transient absorption spectroscopy (fs-TA and ns-TA). The nanosecond transient absorption spectra of both show only one decay component. The excited-state lifetime of $Au_{22}$ is 600 ns, and that of $Au_{16}Cu_6$ is 1.3 μs, which is comparable to their luminescence lifetimes. In contrast to the simple nanosecond dynamics, different femtosecond dynamics were observed in the two types of nanoclusters, and a broad excited-state absorption (ESA) was also observed in both clusters. For $Au_{22}$ (Figure 8.11a), with an excitation wavelength of 380 nm, the ESA peak exhibits a weak ultrafast decay within the first 1 ps but increases between 1 ps and 300 picoseconds. For $Au_{16}Cu_6$ (Figure 8.11b), an ultrafast decay occurs within the first 1 ps, followed by a non-decaying relaxation component that remains constant between 2 ps and 2 ns. The absorption dynamics probed at 700 nm and the different dynamics of the two clusters are shown in Figure 8.11c. The excited-state dynamics reveals that doping increases the rate of intersystem crossing from $S_1$ to $T_1$ by 300 times (Figure 8.11d), thus achieving a nearly 100% photoluminescence quantum yield. This work demonstrates that a nearly 100% photoluminescence quantum yield can be achieved through gold–copper alloy nanoclusters even in solutions at room temperature, which makes applications ranging from biological imaging to light-emitting devices possible. In the future, it will be interesting to study the doping effects of other doped metals, which may lead to unexpected optical properties and also photocatalytic properties that we are going to introduce in the next section.

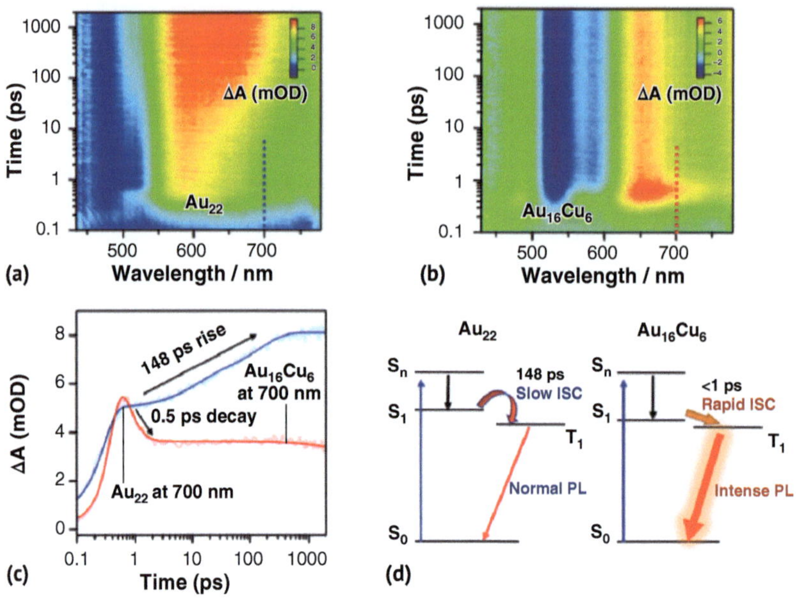


**Figure 8.11** Ultrafast excited-state dynamics results of $Au_{22}$ and $Au_{16}Cu_6$. (a) Fs-TA data map of $Au_{22}$ NCs with 380-nm excitation. (b) Fs-TA data map of $Au_{16}Cu_6$ NCs with 380-nm excitation. (c) TA kinetic traces probed at 700 nm and corresponding fits of two NCs. (d) Excited-state relaxation diagram obtained from the time-resolved data. *Source:* [67]/ with permission of American Association for the Advancement of Science – AAAS.

## 8.2 Metal Nanocluster Photocatalytic Property

Developing and utilizing solar energy-based environmental protection technologies to address the world's environmental and energy challenges is one of the most pressing challenges faced by scientists today. Among these, photocatalysis has attracted significant attention due to its ability to directly convert light energy into chemical energy under mild reaction conditions [68–70]. The essence of photocatalysis lies in photocatalytic reactions mediated by catalysts. When a photocatalytic material is irradiated with light energy exceeding its bandgap, electrons transition, generating electron ($e^-$)–hole ($h^+$) pairs. These pairs then undergo redox reactions with adsorbates on the material's surface under light illumination [71]. Currently, most commonly used photocatalysts are wide-bandgap semiconductor materials, such as $TiO_2$, $SrTiO_3$, $BiVO_4$, $Ta_3N_5$, g-$C_3N_4$, CdS, and $MoS_2$ [72–77]. Although these photocatalysts exhibit advantages such as high stability, low toxicity, mild reaction conditions, and rapid reaction rates, they primarily rely on ultraviolet light sources, leading to low utilization of solar energy. Additionally, their high electron–hole recombination rates significantly reduce the efficiency of photocatalytic reactions.

Metal nanoclusters with atomically precise structures, sized between 1 and 3 nm, exhibit pronounced quantum size effects. Their energy level architectures differ fundamentally from those of larger nanoparticles, where quasi-continuous energy bands decompose into discrete levels [78–80]. This quantization and discretization enable electrons in coinage metal nanoclusters to interact with light through interlevel transitions, facilitating efficient separation of electrons and holes [81]. Additionally, metal nanoclusters possess small energy gaps, typically showing distinct absorption in the visible/NIR spectral regions [82, 83]. Characterized by excellent photostability, these nanoclusters are inherently resistant to oxidation or reduction by photogenerated charge carriers. Collectively, these attributes establish metal nanoclusters as an emerging class of photocatalytic materials garnering increasing attention. Meanwhile, their atomically resolved structures enable investigations of photocatalytic mechanisms at the atomic scale [84, 85], providing theoretical foundations for designing novel photocatalytic materials.

Atomically precise metal nanoclusters exhibit unique electronic structures distinct from both single atoms and nanoparticles (Figure 8.12). Their ultra-small size induces strong quantum size effects, leading to discrete energy levels that enable electron transitions from the HOMO to the LUMO, facilitating electron–hole separation. Additionally, metal nanoclusters possess small energy gaps, typically <2.2 eV, allowing photocatalytic reactions under visible light illumination. As such, metal nanoclusters can be regarded as semiconductor nanomaterials with narrow bandgaps for photocatalytic applications.

In the rational design of photocatalysts, their optical and electronic properties should be firstly considered, as these properties are pivotal in determining the light harvesting abilities and the overall efficiency of the photocatalysts. When the energy of the photon is equal to or greater than the HOMO–LUMO gap of the photocatalyst, an electron is promoted from the HOMO to the LUMO (Figure 8.13, process 1), which results in a charge separation. Then, the generated electron and hole either migrate to the surface for the catalytic reaction (Figure 8.13, processes 2 and 4), or they may recombine at bulk or surface (Figure 8.13, process 3). Therefore, a good photocatalyst should have the properties including light absorption in the visible range for the maximum utilization of sunlight, good electron–hole separation lifetimes, and efficient photogenerated electron and hole transfer to the surface [86].

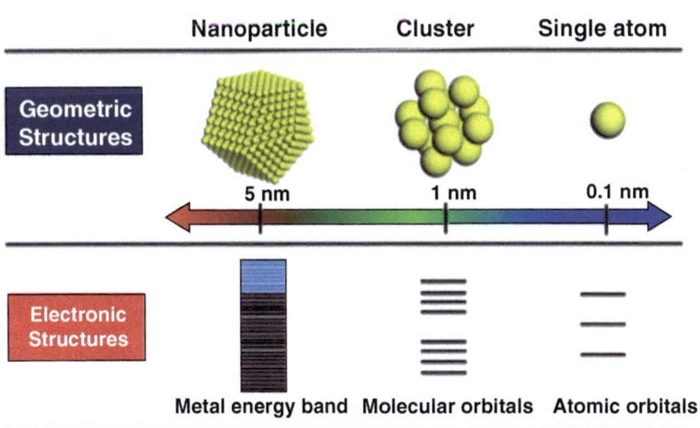


**Figure 8.12** Geometric and electronic structures of single atom, clusters, and nanoparticles. *Source:* [81]/with permission of American Chemical Society.

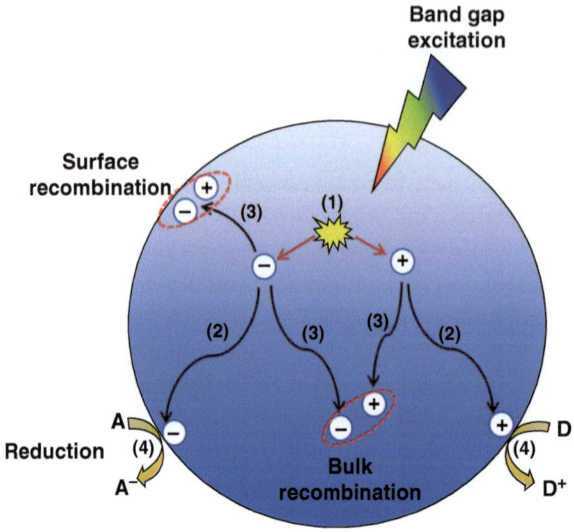


**Figure 8.13** Electronic processes in photocatalytic materials. (1) Photoinduced charge separation, (2) transfer of LUMO electrons to the particle surface, (3) surface or bulk electron–hole recombination, and (4) electrons successfully extracted for photocatalytic redox reactions.

## 8.3 Metal Nanocluster Energy Transfer

Energy transfer refers to the process by which energy in an excited state is transferred from one species (such as metal nanoclusters) to another species (such as reactants or the medium), without involving the actual transfer of electrons. The characteristics of energy transfer include: (i) non-electron transfer: only energy is transferred, without involving the movement of charge; (ii) mode of action: realized through non-radiative mechanisms, such as Förster resonance energy transfer or Dexter energy transfer; (iii) reaction requirements: applicable to systems that require energy input but do not involve redox reactions. The most general and important way of metal nanocluster energy-transfer process is the metal nanocluster-catalyzed generation of singlet oxygen ($^1O_2$).

Kawasaki and coworkers demonstrate that $^1O_2$ can be efficiently produced through the direct photosensitization by anionic $Au_{25}(SR)_{18}^-$ clusters (H-SR = phenylethanethiol or captopril) without using conventional organic photosensitizers under visible/near-IR (532, 650, and 808 nm) irradiation (Figure 8.14a). The $^1O_2$ generation was confirmed by two methods: (i) direct observation of the characteristic phosphorescence of $^1O_2$ at 1270 nm and (ii) using $^1O_2$ specific chemical probes such as organic-soluble 1,3-diphenylisobenzofuran (DPBF) or water-soluble 3,3′-diaminobenzidine (DAB) (Figure 8.14b,c). These results demonstrated that $O_2$ can be readily activated by $Au_{25}(SR)_{18}^-$ under visible or NIR light, which prompted us to carry out photocatalytic applications of $Au_n(SR)_m$ NCs for organic transformation reactions. In this article, catalytic oxidation of organic sulfide to

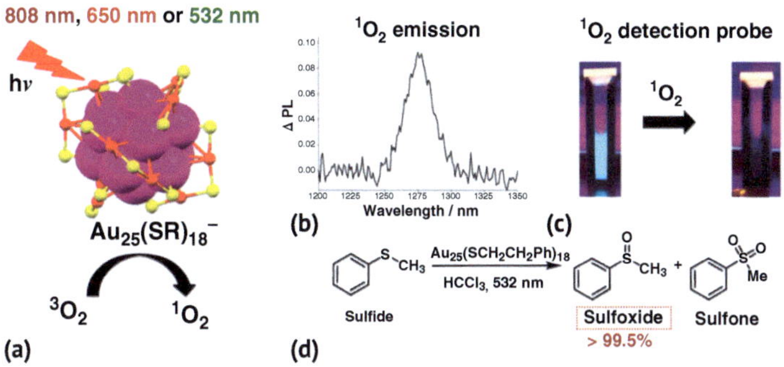


**Figure 8.14** (a) $^1O_2$ generation via energy transfer from photoexcited nanoclusters to $^3O_2$. (b) PL difference spectrum in which the singlet $O_2$ ($^1O_2$) emission is centered around 1276 nm. (c) Photographs of the DPBF-containing DMF solution in the presence of $Au_{25}(PET)_{18}^-$ at excitation of 350 nm before and after irradiation at 532 nm (50 mW) for 2 minutes. (d) Catalytic oxidation of sulfide to sulfoxide by $^1O_2$ over photoexcited $Au_{25}(PET)_{18}^-$. *Source:* [87]/with permission of American Chemical Society.

sulfoxide with nearly 100% selectivity by $^1O_2$ was demonstrated on the photoexcited $Au_{25}(SC_2H_4Ph)_{18}^-$ clusters (Figure 8.14d). It is suggested that the optical gap of $Au_{25}(SR)_{18}$ (~1.3 eV) being larger than the energy of $^1O_2$ (0.97 eV) allows for the efficient energy transfer to $^3O_2$. In addition, the long lifetime of the electronic excited states of $Au_{25}(SR)_{18}$ and the well-defined $O_2$ adsorption sites are the key factors that promote energy transfer from $Au_{25}(SR)_{18}^-$ to molecular oxygen, thus facilitating the formation of $^1O_2$. It was also found that neutral $Au_{25}(SR)_{18}^0$ can also produce $^1O_2$ as efficiently as the anionic $Au_{25}(SR)_{18}^-$ [87].

After a few years, Agrachev and coworkers have provided even more insights into the significance of the charge state of metal NCs on photocatalytic activity, by performing a time-resolved electron paramagnetic resonance (TR-EPR) study on $[Au_{25}(SCH_2CH_2CH_3)_{18}]^-$ NCs in order to understand the photo excitation dynamics [88]. They found that the photoexcitation of this NC causes the generation of a singlet excited state ($S_1$), which then undergoes intersystem crossing to a triplet excited state ($T_1$). This T1 state is then able to generate singlet oxygen ($^1O_2$) from triplet oxygen ($^3O_2$) that could be used for target applications (Figure 8.15). Notably, the authors also found that $Au_{25}(SCH_2CH_2CH_3)_{18}^0$ did not generate any EPR signal in its excited state, which implies that $Au_{25}(SCH_2CH_2CH_3)_{18}^0$ might be relatively inactive in singlet oxygen generation. This data also supports the findings by Rongchao Jin et al., who have also shown that anionic Au nanoclusters performed better than their corresponding neutral counterparts in the photocatalytic generation of singlet oxygen.

Au nanoclusters can be regarded as semiconductors with small bandgaps. Yu and coworkers [89] report the visible light photocatalytic properties of a composite material consisting of $Au_{25}(SR)_{18}$ nanoclusters (R = $CH_2CH_2Ph$) and $TiO_2$ nanocrystals. The effects of $Au_{25}(SR)_{18}$ nanoclusters on the photocatalytic activity of $TiO_2$ nanocrystals were evaluated in the reaction of photocatalytic degradation of

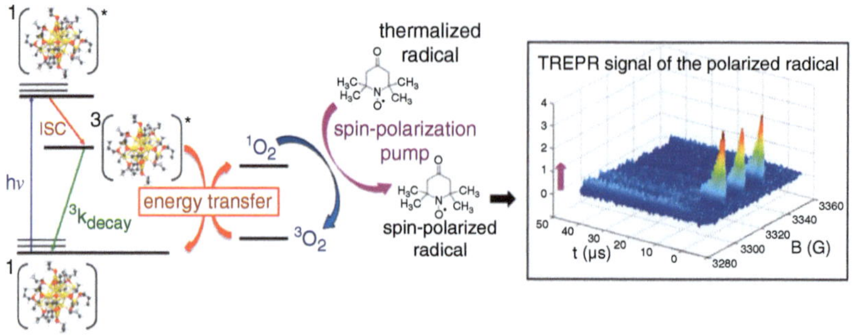


**Figure 8.15** Schematic illustration of the TR-EPR detection of singlet-oxygen generation through the photoexcitation of $[Au_{25}(SCH_2CH_2CH_3)_{18}]^-$. *Source:* [88]/with permission of American Chemical Society.

methyl orange. The loading of $Au_{25}(SR)_{18}$ nanoclusters onto $TiO_2$ results in strong visible light absorption by the composite and, more importantly, 1.6 times increase in visible light photocatalytic activity. The quantum confinement of $Au_{25}(SR)_{18}$ with a HOMO-LUMO gap of $E_g$ ~ 1.3 eV is likely key to enhance the photocatalytic activity [90]. Furthermore, the $Au_{25}(SR)_{18}/TiO_2$ composite nanostructure exhibits high stability in recycling tests. The $Au_{25}(SR)_{18}$ nanoclusters dispersed on the $TiO_2$ surface can act as a small-band-gap semiconductor to absorb visible light, giving rise to electron–hole separation and producing singlet oxygen (Figure 8.16). This work systematically proves that $^1O_2$ generated by $Au_{25}$ nanoclusters (via energy transfer) is directly responsible for methyl orange degradation through probe experiments, control tests, and quenching analyses (Figure 8.16c–e).

Li and coworkers [91] investigated the photosensitive characteristics of $Au_{38}S_2(SAdm)_{20}$ nanometer clusters forming singlet oxygen ($^1O_2$) at visible light wavelengths of 532 and 650 nm. First, the formation of $^1O_2$ was detected by 1,3-diphenylisobenzofuran as the chemical trapping probe as well as direct observation of the characteristic $^1O_2$ emission (ca. 1276 nm). They also compared the photocatalytic amine oxidation performances between the bcc-$Au_{38}S_2(S\text{-}Adm)_{20}$ (S-Adm = 1-adamantanethiolate) and bi-icosahedral $Au_{38}(SR)_{24}$ ($R = C_2H_4Ph$) [92]. For photocatalytic oxidation of methyl phenyl sulfides, $Au_{38}S_2(S\text{-}Adm)_{20}$ gave 57% conversion with 100% selectivity for methyl phenyl sulfoxide, but $Au_{38}(SR)_{24}$ NCs were almost inactive in the sulfoxide reaction (Figure 8.17). The singlet oxygen formed by the energy transfer of light-excited NCs to triplet oxygen was the main active species for this reaction. Thus, higher activity of $Au_{38}S_2(S\text{-}Adm)_{20}$ was attributed to higher $^1O_2$ quantum yield. The two types of $Au_{38}$ NCs were further applied for the photocatalytic oxidation of amines toward imines. The $Au_{38}S_2(S\text{-}Adm)_{20}$ NCs showed 99% conversion of benzyl amines with 99% selectivity, which presented better catalytic activities than 20% conversion of $Au_{38}(SR)_{24}$ (Figure 8.17c). The high catalytic activity of $Au_{38}S_2(S\text{-}Adm)_{20}$ was consistent with the photocatalytic oxidation results of methyl phenyl sulfides, with singlet oxygen as the intermediate of oxidized organic sulfides or benzylamines.

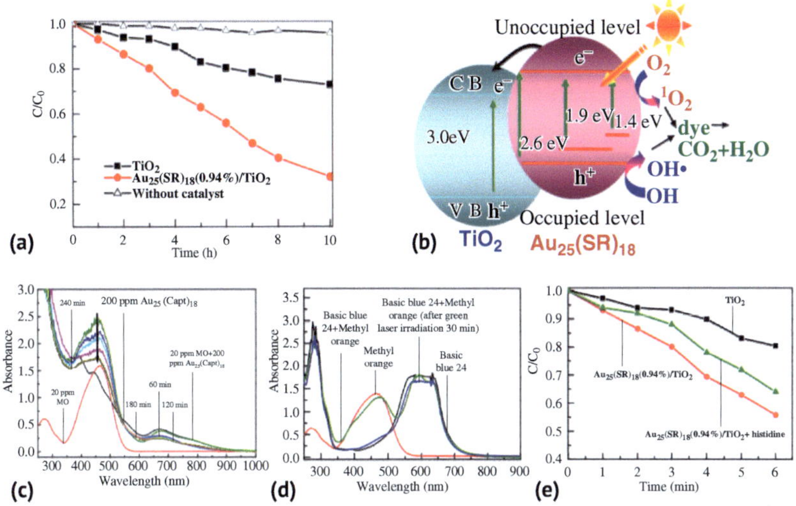


**Figure 8.16** (a) Corresponding concentration changes of methyl orange with reaction time and (b) the photocatalytic mechanisms of $Au_{25}(SR)_{18}/TiO_2$ under visible-light irradiation. (c) UV–vis spectra of methyl orange and $Au_{25}(Capt)_{18}$. Explain: Captopril-protected $Au_{25}(Capt)_{18}$, which is soluble in water and has similar capability as $Au_{25}(SR)_{18}$ to produce singlet oxygen to degrade methyl orange. (d) UV–vis spectra of methyl orange and basic blue 24 ($C_{18}H_{22}ClH_3S{\cdot}xZnCl_2$). Explain: To verify the oxidation capability of $^1O_2$ in decomposing methyl orange, they further used basic blue 24, a known dye for $^1O_2$ production, to produce $^1O_2$. (e) The effect of addition of L-histidine ($^1O_2$ quencher,) on the photocatalytic activity of $Au_{25}(SR)_{18}(0.94\%)/TiO_2$ under visible light irradiation in methyl orange degradation. *Source:* [89]/with permission of American Chemical Society.

In addition, Chong and coworkers [93] achieved 100% conversion of benzylamine to *N*-benzylamine under visible light irradiation by using $Au_{25\text{-}x}Ag_x(SR)_{18}$ NCs. The mechanism involved the oxidation of benzylamine to *N*-benzylidene benzylamine by obtained highly reactive $^1O_2$ from NCs donating energy to triplet oxygen under light irradiation. Specifically, the catalysts of $Au_{25}^-$, $Au_{25}^0$, $Ag_{25}$ , and $Au_{25\text{-}x}Ag_x$ were prepared according to the previous literature [70, 94–96]. As shown in Figure 8.18, those NCs share a common structure, which contains an icosahedral $M_{13}$ core protected by six $M_2S_3$ staple motifs. In the case of $Au_{25-x}Ag_x^-$ NCs, it is interesting that Ag atoms are selectively incorporated in the 12 vertices of the icosahedral core. Other than the common structure, they are all characteristic for the negative valence state. $Au_{25}^-$, $Au_{25}^0$, $Ag_{25}^-$, and $Au_{25-x}Ag_x^-$ all show EPR activity under irradiation, but $Au_{25-x}Ag_x^-$ exhibits the strongest signal. The alloyed $Au_{25-x}Ag_x^-$ NCs catalyst gives better performance than the pure metal NCs due to synergistic effect, with 100% conversion rate and selectivity.

Currently, most photocatalytic reactions proceed under the irradiation of UV–vis light. However, nearly half of the solar spectrum is made up of the much lower energy infrared light. From the energy-saving perspective, the less energy to excite the electron transition is more attractive to researchers [97]. Therefore, developing

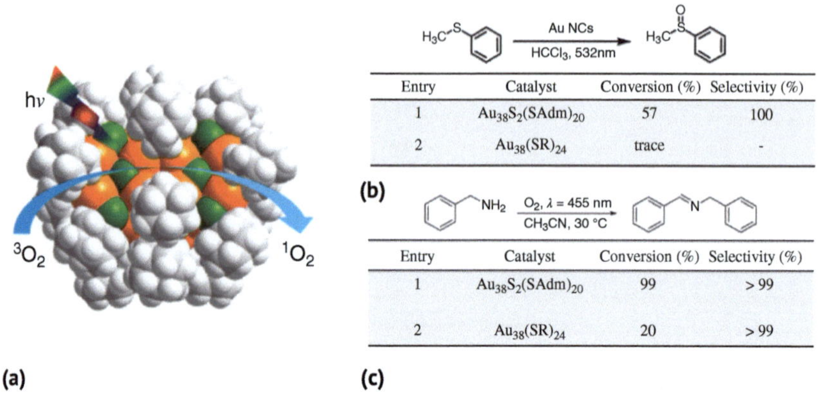


**Figure 8.17** (a) Illustrations of the Generation Process on the $Au_{38}S_2(SAdm)_{20}$ structure, and shown in space-filling mode. Color code: Au, orange; S, green; C, gray; H, white. (b) Catalytic performances of the $Au_{38}S_2(SAdm)_{20}$ and $Au_{38}(SR)_{24}$ NCs in the selective oxidation of methyl phenyl sulfide to sulfoxide under light irradiation with $\lambda = 532$ nm. Reaction conditions: phenyl methyl sulfide (0.05 mmol), $CHCl_3$(2 mL), 0.2 μmol nanoclusters under a $O_2$ flow (5 mL min$^{-1}$) at room temperature for 12 hours under irradiation with $\lambda = 532$ nm. (c) Photocatalytic oxidation of benzylamine to an imine in the presence of $O_2$ and $Au_{38}S_2(SAdm)_{20}$ nanoclusters under LED irradiation ($\lambda \sim 455$ nm) at 30°C. Reaction conditions: benzylamine (0.2 mmol), dodecane (0.1 mmol), 1 mL of toluene, and 0.5 μmol of $Au_{38}S_2(SAdm)_{20}$ or $Au_{38}(PET)_{24}$ nanoclusters in the presence of $O_2$ at 30°C under a LED light centered at $\lambda = 455$ nm for 30 min. *Source:* [91]/with permission of American Chemical Society.

photocatalyst with NIR light response is highly desired. Due to the precisely adjustable number, the type and arrangement of metal atoms, metal NCs exhibit tunable HOMO–LUMO gap, realizing the wide light spectrum utilization from UV to NIR. Wang and coworkers [98] synthesized $Au_{25}(F\text{-}Ph)_{18}$ (F-Ph = 3,4-difluorothiophenol) NCs via surface modification on the $Au_{25}(PET)_{18}$ template (Figure 8.19a). Due to the π–π stacking between the F-Ph ligands, the modified $Au_{25}(F\text{-}Ph)_{18}$ shows improved stability compared to $Au_{25}(F\text{-}Ph)_{18}$, reflecting the almost unchanged UV–vis characteristic peaks of $Au_{25}(F\text{-}Ph)_{18}$ in the presence of $H_2O_2$ and the remaining negatively charged state after irradiation. In addition, $Au_{25}(F\text{-}Ph)_{18}$ displays higher $^1O_2$ production ability, indicating the enhanced catalytic activity than that of $Au_{25}(PET)_{18}$. On the basis of these findings, $Au_{25}(F\text{-}Ph)_{18}$ nanoclusters are used to realize the C–H functionalization of 2-aryl-1,2,3,4-tetrahydroisoquinolines (Figure 8.19b). Under NIR light irradiation, $[Au_{25}]^-$ is photoexcited to $[Au_{25}]^{-*}$, and then $[Au_{25}]^{-*}$ transfers the energy for the formation of $^1O_2$ from $^3O_2$ for the subsequent C-H functionalization process. From the time-dependent synthesis yield curve, it can be seen that when using 850 nm NIR LED instead the 460 nm blue LED light source, the yield of the product is significantly improved from 69% in 48 hours to 99% in 8 hours (Figure 8.19c). A following work by Zhu and coworkers further manifests the advantages of designing efficient metal nanocluster catalysts for photocatalysis involving the energy-transfer process [99]. They synthesized a series of structurally similar $M_{15}$ nanoclusters, including $Au_{15}$, $Au_{15\text{-}x}Ag_x$, $Au_{15\text{-}x}Cu_x$, $Au_{15\text{-}x\text{-}y}Ag_xCu_y$,

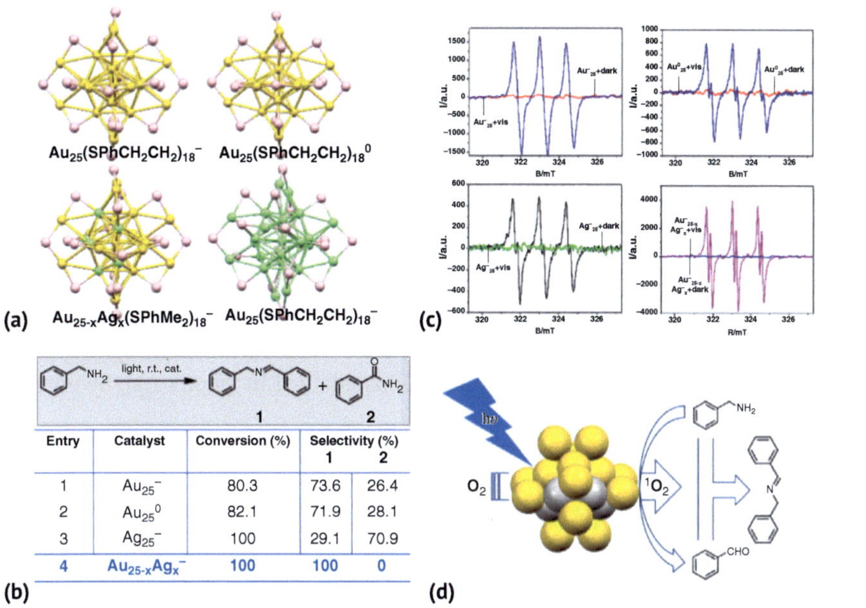


**Figure 8.18** (a) Crystalline structure of $Au_{25}(SPhCH_2CH_2)_{18}^-$, $Au_{25}(SPhCH_2CH_2)_{18}^0$, $Au_{25-x}Ag_x(SPhMe_2)_{18}^-$, and $Ag_{25}(SPhCH_2CH_2)_{18}^-$. The golden and green balls are gold and silver atoms, respectively, whereas the pink are S atoms. (b) Photoinduced catalysis by different catalysts. Reaction condition: 0.2 mL benzylamine mixed with 1 mg NCs blended into 2 mL nitrile under Xe light with 420 nm filter at ~20°C for 5 hours, open air. (c) The EPR spectra of $Au_{25}^-$, $Au_{25}^0$, $Ag_{25}^-$, and $Au_{25-x}Ag_x^-$ obtained from mixture of NCs (1 mg) and 1 mL toluene, 1 mL acetonitrile, 5 mmol spin trap *N-tert*-butyl-α-phenylnitrone (PBN) at 243 K. The light is filtered by 420 nm filter. (d) The proposed mechanism of photocatalysis catalyzed by $Au_{25-x}Ag_x^-$. *Source:* [93]/John Wiley & Sons.

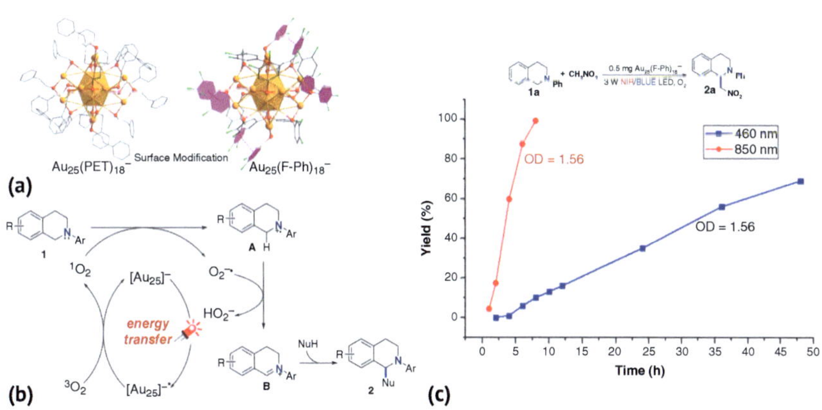


**Figure 8.19** (a) Structure of $Au_{25}(F\text{-}Ph)_{18}$ and $Au_{25}(PET)_{18}$. (b) C–H functionalization of 2-aryl-1,2,3,4-tetrahydroisoquinoline using $Au_{25}(F\text{-}Ph)_{18}^-$ as catalyst. (c) Time-dependent synthesis yield by using $Au_{25}(F\text{-}Ph)_{18}^-$ with NIR and blue LED light. *Source:* [98]/with permission of American Chemical Society.

and $Au_{15-x-y}Ag_xCu_yPt_1$, and systematically investigated their photocatalytic activity toward oxidation of 4-phenylcyclohexan-1-one derived TBS enol ether. Results showed that although these five nanocluster catalysts bear the basically same geometric structures, their catalytic activities were different from each other. Specifically, $Au_{15-x}Cu_x$ showed the best catalytic performance, affording the desired α,β-unsaturated carbonyl product in 83% yield. The other four nanoclusters, namely $Au_{15}$, $Au_{15-x}Ag_x$, $Au_{15-x-y}Ag_xCu_y$, and $Au_{15-x-y}Ag_xCu_yPt_1$ catalyzed the reaction and gave the corresponding product in 75%, 62%, 47%, and 39% yields, respectively. This work indicates that apart from the surface ligand exchange method (Figure 8.19), metal alloying also plays important roles in regulating photocatalytic activities of metal nanoclusters. Both strategies can be applied in the future for designing effective catalysts for other organic transformations enabled by photocatalytic energy-transfer processes.

## 8.4 Metal Nanocluster Electron Transfer

In the photocatalytic process of metal nanoclusters, electron transfer refers to the transition of electrons from the ground state to the excited state under light excitation. These excited electrons then transfer to reactants, active sites on the catalyst surface, or other acceptors through various pathways, while leaving vacancies in the nanocluster. This electron transfer process is a critical step for the photocatalytic reaction, driving the chemical transformation. The actual transfer of electrons is usually accompanied by redox reactions.

### 8.4.1 Photocatalytic Water Splitting

Hydrogen energy, which is clean, efficient, safe, storable and transportable, has been widely as the most ideal pollution-free green energy [69, 100]. The technology of photodissociation of water to produce hydrogen was discovered in 1972 by Fujishima and coworkers [101], who first reported the phenomenon of photocatalytic decomposition of water to produce hydrogen on the surface of $TiO_2$, thus opening up the avenue of photocatalytic water splitting to produce hydrogen. Photocatalytic water splitting is a process that utilizes solar energy to decompose water to produce $H_2$ (Figure 8.20). The theoretical minimum bandgap of photocatalytic materials for water splitting is 1.23 eV [69, 102]. Metal nanoclusters are usually combined with wide bandgap semiconductor materials as cocatalysts to reduce the activation energy of hydrogen evolution reactions and provide active sites, thereby obtaining more efficient composite photocatalysts.

Negishi and coworkers [103] prepared size uniform NC-based catalysts using the front-synthesis approach. They controlled the loading amount (~0.1%) and deposited a series of GSH-protected Au NCs ($Au_{10}(SG)_{10}$, $Au_{15}(SG)_{13}$, $Au_{18}(SG)_{14}$, $Au_{22}(SG)_{16}$, $Au_{25}(SG)_{18}$, $Au_{29}(SG)_{20}$, $Au_{33}(SG)_{22}$, and $Au_{39}(SG)_{24}$) on $BaLa_4Ti_4O_{15}$. After removing the ligand by calcination at 300°C in vacuo, for group 1 ($Au_{10}$, $Au_{15}$, $Au_{18}$, $Au_{25}$, and $Au_{39}$), the size of the Au nanoclusters displayed no significant

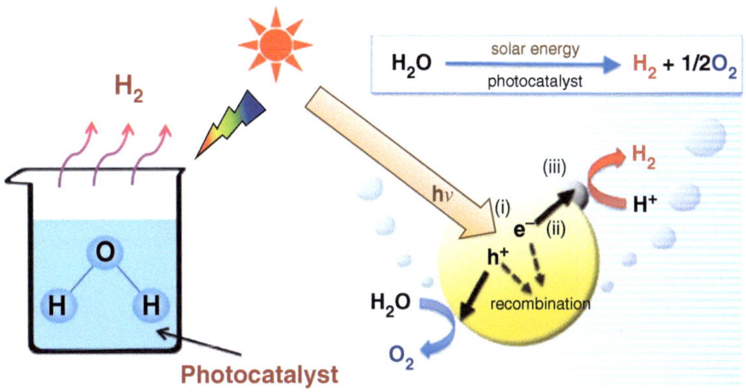


**Figure 8.20** Photocatalytic water splitting as an artificial photosynthesis (inset: principle of water splitting using semiconductor photocatalysts).

change with slight increase, while, for group 2 ($Au_{22}$, $Au_{29}$, and $Au_{33}$), these Au nanoclusters showed clear size increase and a distinct plasmonic peak at ~520 nm. The Au nanocluster-$BaLa_4Ti_4O_{15}$ system with more stable Au nanoclusters (those in group 1) was used to reveal the size effect of Au nanoclusters in photocatalytic water splitting activity. All Au nanocluster-$Bala_4Ti_4O_{15}$ composites produced $H_2$ and $O_2$ in the ratio of 2:1, but the catalytic activity increased with the decrease in nanocluster size, which was due to the sharp increase of surface atoms with the size decrease (Figure 8.21).

Glutathione-protected gold nanoclusters ($Au_x$-GSH NCs) was also discovered to serve as sensitizers [104], injecting electrons into $TiO_2$ nanostructures upon visible light excitation. Moreover, they can generate photocurrents in solar cells, with efficiency comparable to that of CdS QDs (a power conversion efficiency of 2%). Based on this finding, the researchers prepared a photocatalytic system composed of $Au_x$-GSH/Pt/$TiO_2$ nanocomposite particles using $Au_x$-GSH NCs as photosensitizers and conducted photocatalytic reactions under visible light irradiation for water splitting to produce $H_2$ [105]. As shown in Figure 8.22a, after photoexcitation, $Au_x$-GSH NCs can inject electrons into $TiO_2$, and the electrons migrating to the surface can reduce $H^+$ and release $H_2$. By comparing the amount of hydrogen generated by Pt/$TiO_2$ and $Au_x$-GSH/Pt/$TiO_2$ at different times (Figure 8.22b), the amount of $H_2$ released by $Au_x$-GSH/Pt/$TiO_2$ increased linearly within the first 90 minutes. When the individual Pt/$TiO_2$ was irradiated, no $H_2$ production was detected. This result indicates that $Au_x$-GSH NCs dominate the generation of $H_2$. Meanwhile, when ethylene diamine tetra acetic acid (EDTA) was introduced into the aqueous solution as a hole scavenger, the $H_2$ production rate increased significantly. For $Au_x$-GSH-sensitized Pt/$TiO_2$ NPs, the apparent quantum yield (AQY) for $H_2$ production was 0.05%, and it increased to 0.13% after adding EDTA. Although the $H_2$ production efficiency is not high, this system can still produce $H_2$ without an externally applied bias potential, demonstrating the effectiveness of thiol ligand-protected gold nanoclusters in the production of solar fuels. This work shows that

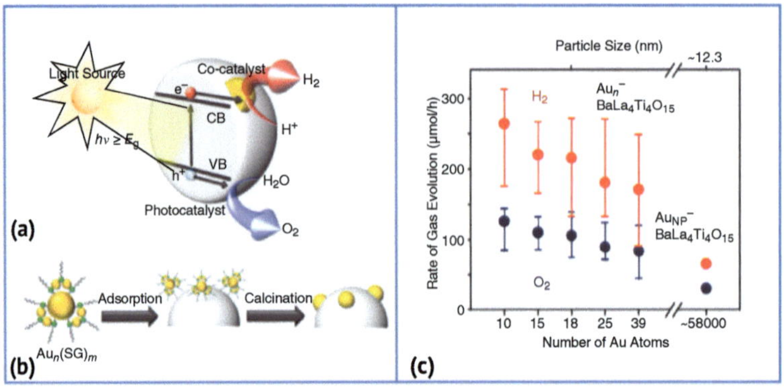


**Figure 8.21** (a) Scheme of photocatalytic water splitting using Au NCs as cocatalysts: CB, conduction band; VB, valence band; $E_g$, bandgap. (b) Schematic of the loading of highly regulated metal clusters onto the photocatalyst surface. (c) Relationship between the size of the Au NCs and the activity of $H_2$ and $O_2$ production. *Source:* [103]/with permission of American Chemical Society.

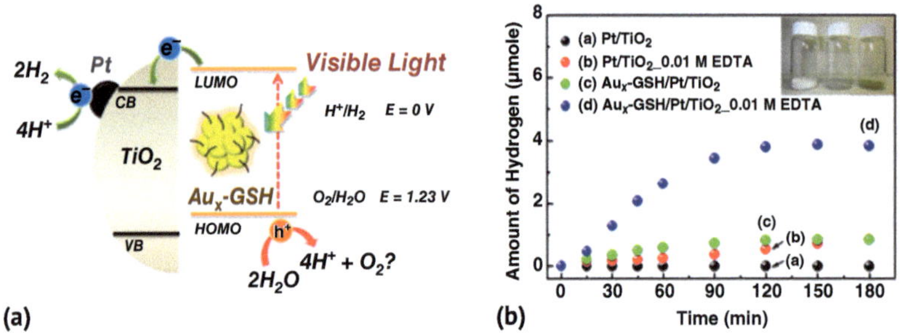


**Figure 8.22** (a) Illustration of the working principle of the $Au_x$-GSH-sensitized Pt/$TiO_2$ NPs photocatalytic system for water splitting reaction under visible-light illumination. (b) Time course of hydrogen evolution following the visible light illumination of an aqueous suspension containing (a) and (b) Pt/$TiO_2$ nanoparticles, and (c) and (d) $Au_x$-GSH-sensitized Pt/$TiO_2$ nanoparticles. The experiments were conducted in aqueous solution (pH 7) (a) and (c) in the absence and (c) and (d) presence of 0.01 M EDTA aqueous solution. *Source:* [105]/with permission of American Chemical Society.

gold nanoclusters are effective in expanding the light response of wide-bandgap semiconductors such as $TiO_2$.

Atomically precise nanoclusters $Ag_{44}(SR)_{30}$ [106] was deposited onto $TiO_2$ and investigated their photocatalytic hydrogen production performance. The test results show that for the $Ag_{44}/TiO_2$ composite system (**1**-$TiO_2$ (p), where **1** represents $Ag_{44}(SR)_{30}$ formed by loading 2% (mass fraction) of $Ag_{44}(SR)_{30}$ onto $TiO_2$, its photocatalytic hydrogen production rate is 10 times that of individual $TiO_2$ nanoparticles. When the light source was changed from visible light to simulated sunlight, the amount of photocatalytic $H_2$ production increased by three orders of magnitude, and the $H_2$ production rate reached 7.4 mmol/(h·g), which is five times higher than the photocatalytic hydrogen production rate of $TiO_2$ modified with plasmonic Ag nanoparticles without precise structure (AgNP-$TiO_2$), as shown in Figure 8.23a. It can even be comparable to $TiO_2$ modified with Pt nanoparticles under similar conditions. To explore the mechanism, the energy level structure of the $Ag_{44}/TiO_2$ composite system was deduced through theoretical calculations, and the charge transfer pathway was revealed by transient absorption spectroscopy. The results show that under visible light irradiation, only the nanoclusters are excited. Due to the forbidden transition dipole of the HOMO–LUMO of the nanoclusters, the electrons in the HOMO of the nanoclusters are excited to LUMO+1 [41]. Since the energy level of LUMO+1 is higher than that of the conduction band of $TiO_2$, the photogenerated electrons spontaneously inject into $TiO_2$ to produce $H_2$, and the holes remain on the nanoclusters for oxidation reactions. In this case, the nanoclusters only act as photosensitizers. When irradiated with UV–vis light, both the nanoclusters and the semiconductor are excited to generate electron–hole pairs. Driven by the internal electric field formed, the photogenerated holes in the valence band of $TiO_2$ can easily transfer to the HOMO energy level of $Ag_{44}(SR)_{30}$, and the electrons of $Ag_{44}(SR)_{30}$ inject into the conduction band of $TiO_2$. The $Ag_{44}/TiO_2$ composite can effectively suppress the recombination of electron–hole pairs and thus has high photocatalytic performance, as shown in Figure 8.23b. Meanwhile, the catalytic efficiency hardly decreases after multiple cycles (Figure 8.23c), indicating that the catalyst has good

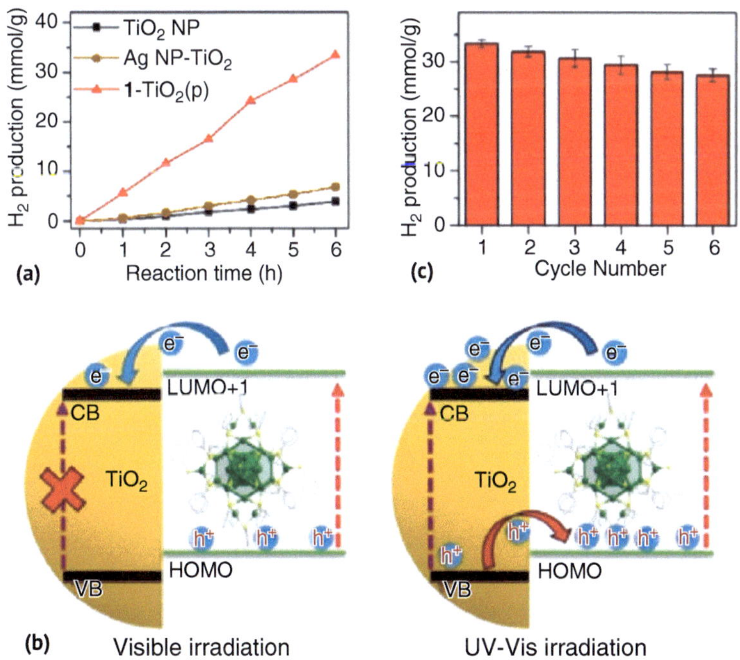


**Figure 8.23** (a) Photocatalytic properties of $TiO_2$ NP, Ag NP-$TiO_2$, and **1**-$TiO_2$ (p) under UV–vis light. (b) Schematic illustration of the charge-transfer pathways in the $Ag_{44}/TiO_2$ photocatalytic system for $H_2$ evolution under irradiation with visible light and UV–vis light. (c) Cycling test over **1**-$TiO_2$ (p) NPs under UV–vis light. *Source:* [106]/John Wiley & Sons/CC BY-NC-ND 4.0.

stability. Therefore, this study demonstrates that the combination of silver nanoclusters and wide-bandgap semiconductors shows a synergistic effect on light collection and charge separation at the binary semiconductor heterojunction. At the same time, the introduction of atomically precise nanoclusters can provide a clear interface for studying charge transfer at the interface, thereby further clarifying the mechanism of photocatalytic hydrogen production.

### 8.4.2 Photocatalytic Degradation of Organic Pollutants

With the rapid development of modern industry and agriculture, organic pollutants such as organic dyes, phenols and their derivatives, antibiotics, surfactants, and pesticides have been released into the environment, posing a serious threat to human sustainable development. Currently, traditional methods for treating organic pollutants in the environment include extraction, adsorption, biodegradation, and ionizing radiation, but these methods are difficult to apply in real-world settings [107]. In contrast, the advantages of photocatalytic degradation of organic pollutants are evident. Under the excitation of an external light source, the photocatalyst undergoes electron transfer, generating photogenerated carriers. These photogenerated carriers can interact with water and air to produce hydroxyl radicals (·OH), superoxide radicals ($\cdot O_2^-$), photogenerated holes ($h^+$), and singlet oxygen ($^1O_2$). These active substances possess strong oxidizing capabilities, capable of decomposing adsorbed organic pollutants into $H_2O$ and $CO_2$, thus completing the degradation process without causing secondary pollution under mild conditions [108, 109].

Yu and coworkers deposited the $Au_{25}(SCH_2CH_2Ph)_{18}$ NCs on $TiO_2$ and applied it into photocatalytic methyl orange degradation, and they found it displayed light-dependent catalytic activity. Due to the introduction of $Au_{25}$, the light absorption range of the composite is extended from ultraviolet to near infrared. Under UV and visible light irradiation, the decomposition ratio (DR) in methyl orange degradation of the composite (with only 0.94% NC loading amount) was 69%, while the DR of bare $TiO_2$ with only 27%, indicating a higher catalytic activity of the composite (Figure 8.24a) [89]. Hu and coworkers reported similar results in their study on the photocatalytic degradation of rhodamine B (RhB) by using $Au_{25}$/porous ZnO as catalysts. In order to balance the separation efficiency of $e^-$–$h^+$ and the active surface area, they optimized the loading amount of $Au_{25}$ in composite. When the loading amount is 1.16% (Figure 8.24c), the catalytic activity of the composite is the highest, and the DR can reach 95% within 15 minutes (Figure 8.24) [110].

Heteroatom doping in coinage metal nanoclusters can significantly alter their electronic structures [111], thereby influencing their photocatalytic properties. Liu and coworkers synthesized a new type of alloy nanocluster protected by adamantanethiol ligands, $[Au_{23-x}Ag_x(S\text{-}Adm)_{15}]$ ($x = 4$–$7$), whose structure is shown in Figure 8.25a. They also investigated its photocatalytic performance in the degradation of pollutants such as rhodamine B and phenol under visible-light irradiation. The $Au_{23-x}Ag_x$ clusters were deposited on a $TiO_2$ support ($Au_{23-x}Ag_x/TiO_2$) for photocatalytic tests. It was found that 0.5% (mass fraction) $Au_{23-x}Ag_x/TiO_2$ exhibited the highest catalytic performance in the photocatalytic degradation of

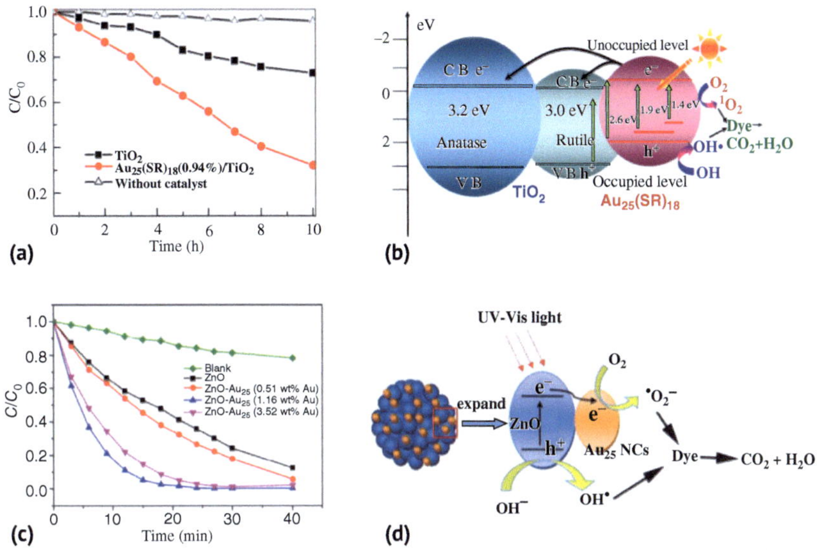


**Figure 8.24** (a) The visible light photocatalytic activity of without the catalyst, pure $TiO_2$, and $Au_{25}(SR)_{18}$ (0.94%)/$TiO_2$. (b) Mechanism of photocatalytic decomposition of organic dyes by $Au_{25}(SR)_{18}/TiO_2$ under visible light irradiation. (c) Photodegradation curves of RhB under UV–vis light. (d) Photocatalytic mechanism of ZnO–$Au_{25}$ nanocomposites. *Source:* Adapted with permission from [89, 110].

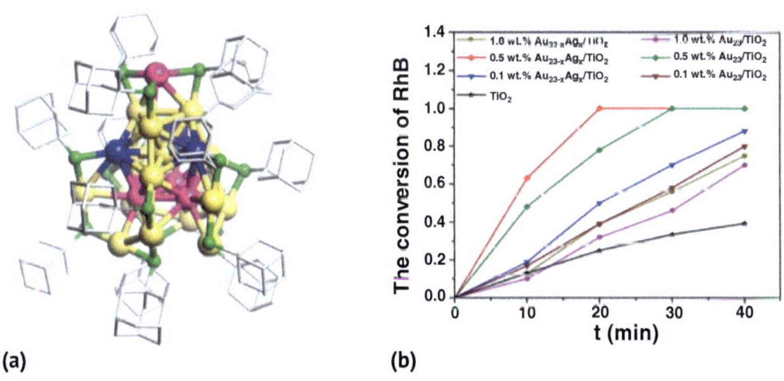


**Figure 8.25** (a) X-ray structure of $[Au_{23-x}Ag_x(S\text{-}Adm)_{15}]$ ($x = 4$–$7$), Au: yellow; Ag with full occupancy: pink; Ag with partial occupancy: blue; S: green; C: gray. (b) The photocatalytic degradation of RhB on $Au_{23-x}Ag_x/TiO_2$, $Au_{23}/TiO_2$, and $TiO_2$ under visible-light irradiation. *Source:* [112]/John Wiley & Sons.

RhB and phenol. Meanwhile, a comparison was made with the $[Au_{23}(S\text{-}Adm)_{16}]$ cluster. Under the same conditions, the catalytic performance of $Au_{23\text{-}x}Ag_x/TiO_2$ under visible-light irradiation was higher than that of $Au_{23\text{-}x}Ag_x/TiO_2$, as shown in Figure 8.25b. The authors attributed the high catalytic activity of $Au_{23\text{-}x}Ag_x$ to the fact that the doping of Ag altered the electronic structure of the nanocluster, making the energy gap of $Au_{23-x}Ag_x(S\text{-}Adm)_{15}$ (1.5 eV) smaller than that of $Au_{23}(S\text{-}Adm)_{16}$ (2.1 eV) [112]. That is, the energy for the electron transition from the occupied orbitals to the unoccupied orbitals in $[Au_{23-x}Ag_x(S\text{-}Adm)_{15}]$ is less than that in $Au_{23}(S\text{-}Adm)_{16}$. Therefore, under visible-light irradiation, $Au_{23-x}Ag_x$ can easily generate more holes and electrons, and more electrons are injected into the conduction band of $TiO_2$, thus enhancing the catalytic performance of the metal nanoclusters.

A copper nanocluster $[Cu_{54}S_{13}O_6(^tBuS)_{20}(^tBuSO_3)_{12}]$ ($Cu_{54}$) (Figure 8.26a) was also synthesized with mixed valence states [13]. The researchers investigated the photocatalytic activity of $Cu_{54}$ nanoclusters on the degradation of phenol in aqueous solution. $Cu_{54}$ has a wide absorption band at 580 nm and an energy gap of 2.05 eV. $Cu_{54}/TiO_2$ catalyst was prepared using $TiO_2$ as the carrier. After loading $Cu_{54}$, the light capture ability of $TiO_2$ nanoparticles was enhanced in the visible light region (Figure 8.26b). In order to evaluate the photocatalytic activity of $Cu_{54}/TiO_2$ composites, a phenol photodegradation reaction driven by visible light ($\lambda > 420$ nm) was carried out. It was found that when the optimal photocatalyst loading was 1.0% (mass fraction) of $Cu_{54}/TiO_2$, the degradation efficiency value of phenol could reach as high as 95% after 1.5 hours. This achieves the same degradation efficiency much faster than other loaded or pure $TiO_2$ samples (Figure 8.26c). The conduction band energy level of $Cu_{54}$ is more negative than that of $TiO_2$. $Cu_{54}$ electrons will be transferred to the conduction band of $TiO_2$ and react with surface $O_2$ to generate $O_2^-$ thereby degrading phenol (Figure 8.26a). Therefore, this catalyst can effectively reduce the electron–hole recombination rate, improve the charge separation performance, and thereby enhance the photocatalytic efficiency. In addition, the holes in the $Cu_{54}$ valence band, due to their oxidizing ability, can also be used to degrade phenol in aqueous solutions.

### 8.4.3 Photocatalytic $CO_2$ Reduction

The excessive emission of $CO_2$ has caused serious environmental problems. The capture of $CO_2$ has been researched for decades and led to some inspiring achievements. However, the effective utilization of $CO_2$ is the ultimate goal of solving $CO_2$ emission. Several works have focused on $CO_2$ hydrogenation at high temperature and pressure without the consideration of hydrogen cost. Electrocatalytic $CO_2$ reduction is a potential alternative strategy, but the required additional electricity is essentially another form of energy. As a green and facile method, photocatalytic $CO_2$ reduction to valuable C-containing products has attracted great interests benefiting from the inexhaustible solar energy [113–116].

The first two steps of photocatalytic $CO_2$ reduction are the same as that of photocatalytic water splitting, except for the final step, due to the complex proton-coupled electron transfer processes and various C-containing products (i.e. $2e^-$ for CO and HCOOH, $6e^-$ for $CH_3OH$, $8e^-$ for $CH_4$ and $CH_3COOH$, $12e^-$ for $C_2H_5OH$ and $C_2H_4$, and $14e^-$ for $C_2H_6$) during $CO_2$ reduction [117, 118]. In the photocatalytic $CO_2$ reduction process of metal nanoclusters, electron transfer is the primary reaction

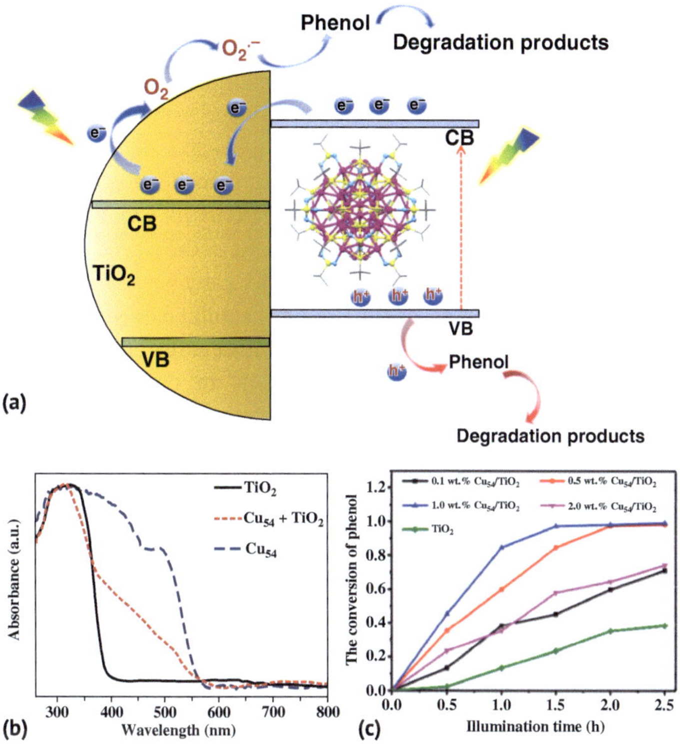


**Figure 8.26** (a) Possible mechanism of photocatalytic phenol degradation and the total crystal structure of $[Cu_{54}S_{13}O_6(^tBuS)_{20}(^tBuSO_3)_{12}]$. Color codes: brown, blue = copper; yellow, purple = sulfur; gray = carbon; red = oxygen. Hydrogen atoms are omitted for clarity. (b) The solid UV–vis absorption spectra of $TiO_2$, $Cu_{54}$ and $Cu_{54}/TiO_2$. (c) The photocatalytic degradation of phenol (5 ppm) on $Cu_{54}/TiO_2$ and $TiO_2$.

mechanism. The electron transfer mechanism not only drives the $CO_2$ reduction reaction but also significantly enhances catalytic performance through synergistic effects and charge separation. This mechanism provides important guidance for designing efficient photocatalysts. In the photocatalytic $CO_2$ reduction process, after the metal nanoclusters absorb photons, electrons transition from the ground state to the excited state, forming electron–hole pairs. These photogenerated electrons are then transferred to the $CO_2$ molecule, driving its reduction reaction. The specific process includes (i) photoelectron transfer: the excited electrons from the conduction band of metal nanoclusters are transferred to $CO_2$ molecules, which reduce them to products such as CO, $CH_4$ and HCOOH. (ii) Charge separation: The high specific surface area and abundant active sites of metal nanoclusters contribute to the effective separation of electron–hole pairs, thus prolonging charge life and improving catalytic efficiency. (iii) Synergistic effect: in some bimetallic nanoclusters, electrons are transferred from one metal to another metal, which further promotes the adsorption and activation of $CO_2$ and significantly improves the reduction efficiency.

In coinage metal nanoclusters, the densely covered ligand layer on the surface leads to a lack of active sites for catalytic reactions, and it is difficult for photogenerated electrons to participate in the reactions. Surface modification strategies can be used to improve the photocatalytic performance of coinage metal nanoclusters [119, 120]. Cui and coworkers [120] used L-cysteine as a bridging ligand to graft metal cations ($Fe^{2+}$, $Co^{2+}$, $Ni^{2+}$, and $Cu^{2+}$) onto inert Au nanoclusters, endowing the gold nanoclusters with visible-light-driven catalytic activity for $CO_2$ reduction. For example, when 4 μmol of $Co^{2+}$ was added to the Au NCs, the conversion rate of $CO_2$ reduction to CO was 3.45 μmol/(h·g), and the selectivity was 65.2%, while the Au nanoclusters without the addition of $Co^{2+}$ did not exhibit catalytic activity. The process of metal cations being bridged to Au nanoclusters for photocatalytic $CO_2$ reduction is shown in Figure 8.27. The metal cations are covalently linked to the S atom on L-cysteine, which promotes the transfer of electrons from the Au nanoclusters to the metal cations, enabling the grafted metal cations to receive photogenerated electrons and serve as catalytic sites for $CO_2$ reduction. DFT calculations indicate that different metal cations have different d-band centers and different binding energies with $CO_2$, thus altering the activity and selectivity of $CO_2$ reduction. This work demonstrates an effective surface modification strategy that can connect the catalytic sites with the light-harvesting centers and provides a model system for the study of the complex effects of catalytic sites in the design of photocatalysts.

Atomically precise coinage metal nanoclusters have received increasing attention in photocatalysis. However, due to their large specific surface area and high surface energy, they are generally thermodynamically unstable and prone to migration and coalescence under light irradiation or high-temperature conditions. This leads to a significant decrease in catalytic activity and selectivity. Therefore, adopting strategies to prevent the aggregation of nanoclusters on the support is crucial for their photocatalytic performance [98, 121–123]. For example, Au nanoclusters were embedded into porous metal–organic framework (MOFs) materials to reduce the aggregation of the nanoclusters. Since the *N*-heterocyclic carbine (NHC) ligand can

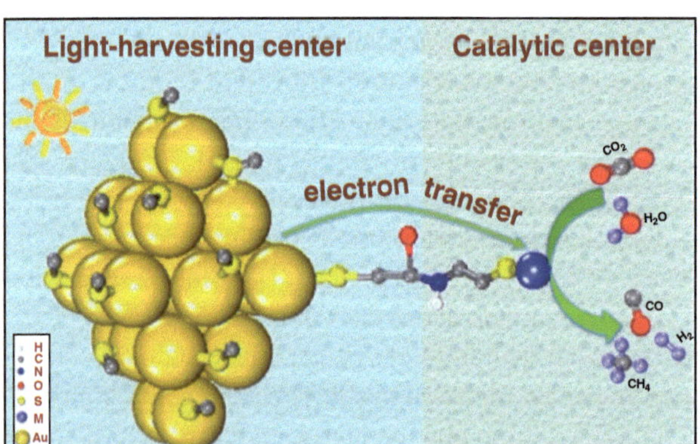


**Figure 8.27** Schematic illustration of the process of modifying Au clusters with L-cys ligands and further grafting them with metal cations for photocatalytic $CO_2$ reduction. *Source:* [120]/with permission of American Chemical Society.

form strong covalent bonds with Au, NHC was used to immobilize Au nanoclusters in MOFs (Au-NC@UiO-68-NHC), as shown in Figure 8.28a, b. It was found that this composite material exhibits a high degree of photostability and chemical stability. Au-NC@UiO-68-NHC has photocatalytic activity for $CO_2$ reduction. As shown in Figure 8.28c, the evolution rate of CO is 57.6 μmol/(g·h), which is more than four times higher than that of the UiO-68-$NH_2$/Au mixture without the stabilization of the NHC ligand, and nearly 20 times higher than that of the pure UiO-68-NHC. Under continuous irradiation by a 300 W xenon lamp for 12 hours, the photocatalytic activity of Au nanocluster@UiO-68-NHC for $CO_2$ reduction still remains above 90%, as shown in Figure 8.28d. In summary, the above results indicate that the porous MOF material provides a versatile platform for studying the photocatalytic activity of Au nanocluster catalysts.

Methane is a widely used fuel and chemical raw material with high value. However, photocatalytic $CO_2$ reduction to $CH_4$ is relatively difficult due to the involvement of the eight-electron transfer process. Atomically precise $Ag_{25}$ nanocluster was utilized to realize nearly 100% selectivity in the photoreduction of

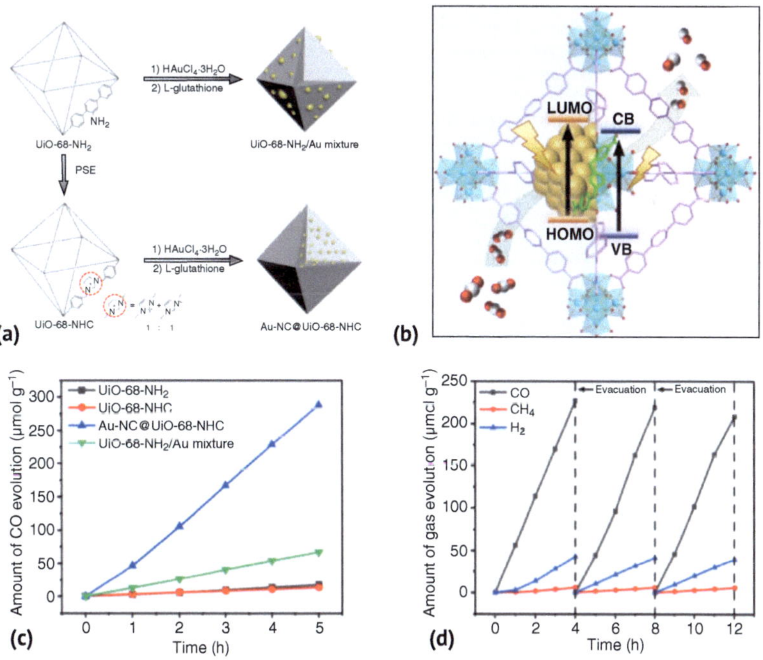


**Figure 8.28** (a) and (b) Synthesis and photocatalytic $CO_2$ reduction mechanism of Au-NC@UiO-68-NHC. (c) Time courses of CO evolution by photocatalytic $CO_2$ reduction using different photocatalysts. (d) Time courses of photocatalytic $CO_2$ reduction on Au-NC@UiO-68-NHC under AM 1.5G irradiation for 12 hours. *Source:* [122]/John Wiley & Sons.

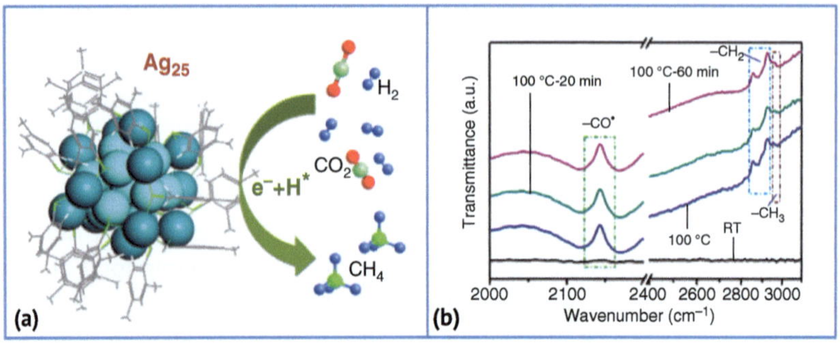


**Figure 8.29** (a) Schematic model of a $Ag_{25}$ cluster for $CO_2$ photoreduction. (b) Time-resolved operando IR spectra of reaction intermediates on $Ag_{25}$ clusters at 100°C for $CO_2$ RR. Reaction conditions: $CO_2$ and $H_2$ (1:3 in volume, 1.5 MPa). *Source:* [124]/with permission of American Chemical Society.

$CO_2$ to $CH_4$ (Figure 8.29a) [124]. Time-resolved operando IR spectroscopy was used to investigate the intermediates generated during $CO_2$ reduction. The bands at 2142, 2854, and 2963 $cm^{-1}$ were assigned to *CO, $-CH_2$, and $-CH_3$, respectively (Figure 8.29b). $CH_4$ was formed through an –H-assisted multielectron reaction pathway. In addition, the adsorption of $CO_2$ on $Ag_{25}$ NCs was energetically favorable, as confirmed by DFT calculation.

The development of stable copper nanoclusters featuring an appropriate band structure and active catalytic sites for the photocatalytic reduction of carbon dioxide remains a formidable challenge. 2-Mercaptopyrimidine (PymSH), existing in thiol and thione tautomeric forms, exhibits a strong affinity toward various metal centers. As such, it is commonly employed as a protective ligand to enhance the stability of nanoclusters. Moreover, the uncoordinated nitrogen atom in PymSH can function as a proton relay, supplying a local proton source that significantly accelerates reaction kinetics. In a recent breakthrough, Dong and coworkers [125] successfully synthesized a novel copper nanocluster, $Cu_6$-NH, which contains locally protonated N–H groups. This newly developed nanocluster not only demonstrates remarkable stability but also possesses semiconducting properties. Under visible light irradiation with a wavelength greater than 400 nm, using water as an electron donor, $Cu_6$-NH achieved an impressive 100% selectivity in the photoreduction of $CO_2$ to CO (Figure 8.30). To gain a comprehensive understanding of the ligand's influence on catalytic activity, the researchers further synthesized isostructural $Cu_6$-N nanoclusters. Through a combination of in situ diffuse reflectance infrared Fourier transform spectroscopy (DRIFTS) measurements and DFT calculations, they determined that the exceptional catalytic performance of $Cu_6$-NH could be attributed to the protonated pyrimidine nitrogen. This unique feature provides a local proton source, effectively lowering the energy barrier for the formation of the crucial *COOH intermediate, a key step in the $CO_2$ reduction process. In other words, the protonated pyrimidine N in $Cu_6$-NH can act as a proton relay station to provide local protons, enabling the formation of the crucial *COOH intermediate during the CO generation process, thus significantly enhancing the photocatalytic performance.

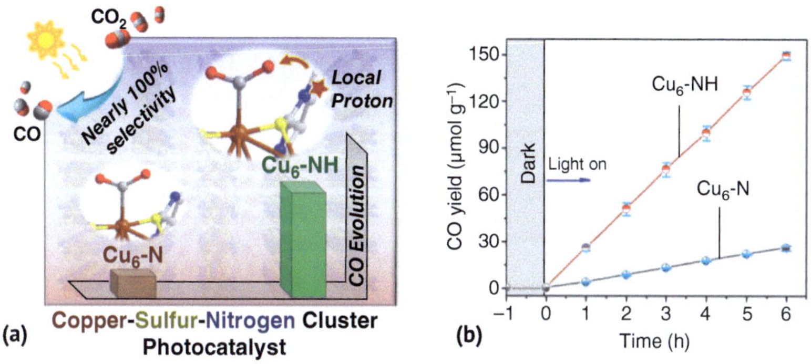


**Figure 8.30** (a) Schematic illustration of the structure-activity relationship in the photocatalytic $CO_2$ reduction process of $Cu_6$ clusters ($Cu_6$-NH and $Cu_6$-N) of the same structure. (b) Time-dependent $CO_2$−CO photoreduction performances of $Cu_6$-NH and $Cu_6$-N. *Source:* [129]/John Wiley & Sons.

### 8.4.4 Photocatalytic Organic Transformation

Imine and its derivatives are important components of synthetic fine chemicals and drugs, and the selective oxidation of amine to the corresponding imine is one of the most important functional group transformations in organic chemistry. Recently, the use of sunlight as a driving force for selective photocatalytic oxidation of organic compounds has attracted more and more attention. It was reported that $Au_{25}$ nanocluster exhibited an excellent activity in the selective oxidation by photocatalysis. Chen and coworkers [126] discovered that atomically precise $Au_{25}(PPh_3)_{10}(SC_3H_6Si(OC_2H_5)_3)_5Cl_2$ ($PPh_3$ = triphenylphosphine, $SC_3H_6Si(OC_2H_5)_3$ – mercaptopropyltriethoxysilane) nanoclusters loaded on $TiO_2$ exhibited excellent selective photocatalytic amine oxidation activity at room temperature. Take benzylamine as an example. The $Au_{25}$ nanoclusters loaded on $TiO_2$ absorb light and can act as narrow bandgap semiconductors, resulting in the separation of electrons and holes. Electrons can be injected into the conduction band of $TiO_2$, and $O_2$ is reduced by electrons to $O_2^-$. Benzylamine can react with photogenerated holes to produce benzylamine cation radicals. The gold atoms exposed on the surface of the nanoclusters seize the α-H of the benzylamine cation radical to generate the carbocation intermediate and the Au-H intermediate. $O_2^-$ reacts with H in the AU-H intermediate and the carbocation intermediate to produce aldehyde amines (Ar–CH=NH) and $H_2O_2$. The unreacted benzylamine undergoes nucleophilic attack on the aldehyde amines, and after deamination, the corresponding imines are produced, as shown in Figure 8.31a. Among them, the turnover frequency (TOF) of 4-methylbenzylamine oxidation is 1522 $h^{-1}$, which is much higher than that of the traditional gold nanoparticle catalytic agent (Figure 8.31b). This work lays the foundation for the visible light photocatalytic selective organic conversion of gold nanoclusters.

Subsequently, Chen and coworkers fabricated $Au_{25}(PPh_3)_{10}(SC_3H_6Si(OC_2H_5)_3)_5Cl_2$ NCs/BiOCl heterostructure via the wet-impregnation method (Figure 8.32) [127]. Because bismuth halide oxide BiOX (X = Cl, Br, I), as a novel layered material with

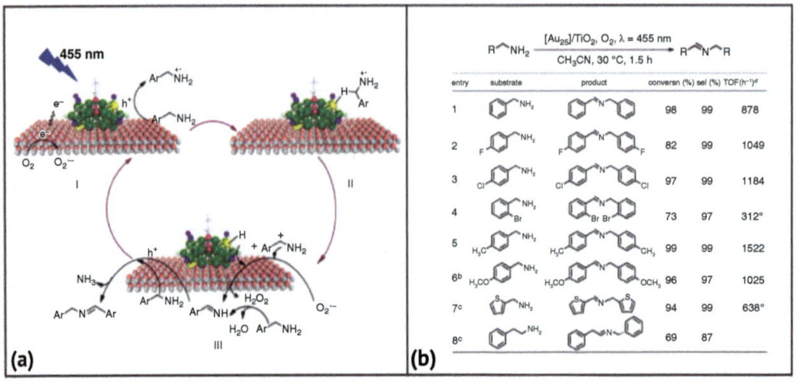


**Figure 8.31** (a) Proposed mechanism for photo-oxidation of benzylamine catalyzed by $TiO_2$-supported $Au_{25}$ clusters. (b) Photocatalytic oxidation of a range of amine substrates over $Au_{25}/TiO_2$ catalyst. *Source:* [126]/with permission of American Chemical Society.

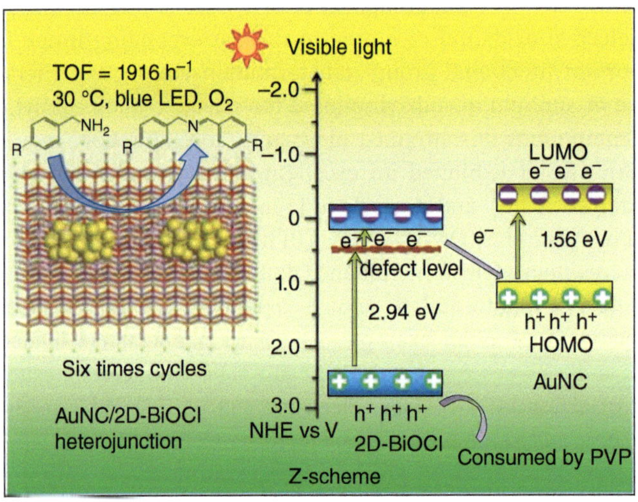


**Figure 8.32** Schematic diagrams of the structural band structure of various amines photocatalytically aerobic oxidized to imines and AuNC and 2D-BiOCl on the AuNC/2D-BiOC structure. *Source:* [127]/with permission of Elsevier.

semiconductor properties, has the following advantages in the field of photocatalysis: (i) its typical BiOCl is composed of an interlaced $[Bi_2O_2]^{2+}$ layer and two layers of Cl ions, featuring a self-built internal electric field that can enhance the separation of carriers. (ii) The layered structure can shorten the transport distance of carriers from the interior to the surface. The 0D/2D heterostructure possessed a large amount of junctions, benefiting the effective charge transfer and separation. Besides this, the heterostructure was conducive to the stabilization of the Au nanoclusters, preventing them from self-oxidation. Based on the analysis of the energy band structure and a series of photoelectrochemical measurements, a Z-scheme charge transfer pathway was confirmed in the $Au_{25}$ nanoclusters/BiOCl heterostructure (Figure 8.32). The $Au_{25}$ nanoclusters/BiOCl realized the photocatalytic oxidative coupling of various amine substrates and showed a high TOF of 1916 $h^{-1}$ for benzylamine (under 100 W LED lamp irradiation in a mixture of ethanol and water system), which were 4.7 and 12,773.3 times higher than that of $Au_{25}$ nanoclusters and 2D-BiOCl, respectively.

Hirai and coworkers utilized the typical icosahedral $M_{13}$ system to investigate the doping effect toward the photocatalytic electron transfer capacity of metal nanoclusters [128]. They synthesized a $Au_{13}(dppe)_5Cl_2$ nanocluster (dppe = 1,2-bis(diphenylphosphino)ethane) and a series of the doping analogues, such as $PdAu_{12}(dppe)_5Cl_2$, $PtAu_{12}(dppe)_5Cl_2$, $RhAu_{12}(dppe)_5Cl_2$ and $IrAu_{12}(dppe)_5Cl_2$. The compositions and structures of these nanoclusters were achieved by ESI-MS, SCXRD, etc. Furthermore, characterizations including UV–vis, XPS, photoluminescence, DPV demonstrated that the heterometal doping significantly affect the redox ability of the excited $MAu_{12}(dppe)_5Cl_2$ nanoclusters (**MAu$_{12}$**) under irradiation (Figure 8.33a). This leads to distinct photocatalytic ability of $Au_{13}(dppe)_5Cl_2$, $PdAu_{12}(dppe)_5Cl_2$, $PtAu_{12}(dppe)_5Cl_2$, $RhAu_{12}(dppe)_5Cl_2$ and $IrAu_{12}(dppe)_5Cl_2$ in a [2+2] bisenone cycloaddition reaction (Figure 8.33c). Notably, the nanocluster catalysts showed dramatically higher selectivity than the conventional catalysts such as $[Ru(bpy)_3]^{2+}$ (bpy = 2,2′-bipyridyl). The thiolate-protected nanocluster $Au_{25}(SR)_{18}$ showed none of catalytic activity under the reaction conditions. A catalytic mechanism was then proposed by Tatsuya Tsukuda et al. representing the redox ability of **MAu$_{12}$** under irradiation (Figure 8.33b). Initially, **MAu$_{12}$** is excited under light irradiation, giving an excited nanocluster ***MAu$_{12}$**. In the meantime, the bisenone substrate reacts with base $LiBF_4$, affording a $[Li\text{-bisenone}]^+$ adduct. This adduct would undergoes an electron transfer process with the excited nanocluster***MAu$_{12}$** to give an oxidized nanocluster, which can be reduced by amine *i*-$Pr_2$NEt to regenerate the original **MAu$_{12}$** catalyst. Notably, the **MAu$_{12}$** nanocluster holds eight free electrons inside the metal kernel. The closed shell structure endows the nanocluster with high stability, which is beneficial for photocatalysis. On the other hand, the eight free electrons are delocalized, thus being readily transformed to the organic substrates under light irradiation. In the whole catalytic cycle, two single electron transfer processes occur. For step i, an electron is transferred from the excited nanocluster ***MAu$_{12}$** to $[Li\text{-bisenone}]^+$ adduct, affording the $[Li\text{-bisenone}]^{\bullet}$ that undergoes a further intramolecular reaction to give the final product. For step ii, an electron is transferred from the reductant amine *i*-$Pr_2$NEt to the oxidized nanocluster **MAu$_{12}$** (7e) to close the catalytic cycle.

Rapid charge recombination and poor stability limit the photocatalytic application of metal nanoclusters. Building a heterostructure with cascade charge transport pathway can effectively solve the dilemma. Two kinds of transition TMC-metal NC-based heterostructure (CdS@PAH@$Ag_9$ NCs and CdSe-$Au_x$-graphene, PAH = poly (allylamine hydrochloride)) were designed via step-by-step

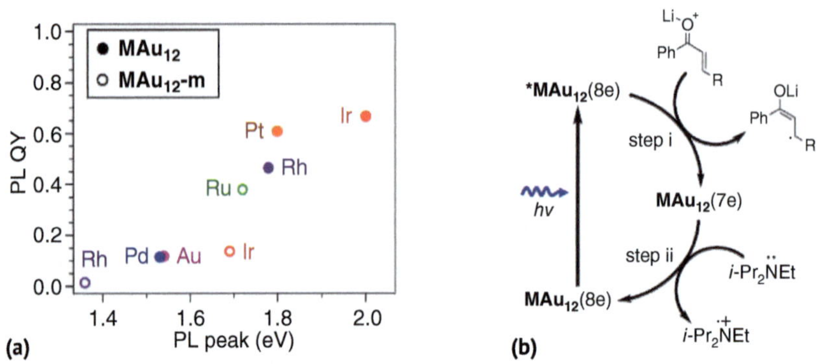


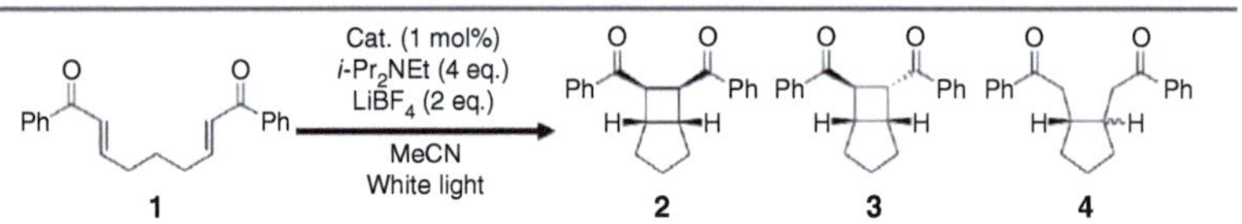


| Catalyst | *t* [min] | Conversion [%] | Selectivity (2:3:4) |
|---|---|---|---|
| **$PtAu_{12}$** | 20 | 52 | 82 : 3 : 15 |
| **$IrAu_{12}$** | 20 | 48 | 88 : 3 : 9 |
| **$RhAu_{12}$** | 20 | 20 | 97 : 0 : 3 |
| **$PdAu_{12}$** | 20 | 8 | 89 : 11 : 0 |
| **$Au_{13}$** | 20 | 6 | 100 : 0 : 0 |
| $[Ru(bpy)_3]^{2+}$ | 1 | 96 | 76 : 2 : 23 |
| **$RuAu_{12}$-m** | 20 | 20 | 99 : 0 : 1 |
| $[Au_{25}(PET)_{18}]^-$ | 20 | trace | |
| Blank | 60 | trace | |

(c)

**Figure 8.33** (a) Optical emission quantum yields of **$MAu_{12}$** and **$MAu_{12}$-m** (dppm-protected $MAu_{12}$ analogues) nanoclusters vs. their energy of their optical emission peaks. (b) Proposed photocatalytic mechanism for **$MAu_{12}$**. (c) Photocatalytic activity of different catalysts in a [2 + 2] bisenone cycloaddition reaction. *Source:* [128]/John Wiley & Sons.

surface modification and electrostatic self-assembly (Figure 8.34a) [129]. In the CdS@PAH@$Ag_9$ nanocluster system, PAH played the dual roles of a surface charge modifier to connect $Ag_9(GSH)_6$ NCs and CdS and an interfacial charge transport mediator to accelerate charge separation. Under light irradiation, the photoinduced electrons generated on the $Ag_9$ nanocluster flowed to CdS with the aid of PAH; thereby the charge lifetime was effectively prolonged, leading to an enhancement in the catalytic activity for the selective anaerobic photocatalytic reduction of nitroaromatics to amino derivatives (Figure 8.34b).

In modern synthesis, bond-forming reactions, especially those under mild conditions, are crucial for new product generation and have broad applications in fine

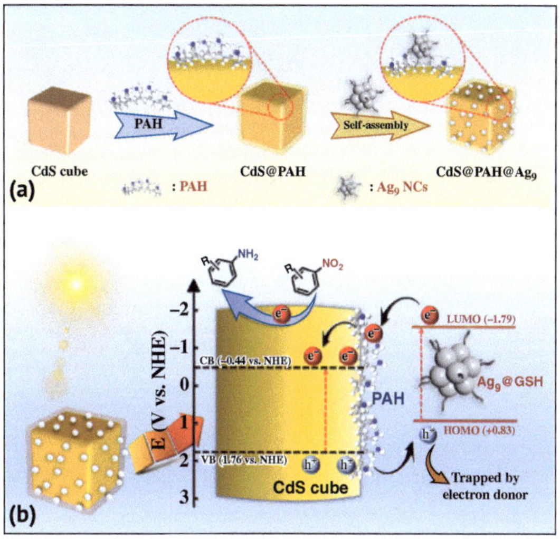


**Figure 8.34** (a) Diagrammatic representation for the electrostatic self-assembly of the CdS@PAH@$Ag_9$ heterostructure. (b) Diagrammatic representation for the photocatalytic mechanism of the CdS@2.0PAH@0.1$Ag_9$ heterostructure. *Source:* [129]/ with permission of American Chemical Society.

chemical synthesis, drug discovery, and sustainable industrial precursor production. Visible light photocatalysis offers advantages over thermal catalysis, but the development of high-efficiency and stable catalysts is essential. Cu-based photocatalysis is a promising, cost-effective alternative to expensive metal-complexes and less stable organic dyes [130]. However, while cross-coupling reactions of aryl bromides and iodides are relatively achievable, those of aryl chlorides still pose a significant challenge. Sagadevan and coworkers and coworkers reported a $[Cu_{61}(S^tBu)_{26}S_6Cl_6H_{14}]$ nanocluster ($Cu_{61}$NC)-based catalyst that can enable C–N bond forming reaction of aryl chlorides under visible-light irradiation at ambient conditions [131]. In the presence of base, at room temperature, the $Cu_{61}$ nanocluster can catalyze the Ullman reaction for het(aryl) very well, and the crystal structure can be found inset (Figure 8.35a). The authors also proposed a plausible mechanism (Figure 8.35b). Specifically, Cu nanocluster reacts with the base forming Cu nanocluster-amine complex A first, then blue LED irradiation leads to the formation of complex B. After a SET process to give an electron to aryl chloride, the oxidized Cu nanocluster-Nu complex, a halide anion, and an arene radical were generated, evidenced by the fluorescence and quenching lifetime tests. After that, the radical attacks the Cu nanocluster-Nu complex, delivering the C–N bond formation and recovering the $Cu_{61}$ nanoclusters. This study introduces atomically precise Cu nanocluster as a new family of photocatalysts for bond-forming reactions under mild conditions. After this work, Nematulloev, Dong, Alamer, Bodiuzzaman, and coworkers developed other useful organic transformations enabled by copper nanocluster photocatalysis [132–135].

Since a series of works of metal nanocluster photocatalytic organic transformations from Osman M. Bakr, Magnus Rueping, and coworkers [131–135], metal nanoclusters-catalyzed organic transformations via photo-induced electron transfer processes have received more and more attentions. The free electrons in their metal kernels and the abundant activation sites on their surface all facilitate the electron transfer process between metal nanoclusters and the organic substrates,

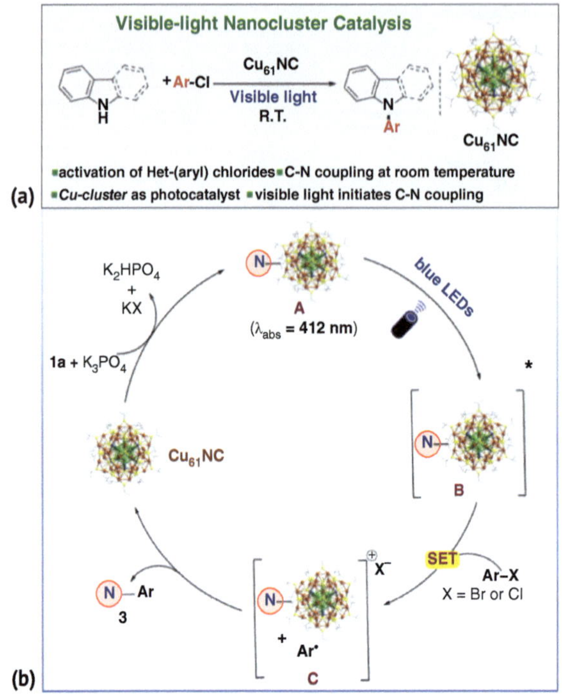


**Figure 8.35** (a) $Cu_{61}$ nanocluster catalyzed Ullmann C–N Coupling. (b) Proposed reaction mechanism. *Source:* [131]/with permission of American Chemical Society.

thus endowing them with high efficiency in photocatalysis. Liu and coworkers designed copper and gold nanocluster photocatalysts and realized nanoclusters-catalyzed multi-component reactions under light irradiation [136, 137]. By using 2,4-dimethylbenzenethiolate (2,4-DMBT) as the protecting ligand, they synthesized a $[Cu_{40}H_{17}(2,4\text{-DMBT})_{24}]^-$ nanocluster, which demonstrates featured emission in the second NIR region (1000–1700 nm) with exited lifetime up to several microsecond. The efficient usage on light irradiation of this nanocluster enables a $Cu_{40}$-catalyzed three-component radical coupling of alkenes, fluoroalkyl iodines, and trimethylsiyl cyanide under blue LED irradiation at room temperature (Figure 8.36a) [136]. This work indicates the viability of using metal nanoclusters for triggering complex organic reactions.

A following work from Liu and coworkers further manifests the promising performance of metal nanoclusters in photocatalysis [137]. They synthesized a series of $[Au_{16}(\text{diphosphine})_6Cl_2]^{2+}$ nanoclusters by employing diphosphines with different substituents and found that one of them, namely $[Au_{16}(C_{30}H_{24}NFP_2)_6Cl_2]^{2+}$ was able to efficiently donating an electron to the *O*-(4-(trifluoromethyl)benzoyl) oxime. The single electron transfer from the $Au_{16}$ nanocluster to *O*-(4-(trifluoromethyl) benzoyl) oxime generates an iminyl radical, which further reacts with 1-isocyano-2-styrylbenzene and aniline to give the final Fukuyama indole reaction products

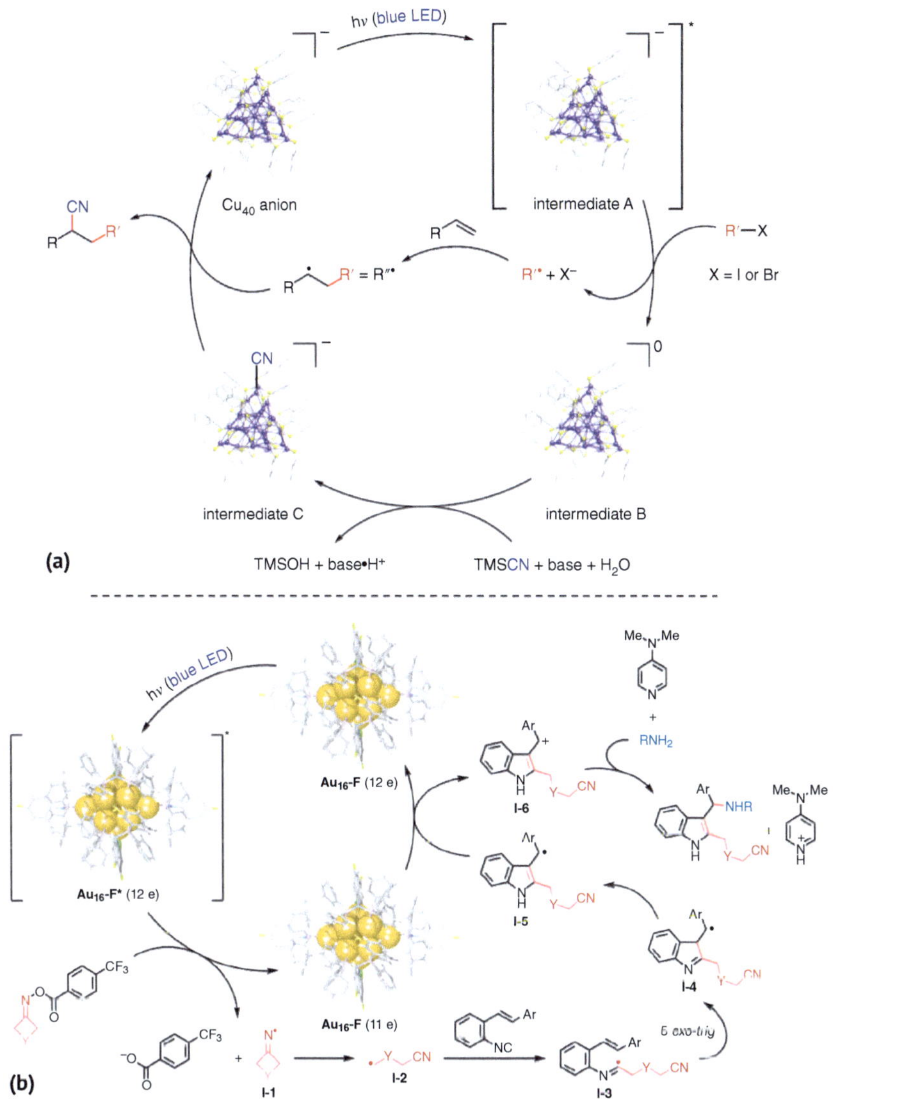


**Figure 8.36** (a) $Cu_{40}$ nanocluster catalyzed three-component: coupling of alkene, alkyl iodine, and TMSCN. (b) $Au_{16}$ nanocluster catalyzed three-component: coupling of 1-isocyano-2-styrylbenzene, *O*-(4-(trifluoromethyl)benzoyl) oxime, and aniline. *Source:* Reproduced with permission from Refs. [136, 137].

(Figure 8.36b). This nanocluster can be readily recovered and recycled, representing a promising avenue for advancing organic synthesis in a sustainable and practical manner. Zhao and coworkers also applied metal nanoclusters to trigger an efficient three-component reaction under light irradiation [138]. They synthesized a $Au_8Cu_6$ nanocluster by their developed photosynthetic method. This synthetic protocol with blue LED enables the generation of the $Au_8Cu_6$ nanocluster with high stability under light irradiation conditions, thus being beneficial for the photocatalysis of this Au-Cu alloyed nanocluster. It was utilized to catalyze a three-component reaction of alkynes, TEMPO and H-phosphonate, furnishing the alkenyl phosphonate with high efficiency.

Zhang and coworkers applied POM clusters for catalyzing organic transformations with high efficiency [139]. For example, they synthesized giant hollow polymolybdate cage $Mo_{240}$, and realized its internal growth and further external growth to give polymolybdate dodecahedral cages, that is, $Mo_{250}$ and $Mo_{260}$ [140]. The regular variations in POM structures and nuclearity endow these clusters with different optical properties and reduction-oxidation capacity. Among them, $Mo_{240}$ and one of $Mo_{260}$ were found to demonstrate exceptional photocatalytic performance in oxidative cross-coupling reaction of unactivated alkanes (e.g. cyclohexane) and N-heteroarenes (e.g. methylisoquinoline-3-carboxylate). The desired oxidative coupling product was obtained in up to 92% yield within one hour (Figure 8.37).

Very recently, Zhang and coworkers designed and synthesized a carbonylmetallated palladium nanocluster $\{Pd_{14}(F\text{-}Ph_3P)_6(CO)_{12}[Co(CO)_4]_2\}$ and realized a visible light-driven carbonylation reaction by using this nanocluster for photocatalysis [141]. There are two features in this work. The first one is the controlled synthesis of the atomically precise palladium nanoclusters, which have been widely considered as a challenging task as a result of the kinetically favored formation of palladium nanoparticles. In this work, $[Co(CO_4)]^-$ was utilized as a protecting ligand to regulate the reduction reaction kinetics. This kind of carbonylmetallate ligand successfully controlled the synthetic process, affording three $Pd_{14}$ nanoclusters with atomically precise structures. The stability of these palladium nanoclusters should be attributed to the electron transfer from $[Co(CO_4)]^-$ to the $Pd_{14}$ kernel,

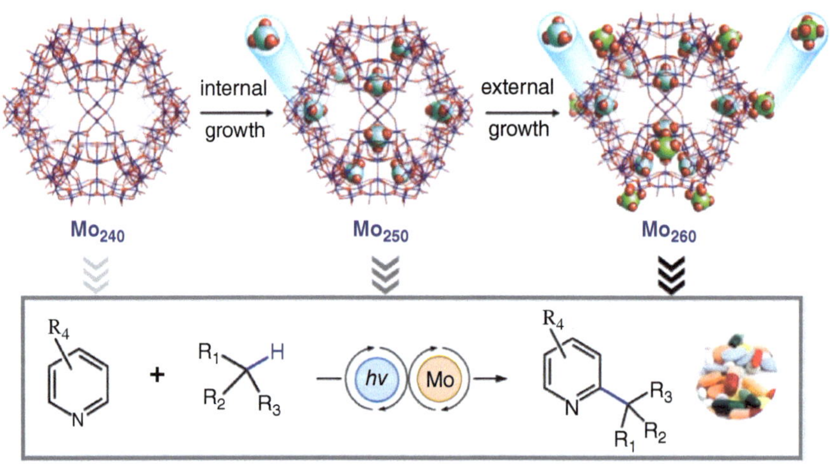


**Figure 8.37** POM-catalyzed oxidative cross-coupling reactions between unactivated alkanes and N-heteroarenes. *Source:* [140]/with permission of American Chemical Society.

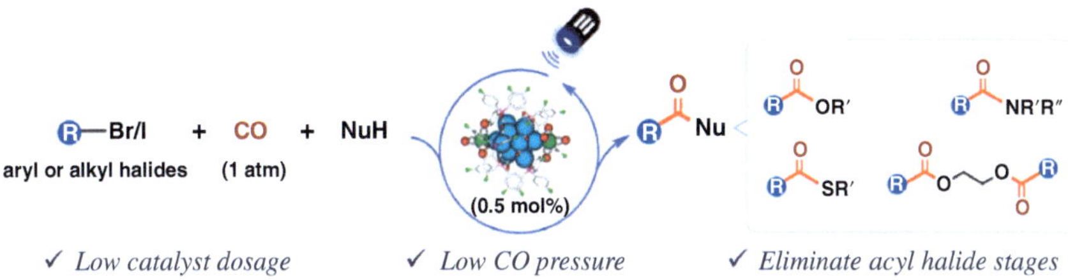


**Figure 8.38** Palladium nanocluster-catalyzed carbonylation reaction of aryl/alkyl halides, CO, and nucleophiles under light irradiation. *Source:* [141]/with permission of American Chemical Society.

which significantly stabilizes the whole nanoclusters. The second feature is the promising performance of a $Pd_{14}Co_2$ nanocluster in photocatalysis involving a CO insertion process (Figure 8.38). Specifically, the $Pd_{14}$ kernel facilitates the oxidative addition of aryl/alkyl halides and then undergoes electron transfer to give aryl/alkyl radicals. The subsequential CO insertion triggers the formation of a metal-acyl bond, which is captured by nucleophiles to afford the final products. Notably, the $[Co(CO_4)]^-$ moieties synergistically activate the nucleophiles, such as amines, alcohols and thiols, playing important roles in the metal nanocluster photocatalysis under mild conditions (1 atm CO, 30 W of 450 nm LED) and promoting the formation of C—N, C—O, and C—S bonds.

### 8.4.5 Photocatalytic Reduction of $N_2$ to $NH_3$

Ammonia ($NH_3$) serves as a pivotal chemical and energy carrier, playing a crucial role in various industrial and energy-related applications. Currently, the traditional Haber–Bosch process, the cornerstone of industrial ammonia synthesis, is confronted with increasingly arduous challenges. These include exorbitant energy consumption, high production costs, and the emission of harmful gases, which not only strain resources but also pose significant environmental threats. In response to these issues, the photocatalytic reduction of nitrogen ($N_2$) to ammonia has emerged as a highly promising alternative, attracting substantial attention and research efforts from the scientific community. This approach shares several remarkable similarities with biological nitrogen fixation. It features a straightforward reaction system, operates under mild reaction conditions, is environmentally friendly, cost-effective, safe, and allows for easy regulation of performance. As a result, it has demonstrated great potential in the field of ammonia synthesis [142]. Distinct from the conventional Haber-Bosch method, photocatalytic nitrogen reduction for ammonia synthesis does not rely on hydrogen ($H_2$) as a reactant. Instead, water ($H_2O$) serves as the proton source. Under ambient conditions, with sunlight as the energy source, ammonia can be directly synthesized from water and nitrogen, as described by the chemical equation: $N_2 + 3H_2O = 2NH_3 + 1.5O_2$ [143]. This innovative process represents a significant step towards more sustainable and efficient ammonia production methods.

Metal nanoclusters with excitonic behavior are capable of generating electrons induced by solar energy and donating electrons to $N_2$ for its reduction. Sun and coworkers utilized $TiO_2$ with oxygen vacancy ($TiO_2$-Ov) to support $Au_4Ru_2(PPh_3)_2(SC_2H_4Ph)_8$ nanoclusters (Figure 8.39) [144]. Under light irradiation, the excited $Au_4Ru_2$ nanoclusters generated photoinduced electron–hole pairs. The photogenerated carriers effectively separated between $Au_4Ru_2$ nanoclusters and $TiO_2$-Ov, in which the electrons transferred to the Ru atoms and then injected to $N_2$ for its activation. $TiO_2$-Ov was responsible for the water splitting to provide the proton for $N_2$ hydrogenation. In comparison with a series of homometal $Au_n$ nanoclusters ($n = 8, 9, 11, 18, 23, 24, 25, 28, 36, 44$), the alloyed $Au_4Ru_2$ nanoclusters can effectively activate $N_2$ (Figure 8.39a). The distance between $N_2$ and nanoclusters in the Au nanoclusters was larger than 3.2 Å, indicating the physical adsorption of $N_2$ on these $Au_n$ nanoclusters. On the contrary, $N_2$ was closer to $Au_4Ru_2$ nanoclusters with a distance of 2.14 Å and 1.84 Å for the side-on and end-on configuration, respectively, suggesting the chemical adsorption of $N_2$ on $Au_4Ru_2$ nanoclusters. $Au_4Ru_2$ nanoclusters/$TiO_2$-Ov showed a $NH_3$ production rate of 44.5 μmol $g^{-1}h^{-1}$ (Figure 8.39b), three times higher than that of Au nanoclusters/$TiO_2$-Ov. The experimental studies combined with theoretical calculations demonstrate that

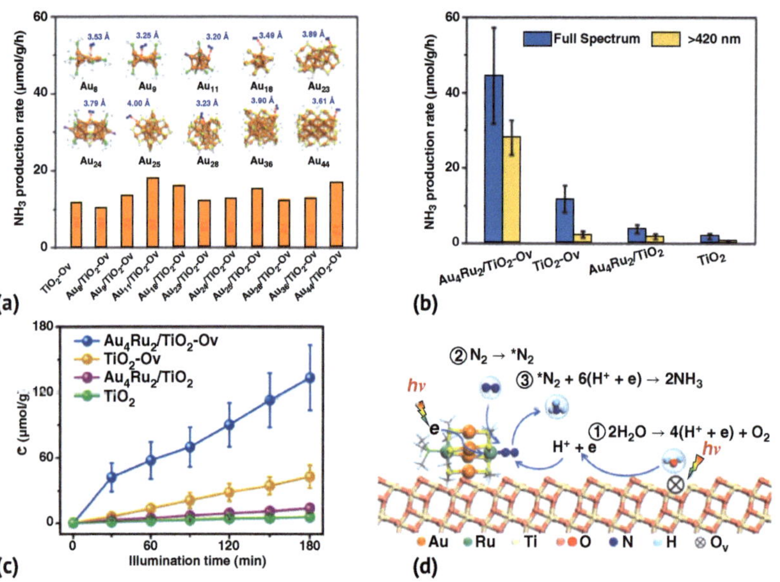


**Figure 8.39** (a) The photocatalytic $N_2$ reduction activity of homometal $Au_n$ NCs ($n$ = 8, 9, 11, 18, 23, 24, 25, 28, 36, and 44) on $TiO_2$-Ov. Inset: the structures of the $Au_n$ NCs, and the distance between $N_2$ and these $Au_n$ NCs. The C, N, S, Au, P, and Cl are shown in gray, blue, yellow, orange, green, and violet colors, respectively. H atoms are omitted for clarity. (b) Photocatalytic $N_2$ reduction mechanism on $Au_4Ru_2$/$TiO_2$-Ov. (c) Time dependent photocatalytic ammonia production over the $Au_4Ru_2$/$TiO_2$-Ov, $Au_4Ru_2$/$TiO_2$, $TiO_2$-Ov, and $TiO_2$ catalysts. (d) Proposed mechanism for the photocatalytic $N_2$ reduction on the $Au_4Ru_2$/$TiO_2$-Ov catalyst. *Source:* [144]/Royal Society of Chemistry.

the cooperative effect between $Au_4Ru_2$ nanocluster and $TiO_2$ substrate with oxygen vacancies leads to an extraordinary activity for light-driven $N_2$ reduction. The electron–hole pairs can be generated from the excited $Au_4Ru_2$ nanocluster. The heterojunction between $Au_4Ru_2$ nanocluster and the $TiO_2$-Ov substrate also facilitates photocarriers separation. Meanwhile, $TiO_2$-Ov induces water splitting to produce hydrogen protons for $N_2$ fixation and conversion on the Ru atoms (Figure 8.39d). Certainly, this work provides deep insights into nonplasmon-induced charge transfer from atomically precise metal nanoclusters and develops a feasible strategy.

## References


1 H. Qian, M. Zhu, Z. Wu, R. Jin, Quantum Sized Gold Nanoclusters with Atomic Precision. *Acc. Chem. Res.* **2012**, *45*, 1470–1479.

2 P. Maity, S. Xie, M. Yamauchi, T. Tsukuda, Stabilized Gold Clusters: From Isolation Toward Controlled Synthesis. *Nanoscale* **2012**, *4*, 4027–4037.

3 D.-E. Jiang, The Expanding Universe of Thiolated Gold Nanoclusters and Beyond. *Nanoscale* **2013**, *5*, 7149–7160.

4 G. Li, R. Jin, Atomically Precise Gold Nanoclusters as New Model Catalysts. *Acc. Chem. Res.* **2013**, *46*, 1749–1758.

5 R. Jin, Atomically Precise Metal Nanoclusters: Stable Sizes and Optical Properties. *Nanoscale* **2015**, *7*, 1549–1565.

6 R. Jin, C. Zeng, M. Zhou, Y. Chen, Atomically Precise Colloidal Metal Nanoclusters and Nanoparticles: Fundamentals and Opportunities. *Chem. Rev.* **2016**, *116*, 10346–10413.

7 M. Zhu, C. M. Aikens, F. J. Hollander, G. C. Schatz, R. Jin, Correlating the Crystal Structure of a Thiol-Protected $Au_{25}$ Cluster and Optical Properties. *J. Am. Chem. Soc.* **2008**, *130*, 5883–5885.

8 S. Chen, S. Wang, J. Zhong, Y. Song, J. Zhang, H. Sheng, Y. Pei, M. Zhu, The Structure and Optical Properties of the $[Au_{18}(SR)_{14}]$ Nanocluster. *Angew. Chem. Int. Ed.* **2015**, *54*, 3145–3149; *Angew. Chem.* **2015**, *127*, 3188–3192.

9 A. Das, C. Liu, H. Byun, K. Nobusada, S. Zhao, N. Rosi, R. Jin, Structure Determination of $[Au_{18}(SR)_{14}]$, *Angew. Chem. Int. Ed.* **2015**, *54*, 3140–3144; *Angew. Chem.* **2015**, *127*, 3183–3187.

10 S. Chen, L. Xiong, S. Wang, Z. Ma, S. Jin, H. Sheng, Y. Pei, M. Zhu, Total Structure Determination of $Au_{21}(S\text{-}Adm)_{15}$ and Geometrical/Electronic Structure Evolution of Thiolated Gold Nanoclusters. *J. Am. Chem. Soc.* **2016**, *138*, 10754–10757.

11 S. Yang, S. Chen, L. Xiong, C. Liu, H. Yu, S. Wang, N. L. Rosi, Y. Pei, M. Zhu, Total Structure Determination of $Au_{16}(S\text{-}Adm)_{12}$ and $Cd_1Au_{14}(S^tBu)_{12}$ and Implications for the Structure of $Au_{15}(SR)_{13}$. *J. Am. Chem. Soc.* **2018**, *140*, 10988–10994.

12 Y. Li, M. J. Cowan, M. Zhou, T.-Y. Luo, Y. Song, H. Wang, N. L. Rosi, G. Mpourmpakis, R. Jin, Atom-by-Atom Evolution of the Same Ligand-Protected $Au_{21}$, $Au_{22}$, $Au_{22}Cd_1$, and $Au_{24}$ Nanocluster Series. *J. Am. Chem. Soc.* **2020**, *142*, 20426–20433.

**13** A. Das, T. Li, K. Nobusada, C. Zeng, N. L. Rosi, R. Jin, Nonsuperatomic $[Au_{23}(SC_6H_{11})_{16}]^-$ Nanocluster Featuring Bipyramidal $Au_{15}$ Kernel and Trimeric $Au_3(SR)_4$ Motif. *J. Am. Chem. Soc.* **2013**, *135*, 18264–18267.

**14** J. Yan, H. Su, H. Yang, S. Malola, S. Lin, H. Hakkinen, N. Zheng, Total Structure and Electronic Structure Analysis of Doped Thiolated Silver $[MAg_{24}(SR)_{18}]^{2-}$ (M = Pd, Pt) Clusters. *J. Am. Chem. Soc.* **2015**, *137*, 11880–11883.

**15** A. Das, T. Li, K. Nobusada, Q. Zeng, N. L. Rosi, R. Jin. Total Structure and Optical Properties of a Phosphine/Thiolate Protected $Au_{24}$ Nanocluster. *J. Am. Chem. Soc.* **2012**, *134*, 20286–20289.

**16** S. Wang, H. Abroshan, C. Liu, T.-Y. Luo, M. Zhu, H. J. Kim, N. L. Rosi, R. Jin, Shuttling Single Metal Atom into and out of a Metal Nanoparticle. *Nat. Commun.* **2017**, *8*, 848.

**17** X. Kang, M. Zhu, Tailoring the Photoluminescence of Atomically Precise Nanoclusters. *Chem. Soc. Rev.* **2019**, *48*, 2422–2457.

**18** Y. Tao, M. Li, J. Ren, X. Qu, Metal Nanoclusters: Novel Probes for Diagnostic and Therapeutic Applications. *Chem. Soc. Rev.* **2015**, *44*, 8636–8663.

**19** H. Yu, B. Rao, W. Jiang, S. Yang, M. Zhu, The Photoluminescent Metal Nanoclusters with Atomic Precision. *Coord. Chem. Rev.* **2019**, *378*, 595–617.

**20** X. Hu, Y. Zheng, J. Zhou, D. Fang, H. Jiang, X. Wang, Silver-Assisted Thiolate Ligand Exchange Induced Photoluminescent Boost of Gold Nanoclusters for Selective Imaging of Intracellular Glutathione. *Chem. Mater.* **2018**, *30*, 1947–1955.

**21** X.-R. Song, N. Goswami, H.-H. Yang, J. Xie, Functionalization of Metal Nanoclusters for Biomedical Applications. *Analyst* **2016**, *141*, 3126–3140.

**22** N. Goswami, K. Zheng, J. Xie, Bio-NCs-the Marriage of Ultrasmall Metal Nanoclusters with Biomolecules. *Nanoscale* **2014**, *6*, 13328–13347.

**23** K. Konishi, M. Iwasaki, M. Sugiuchi, Y. Shichibu, Ligand-Based Toolboxes for Tuning of the Optical Properties of Subnanometer Gold Clusters. *J. Phys. Chem. Lett.* **2016**, *7*, 4267–4274.

**24** Y. Zheng, L. Lai, W. Liu, H. Jiang, X. Wang, Recent Advances in Biomedical Applications of Fluorescent Gold Nanoclusters. *Adv. Colloid Interface Sci.* **2017**, *242*, 1–16.

**25** C. Zeng, H. Qian, T. Li, G. Li, N. L. Rosi, B. Yoon, R. N. Barnett, R. L. Whetten, U. Landman, R. Jin, Total Structure and Electronic Properties of the Gold Nanocrystal $Au_{36}(SR)_{24}$. *Angew. Chem. Int. Ed.* **2012**, *51*, 13114–13118; *Angew. Chem.* **2012**, *124*, 13291–13295.

**26** A. Kim, C. Zeng, M. Zhou, R. Jin, Surface Engineering of $Au_{36}(SR)_{24}$ Nanoclusters for Photoluminescence Enhancement. *Part. Part. Syst. Charact.* **2017**, *34*, 1600388.

**27** X. Kang, S. Wang, Y. Song, S. Jin, G. Sun, H. Yu, M. Zhu, Bimetallic $Au_2Cu_6$ Nanoclusters: Strong Luminescence Induced by the Aggregation of Copper(I) Complexes with Gold(0) Species. *Angew. Chem. Int. Ed.* **2016**, *55*, 3611–3614; *Angew. Chem.* **2016**, *128*, 3675–3678.

**28** X. Kang, X. Li, H. Yu, Y. Lv, G. Sun, Y. Li, S. Wang, M. Zhu, Modulating Photo-Luminescence of $Au_2Cu_6$ Nanoclusters via Ligand-Engineering. *RSC Adv.* **2017**, *7*, 28606–28609.

29 X. Kang, S. Jin, L. Xiong, X. Wei, M. Zhou, C. Qin, Y. Pei, S. Wang, M. Zhu, Nanocluster Growth via "Graft-onto": Effects on Geometric Structures and Optical Properties. *Chem. Sci.* **2020**, *11*, 1691–1697.

30 Q. Li, T.-Y. Luo, M. G. Taylor, S. Wang, X. Zhu, Y. Song, G. Mpourmpakis, N. L. Rosi, R. Jin, Molecular "Surgery" on a 23-Gold-Atom Nanoparticle. *Sci. Adv.* **2017**, *3*, e1603193.

31 M. Zhou, H. Qian, M. Y. Sfeir, K. Nobusada, R. Jin, Effects of Single Atom Doping on the Ultrafast Electron Dynamics of $M_1Au_{24}(SR)_{18}$ (M = Pd, Pt) Nanoclusters. *Nanoscale* **2016**, *8*, 7163–7171.

32 E. B. Guidez, V. Mäkinen, H. Häkkinen, C. M. Aikens, Effects of Silver Doping on the Geometric and Electronic Structure and Optical Absorption Spectra of the $Au_{25\text{-}n}Ag_n(SH)_{18}$- (n = 1, 2, 4, 6, 8, 10, 12) Bimetallic Nanoclusters. *J. Phys. Chem. C.* **2012**, *116*, 20617–20624.

33 K. R. Krishnadas, A. Baksi, A. Ghosh, G. Natarajan, A. Som, T. Pradeep, Interparticle Reactions: An Emerging Direction in Nanomaterials Chemistry. *Acc. Chem. Res.* **2017**, *50*, 1988–1996.

34 Y. Niihori, S. Hossain, S. Sharma, B. Kumar, W. Kurashige, Y. Negishi, Understanding and Practical Use of Ligand and Metal Exchange Reactions in Thiolate-Protected Metal Clusters to Synthesize Controlled Metal Clusters. *Chem. Rec.* **2017**, *17*, 473–484.

35 X. Yuan, X. Dou, K. Zheng, J. Xie, Recent Advances in the Synthesis and Applications of Ultrasmall Bimetallic Nanoclusters. *Part. Part. Syst. Charact.* **2015**, *32*, 613–629.

36 S. Tian, L. Liao, J. Yuan, C. Yao, J. Chen, J. Yang, Z. Wu, Structures and Magnetism of Mono-Palladium and Mono-Platinum Doped $Au_{25}(PET)_{18}$ Nanoclusters. *Chem. Commun.* **2016**, *52*, 9873–9876.

37 B. Zhang, T. Bürgi, Doping Silver Increases the $Au_{38}(SR)_{24}$ Cluster Surface Flexibility. *J. Phys. Chem. C.* **2016**, *120*, 4660–4666.

38 N. Barrabés, B. Zhang, T. Bürgi, Racemization of Chiral $Pd_2Au_{36}(SC_2H_4Ph)_{24}$: Doping Increases the Flexibility of the Cluster Surface. *J. Am. Chem. Soc.* **2014**, *136*, 14361–14364.

39 S. Wang, Y. Song, S. Jin, X. Liu, J. Zhang, Y. Pei, X. Meng, M. Chen, P. Li, M. Zhu, Metal Exchange Method Using $Au_{25}$ Nanoclusters as Templates for Alloy Nanoclusters with Atomic Precision. *J. Am. Chem. Soc.* **2015**, *137*, 4018–4021.

40 S. Wang, Q. Li, X. Kang, M. Zhu, Customizing the Structure, Composition, and Properties of Alloy Nanoclusters by Metal Exchange. *Acc. Chem. Res.* **2018**, *51*, 2784–2792.

41 H. Yang, Y. Wang, H. Huang, L. Gell, L. Lehtovaara, S. Malola, H. Häkkinen, N. Zheng, All-Thiol-Stabilized $Ag_{44}$ and $Au_{12}Ag_{32}$ Nanoparticles with Single-Crystal Structures. *Nat. Commun.* **2013**, *4*, 2422.

42 C. Yao, J. Chen, M.-B. Li, L. Liu, J. Yang, Z. Wu, Adding Two Active Silver Atoms on $Au_{25}$ Nanoparticle. *Nano Lett.* **2015**, *15*, 1281–1287.

43 X. Wang, B. Yin, L. Jiang, C. Yang, Y. Liu, G. Zou, S. Chen, M. Zhu, Ligand-Protected Metal Nanoclusters as Low-Loss, Highly Polarized Emitters For Optical Waveguides. *Science* **2023**, *381*, 784–790.

**44** Z. Wu, R. Jin, On the Ligand's Role in the Fluorescence of Gold Nanoclusters. *Nano Lett.* **2010**, *10*, 2568–2573.

**45** J. Luo, Z. Xie, J. W. Y. Lam, L. Cheng, H. Chen, C. Qiu, H. S. Kwok, X. Zhan, Y. Liu, D. Zhu, B. Z. Tang, Aggregation-Induced Emission of 1-Methyl-1,2,3,4,5-Pentaphenylsilole. *Chem. Commun.* **2001**, *18*, 1740–1741.

**46** J. Mei, Y. Hong, J. W. Y. Lam, A. Qin, Y. Tang, B. Z. Tang, Aggregation-Induced Emission: The Whole is More Brilliant than the Parts. *Adv. Mater.* **2014**, *26*, 5429–5479.

**47** J. Mei, N. L. C. Leung, R. T. K. Kwok, J. W. Y. Lam, B. Z. Tang, Aggregation-Induced Emission: Together We Shine, United We Soar. *Chem. Rev.* **2015**, *115*, 11718–11940.

**48** Z. Luo, X. Yuan, Y. Yu, Q. Zhang, D. T. Leong, J. Y. Lee, J. Xie, From Aggregation-Induced Emission of Au(I)-Thiolate Complexes to Ultrabright Au(0)@Au(I)-Thiolate Core-Shell Nanoclusters. *J. Am. Chem. Soc.* **2012**, *134*, 16662–16670.

**49** X. Dou, X. Yuan, Y. Yu, Z. Luo, Q. Yao, D. T. Leong, J. Xie, Lighting up Thiolated Au@Ag Nanoclusters via Aggregation-Induced Emission. *Nanoscale* **2014**, *6*, 157–161.

**50** M. Sugiuchi, J. Maeba, N. Okubo, M. Iwamura, K. Nozaki, K. Konishi, Aggregation-Induced Fluorescence-to-Phosphorescence Switching of Molecular Gold Clusters. *J. Am. Chem. Soc.* **2017**, *139*, 17731–17734.

**51** T. Chen, S. Yang, J. Chai, Y. Song, J. Fan, B. Rao, H. Sheng, H. Yu, M. Zhu, Crystallization-Induced Emission Enhancement: A Novel Fluorescent Au-Ag Bimetallic Nanocluster with Precise Atomic Structure. *Sci. Adv.* **2017**, *3*, e1700956.

**52** Z. Wu, H. Liu, T. Li, J. Liu, J. Yin, O. F. Mohammed, O. M. Bakr, Y. Liu, B. Yang, H. Zhang, Contribution of Metal Defects in the Assembly Induced Emission of Cu Nanoclusters. *J. Am. Chem. Soc.* **2017**, *139*, 4318–4321.

**53** A. Das, T. Li, G. Li, K. Nobusada, C. Zeng, N. L. Rosib, R. Jin, Crystal Structure and Electronic Properties of a Thiolate-Protected $Au_{24}$ Nanocluster. *Nanoscale* **2014**, *6*, 6458–6462.

**54** C. Yao, S. Tian, L. Liao, X. Liu, N. Xia, N. Yan, Z. Gan, Z. Wu, Synthesis of Fluorescent Phenylethanethiolated Gold Nanoclusters via Pseudo-AGR Method. *Nanoscale* **2015**, *7*, 16200–16203.

**55** Z. Gan, Y. Lin, L. Luo, G. Han, W. Liu, Z. Liu, Chuanhao Yao, L. Weng, L. Liao, J. Chen, X. Liu, Y. Luo, C. Wang, S. Wei, Z. Wu, Fluorescent Gold Nanoclusters with Interlocked Staples and a Fully Thiolate-Bound Kernel. *Angew. Chem. Int. Ed.* **2016**, *55*, 11567–11571; *Angew. Chem.* **2016**, *128*, 11739–11743.

**56** G. Li, Z. Lei, Q.-M. Wang, Luminescent Molecular Ag-S Nanocluster $[Ag_{62}S_{13}(S^{t}Bu)_{32}](BF_4)_4$. *J. Am. Chem. Soc.* **2010**, *132*, 17678–17679.

**57** S. Jin, S. Wang, Y. Song, M. Zhou, J. Zhong, J. Zhang, A. Xia, Y. Pei, M. Chen, P. Li, M. Zhu, Crystal Structure and Optical Properties of the $[Ag_{62}S_{12}(S^{t}Bu)_{32}]_2^{+}$ Nanocluster with a Complete Face-Centered Cubic Kernel. *J. Am. Chem. Soc.* **2014**, *136*, 15559–15565.

**58** Y. Huang, L. Fuksman, J. Zheng, Luminescence Mechanisms of Ultrasmall Gold Nanoparticles. *Dalton Trans.* **2018**, *47*, 6267–6273.

**59** W. Du, S. Jin, L. Xiong, M. Chen, J. Zhang, X. Zou, Y. Pei, S. Wang, M. Zhu, $Ag_{50}(Dppm)_6(SR)_{30}$ and Its Homologue $Au_xAg_{50-x}(Dppm)_6(SR)_{30}$ Alloy Nanocluster: Seeded Growth, Structure Determination, and Differences in Properties. *J. Am. Chem. Soc.* **2017**, *139*, 1618–1624.

**60** X. Kang, L. Xiong, S. Wang, H. Yu, S. Jin, Y. Song, T. Chen, L. Zheng, C. Pan, Y. Pei, M. Zhu, Shape-Controlled Synthesis of Trimetallic Nanoclusters: Structure Elucidation and Properties Investigation. *Chem. Eur. J.* **2016**, *22*, 17145–17150.

**61** G. Soldan, M. A. Aljuhani, M. S. Bootharaju, L. G. AbdulHalim, M. R. Parida, A.-H. Emwas, O. F. Mohammed, O. M. Bakr, Gold Doping of Silver Nanoclusters: A 26-Fold Enhancement in the Luminescence Quantum Yield. *Angew. Chem. Int. Ed.* **2016**, *55*, 5749–5753; *Angew. Chem.* **2016**, *128*, 5843–5847.

**62** S. Wang, X. Meng, A. Das, T. Li, Y. Song, T. Cao, X. Zhu, M. Zhu, R. Jin, A 200-fold Quantum Yield Boost in the Photoluminescence of Silver Doped $Ag_xAu_{25-x}$ Nanoclusters: The 13th Silver Atom Matters. *Angew. Chem., Int. Ed.* **2014**, *53*, 2376–2380; *Angew. Chem.* **2014**, *126*, 2408–2412.

**63** M. Zhou, J. Zhong, S. Wang, Q. Guo, M. Zhu, Y. Pei, A. Xia, Ultrafast Relaxation Dynamics of Luminescent Rod-Shaped, Silver Doped $Ag_xAu_{25-x}$ Clusters. *J. Phys. Chem. C.* **2015**, *119*, 18790–18797.

**64** M. S. Bootharaju, C. P. Joshi, M. R. Parida, O. F. Mohammed, O. M. Bakr, Templated Atom-Precise Galvanic Synthesis and Structure Elucidation of a $[Ag_{24}Au(SR)_{18}]^-$ Nanocluster. *Angew. Chem. Int. Ed.* **2016**, *55*, 922–926; *Angew. Chem.* **2016**, *128*, 934–938.

**65** M. Zhou, C. Zeng, M. Y. Sfeir, M. Cotlet, K. Iida, K. Nobusada, R. Jin, Evolution of Excited-State Dynamics in Periodic $Au_{28}$, $Au_{36}$, $Au_{44}$, and $Au_{52}$ Nanoclusters. *J. Phys. Chem. Lett.* **2017**, *8*, 4023–4030.

**66** F. Hu, Z.-J. Guan, G. Yang, J.-Q. Wang, J.-J. Li, S.-F. Yuan, G.-J. Liang, Q.-M. Wang, Molecular Gold Nanocluster $Au_{156}$ Showing Metallic Electron Dynamics. *J. Am. Chem. Soc.* **2021**, *143*, 17059–17067.

**67** W.-Q. Shi, L. Zeng R.-L. He, X.-S. Han, Z.-J. Guan, M. Zhou, Q.-M. Wang, Near-unity NIR Phosphorescent Quantum Yield from a Room-Temperature Solvated Metal Nanocluster. *Science* **2024**, *383*, 326–330.

**68** S. Ardo, G. J. Meyer, Photodriven Heterogeneous Charge Transfer with Transition-Metal Compounds Anchored to $TiO_2$ Semiconductor Surfaces. *Chem. Soc. Rev.* **2009**, *38*, 115–164.

**69** A. Kudo, Y. Miseki, Heterogeneous Photocatalyst Materials for Water Splitting. *Chem. Soc. Rev.* **2009**, *38*, 253–278.

**70** H. Chen, C. E. Nanayakkara, V. H. Grassian, Titanium Dioxide Photocatalysis in Atmospheric Chemistry. *Chem. Rev.* **2012**, *112*, 5919–5948.

**71** Q. Guo, C. Zhou, Z. Ma, X. Yang, Fundamentals of $TiO_2$ Photocatalysis: Concepts, Mechanisms, and Challenges. *Adv. Mater.* **2019**, *31*, 1901997.

**72** F. Opoku, K. K. Govender, C. G. C. E. van Sittert, P. P. Govender, Enhancing Charge Separation and Photocatalytic Activity of Cubic $SrTiO_3$ with Perovskite-Type Materials $MTaO_3$ (M = Na,K) for Environmental Remediation: A First-Principles Study. *Chemistryselect* **2017**, *2*, 6304–6316.

**73** A. Malathi, J. Madhavana, M. Ashokkumar, P. Arunachalam, A Review on $BiVO_4$ Photocatalyst: Activity Enhancement Methods for Solar Photocatalytic Applications. *Appl. Catal. A: Gen.* **2018**, *555*, 47–74.

**74** P. Zhang, J. Zhang, J. Gong, Tantalum-Based Semiconductors for Solar Water Splitting. *Chem. Soc. Rev.* **2014**, *43*, 4395–4422.

**75** S. Cao, J. Low, J. Yu, M. Jaroniec, Polymeric Photocatalysts Based on Graphitic Carbon Nitride. *Adv. Mater.* **2015**, *27*, 2150–2176.

**76** L. Cheng, Q. Xiang, Y. Liao, H. Zhang, CdS-Based Photocatalysts. *Energy Environ. Sci.* **2018**, *11*, 1362–1391.

**77** Z. Li, X. Meng, Z. Zhang, Recent Development on $MoS_2$-Based Photocatalysis: A review. *J. Photochem. Photobiol. C.* **2018**, *35*, 39–55.

**78** J, Chen, Q, -F. Zhang, T. A. Bonaccorso, P. G. Williard, L.-S. Wang, Controlling Gold Nanoclusters by Diphospine Ligands. *J. Am. Chem. Soc.* **2014**, *136*, 92–95.

**79** Q. Zhu, X. Huang, Y. Zeng, K. Sun, L. Zhou, Y. Liu, L. Luo, S. Tian, X. Sun, Controllable Synthesis and Electrocatalytic Applications of Atomically Precise Gold Nanoclusters. *Nanoscale Adv.* **2021**, *3*, 6330–6341.

**80** J. Chen, Q.-F. Zhang, P. G. Williard, L.-S. Wang, Synthesis and Structure Determination of a New $Au_{20}$ Nanocluster Protected by Tripodal Tetraphosphine Ligands. *Inorg. Chem.* **2014**, *53*, 3932–3934.

**81** L. Liu, A. Corma, Metal Catalysts for Heterogeneous Catalysis: From Single Atoms to Nanoclusters and Nanoparticles. *Chem. Rev.* **2018**, *118*, 4981–5079.

**82** I. Chakraborty, T. Pradeep, Atomically Precise Clusters of Noble Metals: Emerging Link between Atoms and Nanoparticles. *Chem. Rev.* **2017**, *117*, 8208–8271.

**83** H. Lu, B. Chen, Y. Li, J. Shi, J. Li, L. Wang, S. Luo, C. Fan, J. Shen, J. Chen, Benzyl-Rich Ligand Engineering of the Photostability of Atomically Precise Gold Nanoclusters. *Chem. Commun.* **2022**, *58*, 2395–2398.

**84** J. Fang, B. Zhang, Q. Yao, Y. Yang, J. Xie, N. Yan, Recent Advances in the Synthesis and Catalytic Applications of Ligand-Protected, Atomically Precise Metal Nanoclusters. *Coord. Chem. Rev.* **2016**, *322*, 1–29.

**85** O. Jin H. Chai, Z. Liu, T. Chen, J. Xie, Engineering Ultrasmall Metal Nanoclusters for Photocatalytic and Electrocatalytic Applications. *Nanoscale* **2019**, *11*, 20437–20448.

**86** C. C. Nguyen, N. N. Vu, T.-O. Do, Recent Advances in the Development of Sunlight-Driven Hollow Structure Photocatalysts and their Applications. *J. Mater. Chem. A.* **2015**, *3*, 18345–18359.

**87** H. Kawasaki, S. Kumar, G. Li, C. Zeng, D. R. Kauffman, J. Yoshimoto, Y. Iwasaki, R. Jin, Generation of Singlet Oxygen by Photoexcited $Au_{25}(SR)_{18}$ Clusters. *Chem. Mater.* **2014**, *26*, 2777–2788.

**88** M. Agrachev, M. Ruzzi, A. Venzo, F. Maran, Nuclear and Electron Magnetic Resonance Spectroscopies of Atomically Precise Gold Nanoclusters. *Acc. Chem. Res.* **2019**, *52*, 44–52.

**89** C. Yu, G. Li, S. Kumar, H. Kawasaki, R. Jin, Stable $Au_{25}(SR)_{18}/TiO_2$ Composite Nanostructure with Enhanced Visible Light Photocatalytic Activity. *J. Phys. Chem. Lett.* **2013**, *4*, 2847–2852.

**90** T. D. Green, K. L. Knappenberger, Relaxation Dynamics of $Au_{25}L_{18}$ Nanoclusters Studied by Femtosecond Time-Resolved Near Infrared Transient Absorption Spectroscopy. *Nanoscale* **2012**, *4*, 4111–4118.

**91** Z. Li, C. Liu, H. Abroshan, D. R. Kauffman, G. Li, $Au_{38}S_2(SAdm)_{20}$ Photocatalyst for One-Step Selective Aerobic Oxidations. *ACS Catal.* **2017**, *7*, 3368–3374.

**92** H. Qian, W. T. Eckenhoff, Y. Zhu, T. Pintauer, R. Jin, Total Structure Determination of Thiolate-Protected $Au_{38}$ Nanoparticles. *J. Am. Chem. Soc.* **2010**, *132*, 8280–8281.

**93** H. Chong, G. Gao, J. Chai, S. Yang, B. Rao, G. Li, M. Zhu, Photoinduced Oxidation Catalysis by $Au_{25-x}Ag_x(SR)_{18}$ Nanoclusters. *ChemNanoMat* **2018**, *4*, 482–486.

**94** M. Zhu, C. M. Aikens, M. P. Hendrich, R. Gupta, H. Qian, G. C. Schatz, R. Jin, Reversible Switching of Magnetism in Thiolate-Protected $Au_{25}$ Superatoms. *J. Am. Chem. Soc.* **2009**, *131*, 2490–2492.

**95** C. P. Joshi, M. S. Bootharaju, M. J. Alhilaly, O. M. Bakr, $[Ag_{25}(SR)_{18}]^-$: The "Golden" Silver Nanoparticle. *J. Am. Chem. Soc.* **2015**, *137*, 11578–11581.

**96** C. Kumara, C. M. Aikens, A. Dass, X-ray Crystal Structure and Theoretical Analysis of $Au_{25-x}Ag_x(SCH_2CH_2Ph)_{18}{}^-$ Alloy. *J. Phys. Chem. Lett.* **2014**, *5*, 461–466.

**97** C. Han, B. K. Kundu, Y. Liang, Y. Sun, Near-Infrared Light-Driven Photocatalysis with an Emphasis on Two-Photon Excitation: Concepts, Materials, and Applications. *Adv. Mater.* **2024**, *36*, 2307759.

**98** S. Wang, L. Tang, B. Cai, Z. Yin, Y. Li, L. Xiong, X. Kang, J. Xuan, Y. Pei, M. Zhu, Ligand Modification of $Au_{25}$ Nanoclusters for Near-Infrared Photocatalytic Oxidative Functionalization. *J. Am. Chem. Soc.* **2022**, *144*, 3787–3792.

**99** C. Zhu, Z.-L. Chen, H. Li, L. Lu, K. Xi, J. Xuan, M. Zhu, Rational Design of Highly Phosphorescent Nanoclusters for Efficient Photocatalytic Oxidation. *J. Am. Chem. Soc.* **2024**, *146*, 23212–23220.

**100** X. Chen, S. Shen, L. Guo, S. S. Mao, Semiconductor-Based Photocatalytic Hydrogen Generation. *Chem. Rev.* **2010**, *110*, 6503–6570.

**101** A. Fujishima, K. Honda, Electrochemical Photolysis of Water at a Semiconductor Electrode. *Nature* **1972**, *238*, 37–38.

**102** T. Wang, J. Gong, Single-Crystal Semiconductors with Narrow Band Gaps for Solar Water Splitting. *Angew. Chem. Int. Ed.* **2015**, *54*, 10718–10732; Einkristalline Halbleiter mit kleinen Bandlücken für die solare Wasserspaltung. *Angew. Chem.* **2015**, *127*, 10866–10881.

**103** Y. Negishi, Y. Matsuura, R. Tomizawa, W. Kurashige, Y. Niihori, T. Takayama, A. Iwase, A. Kudo, Controlled Loading of Small Aun Clusters (n = 10-39) onto $BaLa_4Ti_4O_{15}$ Photocatalysts: Toward an Understanding of Size Effect of Cocatalyst on Water-Splitting Photocatalytic Activity. *J. Phys. Chem. C.* **2015**, *119*, 11224–11232.

**104** Y.-S. Chen, H. Choi, P. V. Kamat, Metal-Cluster-Sensitized Solar Cells. A New Class of Thiolated Gold Sensitizers Delivering Efficiency Greater than 2%. *J. Am. Chem. Soc.* **2013**, *135*, 8822–8825.

**105** Y.-S. Chen, P. V. Kamat, Glutathione-Capped Gold Nanoclusters as Photosensitizers. Visible Light-Induced Hydrogen Generation in Neutral Water. *J. Am. Chem. Soc.* **2014**, *136*, 6075–6082.

**106** Y. Wang, X.-H. Liu, Q. Wang, M. Quick, S. A. Kovalenko, Q.-Y. Chen, N. Koch, N. Pinna, Insights into Charge Transfer at an Atomically Precise Nanocluster/Semiconductor Interface. *Angew. Chem. Int. Ed.* **2020**, *59*, 7748–7754; *Angew. Chem.* **2020**, *132*, 7822–7828.

**107** Y. Zhao, Y. Li, L. Sun, Design and Construction Strategies to Improve Covalent Organic Frameworks Photocatalyst's Performance for Degradation of Organic Pollutants. *Chemosphere* **2022**, *286*, 131646.

**108** H. Dong, G. Zeng, L. Tang, C. Fan, C. Zhang, X. He, Y. He, An Overview on Limitations of $TiO_2$-Based Particles for Photocatalytic Degradation of Organic Pollutants and the Corresponding Countermeasures. *Water Res.* **2015**, *79*, 128–146.

**109** Y. Zhao, Y. Li, L. Sun, Recent Advances in Photocatalytic Decomposition of Water and Pollutants for Sustainable Application. *Chemosphere* **2021**, *276*, 130201.

**110** J. Hu, H. Yuan, P. Li, J. Wang, Q. Liu, H. Wang, Q. Wang, X. Yuan, Synthesis and Photocatalytic Activity of ZnO-$Au_{25}$ Nanocomposites. *Sci. China Chem.* **2016**, *59*, 277–281.

**111** X. Liu, G. Yao, X. Cheng, J. Xu, X. Cai, W. Hu, W. W. Xu, C. Zhang, Y. Zhu, Cd-driven Surface Reconstruction and Photodynamics in Gold Nanoclusters. *Chem. Sci.* **2021**, *12*, 3290–3294.

**112** C. Liu, X. Ren, F. Lin, X. Fu, X. Lin, T. Li, K. Sun, J. Huang, Structure of the $Au_{23-x}Ag_x(S\text{-}Adm)_{15}$ Nanocluster and Its Application for Photocatalytic Degradation of Organic Pollutants. *Angew. Chem. Int. Ed.* **2019**, *58*, 11335–11339; *Angew. Chem.* **2019**, *131*, 11457–11461.

**113** L.-J. Liu, J.-W. Zhang, M. Asad, Z.-Y. Wang, S.-Q. Zang, Thomas C. W. Mak, A High-Nuclearity CuI /CuII Nanocluster Catalyst for Phenol Degradation. *Chem. Commun.* **2021**, *57*, 5586–5589.

**114** S. Fang, M. Rahaman, J. Bharti, E. Reisner, M. Robert, G. A. Ozin, Y. H. Hu, Photocatalytic $CO_2$ Reduction. *Nat. Rev. Methods Primers.* **2023**, *3*, 62.

**115** J. Fu, K. Jiang, X. Qiu, J. Yu, M. Liu, Product Selectivity of Photocatalytic $CO_2$ Reduction Reactions. *Mater. Today.* **2020**, *32*, 222–243.

**116** Z. Jiang, H. Sun, T. Wang, B. Wang, W. Wei, H. Li, S. Yuan, T. An, H. Zhao, J. Yu, P. K. Wong, Nature-Based Catalyst for Visible-Light-Driven Photocatalytic $CO_2$ Reduction. *Energy. Environ. Sci.* **2018**, *11*, 2382–2389.

**117** D. Li, M. Kassymova, X. Cai, S.-Q. Zang, H. L. Jiang, Photocatalytic $CO_2$ Reduction Over Metal-Organic Framework-Based Materials. *Coord. Chem. Rev.* **2020**, *412*, 213262.

**118** E. Nikoloudakis, I. Lo′pez-Duarte, G. Charalambidis, K. Ladomenou, M. Ince, A. G. Coutsolelos, Porphyrins and Phthalocyanines as Biomimetic Tools for Photocatalytic $H_2$ Production and $CO_2$ Reduction. *Chem. Soc. Rev.* **2022**, *51*, 6965–7045.

**119** K. Sun, Y. Qian, H.-L. Jiang, Metal-Organic Frameworks for Photocatalytic Water Splitting and $CO_2$ Reduction. *Angew. Chem. Int. Ed.* **2023**, *62*, e202217565; *Angew. Chem.* **2023**, *135*, e202217565.

**120** X. Cui, J. Wang, B. Liu, S. Ling, R. Long, Y. Xiong, Turning Au Nanoclusters Catalytically Active for Visible-Light-Driven $CO_2$ Reduction through Bridging Ligands. *J. Am. Chem. Soc.* **2018**, *140*, 16514–16520.

**121** T. Kawawaki, Y. Kataoka, M. Hirata, Y. Akinaga, R. Takahata, K. Wakamatsu, Y. Fujiki, M. Kataoka, S. Kikkawa, A. S. Alotabi, S. Hossain, D. J. Osborn, T. Teranishi, G. G. Andersson, G. F. Metha, S. Yamazoe, Y. Negishi, Creation of High-Performance Heterogeneous Photocatalysts by Controlling Ligand Desorption and Particle Size of Gold Nanocluster. *Angew. Chem. Int. Ed.* **2021**, *60*, 21340–21350; *Angew. Chem.* **2021**, *133*, 21510–21520.

**122** Y. Jiang, Y. Yu, X. Zhang, M. Weinert, X. Song, J. Ai, L. Han, H. Fei, N-Heterocyclic Carbene-Stabilized Ultrasmall Gold Nanoclusters in a Metal-Organic Framework for Photocatalytic $CO_2$ Reduction. *Angew. Chem. Int. Ed.* **2021**, *60*, 17388–17393; *Angew. Chem.* **2021**, *133*, 17528–17533.

**123** X.-L. Pei, Z.-J. Guan, Z.-A. Nan, Q.-M. Wang, Heterometallic Coinage Metal Acetylenediide Clusters Showing Tailored Thermochromic Luminescence. *Angew. Chem. Int. Ed.* **2021**, *60*, 14381–14384; *Angew. Chem.* **2021**, *133*, 14502–14505.

**124** Y. Xiong, H. Chen, Y. Hu, S. Yang, X. Xue, L. He, X. Liu, J. Ma, Z. Jin, Photodriven Catalytic Hydrogenation of $CO_2$ to $CH_4$ with Nearly 100% Selectivity over $Ag_{25}$ Clusters. *Nano Lett.* **2021**, *21*, 8693–8700.

**125** J.-P. Dong, Y. Xu, X.-G. Zhang, H. Zhang, L. Yao, R. Wang, S.-Q. Zang, Copper-Sulfur-Nitrogen Cluster Providing a Local Proton for Efficient Carbon Dioxide Photoreduction. *Angew. Chem. Int. Ed.* **2023**, *62*, e202313648; *Angew. Chem.* **2023**, *135*, e202313648.

**126** H. Chen, C. Liu, M. Wang, C. Zhang, N. Luo, Y. Wang, H. Abroshan, G. Li, F. Wang, Visible Light Gold Nanocluster Photocatalyst: Selective Aerobic Oxidation of Amines to Imines. *ACS Catal.* **2017**, *7*, 3632–3638.

**127** H. Chen, L. Peng, Y. Bian, X. Shen, J. Li, H.-C. Yao, S.-Q. Zang, Z. Li, Exerting Charge Transfer to Stabilize Au Nanoclusters for Enhanced Photocatalytic Performance Toward Selective Oxidation of Amines. *Appl. Catal. B.* **2021**, *284*, 119704.

**128** H. Hirai, S. Takano, T. Nakashima, T. Iwasa, T. Taketsugu, T. Tsukuda, Doping-Mediated Energy-Level Engineering of $M@Au_{12}$ Superatoms (M=Pd, Pt, Rh, Ir) for Efficient Photoluminescence and Photocatalysis. *Angew. Chem. Int. Ed.* **2022**, *61*, e202207290; *Angew. Chem.* **2022**, *134*, e202207290.

**129** Q. Chen, X.-Z. Ge, L. Yu, F.-X. Xiao, Atomically Precise Metal Nanocluster Photosystem: Electron Relay Boosts Photocatalytic Organic Transformation. *Inorg. Chem.* **2023**, *62*, 19358–19365.

**130** O. Reiser, Shining Light on Copper: Unique Opportunities for Visible-Light-Catalyzed Atom Transfer Radical Addition Reactions and Related Processes. *Acc. Chem. Res.* **2016**, *49*, 1990–1996.

**131** A. Sagadevan, A. Ghosh, P. Maity, O. F. Mohammed, O. M. Bakr, M. Rueping, Visible-Light Copper Nanocluster Catalysis for the C-N Coupling of Aryl Chlorides at Room Temperature. *J. Am. Chem. Soc.* **2022**, *144*, 12052–12061.

**132** S. Nematulloev, A. Sagadevan, B. Alamer, A. Shkurenko, R. Huang, J. Yin, C. Dong, P. Yuan, K. E. Yorov, A. A. Karluk, W. J. Mir, B. E. Hasanov, M. N. Hedhili, N. M. Halappa, M. Eddaoudi, O. F. Mohammed, M. Rueping, O. M. Bakr, Atomically Precise Defective Copper Nanocluster Catalysts for Highly Selective C-C Cross-Coupling Reactions. *Angew. Chem. Int. Ed.* **2023**, *62*, e202303572.

**133** C. Dong, R.-W. Huang, A. Sagadevan, P. Yuan, L. Gutiérrez-Arzaluz, A. Ghosh, S. Nematulloev, B. Alamer, O. F. Mohammed, I. Hussain, M. Rueping, O. M. Bakr, Isostructural Nanocluster Manipulation Reveals Pivotal Role of One Surface Atom in Click Chemistry. *Angew. Chem. Int. Ed.* **2023**, *62*, e202307140.

**134** B. Alamer, A. Sagadevan, M. Bodiuzzaman, K. Murugesan, S. Alsharif, R.-W. Huang, A. Ghosh, M. H. Naveen, C. Dong, S. Nematulloev, J. Yin, A. Shkurenko, M. Abulikemu, X. Dong, H. Han, M. Eddaoudi, M. Rueping, O. M. Bakr, Planar Core and Macrocyclic Shell Stabilized Atomically Precise Copper Nanocluster Catalyst for Efficient Hydroboration of C–C Multiple Bond. *J. Am. Chem. Soc.* **2024**, *146*, 16295–16305.

**135** M. Bodiuzzaman, K. Murugesan, P. Yuan, B. Maity, A. Sagadevan, N. Malenahalli H, S. Wang, P. Maity, M. F. Alotaibi, D.-e. Jiang, M. Abulikemu, O. F. Mohammed, L. Cavallo, M. Rueping, O. M. Bakr, Modulating Decarboxylative Oxidation Photocatalysis by Ligand Engineering of Atomically Precise Copper Nanoclusters. *J. Am. Chem. Soc.* **2024**, *146*, 26994–27005.

**136** L.-J. Liu, M.-M. Zhang, Z. Deng, L.-L. Yan, Y. Lin, D. L. Phillips, V. W.-W. Yam, J. He, NIR-II Emissive Anionic Copper Nanoclusters with Intrinsic Photoredox Activity in Single-Electron Transfer. *Nat. Commun.* **2024**, *15*, 4688.

**137** L.-J. Liu, X. Lei, J. Guo, X. Mo, Y. Lao, S. Zhuang, H. Zeng, S. Yang, Y. Zhao, W. W. Xu, J. He, Johnson Solid Gold Nanoclusters as Heterogeneous Visible-Light Photocatalysts for Fukuyama Indole Synthesis. *J. Am. Chem. Soc.* **2025**, *147*, 27981–27991.

**138** Y. Zhao, Z.-M. Zhu, W. Fan, W. Zhu, J.-J. Yang, Y. Tao, W. Fei, H. Bi, S. Zhang, M.-B. Li, Photosynthesis of $Au_8Cu_6$ Nanocluster for Photocatalysis in Oxidative Functionalization of Alkynes. *Nat. Commun.* **2024**, *15*, 9632.

**139** L. Zhang, J. Liu, Y.-Q. Lan, Hetero-Motif Molecular Junction Photocatalysts: A New Frontier in Artificial Photosynthesis. *Acc. Chem. Res.* **2024**, *57*, 870–883.

**140** J.-M. Lin, Z.-B. Mei, C. Guo, J.-R. Li, Y. Kuang, J.-W. Shi, J.-J. Liu, X. Li, S.-L. Li, J. Liu, Y.-Q. Lan, Synthesis of Isotypic Giant Polymolybdate Cages for Efficient Photocatalytic C–C Coupling Reactions. *J. Am. Chem. Soc.* **2024**, *146*, 22797–22806.

**141** L.-M. Zhang, H.-Z. Wei, H. Zhang, X.-H. Zhong, S.-F. Yuan, B. Wang, J. Wei, Q. Bai, Z. Zhang, K.-G. Liu, T. Wu, D. Li, Carbonylmetallated Palladium Nanoclusters for Visible-Light-Driven Carbonylation under 1 atm of CO. *J. Am. Chem. Soc.* **2025**, *147*, 28073–28084.

**142** B. Sun, S. Lu, Y. Qian, X. Zhang, J. Tian, Recent Progress in Research and Design Concepts for the Characterization, Testing, and Photocatalysts for Nitrogen Reduction Reaction. *Carbon Energy* **2023**, *5*, e305.

**143** S. Lin, X. Zhang, L. Chen, Q. Zhang, L. Ma, J. Liu, A Review on Catalysts for Electrocatalytic and Photocatalytic Reduction of $N_2$ to Ammonia. *Green Chem.* **2022**, *24*, 9003–9026.

**144** Y. Sun, W. Pei, M. Xie, S. Xu, S. Zhou, J. Zhao, K. Xiao, Y. Zhu, Excitonic $Au_4Ru_2(PPh_3)_2(SC_2H_4Ph)_8$ Cluster for Light-Driven Dinitrogen Fixation. *Chem. Sci.* **2020**, *11*, 2440–2447.

# 9

# Metal Nanocluster Electrocatalysis

## 9.1 Metal Nanocluster Electrochemical Property

Metal nanocluster electrocatalysis enables electron(s) transfer between the substrates and the metal nanocluster catalysts, thus activating substrates and realizing their further high value-added conversion. Being different from the metal nanocluster photocatalysis, metal nanocluster electrocatalysis cannot underdo with energy transfer. Therefore, the electrochemical property of metal nanoclusters, especially their reversible electron-donating and electron-accepting ability is crucial for evaluating electrocatalytic performance of metal nanoclusters. Before we discuss metal nanocluster electrocatalysis, we herein talk about the electrochemical property of metal nanoclusters first of all.

Metal nanoclusters encompass the bulk-to-molecule transition region, where the discrete electronic states start to appear. Meanwhile, the electronic band energetics bring the quantum confinement effect that leads to a significant electrochemical property of metal nanoclusters compared to metal complexes and metal nanoparticles [1–5]. Considering that organic ligand-protected metal nanoclusters are good model for investigating the structure, catalytic activity, and structure–property correlation of metal nanoclusters, we mainly focus on this kind of species as a representative in the following to showcase the unique electrochemical property and electrocatalytic performance of metal nanoclusters. Voltammetry is the most important method for characterizing the electronic structures and reversible electron-donating/accepting ability of metal nanoclusters. Especially, due to the fact that these ultra-small metal species are difficult to be prepared in large-scale and they are also sensitive toward high current electrolysis, square wave voltammetry (SWV) or differential pulse voltammetry (DPV) is more frequently employed than cyclic voltammetry (CV) to investigate the electrochemical property of metal nanoclusters [6]. The as-obtained spectra are helpful to correlate the electronic structures of metal nanoclusters with their optical property and energy gap (HOMO–LUMO gap), thus describing a full electrocatalytic property of metal nanoclusters.

*Metal Nanocluster Catalysis: From Fundamentals to Applications*, First Edition. Manzhou Zhu and Man-Bo Li. © 2026 Wiley-VCH GmbH. All rights reserved, including rights for text and data mining and training of artificial intelligence technologies or similar technologies. Published 2026 by Wiley-VCH GmbH.

### 9.1.1 Size-dependent Electrochemical Property of Metal Nanoclusters

As we mentioned above, the metal nanocluster falls in the intermediate size between the single metal atom and the metal nanocrystal. Therefore, their electrochemical property is very sensitive to the nanocluster size. Taking the thiolate ligand-protected gold nanoclusters as a typical example, their mass information, UV–vis absorption spectra and SWV spectra are shown in Figure 9.1 [7]. From the MALDI-TOF-MS spectra, obvious peaks around $m/z$ = 7034, 10,298, 17,300, 25,249, 35,397, and 74,852 Da were observed for the thiolate ligand-protected gold nanoclusters $Au_{25}(SC_6H_{13})_{18}$, $Au_{38}(SC_6H_{13})_{24}$, $Au_{67}(SC_6H_{13})_{35}$, $Au_{102}(SC_6H_{13})_{44}$, $Au_{144}(SC_6H_{13})_{60}$, and $Au_{333}(SC_6H_{13})_{79}$. The mass information on the one hand verifies the compositions of these gold nanoclusters, and on the other hand indicates that the samples are pure because none of other peaks were observed in the spectra.

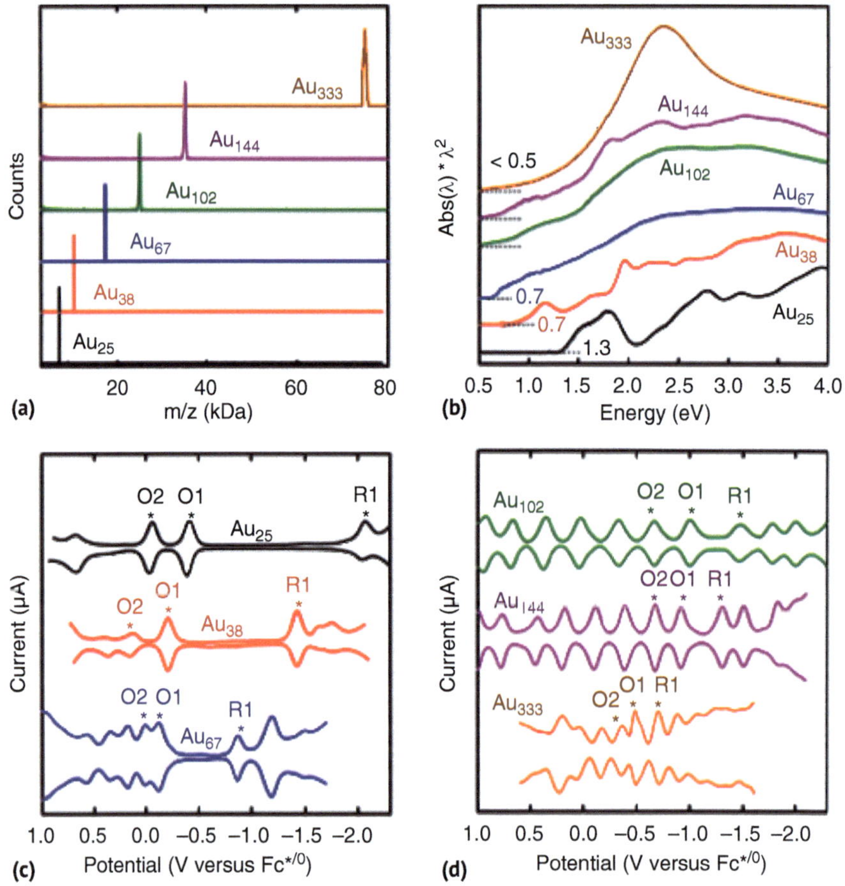


**Figure 9.1** MALDI-TOF-MS (a), UV–vis (b), and SWV (c, d) spectra of thiolate ligand-protected gold nanoclusters $Au_{25}(SC_6H_{13})_{18}$, $Au_{38}(SC_6H_{13})_{24}$, $Au_{67}(SC_6H_{13})_{35}$, $Au_{102}(SC_6H_{13})_{44}$, $Au_{144}(SC_6H_{13})_{60}$, and $Au_{333}(SC_6H_{13})_{79}$. *Source:* [6]/with permission of American Chemical Society.

The UV–vis spectra of these six gold nanoclusters give their optical gap information. As shown in Figure 9.1b, the optical gaps of $Au_{25}(SC_6H_{13})_{18}$, $Au_{38}(SC_6H_{13})_{24}$, and $Au_{67}(SC_6H_{13})_{35}$ were estimated as ~1.3, 0.9, and 0.7 eV, respectively. The optical gaps of $Au_{102}(SC_6H_{13})_{44}$, $Au_{144}(SC_6H_{13})_{60}$, and $Au_{333}(SC_6H_{13})_{79}$ were all smaller than 0.5 eV from the UV–vis spectra. This observation is consistent with the previously revealed results, that is, the energy gap (which is proportional to optical gap) decreases significantly with the size increasement of metal nanoclusters. This result is also in line with the common sense, because molecules generally display clearly discrete energy levels while bulk metals and metal nanocrystals present continuous energy bands. Increasement of the gold nanoclusters' size will finally trigger their evolution to FCC gold nanocrystals and then bulk metals, which do not have the discrete energy levels anymore. In contrary, decrease of the gold nanoclusters' size will finally result in the generation of metal complexes, which bear discrete energy levels and large HOMO–LUMO gaps.

Being similar to the decrease order of the optical gaps of the six gold nanoclusters with size increased, the electrochemical gaps of $Au_{25}(SC_6H_{13})_{18}$, $Au_{38}(SC_6H_{13})_{24}$, $Au_{67}(SC_6H_{13})_{35}$, $Au_{102}(SC_6H_{13})_{44}$, $Au_{144}(SC_6H_{13})_{60}$, and $Au_{333}(SC_6H_{13})_{79}$ also decrease sequentially (Figure 9.1c,d). The electrochemical gap between the first reduction potential (R1) and first oxidation potential (O1) of $Au_{25}(SC_6H_{13})_{18}$ was determined to be 1.65 V from the SWV spectrum, while the value for $Au_{333}(SC_6H_{13})_{79}$ was determined to be 0.22 V. Meanwhile, the oxidation and reduction states for the $Au_{333}(SC_6H_{13})_{79}$ significantly increase compared to that of $Au_{25}(SC_6H_{13})_{18}$ nanocluster. This phenomenon can be well explained by the increased free electrons with the size growth of metal nanoclusters. The size-dependent electrochemical property results in very small O1-R1 gaps of metal nanoclusters with large metal kernels. Therefore, these nanoclusters can be recognized as quantum capacitors, that is, their double layer capacitances become small and quantized double layer charging behavior can be experimentally observed. Compared to the charging behavior of metal nanoclusters with smaller sizes, namely $Au_{25}(SC_6H_{13})_{18}$, $Au_{38}(SC_6H_{13})_{24}$, and $Au_{67}(SC_6H_{13})_{35}$, the metal nanoclusters with large sizes and very small HOMO–LUMO gaps are more like molecular capacitor and thus become quantized double layer charging behavior.

The quantized double layer charging observed in SWV spectra of thiolate ligand-protected gold nanoclusters suggests a molecule-to-metal transition in metal nanoclusters. $Au_{102}(SC_6H_{13})_{44}$, $Au_{144}(SC_6H_{13})_{60}$, and $Au_{333}(SC_6H_{13})_{79}$ possess metallic electrochemical property while $Au_{25}(SC_6H_{13})_{18}$, $Au_{38}(SC_6H_{13})_{24}$, and $Au_{67}(SC_6H_{13})_{35}$ are more molecule-like. The evolution from molecular state to metallic state of metal nanoclusters has also been observed by Zhou and coworkers by employing the femtosecond and nanosecond transient absorption (fs- and ns-TA) spectroscopy [8–12]. Kwak and coworkers also observed that the exciton dynamics of the thiolate ligand-protected gold nanoclusters are opposite to their electron dynamics, and their exciton lifetimes increase with increased HOMO–LUMO gaps (i.e. with decreased sizes of metal nanocluster) [7]. It should be noted that the surface ligands also significantly affect the electrochemical property of metal nanoclusters. Compared to the thiolate ligand-protected gold nanocluster $Au_{25}(SC_6H_{13})_{18}$, the thiolate and phosphine ligands co-protected gold nanocluster $Au_{25}(PPh_3)_{10}(SC_6H_{13})_5Cl_2$ exhibits a smaller electrochemical gap of 1.54 V, although both nanoclusters possess the same number of gold atoms and the similar size [13]. Furthermore, the O1 and R1 peaks of $Au_{25}(PPh_3)_{10}(SC_6H_{13})_5Cl_2$ are both positively shifted relative to $Au_{25}(SC_6H_{13})_{18}$. The ligand-modified electrochemical property of metal nanoclusters will be discussed in detail in Section 9.1.3.

### 9.1.2 Metal Doping Effect of Metal Nanoclusters in Electrochemical Property

Even if two metal nanoclusters have the similar size, surface ligands and compositions, their electrochemical properties can be distinct from each other, to a large extent. This factor is that we are going to discuss in this part, that is, metal doping effect [14–20]. In the year of 2015, the research group of Zhikun Wu and Dongil Lee successfully synthesized four single metal atom-doped $MAu_{24}(SR)_{18}$ nanoclusters. Yao and coworkers obtained $CdAu_{24}(SR)_{18}$ and $HgAu_{24}(SR)_{18}$ nanoclusters by their developed anti-galvanic reaction method [21, 22] (for details, see the corresponding description in Section 4.1), while Kwak and coworkers obtained $PdAu_{24}(SR)_{18}$ and $PtAu_{24}(SR)_{18}$ nanoclusters by the in situ synthesis method [23] (for details, see the corresponding description in Section 4.1). Based on the theoretical calculations and characterizations such as XPS, the Pd or Pt atom should be in the center of the $PdAu_{24}(SC_6H_{13})_{18}$ and $PtAu_{24}(SC_6H_{13})_{18}$ nanoclusters (Figure 9.2a). Indeed, the later research verified the geometric structures of $PdAu_{24}(SC_6H_{13})_{18}$ and $PtAu_{24}(SC_6H_{13})_{18}$ proposed on that time by Dongil Lee, De-en Jiang et al. Yao and

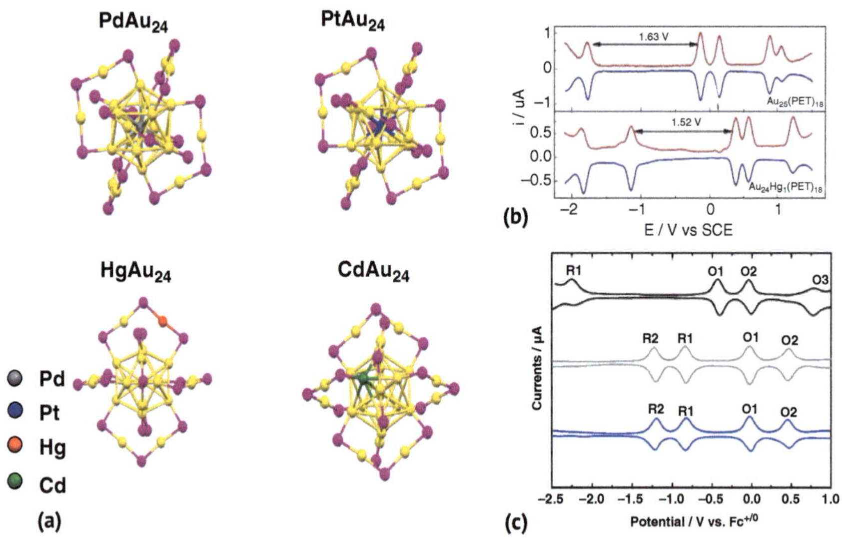


**Figure 9.2** (a) Calculated structures of $PdAu_{24}(SR)_{18}$ and $PtAu_{24}(SR)_{18}$ nanoclusters, and the SCXRD revealed structures of $CdAu_{24}(SR)_{18}$ and $HgAu_{24}(SR)_{18}$ nanoclusters. (b) DPV spectra of $HgAu_{24}(SCH_2CH_2Ph)_{18}$ and $Au_{25}(SCH_2CH_2Ph)_{18}$ nanoclusters. (c) SWV spectra of $PdAu_{24}(SC_6H_{13})_{18}$, $PtAu_{24}(SC_6H_{13})_{18}$ and $Au_{25}(SC_6H_{13})_{18}$ nanoclusters.

coworkers revealed the geometric structures of $CdAu_{24}(SR)_{18}$ and $HgAu_{24}(SR)_{18}$ nanoclusters by SCXRD. Interestingly, the location of Cd and Hg atoms are significantly different from that of Pd and Pt atoms. Specifically, Cd is determined to locate at the outer layer of the $MAu_{13}$ kernel of $CdAu_{24}(SR)_{18}$ nanocluster [21], while the Hg stays further way from the metal kernel and it locates at the "–SR–Au–SR–Au–SR–" staples of $HgAu_{24}(SR)_{18}$ nanocluster, that is, the outermost layer of the whole nanocluster [22].

The MALDI–TOF–MASS spectra and ESI-MS spectrum confirm the pureness of these four $MAu_{24}(SR)_{18}$ nanoclusters, and their electrochemical properties were investigated based on these pure samples. Before we discuss their DPV and SWV spectra, it should be mentioned that $CdAu_{24}(SR)_{18}$, $HgAu_{24}(SR)_{18}$, $PdAu_{24}(SR)_{18}$ and $PtAu_{24}(SR)_{18}$ were all obtained as charge-neutral metal nanoclusters. Therefore, compared to the open-shell $[Au_{25}(SR)_{18}]^0$ nanocluster (7e system), the free electrons (i.e. number of superatomic electrons) in $[CdAu_{24}(SR)_{18}]^0$, $[HgAu_{24}(SR)_{18}]^0$, $[PdAu_{24}(SR)_{18}]^0$, and $[PtAu_{24}(SR)_{18}]^0$ nanoclusters are determined to be 8, 8, 6 and 6 based on their compositions and the equation of $n^* = NV_A - L- z$ ($n^*$ is the number of superatomic electrons of a metal nanocluster; N and L are the numbers of metal atoms and surface thiolate ligands, respectively, of a metal nanocluster; $V_A$ is the valence of the metal; z is the charge state of a metal nanocluster). Results from Liao and coworkers' work demonstrates that after Hg doping, the electrochemical gap (1.52 V) of $HgAu_{24}(SCH_2CH_2Ph)_{18}$ decreases compared to the value (1.63 V) of $Au_{25}(SCH_2CH_2Ph)_{18}$ [22]. Additionally, the positive shifts were observed for the first reduction and oxidation potentials of $HgAu_{24}(SCH_2CH_2Ph)_{18}$ (Figure 9.2b). The DPV spectrum of $CdAu_{24}(SCH_2CH_2Ph)_{18}$ was not given by the authors.

The results from Kank and coworkers also indicate that after the foreign metal doping, the as-obtained $PdAu_{24}(SC_6H_{13})_{18}$ and $PtAu_{24}(SC_6H_{13})_{18}$ nanoclusters bear a significantly decreased electrochemical gap [23], compared to that of $Au_{25}(SC_6H_{13})_{18}$ nanocluster. Specifically, the O1-R1 gap of $PdAu_{24}(SC_6H_{13})_{18}$ and $PtAu_{24}(SC_6H_{13})_{18}$ were determined to be 0.75 and 0.73 v, which are smaller than the 1.65 V of $Au_{25}(SC_6H_{13})_{18}$ Moreover, the similar positive shifts of the first reduction and oxidation potentials of $PdAu_{24}(SC_6H_{13})_{18}$ and $PtAu_{24}(SC_6H_{13})_{18}$ were observed as well (Figure 9.2c). Notably, the experimental phenomena of the decreased electrochemical gap and positively shifted O1 and R1 after the foreign metal doping are only observed in gold nanoclusters. For the silver nanoclusters, the change law is not applicable. For example, $Ag_{25}(SR)_{18}$ bears the similar core-shell structure of $Au_{25}(SR)_{18}$ [24]. The O1-R1 gap of $Ag_{25}(SPhMe_2)_{18}$ was experimentally determined as 1.82 V by the SWV spectrum. However, the Pt or Pd doped $PtAg_{24}(SR)_{18}$ and $PdAg_{24}(SR)_{18}$ present the similar SWV spectra to $Ag_{25}(SR)_{18}$, and their electrochemical gaps range from 1.71 to 1.96 V [25]. These observations demonstrate that metal nanoclusters with the similar structure while different metal types possess completely different electrochemical properties.

The changed electrochemical properties of thiolate ligand-protected gold nanoclusters after metal atom-doping can be attributed to the Jahn–Teller distortion effect. For example, $[Au_{25}(SC_6H_{13})_{18}]^-$ nanocluster is a stable 8e system ($1S^21P^6$), consisting of eight free electrons. DFT calculations revealed that the HOMO level of $[Au_{25}(SC_6H_{13})_{18}]^-$ is triply degenerate 1P orbitals while its LUMO level is doubly degenerate 1D orbitals. In sharp contrast, $PdAu_{24}(SC_6H_{13})_{18}$ and $PtAu_{24}(SC_6H_{13})_{18}$ nanoclusters are 8e systems ($1S^21P^4$), consisting of six free electrons. The triply degenerate HOMO levels of $PdAu_{24}(SC_6H_{13})_{18}$ and $PtAu_{24}(SC_6H_{13})_{18}$ split into a doubly degenerate HOMO, and the LUMO levels are symmetrically occupied by electrons. The dramatically reduced HOMO–LUMO gaps (and also the electrochemical gaps) are resulted from the Jahn–Teller distortion of the 6e $[MAu_{24}(SR)_{18}]^0$ nanoclusters, where splitting of the 1P orbitals accompanies concurrently. This kind of doping-induced Jahn–Teller distortion effect of thiolate ligand-protected gold nanoclusters has also been verified in the $Au_{38}(SR)_{24}$ and $M_2Au_{36}(SR)_{24}$ nanoclusters system [26, 27]. The HOMO–LUMO energy gaps of $[Au_{38}(SR)_{24}]^0$ and $[Pd_2Au_{36}(SR)_{24}]^0$ were determined to be 0.86 and 0.26 eV based on the analysis of their SWV spectra. The decreased HOMO–LUMO (electrochemical) gaps after doping can be well explained by the Jahn–Teller distortion effect as well. DFT calculations predict that $[Au_{38}(SR)_{24}]^0$ has a 14e system ($1S^21P^61D^6$). In contrast, $[Pd_2Au_{36}(SR)_{24}]^0$ has 12 free electrons and possesses a 12e system ($1S^21P^61D^4$). The Jahn–Teller distortion leads to the orbital splitting of $[Pd_2Au_{36}(SR)_{24}]^0$. That is, the doubly degenerate HOMO level would break into HOMO and LUMO of $[Pd_2Au_{36}(SR)_{24}]^0$, thus resulting in a significantly smaller HOMO–LUMO (electrochemical) gap of the nanocluster after Pd doping.

### 9.1.3 Ligand Modification Regulate Electrochemical Property of Metal Nanoclusters

Apart from the size-dependent and doping-affected electrochemical property, as aforementioned, the surface ligands also play important role in regulating the electrochemical property of metal nanoclusters. In Section 9.1.1, we mentioned that the thiolate ligand-protected gold nanocluster $Au_{25}(SC_6H_{13})_{18}$ bears a completely different SWV spectra with the thiolate and phosphine ligands co-protected gold nanocluster $Au_{25}(PPh_3)_{10}(SC_6H_{13})_5Cl_2$ even if they have the same number of gold atoms and the similar size. In this case, both nanoclusters have very different geometric structures. Actually, metal nanoclusters with the similar geometric structure but different surface ligands also exhibit quite different electrochemical properties. Kank and coworkers investigated the electronic structure of water-soluble gold nanoclusters. Although the attempt for revealing the electrochemical gap of glutathione-protected water-soluble $Au_{25}(SG)_{18}$ was failed and only a broad current peak for the oxidation of the nanocluster was observed [28], Kank and coworkers successfully described the electrochemical property of a water-soluble $Au_{25}(SR)_{18}$ nanocluster protected by (3-mercaptopropyl)sulfonate by a phase transfer method (Figure 9.3) [29].

The (3-mercaptopropyl)sulfonate-protected $Au_{25}(SR)_{18}$ nanocluster was originally dissolved in water and the information on its electrochemical property is difficult to be acquired under the water solution. Dongil Lee et al. developed an ion-pairing strategy and transferred the water-soluble $Au_{25}(SR)_{18}$ nanocluster from water to toluene by using the phase transfer reagent tetraoctylammonium cation ($TOA^+$).

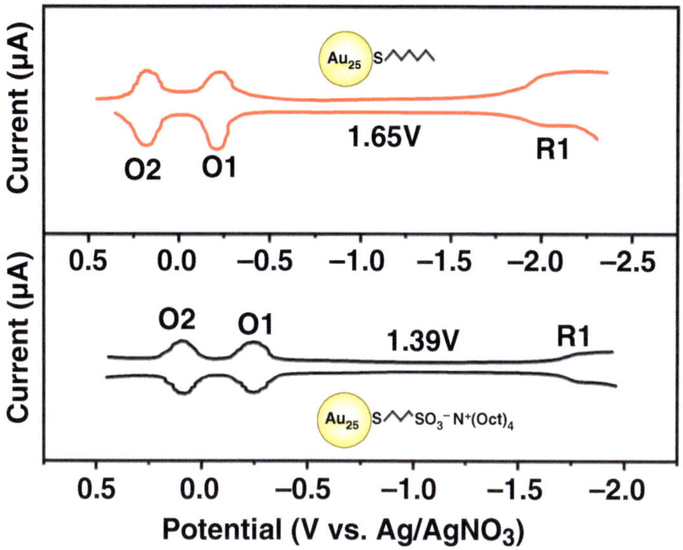


**Figure 9.3** SWV spectra of $Au_{25}(SR)_{18}$ nanoclusters with different surface thiolate ligands.

High-quality oil-soluble $Au_{25}(SR)_{18}$-TOA nanocluster was prepared by the ion-pairing strategy. Its composition and structure were confirmed by ESI-MS, UV–vis, and other characterizations. Results indicate that the phase transfer process does not have influence on the structure of the nanocluster. After transferring into the oil phase, the electrochemical property of the $Au_{25}(SR)_{18}$-TOA nanocluster was readily obtained and the O1-R1 gap was determined to be 1.39 V by SWV spectrum. This value is dramatically smaller than that (1.65 V) of $Au_{25}(SC_6H_{13})_{18}$ nanocluster. Detailed analysis revealed that the decrease of the electrochemical gap of $Au_{25}(SR)_{18}$-TOA is ascribed to the positive shift of its first reduction potential (Figure 9.3). In other words, the first oxidation potential of $Au_{25}(SR)_{18}$-TOA and $Au_{25}(SC_6H_{13})_{18}$ are basically the same. It is the positive shift of the first reduction potential of $Au_{25}(SR)_{18}$-TOA that results in its smaller O1-R1 gap compared to $Au_{25}(SC_6H_{13})_{18}$. The positive shift of the first reduction potential of $Au_{25}(SR)_{18}$-TOA can be well explained by the electrostatic field effect of the (3-mercaptopropyl)sulfonate ligands on the surface of the $Au_{25}$ kernel. That is, the polarized $SO_3^-$ anions would induce more positive charge on the gold kernel, resulting in an $Au_{25}(SR)_{18}$ nanocluster that is more easily to be reduced.

The ligand modification strategy is able to regulate the electrochemical property of metal nanoclusters in many aspects. It can further affect the electrocatalytic performance of metal nanoclusters. For a comprehensive understanding of fundamentals and challenges of ligand modification strategy in regulating electrochemical property and electrocatalytic performance of metal nanoclusters, a review from Kong and coworkers would be recommended [30]. Various ligand modification strategies for efficient electrocatalysis are involved in this review, which is highly inspiring for designing metal nanocluster electrodes and electrocatalysts.

Although the metal nanocluster electrode and the metal nanocluster electrocatalyst are not distinguished in many publications, we prefer introducing them respectively in the following two sections for a clearer understanding on the metal nanocluster electrocatalytic mechanism. Actually, for organic chemists, they generally describe the electrocatalysis from the aspect of electrocatalysts. They design catalysts that can be well dissolved in the electrolyte solution and act as homogeneous catalysts, and thus developing organic electrochemical synthesis methods based on that. For inorganic chemists, they are more frequently involved in designing of electrodes, in which the dissolution process of the electrocatalysts should be avoided because this process would cause the deactivation of the electrode. Thus, the electrocatalysts are generally heterogeneous catalysts during the catalysis. The catalytic mechanisms of homogeneous electrocatalysis and heterogeneous electrocatalysis are distinct from each other in many cases. We take the electrochemical hydrogen evolution reaction (HER) as an example [31]. For the heterogeneous catalytic pathway, three possible elementary steps are generally involved in the electrochemical HER process of $H_2O$ (Figure 9.4a). The first step is a single electron reduction of proton to give an adsorbed hydrogen atom (H*) on the metal surface (Volmer reaction). Subsequently, the H* is transformed into $H_2$ via a second step (Heyrovsky reaction or Tafel reaction). In the Heyrovsky process, another proton diffuses to the H* and reacts with the second electron to release $H_2$. Alternatively, two adjacent H* combine directly to give $H_2$ in the Tafel reaction. For the homogeneous catalytic pathway, the HER mechanisms are diversified, depending on different structures of homogeneous metal catalysts. Nevertheless, metal-hydrides are commonly regarded as key intermediates in the catalytic cycle and the typical catalytic mechanisms can be distinguished as homolytic and heterolytic pathways (Figure 9.4b). In the homolytic pathway, the homogeneous metal catalyst $[M]^n$ is initially reduced to lower valent species [M]n-1, which rapidly combines with proton to give metal hydride species $[M]^{n+1}$–H. With a weak metal–hydrogen bond in the homolytic pathway, $[M]^{n+1}$–H preferentially produce hydrogen via a bimolecular homocoupling. Instead, a strong metal-hydrogen bond would lead to a sluggish homolytic pathway, and the metal hydride species $[M]^{n+1}$–H can be further reduced

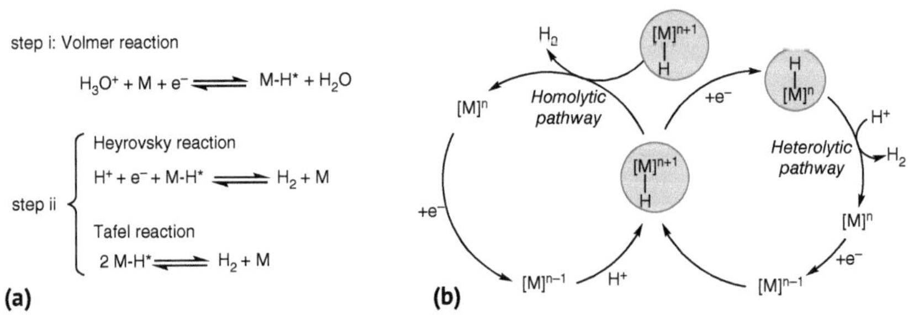


**Figure 9.4** Heterogeneous and homogeneous catalytic pathways involved in HER. M denotes the heterogeneous metal catalysts, $[M]^n$ denotes the homogeneous metal catalysts, H* denotes the adsorbed hydrogen atoms. (a) Heterogeneous HER catalytic mechanism and (b) homogeneous HER catalytic mechanism.

to reactive $[M]^n$–H. $[M]^n$–H reacts with a proton in the heterolytic pathway to afford hydrogen and regenerate the homogeneous catalyst $[M]^n$.

As a bridge between homogeneous and heterogeneous metal catalysis, metal nanocluster catalysis features both molecular and crystalline characteristics. They can be utilized as molecular electrocatalysts for homogeneous organic electrochemical synthesis. Meanwhile, they can be designed as electrode materials for many heterogeneous catalytic processes. Thus, we introduce metal nanocluster electrocatalysis from the two aspects in Sections 9.2 and 9.3, that is, metal nanocluster heterogeneous electrode and metal nanocluster homogeneous catalyst.

## 9.2 Metal Nanocluster Heterogeneous Electrode

Most of the reports on metal nanocluster electrocatalysis are based on metal nanocluster heterogeneous electrode. That is, metal nanoclusters are used as modifiers and loaded on supports. The modification methods of the metal nanocluster electrode are basically the same. Herein, we take a typical example from Zhuang and coworkers for illustration [32]. In this work, they synthesized a new Au-Cd alloyed nanocluster $Au_{47}Cd_2(TBBT)_{31}$ (TBBTH = 4-*tert*-butylbenzenelthiol) by the antigalvanic reduction method that they developed. The synthesis is based on a mother nanocluster $Au_{44}(TBBT)_{28}$ [32]. $Au_{44}(TBBT)_{28}$ and excess TBBTH were dissolved in toluene to form an organic phase. The dopant $Cd(NO_3)_2$ was dissolved in water to form a water phase. The antigalvanic reduction process between the gold nanocluster and the Cd salt occurred between the as-obtained organic phase and the water phase at 70 °C for 24 hours, and the new nanocluster $Au_{47}Cd_2(TBBT)_{31}$ was obtained in relatively high yield (60%). In the following electrocatalysis investigation, they compared the catalytic activity of the newly prepared $Au_{47}Cd_2(TBBT)_{31}$ nanocluster and the mother nanocluster $Au_{44}(TBBT)_{28}$. The metal nanocluster electrodes were prepared as following [32]: "To loading the $Au_{47}Cd_2(TBBT)_{31}$ and $Au_{44}(TBBT)_{28}$ on a Vulcan XC-72 carbon support, a calculated amount of carbon powder was added to the n-hexane solution of the clusters. After the mixture was stirred for six hours, the carbon supported Au nanoclusters were collected by centrifugation and dried at room temperature in a vacuum. A determined amount of these catalysts was dissolved in aqua regia for 12 hours. Then, the obtained mixture was centrifuged, and the centrifugate was diluted with deionized water and analyzed by ICP-AES. The accurate content of Au in the loaded catalyst was measured to be 14.3 wt% for $Au_{47}Cd_2(TBBT)_{31}$ and 13.2 wt% for $Au_{44}(TBBT)_{28}$ by ICP-AES." "The working electrode was a thin layer of a Nafion-impregnated catalyst cast on a glassy carbon disk. To prepare the working electrode, 5 mg of the carbon-supported Au nanoclusters was ultrasonically dispersed in 1 mL of ethanol containing 0.05 mL of Nafion solution to form an ink. Then, 40 μL of the ink was dispensed onto a $5 \times 5\ mm^2$ glassy carbon electrode (GCE) with two sides, and then, the GCE was subsequently dried in a stream of warm air at 70 °C for one hour. The content of Au on the working electrode was calculated using ICP-AES measurements."

Based on the metal nanocluster electrodes prepared above, they further tested the electrocatalytic performance of both nanoclusters by employing the $CO_2$ electroreduction as a model reaction. The results showed that $Au_{47}Cd_2(TBBT)_{31}$ exhibited much higher activity than $Au_{44}(TBBT)_{28}$ in the reduction reaction from $CO_2$ to CO (Figure 9.5a). Specifically, the $Au_{47}Cd_2(TBBT)_{31}$ nanocluster showed 96% of Faradaic efficiency at −0.57 V, compared to the 83% Faradaic efficiency of $Au_{44}(TBBT)_{28}$ nanocluster under the same reaction conditions. Meanwhile, the partial current density of CO was determined to be $-3.2\ mA\ cm^{-2}$ for $Au_{47}Cd_2(TBBT)_{31}$ at −0.57 V, which is more negative than that of the $Au_{44}(TBBT)_{28}$ nanocluster at −0.57 V ($-1.6\ mA\ cm^{-2}$). Additionally, the mass partial current density of CO for $Au_{47}Cd_2(TBBT)_{31}$ is calculated to be −55.6 A/g, which is much denser than that of CO for $Au_{44}(TBBT)_{28}$ (−30.8 A/g). In other words, $Au_{47}Cd_2(TBBT)_{31}$ nanocluster demonstrates higher Faradaic efficiency of CO and mass activity than $Au_{44}(TBBT)_{28}$ nanocluster. The electrocatalytic performance of $Au_{47}Cd_2(TBBT)_{31}$ even exceeds most previously reported state-of-the-art catalysts.

The $Au_{47}Cd_2(TBBT)_{31}$ nanocluster features a hard-sphere random close-packed structure, which has not been observed in other metal nanoclusters reported previously. This kind of unique structure is beneficial for the exposure of the active Cd sites, which play an important role in adsorption and stabilization of the key intermediate COOH*. The binding of Cd sites on $Au_{47}Cd_2(TBBT)_{31}$ toward the oxygen atom gives a release of COOH* 0.39 eV in free energy, while the value on $Au_{44}(TBBT)_{28}$ increases with 0.17 eV. Due to the fact that the COOH* formation process is the rate-determining step, $Au_{47}Cd_2(TBBT)_{31}$ shows the best electrocatalytic performance in the whole reaction even though it is not the most favorable catalyst in the following elementary steps.

Another elegant work from Zhuang and coworkers demonstrates that in addition to the surface metal atoms, the kernelled metal atoms also affect the electrocatalytic performance of metal nanoclusters. They prepared two structurally similar metal nanoclusters, namely $Au_4Pd_6(TBBT)_{12}$ and $Au_3AgPd_6(TBBT)_{12}$. Both nanoclusters possess the similar structures [33]. The only difference comes from the kernelled metal atom. In $Au_4Pd_6(TBBT)_{12}$, the kernelled metal atom is Au,

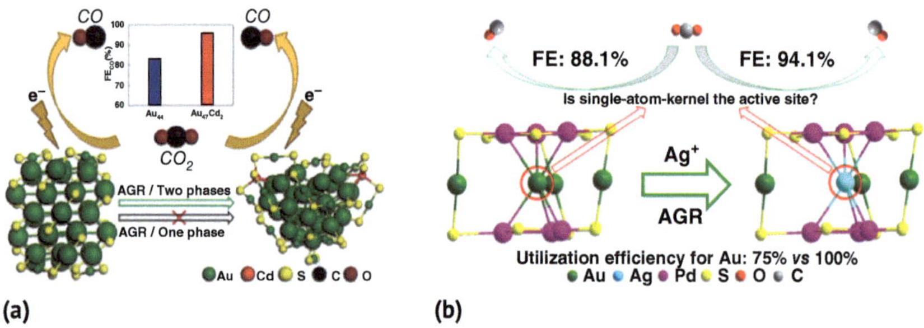


**Figure 9.5** (a) Surface metal-doped $Au_{47}Cd_2(TBBT)_{31}$, and its electrocatalytic performance comparison with $Au_{44}(TBBT)_{28}$ in $CO_2$ reduction reaction. (b) Kernel metal-doped $Au_3AgPd_6(TBBT)_{12}$, and its electrocatalytic performance comparison with $Au_4Pd_6(TBBT)_{12}$ in $CO_2$ reduction reaction.

while in $Au_3AgPd_6(TBBT)_{12}$, the kernelled metal is Ag (Figure 9.5b). The surface environment of both nanoclusters is completely the same. However, they demonstrate different catalytic activities in the $CO_2$ electroreduction reaction. Specifically, $Au_3AgPd_6(TBBT)_{12}$ exhibited Faradaic efficiency of 94.1% and $Au_4Pd_6(TBBT)_{12}$ exhibited Faradaic efficiency of 88.1%, at −0.57 V. Further investigations with the help of theoretical calculations revealed that the kernelled gold or silver atom in both $Au_4Pd_6(TBBT)_{12}$ and $Au_3AgPd_6(TBBT)_{12}$ nanocluster were not the activation sites for electrocatalysis while exerted an influence on their electronic structure and thus effected their electrocatalytic performance. XPS spectra reveal that the binding energies of Au 4f and Pd 3d in $Au_3AgPd_6(TBBT)_{12}$ are dramatically higher than that in $Au_4Pd_6(TBBT)_{12}$. Moreover, the d-band center in $Au_3AgPd_6(TBBT)_{12}$ also decreases significantly, which results in a weakened adsorption of the intermediate CO* on the surface activation sites of $Au_3AgPd_6(TBBT)_{12}$. Apart from the influence on adsorption of intermediates, the electronic changes resulted from the kernelled metal atom effectively suppress competitive HER, and the more difficult desorption of H* over $Au_3AgPd_6(TBBT)_{12}$ enables its highly selective electrocatalysis for the transformation from $CO_2$ to CO.

The foreign metal doping effect in $CO_2$ electroreduction reaction was also investigated by Deng and coworkers [34]. They revealed an Au–Cu synergistic effect that highly affected the electrocatalytic performance of metal nanoclusters. In their study, an Au–Cu-alloyed nanocluster $[Au_{15}Cu_4(DPPM)_6Cl_4(C{\equiv}CR)]^{2+}$ was synthesized. This nanocluster consists of a $Au_{15}Cu_4$ kernel that comprises two interpenetrating incomplete, centered icosahedra $Au_9Cu_2$ and $Au_8Cu_3$. The Au–Cu kernel is further protected by six DPPM, four halide, and one alkynyl ligand. Compared to a structurally similar Au nanocluster $[Au_{18}(DPPM)_6Br_4]^{2+}$ that only showed 60% CO Faradaic efficiency at −3.75 V [35], the Au-Cu alloyed nanocluster $[Au_{15}Cu_4(DPPM)_6Cl_4(C{\equiv}CR)]^{2+}$ exhibited a higher CO Faradaic efficiency of more than 90%. In addition, $[Au_{15}Cu_4(DPPM)_6Cl_4(C{\equiv}CR)]^{2+}$ exhibited an industrial-level CO partial current density of up to −413 mA $cm^{-2}$ at −3.75 V with the gas diffusion electrode based membrane electrode assembly. The value is twofold higher than that of un-doped $[Au_{18}(DPPM)_6Br_4]^{2+}$. Experimental combined with the theoretical investigations revealed that the unique pair of Au–Cu dual sites after stripping of an alkynyl and Cl ligands from the intact nanocluster is critical in the electroreduction transformation from $CO_2$ to CO.

Given the importance of the $CO_2$ electroreduction process, many research groups have been involved in metal nanocluster-catalyzed $CO_2$ electroreduction reaction [36–42]. Su [36], Gao [37], and Liu [38] respectively synthesized atomically precise $Au_8Cu_1$ and $Au_8Ag_1$ nanoclusters, $[Au_{22}H_3(dppe)_3(PPh_3)_8]^{3+}$ nanocluster, and three isostructural $Cu_8$ nanoclusters. These metal nanoclusters were supported on different materials and employed as electrodes for investigating their electrocatalytic properties. The as-obtained results are helpful to reveal the metal-doping effect, hydride-doping effect and isostructural effect of metal nanoclusters in electrocatalysis. Corresponding analysis indicates that copper nanoclusters or copper-doping in other metal nanoclusters are beneficial for the $CO_2$ electroreduction. Meanwhile, hydrides significantly affect the electronic structure of the coordinated metal activation sites and are also beneficial for the formation of COOH* intermediate, which is a key species for the further conversion into CO.

Concerning the metal nanocluster-catalyzed $CO_2$ electrochemical reduction, a breakthrough comes from Tan and coworkers [43]. By utilizing a copper-doped gold nanocluster $Au_{13}Cu_3$ as the electrocatalyst, Tan and coworkers realized an excellent electrochemical reduction of $CO_2$, with the current density as high as 70 mA $cm^{-2}$ (−0.97 V) and the maximum CO Faraday efficiency ($FE_{CO}$) of 99% at −0.58 V. This work also manifests the copper-doping effect in metal nanocluster electrocatalysis. They synthesized four copper-doped $Au_{13}$ icosahedral nanoclusters, which are abbreviated as $Au_{13}Cu_1$, $Au_{13}Cu_2$, $Au_{13}Cu_3$, and $Au_{13}Cu_4$. With the addition of undoped $Au_{13}$, they systematically compared the atomically precise structures and catalytic performance of these five nanoclusters in electrochemical $CO_2$ reduction reaction toward the production of CO. From Figure 9.6, we can clearly observe that $Au_{13}Cu_3$ adapts the most unsymmetric structure. Specifically, $Au_{13}Cu_1$ possesses a $C_3$-symmetric structure. Six thiolate ligands and three phosphine ligands completely occupy twelve gold atoms on the surface of the $Au_{13}$ kernel. The doped copper atom coordinates with three thiol ligands on a triangular surface to form a $Cu_1S_3$ coordination mode. $Au_{13}Cu_2$ bears a centrosymmetric structure with six phosphine

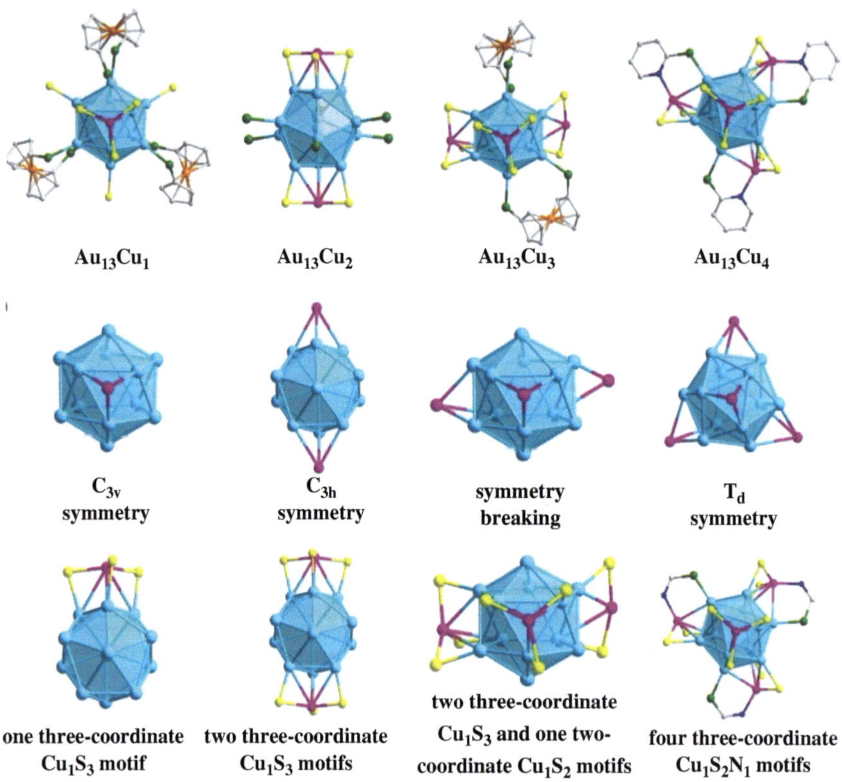


**Figure 9.6** Structural features of $Au_{13}Cu_1$, $Au_{13}Cu_2$, $Au_{13}Cu_3$, and $Au_{13}Cu_4$ nanoclusters. *Source:* [43]/with permission of John Wiley & Sons.

ligands coordinated at the waist of the $Au_{13}$ kernel. Being similar with the substructure in $Au_{13}Cu_1$ nanocluster, two doped copper atoms coordinate with six thiolate ligands to form two $Cu_1S_3$ motifs. $Au_{13}Cu_4$ is also a $C_3$-symmetric structure. Four doped copper atoms are bisected on the surface of the icosahedron $Au_{13}$ kernel, each occupying three surface gold atoms. The as-formed surface motifs are symmetrically arranged on the four triangular faces of the $Au_{13}$ kernel. For $Au_{13}Cu_3$ nanocluster, three Cu atoms covered on the surface of the $Au_{13}$ icosahedron kernel while they cannot be equally distributed on the surface, thus forming an asymmetric surface structure. Meanwhile, the eight sulfur atoms cannot be equally distributed to the three copper atoms either. Both structural features result in an unsymmetric structure of $Au_{13}Cu_3$ with $C_1$-symmetry. In a word, the incorporation of three doped copper atoms on the surface brings a symmetry-breaking structure to the icosahedron $Au_{13}$ kernel.

In accordance with the symmetry rules, these metal nanoclusters demonstrated increased catalytic activity in $CO_2$ electrochemical reduction reaction with the decrease of their symmetry (Figure 9.7). For example, $Au_{13}Cu_1$, $Au_{13}Cu_2$, and $Au_{13}Cu_4$ with lower symmetry exhibited slightly higher catalytic activity than $Au_{13}$, which possesses the highest symmetry among the five nanocluster catalysts. $Au_{13}Cu_3$ with unsymmetric structure demonstrated much higher activity, compared to the other four nanoclusters. It should be noted that these metal nanoclusters were loaded on activated carbon to form metal nanocluster electrodes. TEM analysis revealed that all of the nanoclusters were uniformly dispersed on the support, presenting 1–2 nm sizes. Their catalytic performance was investigated based on these metal nanocluster electrodes. CO was detected as the major product in the $CO_2$ electrochemical reduction reaction by using these nanocluster electrodes. The $Au_{13}Cu_3$ electrode maintains a $FE_{CO}$ of more than 95% with the potential from −0.5 to −0.8 V, and acquires the maximum $FE_{CO}$ of 99% at −0.58 V. The other metal nanocluster electrodes showed significantly lower activities than symmetry-breaking $Au_{13}Cu_3$.

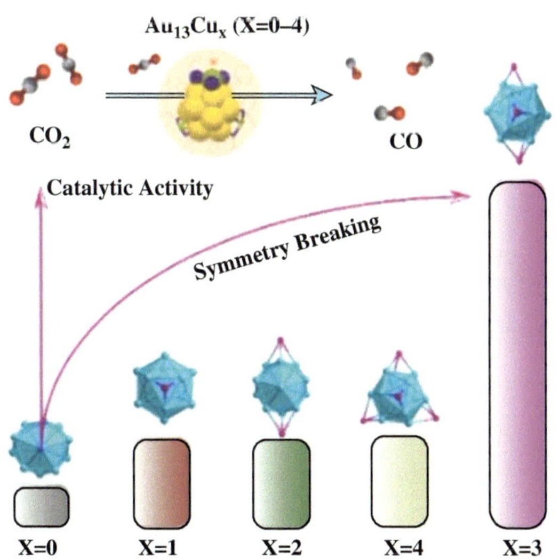


**Figure 9.7** Catalytic performance of $Au_{13}$, $Au_{13}Cu_1$, $Au_{13}Cu_2$, $Au_{13}Cu_3$, and $Au_{13}Cu_4$ nanoclusters in $CO_2$ electrochemical reduction reaction. *Source:* [43]/with permission of John Wiley & Sons.

Based on the atomically precise structures of structurally symmetric $Au_{13}Cu_1$ and structurally unsymmetric $Au_{13}Cu_3$, they conducted DFT calculation concerning the catalytic mechanisms of both nanoclusters (Figure 9.8). Results showed that the symmetry-breaking not only contributes to more active $CuS_3$ sites that are thermodynamically more feasible for $CO_2$ adsorption but also leads to easier $CO_2$ protonation and the subsequent reduction process and CO release process, thus resulting in higher catalytic activity of $Au_{13}Cu_3$. This detailed analysis on catalytic mechanism of metal nanocluster electrodes provides structural foundation for designing electrocatalysts with symmetry-breaking sites to efficiently activate linear $CO_2$ molecule.

Apart from the direct reduction of $CO_2$ to afford CO, the catalytic hydrogenations of $CO_2$ under electrochemical conditions were also acquired by using metal nanocluster electrodes. As we mentioned in Chapter 3, the research group of Shuang-Quan Zang developed a series of metal nanocluster-based MOF materials, which have significantly improved stability and properties compared to the single clusters. In 2025, Dong and coworkers synthesized two structurally similar cluster-based COF materials, namely 2Cu-NiPc-DHDA-COF and La-2Cu-NiPc-DHDA-COF. The only difference between both COF materials is that two copper atoms were located in the structural unit of 2Cu-NiPc-DHDA-COF while two cooper atoms and another La atom were located in the structural unit of La-2Cu-NiPc-DHDA-COF [44]. Both COF materials have clear structures and simple coordination environments (Figure 9.9), which are ideal models for investigating structure-property correlation. By loading the metal nanocluster-based COF on hydrophobic carbon paper, they studied the electrocatalytic $CO_2$ hydrogenation reaction of both electrodes. The results showed intriguing structure-property correlations. 2Cu-NiPc-DHDA-COF catalyzed the reaction, giving $C_2H_4$ as the major product (FE = 40.1%). In sharp contrast, La-2Cu-NiPc-DHDA-COF catalyzed the reaction and afforded $CH_4$ as the major product (FE = 57.1%). The comparison indicates that La element provides chances for manipulating the catalytic selectivity of the metal nanocluster-based MOF materials in electrocatalytic $CO_2$ hydrogenation reaction. This kind of oxophilic element doping effect can be explained by the fact that La regulates

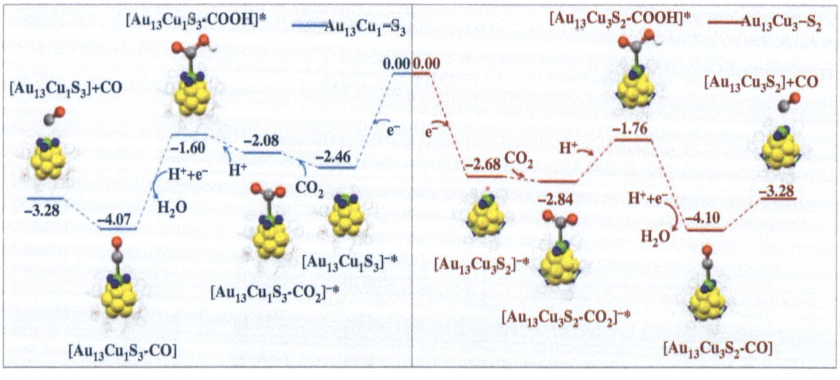


**Figure 9.8** Catalytic mechanisms of $Au_{13}Cu_1$ and $Au_{13}Cu_3$ catalyzed $CO_2$ electrochemical reduction reaction. *Source:* [43]/with permission of John Wiley & Sons.

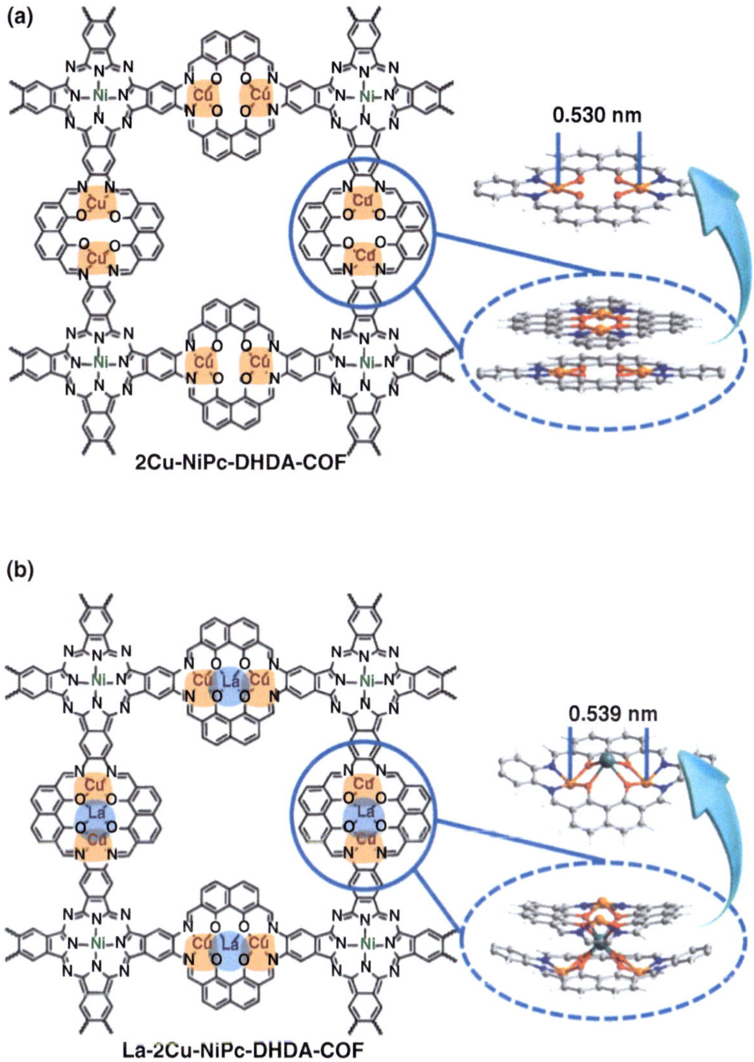


**Figure 9.9** Structural features of metal nanocluster-based MOF electrodes 2Cu-NiPc-DHDA-COF and La-2Cu-NiPc-DHDA-COF. *Source:* [44]/with permission of John Wiley & Sons.

the surface charge distribution of Cu activation sites, thus reducing the formation energy barriers of *COOH and *H and improving the local coverage of *CO and *H. Meanwhile, the La doping is able to regulate the formation energy barriers of $*CH_2O$ and *COCHO, directing the reaction pathway to the formation of methane.

By purposely designing a pure copper nanocluster, Biswas and coworkers realized electrocatalytic $CO_2$ reduction toward the formation of HCOOH [45]. Generally, the copper nanoclusters result in CO as the final product in the electrocatalytic $CO_2$ reduction reactions, because the stable copper nanoclusters are limited to Cu(I) activation sites, which are beneficial for the production of CO (Figure 9.10). Herein, Yuichi Negushi et al. applied a mixed ligand protecting strategy and synthesized a

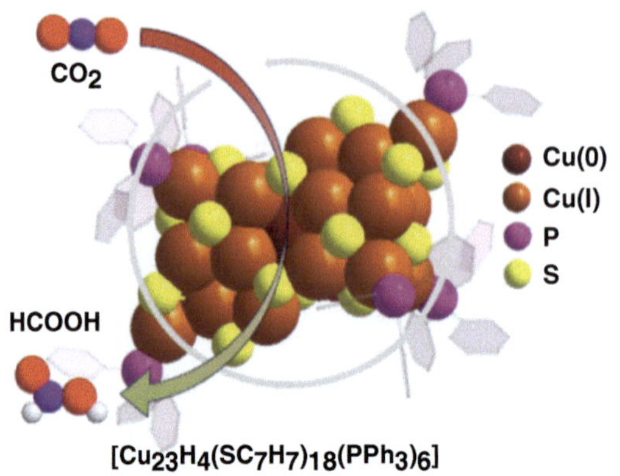


**Figure 9.10** Electrocatalytic $CO_2$ reduction toward the formation of HCOOH enabled by $[Cu_{23}H_4(SC_7H_7)_{18}(PPh_3)_6]$ nanocluster electrode. *Source:* [45]/with permission of American Chemical Society.

$[Cu_{23}H_4(SC_7H_7)_{18}(PPh_3)_6]$ nanocluster. Cu(0) activation sites are found to be located on the surface of the nanocluster, which are stabilized by the additional Cu(I) units, the thiolate ligands and the interstitial hydrides. The existence of Cu(0) activations sites leads to the lower limiting potential for the formation of the key intermediate *HCOO, which can be further transformed to the final HCOOH product. Thus, the selectivity of electrocatalytic $CO_2$ reduction reaction was switched under $[Cu_{23}H_4(SC_7H_7)_{18}(PPh_3)_6]$ nanocluster electrode, giving HCOOH as the major product with a maximum $FE_{HCOOH}$ of 26% at an applied potential of −1.2 V. The TOF for HCOOH formation was calculated to be 113.4 $h^{-1}$, indicating a significantly higher rate of product generation per active site compared to the formation of CO, which shows a much lower TOF of only 6.54 $h^{-1}$. The Cu(0) center-containing copper nanocluster electrode possesses robust framework, and maintains its structure during the electrocatalysis.

ORR, HER and the nitrogen/nitrate reduction to ammonia reactions (NRR) are the most frequently investigated electrochemical transformations as a result of their applicability. We have described these processes in Section 3.1 "Carbon nanomaterial immobilized metal nanocluster." Herein, we briefly introduce ORR, HER, and NRR electrochemical processes occurred on organic ligand-protected atomically precise metal nanocluster electrodes. Wang and coworkers systematically studied ORR process on platinum-group metal nanocluster electrodes [46, 47]. Notably, the synthesis of platinum-group metal nanoclusters has been recognized as a major challenge. By employing a mixed ligand protecting strategy and modifying the reduction process, Wang and coworkers successfully synthesized the $Pd_8B(PPh_2)_3(PPh_3)_2(2,4\text{-DMBT})_6$ and $[Pt_{23}(XylNC)_{20}(CO)_5Cl_2]^{2+}$ nanoclusters, where XylNC represents 2,6-dimethylphenyl isocyanide. The platinum-group metal nanoclusters demonstrate high activity toward ORR not only under acidic reaction conditions but also in basic reaction conditions. For example, based on the CV curves, the electrochemical activation surface areas of $[Pt_{23}(XylNC)_{20}(CO)_5Cl_2]^{2+}$ was obtained as 73.3 $m^2\ g^{-1}$ in an $N_2$-saturated 0.1 M $HClO_4$ solution [46]. This value is higher than that (86.5 $m^2\ g^{-1}$) of commercially available Pt/C catalyst. The peak potential (0.75 V) of the $[Pt_{23}(XylNC)_{20}(CO)_5Cl_2]^{2+}$ nanocluster is also more positive than that (0.72 V) of the Pt/C catalyst. the specific activity of $[Pt_{23}(XylNC)_{20}(CO)_5Cl_2]^{2+}$ was determined to be 1.04 mA $cm^{-2}$, which was significantly higher than that

(0.13 mA $cm^{-2}$) of the Pt/C catalyst. In addition, the mass activity of $[Pt_{23}(XylNC)_{20}(CO)_5Cl_2]^{2+}$ at 0.9 V also increased to 0.9 A $mg^{-1}$, which was eight times higher than that (0.1 A $mg^{-1}$) of the Pt/C catalyst under the same conditions. The high electrocatalytic activity of the $[Pt_{23}(XylNC)_{20}(CO)_5Cl_2]^{2+}$ nanocluster should be ascribed to its significantly reduced absorption energy of oxygen intermediates. The d band center of $[Pt_{23}(XylNC)_{20}(CO)_5Cl_2]^{2+}$ was calculated to be 3.014 eV, which is lower than that (2.423 eV) of bulk Pt, indicating that the oxygen-reduction-intermediate adsorption of the Pt nanocluster is weaker than that of bulk Pt. This electronically structural feature is responsible for the outstanding catalytic ORR performance of the platinum-group metal nanocluster electrodes in acidic reaction conditions. Under basic reaction conditions of in $O_2$-saturated 0.1 M KOH solution, the platinum-group metal nanocluster $Pd_8B(PPh_2)_3(PPh_3)_2(2,4\text{-DMBT})_6$ exhibited the specific activity of −2.72 mA $cm^{-2}$ at 0.9 V for the ORR process [47]. This value is also higher than that (1.61 mA $cm^{-2}$) of commercially available Pt/C catalyst. Among the investigated Pd catalysts, including Pd complexes, Pd nanoparticles and the commercially available Pt/C catalyst, the Tafel slope for the $Pd_8B(PPh_2)_3(PPh_3)_2(2,4\text{-DMBT})_6$ nanocluster was determined to be the smallest as 66.5 mV $dec^{-1}$, suggesting the fastest ORR kinetics on the surface of the platinum-group metal nanocluster. The high performance of the $Pd_8B(PPh_2)_3(PPh_3)_2(2,4\text{-DMBT})_6$ nanocluster in ORR can also be explained by the reduced absorption energy of oxygen intermediates, which is the rate-determining step.

Li and coworkers investigated the HER activity of metal nanocluster electrodes [48]. Although crystalline gold and silver including the ligand-protected Au or Ag nanoparticles are catalytically inert toward the HER process, Rongchao Jin designed and synthesized a $Au_{36}Ag_2(SR)_{18}$ nanocluster, where the RSH is $C_{10}H_{15}SH$. This nanocluster possesses a low coverage of ligands and a trimeric structure composed of three icosahedral units. The surface activation sites trigger high electrocatalytic activity of $Au_{36}Ag_2(SR)_{18}$ in HER at lower overpotentials. The current density of $Au_{36}Ag_2(SR)_{18}$ at −0.3 V (vs. RHE) is 3.8, which is 5.1 times higher than that of $Au_{25}(SR)_{18}$ and $Au_{38}(SR)_{24}$ nanoclusters. Apart from the low ligand-to-metal ratio, low-coordinated Au atoms and unfilled superatomic orbitals, density functional theory (DFT) calculations also reveal that the lower hydrogen binding energy and higher electron affinity of $Au_{36}Ag_2(SR)_{18}$ are responsible for an energetically feasible HER pathway.

Han and coworkers prepared the nickel-doped gold and silver nanoclusters, namely the $M_4Ni_2(SR)_8$ nanoclusters, where M = Au/Ag, SR = 2,4-dimethylthiophenol [49]. Compared to the pure gold or silver nanoclusters with the similar composition and structures, $Au_4Ni_2(SR)_8$ and $Ag_4Ni_2(SR)_8$ electrodes exhibit improved electrochemical nitrogen reduction reaction activity, probably due to the fact that introduction of the nickel atoms significantly suppresses the hydrogen gas generation competition reaction. Meanwhile, owing to the higher $N_2$ chemisorption ability and lower rate-determining barrier, $Ag_4Ni_2(SR)_8$ exhibit the better NRR activity at −0.2 V (vs. RHE) with a $NH_3$ yield of 23.32μg$mg^{-1}h^{-1}$ and a faradic efficiency of 78.97%.

Nitrate is another important nitrogen source apart from $N_2$ for the production of $NH_3$. It has a low N=O bond dissociation energy of 204 kJ $mol^{-1}$, which is significantly lower than that (946 kJ/mol) of the N≡N bond in $N_2$. Meanwhile it is a commonly environmental pollutant and has high solubility in water. Therefore, using nitrate as the nitrogen source for electrochemical $NH_3$ synthesis meets the requirement of energy economics and environmental restoration. Liu and coworkers made great contribution in this research area by applying metal nanocluster electrocatalysis. They prepared a $[Ag_9(mba)_8H_8]^+$ nanocluster/$Ti_3C_2$ MXene mixed material ($H_2$mba = thiosalicylic acid), and loaded this material on carbon paper to form a working electrode. This kind of metal nanocluster/MXene-based heterogeneous electrode enables a promising catalytic performance for the nitrate reduction toward the production of $NH_3$, with $FE_{NH4+}$ of 80.2% at −0.95 V [50]. Only a small amount of byproduct $NO_2^-$ was observed. Moreover, stability tests conducted at −0.95 V demonstrated that the current density performed no decay after 108 hours for the $[Ag_9(mba)_8H_8]^+$ nanocluster/$Ti_3C_2$ MXene electrode, indicating the structural stability and catalytic durability of this material. A series of in situ characterizations revealed that the interaction between the composite structure of $[Ag_9(mba)_8H_8]^+$ nanocluster and $Ti_3C_2$ MXene played important roles in affecting the rate-determining step of the nitrate electrochemical reduction reaction. $[Ag_9(mba)_8H_8]^+$ is conducive to promoting the conversion of $NO_3^-$ to $NO_2^-$, and the enriched and covered $NO_2^-$ on the active centers of MXene promotes the sustained reaction of $NO_2^-$ to $NH_4^+$. The accumulation of the key intermediate *NO on the $[Ag_9(mba)_8H_8]^+$ nanocluster/$Ti_3C_2$ MXene is significantly weakened, resulting in the increased selectivity and Faraday efficiency of the tandem process from nitrate to $NH_3$ (Figure 9.11).

To further reveal the impact of microenvironment toward the electrocatalytic performance of the metal kernel, Chen and coworkers synthesized a series of copper nanoclusters and investigated their structure-based catalytic performance in nitrate reduction reactions. They synthesized two structurally similar $Cu_4$ nanoclusters, that is, $Cu_4(BTT)_4$ (BTT = $C_{10}H_{10}NS_2$) and $Cu_4(PTI)_4$ (PTI = $C_{12}H_{13}N_2OS$). The two $Cu_4$ nanoclusters have the identical metallic kernel structures but distinct ligand architectures (Figure 9.12a,b) [51]. Especially, $Cu_4(PTI)_4$ bears the stronger electron-withdrawing ligands on the surface than $Cu_4(BTT)_4$. This causes their different HOMO–LUMO

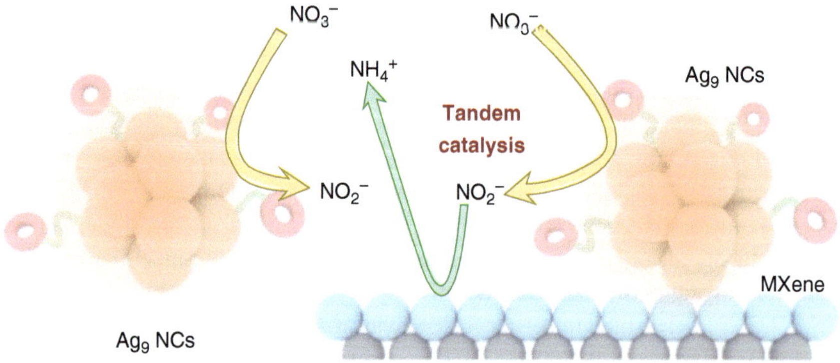


**Figure 9.11** $[Ag_9(mba)_8H_8]^+$ nanocluster/$Ti_3C_2$ MXene composite system for highly efficient nitrate reduction. *Source:* [50]/with permission of John Wiley & Sons.

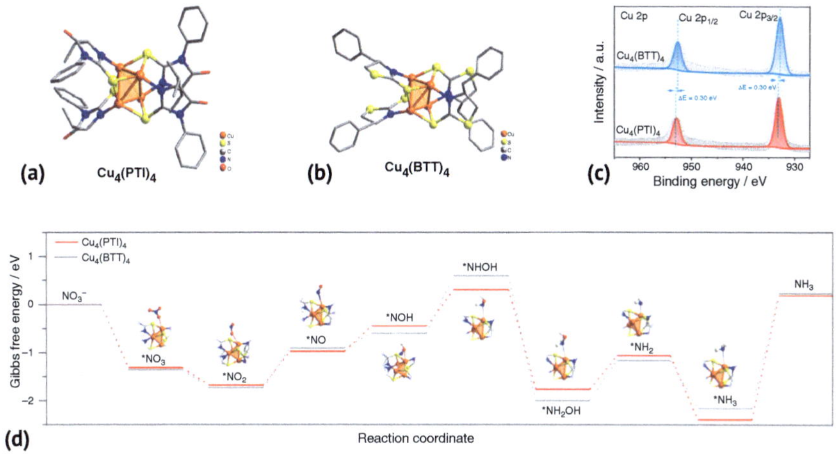


**Figure 9.12** (a, b) Atomically precise structures of $Cu_4(BTT)_4$ and $Cu_4(PTI)_4$ nanoclusters. (c) Cu 2p XPS spectra of $Cu_4(BTT)_4$ and $Cu_4(PTI)_4$ nanoclusters. (d) Proposed catalytic pathways of $Cu_4(BTT)_4$ and $Cu_4(PTI)_4$ nanoclusters in electrochemical nitrate reduction. *Source:* [51]/with permission of John Wiley & Sons.

gaps, absorption characteristics, and even slightly different valence states of the copper atoms in the metal kernels (Figure 9.12c). These structural features result in their distinct catalytic activity in electrochemical nitrate reduction. $Cu_4(PTI)_4$ gave nearly 100% FE toward the production of $NH_3$ at optimal potentials, while $Cu_4(BTT)_4$ only demonstrated a maximum FE of 75%. Experimental results combined calculations revealed that the surface ligands significantly affected the electronic structures of both $Cu_4$ nanoclusters. $Cu_4(PTI)_4$ with the more electron-withdrawing ligands on the surface is beneficial for adsorption of $NH_2OH$, which is a crucial intermediate during the ammonia synthesis process. Meanwhile, the rate-determining step, namely the transformation from $^*NOH$ to $^*NH_2OH$ bears a lower energy barrier for $Cu_4(PTI)_4$ catalysis compared to $Cu_4(BTT)_4$ catalysis (Figure 9.12d).

Even bearing the same organic ligands on the surface, three structurally similar copper nanoclusters, that is, $CuI_2(BINAP)_2$, $CuBr_2(BINAP)_2$, and $CuCl_2(BINAP)_2$, show significantly different catalytic performance in electrochemical nitrate reduction. This interesting phenomenon was revealed recently by Zheng and coworkers [52]. They conducted charge density analysis and found that the halides (i.e. Cl, Br, I) are negatively charged whereas the copper atoms are positively charged. Meanwhile, as a result of the higher electronegativity of Cl, a large electron cloud exists around Cl compared to that of I and Br. In other words, more charge transfer between Cu and Cl occurs compared to Cu–Br and Cu–I. This kind of internal charges redistribution originated from electronegative halides on the surface of the dimetallic copper nanoclusters was found to be beneficial for their electrocatalytic nitrate reduction. Especially, 94.0% of $FE_{NH4+}$ at −0.946 V was achieved for $CuCl_2(BINAP)_2$, which is significantly higher than that of $CuI_2(BINAP)_2$ and $CuBr_2(BINAP)_2$. In situ experimental results combined with theoretical calculations revealed that the charge transfer from copper atoms to halide atoms benefits the approach of alkali metal-ionized water to the surface of the nanocluster catalysts and subsequent dissociation process. Therefore, localized H* intermediates are enriched on the surface of the nanocluster electrodes, promoting the sequential electrocatalytic hydrogenation processes of nitrates shown in Figure 9.13.

After the detailed investigation on atomically precise metal nanocluster-catalyzed electrochemical nitrate reduction, Wei and coworkers developed a practical method for the production of ammonia under industrial-level current density and very low overpotential in a neutral medium [53]. They designed a copper foam (CF) immobilized CoNi bimetallic electrode with rich metal atom-terminated sites (MAT-CoNi/CF) based on the previous results obtained from metal nanocluster electrocatalysis and the further guidance of theoretical calculation. The maintained metal atom-terminated surface through interaction with copper atoms was confirmed to play important roles in reducing overpotential and achieving high current density during electrocatalytic nitrate reduction processes. A nearly 100% FE of ammonia was achieved at low overpotential (−0.15 V) and industrially relevant current density. Owing to the advantage of the reaction conduced under neutral conditions, a home-made gas stripping and absorption device was constructed and the complete conversion process for high-purity $NH_4NO_3$ was realized (Figure 9.14). This work suggests the practical application of the electrocatalytic nitrate reduction based on metal nanoclusters and their analogous electrodes.

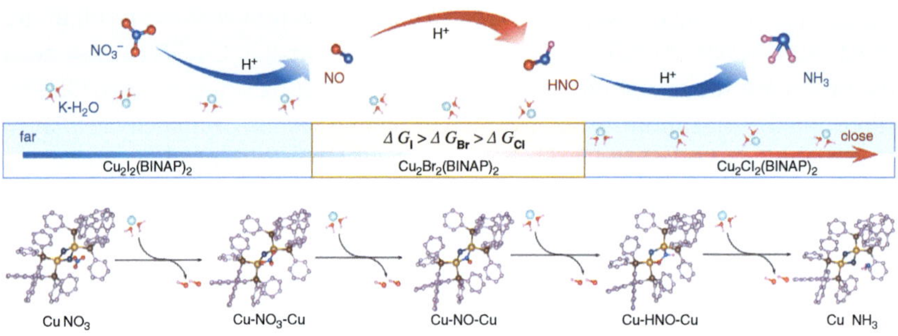


**Figure 9.13** The sequential electrocatalytic hydrogenation processes of nitrate on the surface of $CuI_2(BINAP)_2$, $CuBr_2(BINAP)_2$ and $CuCl_2(BINAP)_2$, respectively. *Source:* [52]/with permission of John Wiley & Sons.

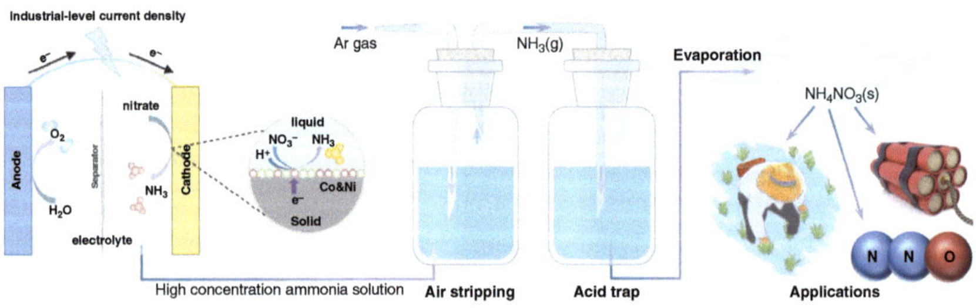


**Figure 9.14** Nitrate electrochemical reduction for the production of ammonia under industrial-level current density and very low overpotential in a neutral medium that was enabled by a MAT-CoNi/CF electrode. *Source:* [53]/with permission of John Wiley & Sons.

In addition to Shuang-Quan Zang, Jinmeng Cai, et al., other research groups made progress in the research area of metal nanocluster-catalyzed nitrate/nitrite electrochemical reduction. Dong and coworkers prepared a N-heterocyclic carbene-stabilized $Cu_3$ nanocluster, which features the mixed valence states of the copper element [54]. SCXRD revealed the atomically precise structure of this copper nanocluster and the featured Cu–Cu bond that is shorter than the typically observed mixed-valence copper complexes. The obvious Cu–Cu interaction triggers electron delocalization across the metal centers. Detailed characterizations indicate that the mixed valence states of the copper element and the as-revealed Robin-Day Class-III mixed-valence system are very crucial for the electrocatalytic activity of the $Cu_3$ nanocluster (Figure 9.15). The terminal two Cu centers are confirmed to serve as the activation sites for the activation of nitrite and the redox ability of the central copper atom plays important roles in the electrocatalysis. Based on this unique catalytic system, the $NO_2^-$ was selectively reduced to $NH_3$ with high Faradaic efficiency up to 95%. Meanwhile, the hydrogen evolution process was significantly suppressed under the $Cu_3$ catalytic conditions.

By immobilization the $Ni_6(SC_2H_4COOH)_{12}$ nanocluster on the surface of a layered doule hydroxides (CuFe-LDH) support, Gu and coworkers prepared a composite metal nanocluster-based electrode $Ni_6$@CuFe-LDH, which catalyzes the conversion from nitrate to ammonia with Faraday efficiency of 97% and the $NH_3$ productivity of 0.91 mmol $mg^{-1}$ $h^{-1}$. The catalytic performance is much higher than the separate $Ni_6(SC_2H_4COOH)_{12}$ nanocluster or CuFe-LDH support [55]. Mechanism studies revealed that the high electrocatalytic activity of the composite material should be ascribed to the synergy between the $Ni_6(SC_2H_4COOH)_{12}$ nanocluster and the CuFe-LDH support, which alters the rete-determining step of the $^*HN_3$ intermediate

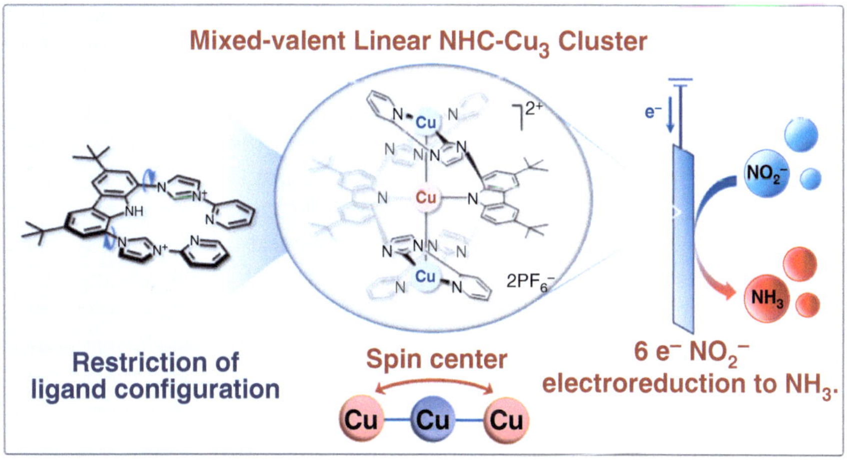


**Figure 9.15** A N-heterocyclic carbene-stabilized $Cu_3$ nanocluster with Robin-Day Class-III mixed-valence system triggers efficient nitrite reduction toward the generation of ammonia. *Source:* [54]/with permission of American Chemical Society.

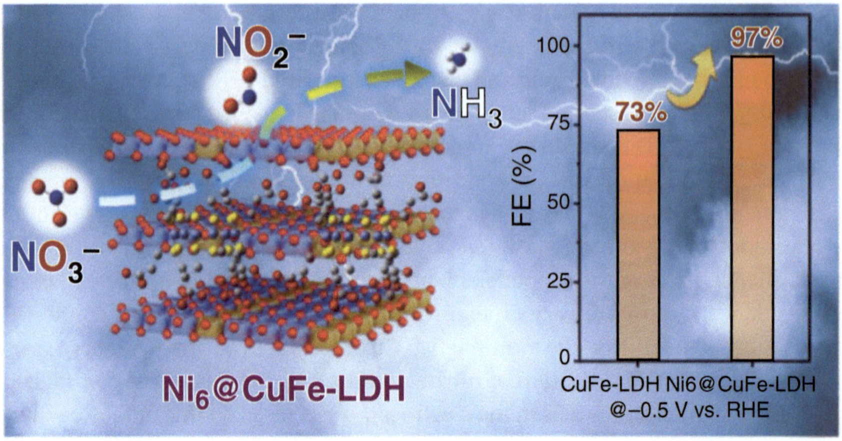


**Figure 9.16** Metal nanocluster-based electrode $Ni_6$@CuFe-LDH for highly efficient nitrate electrochemical reduction. *Source:* [55]/American Chemical Society/CC BY 4.0.

desorption to the transformation from $*NO_3$ to $*NO_2$ (Figure 9.16). The switch on the rate-determining step of the nitrate electrochemical reduction promotes the conversion of the key intermediate nitrite to ammonia as the deeply hydrogenated product.

Combining the aforementioned metal nanocluster-catalyzed $CO_2$ electrochemical reduction process and $NO_3^-$ electrochemical reduction process, Chen and coworkers realized an interesting urea electrochemical synthesis [56]. The icosahedral $M_{13}$-kernelled $Ag_{14}Pd(PTFE)_6(TPP)_8$ and $Ag_{13}Au_5(PTFE)_{10}(DPPP)_4$ nanoclusters (PTFE = pentafluorothiophenol, TPP = triphenylphosphine, DPPP = 1,3-bis(diphenylphosphino)propane) were applied for the featured electrochemical transformation involving a crucial C–N coupling process (Figure 9.17). The bimetallic structures of both nanoclusters induce charge polarization at the activation sites, thus triggering the formation of the key intermediates of *NOH and *COOH at the rate-determining steps. A subsequential C–N coupling between the two key intermediates furnishes the final $CO(NH_2)_2$ with the production rates of $Ag_{14}Pd(PTFE)_6(TPP)_8$ and $Ag_{13}Au_5(PTFE)_{10}(DPPP)_4$ at the maximum Faradaic efficiency of 143.3 and 82.3 mg $h^{-1}g_{cat}^{-1}$, respectively.

Apart from catalyzing small molecules electrochemical transformations, metal nanoclusters were employed for organic synthesis as well. Zhang and coworkers demonstrated such electrochemical transformation recently [57]. They synthesized a 6-mercaptohexanoic acid (MHA) ligand protected silver nanocluster, namely $Ag_{25}(MHA)_{18}$ and prepared the metal nanocluster-based electrode. The composition and structure of the metal nanocluster were verified by several characterizations such as ESI-MS, XPS, etc. In situ analysis revealed that the featured structure of $Ag_{25}(MHA)_{18}$ significantly regulated its catalytic performance in electrochemical semi-hydrogenation of alkynes. The regulation can be summarized into four points (Figure 9.18). (i) The amphophilic MHA ligands on the surface of the nanocluster are able to preconcentrate the water and alkyne molecules near the interface of the

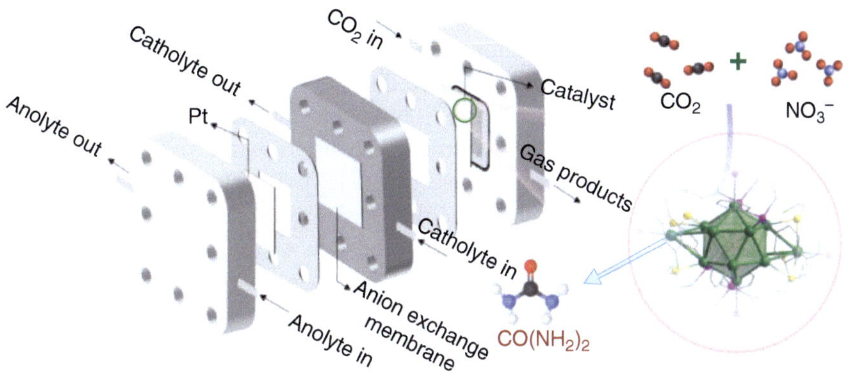


**Figure 9.17** The icosahedral $M_{13}$-kernelled $Ag_{14}Pd(PTFE)_6(TPP)_8$ and $Ag_{13}Au_5(PTFE)_{10}(DPPP)_4$ nanoclusters for featured $CO_2$ and $NO_3^-$ electrochemical transformation involving a crucial C–N coupling process. *Source:* [56]/Oxford University Press/CC BY 4.0.

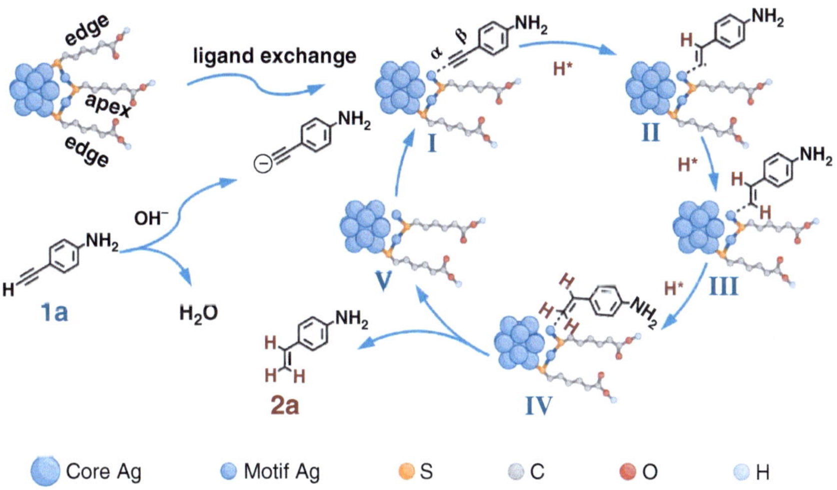


**Figure 9.18** $Ag_{25}(MHA)_{18}$-catalyzed semi-hydrogenation of alkynes under electrolysis. *Source:* [57]/with permission of John Wiley & Sons.

nanocluster catalysts. (ii) The long-chain ligands are beneficial to the disruption of the hydrogen bond network of substrates at the interface. (iii) The negatively charged silver kernel of $Ag_{25}(MHA)_{18}$ can attract the weakly hydrogen-bonded water molecules via counterions and promote the formation of active hydrogen intermediates (H*). (iv) The enzyme-like pockets on the surface of $Ag_{25}(MHA)_{18}$ can recognize the alkyne molecules via σ-bonding, thereby activating alkynes and triggering the further semi-hydrogenation process in the presence of H*. Bearing these structural advantages, the $Ag_{25}(MHA)_{18}$ electrode efficiently catalyzes semi-hydrogenation of alkynes under electrolysis, affording alkenes with 98% conversion, 99% selectivity, and 85% FE in a broad substrate pool.

It should be noted that due to the sensitivity of the metal nanoclusters toward the electrolysis conditions, sometimes the metal nanoclusters face the problem of structural aggregation or decomposition during electrocatalysis. To further stabilize the catalytically active metal nanocluster electrodes, Wang and coworkers developed an electropolymerization strategy to improve the catalytic efficiency and durability of metal nanocluster electrodes [58]. They designed and synthesized a series of carbazole-based ligand-protected metal nanoclusters, that is, the $[Cu_{14}(cba)_6(S^tBu)_8]$, $[Cu_6Au_6(cbz)_6(S^tBu)_6]$, and $[Cu_6Ag_4(cbz)_{10}(PPN)_2]$, in which Hcba is 5-(9*H*-carbazol-9-yl)pentanoic acid, Hcbz is 9-(prop-2-yn-1-yl)-9*H*-carbazole, and PPN is 2,6-bis(diphenylphosphino)pyridine. The carbazole units can serve as polymeric active sites for electropolymerization of these metal nanoclusters. All three nanoclusters were polymerized into hybrid materials by the electropolymerization strategy. Taking the electropolymerization of $[Cu_{14}(cba)_6(S^tBu)_8]$ as an example, $[Cu_{14}(cba)_6(S^tBu)_8]$ was carried out in dichloromethane by using 0.1 M tetrabutylammonium tetrafluoroborate (TBTA) as electrolyte via multicycled CV methods. This metal nanocluster polymerized on various working electrodes, including indium-doped tin oxide (ITO) glass, carbon paper (CP), and nickel foam (NF), to form the metal nanocluster electrodes. The electropolymerization strategy improve the metal nanocluster electrode from two aspects. (i) Electropolymerization greatly enhances catalytic efficiency because the polymerized extended network facilitates electron transport to the metal nanocluster catalytic sites. (ii) Electropolymerization highly enhances catalytic performance because the polymerized extended network effectively protects the metal nanoclusters from aggregation or decomposition during the electrocatalytic process. Owing to the two factors, the polymerized metal nanoclusters showed dramatically improved electrocatalytic performance than the isolated ones. As a representative, the polymerized $[Cu_{14}(cba)_6(S^tBu)_8]$ showed about fourfold $NH_3$ yield rate and about twofold Faraday efficiency enhancement compared to the isolated unpolymerized $[Cu_{14}(cba)_6(S^tBu)_8]$ nanocluster in the electroreduction process from nitrate to ammonia. Moreover, the polymerized $[Cu_{14}(cba)_6(S^tBu)_8]$ material can be well recovered and recycled, thus demonstrating good durability. Figure 9.19 illustrates the significant improvement of charge transfer ability of the metal nanocluster-based polymers, which is mainly ascribed to the formation of the biscarbazole network. For the discrete metal nanocluster, the catalytic sites are only electro-activated at or near the surface of a conductive electrode.

The last example we introduce in this section is a metal complex-anchored metal nanocluster electrode (Figure 9.20). This elegant work comes from Zhao and coworkers [59]. They successfully anchored the metal complex on $Au_{25}(SR)_{18}$ nanocluster by a ligand exchange method (for detailed description of the ligand exchange method, see Section 4.1). Specifically, $Au_{25}(SR)_{18}$ nanocluster reacts with the prepared functional ligand HSR', in which the foreign metal such as Ni, Ru, Fe, and Co was located on the ligand HSR' in advance. The ligand exchange process underwent smoothly, producing the metal complex-anchored metal nanocluster $Au_{25}(SR)_{17}$-(M complex) efficiently. ESI-MS spectra confirmed the successful anchoring of the

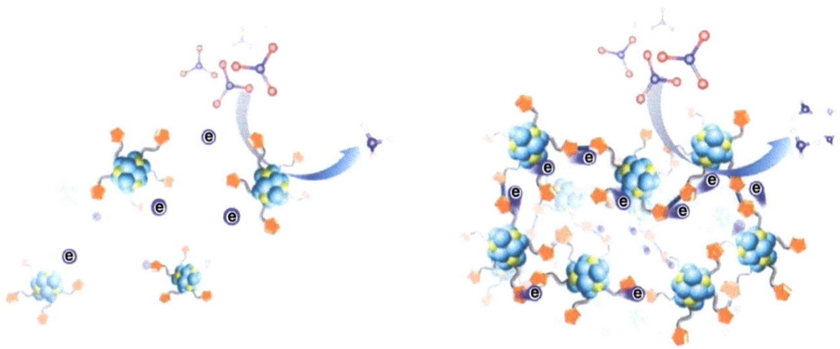

Figure 9.19 Electropolymerization strategy for improving the catalytic efficiency and durability of metal nanocluster electrodes. *Source:* [58]/with permission of John Wiley & Sons.

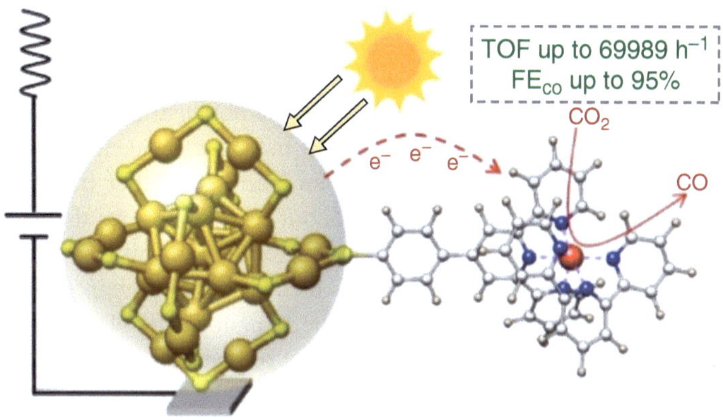


Figure 9.20 Metal complex-anchored $Au_{25}(SR)_{17}$-(M complex) nanocluster electrode for efficient photocoupled electrocatalytic $CO_2$ reduction. *Source:* [59]/with permission of John Wiley & Sons.

metal complex on the surface of the gold nanocluster. The $Au_{25}(SR)_{17}$-(M complex) nanocluster with modified surface contains both photo-synthesizer (the gold kernel) and accessible activation center (the metal complex), which was employed as a promising catalysts for $CO_2$ reduction. The Faradaic efficiency of CO exceeds 90% in a wide potential range by using the $Au_{25}(SR)_{17}$-(M complex) nanocluster electrode. This value is two times higher than that of $Au_{25}(SR)_{18}$ nanocluster electrode. $Au_{25}(SR)_{17}$-(M complex) demonstrated extremely large turnover frequencies (TOF = 63,012 $h^{-1}$ at −0.97 V for $Au_{25}(SR)_{17}$-(Ru complex), and TOF = 69,989 $h^{-1}$ at −1.07 V for $Au_{25}(SR)_{17}$-(Ni complex)), which is also much higher than that of the $Au_{25}(SR)_{18}$ nanocluster. Moreover, stability test indicates the excellent long-term stability of the metal complex-anchored metal nanocluster electrode in contrast with the unmodified $Au_{25}(SR)_{18}$ nanocluster electrode.

## 9.3 Metal Nanocluster Homogeneous Electrocatalyst

Employment of metal nanoclusters as electrocatalysts that are dispersed in the electrolyte is rarely reported, because it requires a highly stable structure of the metal nanocluster to avoid any structural decomposition under the reaction conditions. Herein, we list two examples.

Hu and coworkers developed an efficient Pt-based electrocatalyst from a commercially available $PtCl_2$ salt. A 3.0 mol% of $PtCl_2$ triggers the electrohydrogenation of olefins at 300 mA, giving the hydrogenated product in 97% yield with current density of 133 mA $cm^{-2}$ and Faradaic efficiency of 48% [60]. This catalytic system is highly efficient, and various conjugated or unconjugated olefins are smoothly transformed to the hydrogenated products quantitatively. Mechanism studies suggested that ultra-small-sized Pt nanoclusters were formed during the electrocatalysis process. Qingquan Lu et al. confirmed the existence of Pt nanoclusters by several experiments. First, the electrohydrogenation reaction was completely shut down in the mercury poisoning test, ligand poisoning test or hot filtration tests. These control experiments suggest the formation of Pt aggregates during the reaction. Second, the electrolyte solution on the cathode were then collected and sent for transmission electron microscopy characterization, which give the direct evidence of the existence of Pt nanoclusters with average size less than 2 nm. The lattice fringe of as-obtained Pt nanoclusters was determined to be about 0.23 nm. Third, the XPS analysis also confirmed the formation of Pt(0) species, which should be attributed to the reduction of Pt(II) salts on the cathode and their further aggregation into catalytically active Pt nanoclusters. The high activity of the Pt nanocluster electrocatalyst can be explained by their capacity for the adsorption of hydrides, which is the crucial species for the hydrogenation of olefins. The $PtCl_2$ catalyst precursors were firstly reduced and aggregated to Pt nanoclusters at a low reductive potential on the cathode. Subsequentially, the in situ-generated Pt nanoclusters interact with the protons and electrons, triggering the formation of Pt-H species. This is the real catalytic species and catalyzes the hydrogenation of olefins, affording the hydrogenated products that adsorbed on the surface of the Pt nanoclusters. Finally, desorption of the products regenerates the Pt nanocluster electrocatalyst that can be involved in the next catalytic cycle (Figure 9.21).

Notably, the in situ generated Pt nanocluster-catalyzed olefins electrohydrogenation process can be switched to a homogeneous Pt-complex electrocatalysis by the addition of organic ligands. The authors found that 2,2′-bipyridine coordinated with the $PtCl_2$ precursors effectively, giving the corresponding Pt complexes that also catalyze the hydrogenation of olefins under electrolysis conditions. Although the catalytic yields are slightly lower, this catalytic process demonstrates high selectivity toward specific olefin substrates. Thus, by selectively using 2,2′-bipyridine, the authors can switch the electrocatalytic processes, affording desired hydrogenated product via the in situ generated Pt nanocluster electrocatalysis or the Pt complex electrocatalysis.

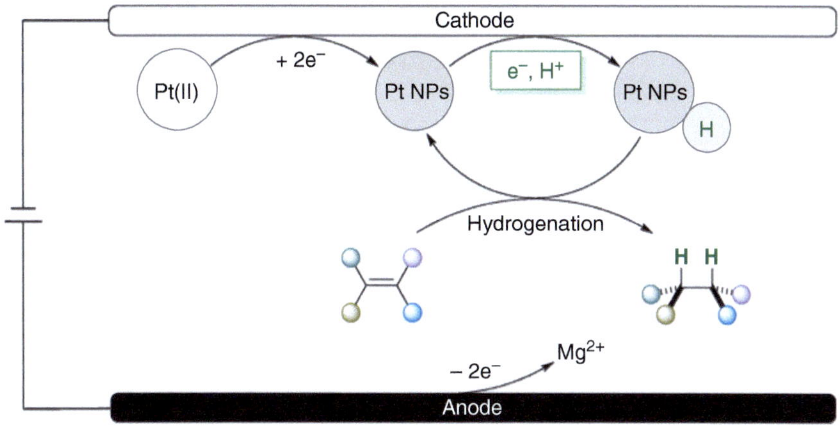


**Figure 9.21** Catalytic mechanism of the in situ-generated Pt nanoclusters-catalyzed efficient electrohydrogenation of olefins. *Source:* [49]/with permission of John Wiley & Sons.

Another example on using the metal nanoclusters as electrocatalysts is reported by Liang and coworkers [61]. They prepared a binuclear cobalt nanocluster electrocatalyst by using amino acids as the protecting ligands (Figure 9.22). The molecular formula of the Co nanocluster is determined to be $Co_2(OH)_2$(amino acid)$_4$, where the amino acid can be selectively changed to give different $Co_2$ nanoclusters. Their structures are basically the same, consisting of a kernel in which two cobalt atoms are bridged by two OH groups. Each cobalt atom is surrounded by two amino acids, forming the total structure of the $Co_2(OH)_2$(amino acid)$_4$ nanocluster. Owing to the fact that the hydrogen atoms in the nitrogen sites still exist while the hydrogen atoms in the acid sites are removed, each cobalt atom should be +3 charged.

This nanocluster can be well-dissolved in DMF, and catalyze the selective C-H functionalization of allenes under electrolysis conditions (Figure 9.22). Specifically, 1,3-disubstituted allenes react with alcohols in the presence of the $Co_2(OH)_2$ (amino acid)$_4$ nanocluster electrocatalyst, giving product enallenols efficiently under 20 mA electrolysis. The C—H functionalization of allenes is highly selective. Only the C—H bond near the aryl substituent is functionalized, while the C—H bond near the alkyl substituent is kept. This selectivity should be attributed to the slightly different acidity of the two C—H sites, which can be well recognized by the $Co_2(OH)_2$(amino acid)$_4$ nanocluster electrocatalyst.

With the development of the synthetic chemistry on metal nanoclusters, it is believed that more and more metal nanocluster electrocatalysts will be discovered in the future. Stability of this kind electrocatalyst during reaction is a major challenge because it is always in the electrolyte solution. Organic ligand-protected metal nanoclusters would be a suitable candidate, because they are the one hand highly dissolvable in many solvents and on the other hand demonstrate abundant charge states that enable efficient electron transfers between the metal nanocluster electrocatalysts and the substrates.

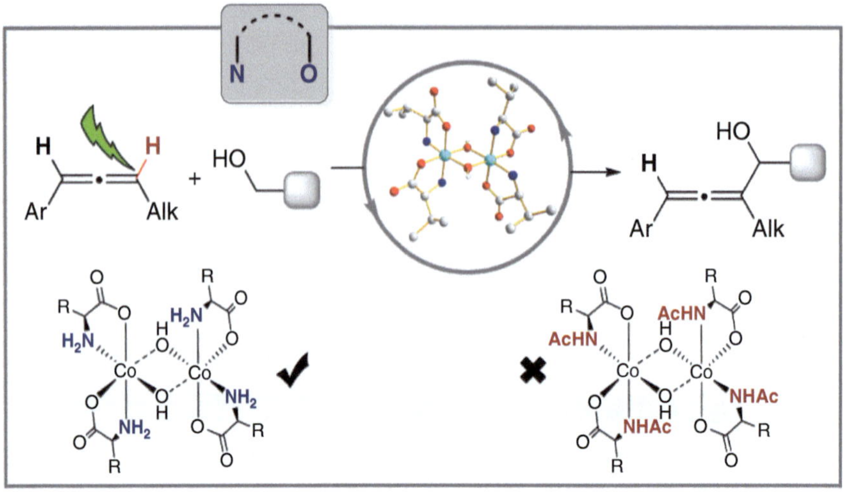


**Figure 9.22** Binuclear cobalt nanocluster electrocatalyst and its catalytic performance in selective C–H functionalization of allenes.

## References


**1** Q. Liu, X. Wang, X. Wang, Sub-1 nm Materials Chemistry: Challenges and Prospects. *J. Am. Chem. Soc.* **2024**, *146*, 26587–26602.

**2** P. Yang, U. Kortz, Discovery and Evolution of Polyoxopalladates. *Acc. Chem. Res.* **2018**, *51*, 1599–1608.

**3** A. W. Cook, T. W. Hayton, Case Studies in Nanocluster Synthesis and Characterization: Challenges and Opportunities. *Acc. Chem. Res.* **2018**, *51*, 2456–2464.

**4** X. Kang, H. Chong, M. Zhu, $Au_{25}(SR)_{18}$: The Captain of the Great Nanocluster Ship. *Nanoscale* **2018**, *10*, 10758–10834.

**5** M. Agrachev, M. Ruzzi, A. Venzo, F. Maran, Nuclear and Electron Magnetic Resonance Spectroscopies of Atomically Precise Gold Nanoclusters. *Acc. Chem. Res.* **2019**, *52*, 44–52.

**6** K. Kwak, D. Lee, Electrochemistry of Atomically Precise Metal Nanoclusters. *Acc. Chem. Res.* **2019**, *52*, 12–22.

**7** K. Kwak, V. D. Thanthirige, K. Pyo, D. Lee, G. Ramakrishna, Energy Gap Law for Exciton Dynamics in Gold Cluster Molecules. *J. Phys. Chem. Lett.* **2017**, *8*, 4898–4905.

**8** M. Zhou, X. Du, H. Wang, R. Jin, The Critical Number of Gold Atoms for a Metallic State Nanocluster: Resolving a Decades-Long Question. *ACS Nano.* **2021**, *15*, 13980–13992.

**9** M. Zhou, C. Zeng, Y. Chen, S. Zhao, M. Y. Sfeir, M. Zhu, R. Jin, Evolution from the Plasmon to Exciton State in Ligand-Protected Atomically Precise Gold Nanoparticles. *Nat. Commun.* **2016**, *7*, 13240.

10 M. Zhou, T. Higaki, G. Hu, M. Y. Sfeir, Y. Chen, D.-E. Jiang, R. Jin, Three-Orders-of-Magnitude Variation of Carrier Lifetimes with Crystal Phase of Gold Nanoclusters. *Science* **2019**, *364*, 279–282.

11 M. Zhou, C. Zeng, Y. Song, J. W. Padelford, G. Wang, M. Y. Sfeir, T. Higaki, R. Jin, On the Non-Metallicity of 2.2 nm $Au_{246}(SR)_{80}$ Nanoclusters. *Angew. Chem. Int. Ed.* **2017**, *56*, 16257–16261; *Angew. Chem.* **2017**, *129*, 16475–16479.

12 M. Zhou, T. Higaki, Y. Li, C. Zeng, Q. Li, M. Y. Sfeir, R. Jin, Three-Stage Evolution from Nonscalable to Scalable Optical Properties of Thiolate-Protected Gold Nanoclusters. *J. Am. Chem. Soc.* **2019**, *141*, 19754–19764.

13 S. Park, D. Lee, Synthesis and Electrochemical and Spectroscopic Characterization of Biicosahedral $Au_{25}$ Clusters. *Langmuir* **2012**, *28*, 7049–7054.

14 Z. Gan, N. Xia, Z. Wu, Discovery, Mechanism, and Application of Antigalvanic Reaction. *Acc. Chem. Res.* **2018**, *51*, 2774–2783.

15 X. Kang, Y. Li, M. Zhu, R. Jin, Atomically Precise Alloy Nanoclusters: Syntheses, Structures, and Properties. *Chem. Soc. Rev.* **2020**, *49*, 6443–6514.

16 A. Ghosh, O. F. Mohammed, O. M. Bakr, Atomic-Level Doping of Metal Clusters. *Acc. Chem. Res.* **2018**, *51*, 3094–3103.

17 S. Hossain, Y. Niihori, L. V. Nair, B. Kumar, W. Kurashige, Y. Negishi, Alloy Clusters: Precise Synthesis and Mixing Effects. *Acc. Chem. Res.* **2018**, *51*, 3114–3124.

18 N. Xia, Z. Wu, Controlling Ultrasmall Gold Nanoparticles with Atomic Precision. *Chem. Sci.* **2021**, *12*, 2368–2380.

19 X. Liu, X. Cai, Y. Zhu, Catalysis Synergism by Atomically Precise Bimetallic Nanoclusters Doped with Heteroatoms. *Acc. Chem. Res.* **2023**, *56*, 1528–1538.

20 G. X. Pei, L. Zhang, X. Sun, Recent Advances of Bimetallic Nanoclusters with Atomic Precision for Catalytic Applications. *Coord. Chem. Rev.* **2024**, *506*, 215692.

21 C. Yao, Y.-J. Lin, J. Yuan, L. Liao, M. Zhu, L.-H. Weng, J. Yang, Z. Wu, Mono-Cadmium vs Mono-Mercury Doping of $Au_{25}$ Nanoclusters. *J. Am. Chem. Soc.* **2015**, *137*, 15350–15353.

22 L. Liao, S. Zhou, Y. Dai, L. Liu, C. Yao, C. Fu, J. Yang, Z. Wu, Mono-Mercury Doping of $Au_{25}$ and the HOMO/LUMO Energies Evaluation Employing Differential Pulse Voltammetry. *J. Am. Chem. Soc.* **2015**, *137*, 9511–9514.

23 K. Kwak, Q. Tang, M. Kim, D.-E. Jiang, D. Lee, Interconversion Between Superatomic 6-Electron and 8-Electron Configurations of $M@Au_{24}(SR)_{18}$ Clusters (M = Pd, Pt). *J. Am. Chem. Soc.* **2015**, *137*, 10833–10840.

24 C. P. Joshi, M. S. Bootharaju, M. J. Alhilaly, O. M. Bakr, $[Ag_{25}(SR)_{18}]^-$: The “Golden” Silver Nanoparticle. *J. Am. Chem. Soc.* **2015**, *137*, 11578–11581.

25 X. Kang, S. Chen, S. Jin, Y. Song, Y. Xu, H. Yu, H. Sheng, M. Zhu, Heteroatom Effects on the Optical and Electrochemical Properties of $Ag_{25}(SR)_{18}$ and Its Dopants. *ChemElectroChem* **2016**, *3*, 1261–1265.

26 H. Qian, W. T. Eckenhoff, Y. Zhu, T. Pintauer, R. Jin, Total Structure Determination of Thiolate-Protected $Au_{38}$ Nanoparticles. *J. Am. Chem. Soc.* **2010**, *132*, 8280–1.

27 M. Kim, Q. Tang, A. V. N. Kumar, K. Kwak, W. Choi, D.-E. Jiang, D. Lee, Dopant-Dependent Electronic Structures Observed for $M_2Au_{36}(SC_6H_{13})_{24}$ Clusters (M = Pt, Pd). *J. Phys. Chem. Lett.* **2018**, *9*, 982–989.

**28** K. Kwak, S. S. Kumar, D. Lee, Selective Determination of Dopamine Using Quantum-Sized Gold Nanoparticles Protected with Charge Selective Ligands. *Nanoscale* **2012**, *4*, 4240–4246.

**29** K. Kwak, D. Lee, Electrochemical Characterization of Water-Soluble $Au_{25}$ Nanoclusters Enabled by Phase-Transfer Reaction. *J. Phys. Chem. Lett.* **2012**, *3*, 2476–2481.

**30** X. Kong, J. Zhu, Z. Xu, Z. Geng, Fundamentals and Challenges of Ligand Modification in Heterogeneous Electrocatalysis. *Angew. Chem. Int. Ed.* **2025**, *64*, e202417562; *Angew. Chem.* **2025**, *137*, e202417562.

**31** S. Zhang, M.-B. Li, Repurposing HER Catalysis toward Metal Hydride-Mediated Electro-Reductive Transformations. *Tetrahedron Chem* **2024**, *11*, 100080.

**32** S. Zhuang, D. Chen, L. Liao, Y. Zhao, N. Xia, W. Zhang, C. Wang, J. Yang, Z. Wu, Hard-Sphere Random Close-Packed $Au_{47}Cd_2(TBBT)_{31}$ Nanoclusters with a Faradaic Efficiency of Up to 96% for Electrocatalytic $CO_2$ Reduction to CO. *Angew. Chem. Int. Ed.* **2020**, *59*, 3073–3077; *Angew. Chem.* **2020**, *132*, 3097–3101.

**33** S. Zhuang, D. Chen, W. Fan, J. Yuan, L. Liao, Y. Zhao, J. Li, H. Deng, J. Yang, J. Yang, Z. Wu, Single-Atom-Kernelled Nanocluster Catalyst. *Nano Lett.* **2022**, *22*, 7144–7150.

**34** G. Deng, H. Yun, M. S. Bootharaju, F. Sun, K. Lee, X. Liu, S. Yoo, Q. Tang, Y. J. Hwang, T. Hyeon, Copper Doping Boosts Electrocatalytic $CO_2$ Reduction of Atomically Precise Gold Nanoclusters. *J. Am. Chem. Soc.* **2023**, *145*, 27407–27414.

**35** S. Jin, W. Du, S. Wang, X. Kang, M. Chen, D. Hu, S. Chen, X. Zou, G. Sun, M. Zhu, Thiol-Induced Synthesis of Phosphine-Protected Gold Nanoclusters with Atomic Precision and Controlling the Structure by Ligand/Metal Engineering. *Inorg. Chem.* **2017**, *56*, 11151–11159.

**36** S. Su, Y. Zhou, L. Xiong, S. Jin, Y. Du, M. Zhu, Structure-Activity Relationships of the Structural Analogs $Au_8Cu_1$ and $Au_8Ag_1$ in the Electrocatalytic $CO_2$ Reduction Reaction. *Angew. Chem. Int. Ed.* **2024**, *63*, e202404629; *Angew. Chem.* **2024**, *136*, e202404629.

**37** Z.-H. Gao, K. Wei, T. Wu, J. Dong, D.-E. Jiang, S. Sun, L.-S. Wang, A Heteroleptic Gold Hydride Nanocluster for Efficient and Selective Electrocatalytic Reduction of $CO_2$ to CO. *J. Am. Chem. Soc.* **2022**, *144*, 5258–5262.

**38** L.-J. Liu, Z.-Y. Wang, Z.-Y. Wang, R. Wang, S.-Q. Zang, T. C. W. Mak, Mediating $CO_2$ Electroreduction Activity and Selectivity over Atomically Precise Copper Clusters. *Angew. Chem. Int. Ed.* **2022**, *61*, e202205626; *Angew. Chem.* **2022**, *134*, e202205626.

**39** D. Yang, W. Pei, S. Zhou, J. Zhao, W. Ding, Y. Zhu, Controllable Conversion of $CO_2$ on Non-Metallic Gold Clusters. *Angew. Chem. Int. Ed.* **2020**, *59*, 1919–1924; *Angew. Chem.* **2020**, *132*, 1935–1940.

**40** X. Cai, W. Hu, S. Xu, D. Yang, M. Chen, M. Shu, R. Si, W. Ding, Y. Zhu, Structural Relaxation Enabled by Internal Vacancy Available in a 24-Atom Gold Cluster Reinforces Catalytic Reactivity. *J. Am. Chem. Soc.* **2020**, *142*, 4141–4153.

**41** S. Zhao, N. Austin, M. Li, Y. Song, S. D. House, S. Bernhard, J. C. Yang, G. Mpourmpakis, R. Jin, Influence of Atomic-Level Morphology on Catalysis: The Case of Sphere and Rod-Like Gold Nanoclusters for $CO_2$ Electroreduction. *ACS Catal.* **2018**, *8*, 4996–5001.

**42** H. Seong, Y. Jo, V. Efremov, Y. Kim, S. Park, S. M. Han, K. Chang, J. Park, W. Choi, W. Kim, C. H. Choi, J. S. Yoo, D. Lee, Transplanting Gold Active Sites into Non-Precious-Metal Nanoclusters for Efficient $CO_2$-to-CO Electroreduction. *J. Am. Chem. Soc.* **2023**, *145*, 2152–2160.

**43** Y. Tan, G. Sun, T. Jiang, D. Liu, Q. Li, S. Yang, J. Chai, S. Gao, H. Y, M. Zhu, Symmetry Breaking Enhancing the Activity of Electrocatalytic $CO_2$ Reduction on an Icosahedron-Kernel Cluster by Cu Atoms Regulation. *Angew. Chem. Int. Ed.* **2024**, *63*, e202317471.

**44** X.-Y. Dong, H. Chen, S. Wang, R.-Y. Zou, S.-Q. Zang, J. Cai, Introducing La into a Customized Dual Cu Covalent Organic Framework to Steer $CO_2$ Electroreduction Selectivity from $C_2H_4$ to $CH_4$. *Adv. Mater.* **2025**, *37*, 2413710.

**45** S. Biswas, Y. Shingyouchi, M. Kamiyama, M. Ogami, H. Song, B. Li, S. Wang, T. Kawawaki, D.-e. Jiang, Y. Negishi, Atomically Precise $[Cu_{23}H_4(SC_7H_7)_{18}(PPh_3)_6]$ Nanocluster: Structural Integration of Johnson Solids through a Cu(0) Center and Electrocatalytic Functionality. *J. Am. Chem. Soc.* **2025**, *147*, 23733–23742.

**46** R. Wang, D. Chen, L. Fang, W. Fan, Q. You, G. Bian, Y. Zhou, W. Gu, C. Wang, L. Bai, J. Li, H. Deng, L. Liao, J. Yang, Z. Wu, Atomically Precise Nanometer-Sized Pt Catalysts with an Additional Photothermy Functionality. *Angew. Chem. Int. Ed.* **2024**, *63*, e202402565; *Angew. Chem.* **2024**, *136*, e202402565.

**47** S. Zhuang, D. Chen, Q. You, W. Fan, J. Yang, Z. Wu, Thiolated, Reduced Palladium Nanoclusters with Resolved Structures for the Electrocatalytic Reduction of Oxygen. *Angew. Chem. Int. Ed.* **2022**, *61*, e202208751; *Angew. Chem.* **2022**, *134*, e202208751.

**48** Y. Li, S. Li, A. V. Nagarajan, Z. Liu, S. Nevins, Y. Song, G. Mpourmpakis, R. Jin, Hydrogen Evolution Electrocatalyst Design: Turning Inert Gold into Active Catalyst by Atomically Precise Nanochemistry. *J. Am. Chem. Soc.* **2021**, *143*, 11102–11108.

**49** M. Han, M. Guo, Y. Yun, Y. Xu, H. Sheng, Y. Chen, Y. Du, K. Ni, Y. Zhu, M. Zhu, Effect of Heteroatom and Charge Reconstruction in Atomically Precise Metal Nanoclusters on Electrochemical Synthesis of Ammonia. *Adv. Funct. Mater.* **2022**, *32*, 2202820.

**50** L. Liu, S.-J. Zheng, H. Chen, J. Cai, S.-Q. Zang, Tandem Nitrate-to-Ammonia Conversion on Atomically Precise Silver Nanocluster/MXene Electrocatalyst. *Angew. Chem. Int. Ed.* **2024**, *63*, e202316910.

**51** H. Chen, K.-S. Qi, X.-Y. Dong, Y.-N. Pei, Y.-J. Jin, H. Zhang, S. Li, J. Wu, J. Cai, S.-Q. Zang, Ligand-Mediated Activity of $Cu_4$ Clusters Boosts Electrocatalytic Nitrate Reduction. *Angew. Chem. Int. Ed.* **2025**, *64*, e202510429.

**52** S.-J. Zheng, X.-Y. Dong, H. Chen, R.-W. Huang, J. Cai, S.-Q. Zang, Unveiling Ionized Interfacial Water-Induced Localized H* Enrichment for Electrocatalytic Nitrate Reduction. *Angew. Chem. Int. Ed.* **2025**, *64*, e202413033.

**53** Y. Wei, J. Huang, H. Chen, S.-J. Zheng, R.-W. Huang, X.-Y. Dong, L.-K. Li, A. Cao, J. Cai, S.-Q. Zang, Electrocatalytic Nitrate Reduction on Metallic CoNi-Terminated catalyst with Industrial-Level Current Density in Neutral Medium. *Adv. Mater.* **2024**, *36*, 2404774.

**54** H. Dong, X. Wang, P. Shi, N. Chen, Z. Zhang, J. Yan, A. Lei, H. Yi, N-Heterocyclic Carbene-Ligated Mixed-Valent Linear $Cu_3$ Cluster for Enhanced Electrocatalytic Nitrite Reduction. *J. Am. Chem. Soc.* **2025**, https://doi.org/10.1021/jacs.5c09431.

**55** X. Gu, J. Zhang, S. Guo, Y. Zhang, L. Xu, R. Jin, G. Li, Tiara Ni Clusters for Electrocatalytic Nitrate Reduction to Ammonia with 97% Faradaic Efficiency. *J. Am. Chem. Soc.* **2025**, *147*, 22785–22795.

**56** H. Chen, L. Liu, X.-H. Ma, S.-J. Zheng, X.-Y. Dong, R.-W. Huang, Z.-Y. Wang, J. Cai, S.-Q. Zang, Atomically Precise Silver-Based Bimetallic Clusters for Electrocatalytic Urea Synthesis. *Natl. Sci. Rev.* **2025**, *12*, nwae440.

**57** Z. Zhang, R. Yin, Z. Song, M. Zhang, B. Zhang, S. Lu, Q. Yao, D.-e. Jiang, J. Xie, W. Hu, Efficient Electrocatalytic Semi-Hydrogenation of Alkynes by Interfacial Engineering of Atomically Precise Silver Nanoclusters. *Angew. Chem. Int. Ed.* **2025**, *64*, e202500389.

**58** Y.-M. Wang, J. Cai, Q.-Y. Wang, Y. Li, Z. Han, S. Li, C.-H. Gong, S. Wang, S.-Q. Zang, T. C. W. Mak, Electropolymerization of Metal Clusters Establishing a Versatile Platform for Enhanced Catalysis Performance. *Angew. Chem. Int. Ed.* **2022**, *61*, e202114538; *Angew. Chem.* **2022**, *134*, e202114538.

**59** J. Zhao, A. Ziarati, A. Rosspeintner, T. Bürgi, Anchoring of Metal Complexes on $Au_{25}$ Nanocluster for Enhanced Photocoupled Electrocatalytic $CO_2$ Reduction. *Angew. Chem. Int. Ed.* **2024**, *63*, e202316649; *Angew. Chem.* **2024**, *136*, e202316649.

**60** P. Hu, W. Xu, L. Tian, H. Zhu, F. Li, X. Qi, Q. Lu, Electrocatalytic Hydrogenation of Olefins. *Angew. Chem. Int. Ed.* **2025**, *64*, e202501215; *Angew. Chem.* **2025**, *137*, e202501215.

**61** Y. Liang, J. Feng, H. Li, X. Wang, Y. Zhang, W. Fan, S. Zhang, M.-B. Li, A Hydrogen Evolution Catalyst [$CO_2O_2$] Metallacycle Enables Regioselective Allene C($sp^2$)-H Functionalization. *Angew. Chem. Int. Ed.* **2024**, *63*, e202400938; *Angew. Chem.* **2024**, *136*, e202400938.

# 10

# Metal Nanocluster Enzyme-catalysis

## 10.1 Nanozyme and Clusterzyme

Enzymes widely exist in the nature. They catalyze various transformations efficiently, mildly, and selectively, thus constructing our current ecosystem [1, 2]. Enzyme-mimicking catalysis is always a hot topic in the research field of catalysis. Taking the oxidation process as a typical example, nature has found a way to realize controlled aerobic oxidations under highly mild conditions with the molecular oxygen as the oxidant. Here is an interesting thing that $O_2$ is a molecule possessing the triplet ground state. As we mentioned in Chapter 8, the vast majority of molecules have the singlet ground state. Under excitation (e.g. light irradiation), these molecules are promoted to the excited state, that is, the triplet state and thus involved in the subsequential processes of the transformations. Very few molecules have the triplet ground state. It is not a coincidence that $O_2$ possesses the triplet ground state but a way that the nature protects organic compounds from destructive oxidation as a result of the high energy barrier for electron transfer process from the organic compounds to the molecular oxygen. The enzyme-mediated mild oxidation process is quite complicated in the organism and contains a variety of redox cofactors. We simplify this process in Figure 10.1. An alcohol is dehydrogenated by nicotinamide adenine dinucleotide (NAD) to give a ketone or aldehyde. The generated NADH is an energy-rich species and can be involved in transferring energy. Its oxidation to NAD transfers electrons into the enzyme I, which is a NADH dehydrogenase enzyme. Meanwhile, the enzyme II, namely the succinate dehydrogenase enzyme triggers the transformation from succinate to fumarate, producing electrons as well. The mobile coenzymes ubiquinone (coenzyme Q) and the cytochrome c act as linkers to accept electrons and then transfer electrons from enzyme III (cytochrome $bc_1$) to enzyme IV (cytochrome c oxidase), which carries these electrons to $O_2$ and reduces it to water.

By mimicking this process, many efficient oxidations and oxidation-mediated organic transformations have been realized under mild reaction conditions [3].

*Metal Nanocluster Catalysis: From Fundamentals to Applications*, First Edition. Manzhou Zhu and Man-Bo Li. © 2026 Wiley-VCH GmbH. All rights reserved, including rights for text and data mining and training of artificial intelligence technologies or similar technologies. Published 2026 by Wiley-VCH GmbH.

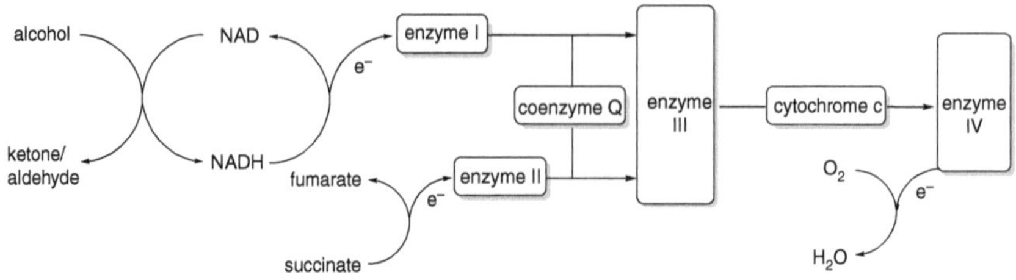


**Figure 10.1** Enzyme-mediated mild oxidation process in organism.

The as-developed redox catalytic species range from small single metal atoms (e.g. metal complexes) to big nanoparticles and other more microscopic materials. Nanozyme [4] is one of promising candidates, featuring stable and recoverable structures, regulable surface, and exceptional biomimetic performance. It starts to attract widespread attentions from 2007, in which year Gao and coworkers discovered the intrinsic peroxidase-like activity of the ferromagnetic nanoparticle [5]. In this study, the $Fe_3O_4$ nanoparticles ranging from 30 to 300 nm all demonstrated intrinsic peroxidase-like activity. Compared to natural peroxidase enzymes such as the horseradish peroxidase (HRP), the $Fe_3O_4$ nanozyme is also oxidant-, pH-, and temperature-dependent. Meanwhile, the $Fe_3O_4$ nanozyme also shows a typical Michaelis–Menten kinetics, demonstrating the enzyme-like property of the nanoparticles. Moreover, the $Fe_3O_4$ nanozyme exhibits additional advantages than the natural HRP enzyme. First, the $Fe_3O_4$ nanozyme is much more robust than HRP during catalysis. The enzymic activity of natural HRP decreases significantly after its exposure to harsher pH or higher temperature. In sharp contrast, the enzymic activity of the $Fe_3O_4$ nanozyme could be basically maintained under harsher reaction conditions. Second, the $Fe_3O_4$ nanozyme displays higher binding affinity toward the substrates such as the 3,3,5,5-tetramethylbenzidine (TMB) than HRP, and demonstrates a 40-fold higher level of enzymic activity than HRP with the same amount of catalyst loading. Third, the $Fe_3O_4$ nanozyme has magnetic property, which can be utilized for efficient catalyst recycling. Meanwhile, the preparation of the $Fe_3O_4$ nanozyme is much easier and low-cost than the natural HRP enzyme. In a word, the $Fe_3O_4$ nano-enzyme demonstrates more stable structure, higher enzymic activity and lower cost than the natural enzyme. This is why nanozyme is so attracting for chemists, biologists, and material scientists.

Since the discovery of the promising peroxidase-like activity of the $Fe_3O_4$ nanoparticles, various nanomaterials, such as metal nanoparticles, metal oxide nanoparticles, polymetallic sulfide nanomaterials, and nanocarbon materials have been reported successively to have certain enzymic activities [6–11]. The term "nanozyme" was proposed by Wei and coworkers for the first time in 2013 [12]. They summarized the nanomaterials that have enzyme-like characteristics in a review paper and named these nanomaterials as nanozymes, which was also called "next-generation artificial enzymes" in this review. After six years, Wu and coworkers authored another review paper entitled "Nanomaterials with enzyme-like characteristics (nanozymes): next-generation artificial enzymes (II)" [13].

They highlighted the new achievements in the field of nanozymes along with the significant advances in nanotechnology, biotechnology, catalysis, and computational science since the publication of the review paper in 2013. Among the developed nanozymes, the ones exhibiting reactive oxygen species (ROS) activity are the most investigated. The oxidoreductase-like nanozymes can activate the oxygen molecule under either environmental conditions or excited conditions (e.g. light irradiation) to produce ROS, including the singlet oxygen, superoxide radical, hydroxyl radical, etc. via energy transfer process or electron transfer process. By utilizing this enzyme-like activity, the nanozymes have been widely applied in antineoplastic therapy, antibacterial and antiviral therapy, and environmental monitoring. The corresponding biomimetic activities of these nanozymes have been found to be highly dependent on their composition, geometric structures, and electronic structures. On the other hand, as a result of their poly-dispersity and ill-defined structures, the enzymic activity of most of nanozymes cannot be well explained and correlated with their structural features. The catalytic mechanism is also ambiguous in most cases.

The concept of clusterzyme emerged at this moment. Metal nanoclusters generally have smaller size than the conventional metal nanoparticles. They are also more catalytically active than metal nanoparticles. Most importantly, the atomically precise structures of metal nanoclusters endow them with ideal models for correlating the enzymic activity with compositions and structures of nanomaterials. The history of clusterzyme is not long. The first report traces back to 2020. Zhao and coworkers developed a composite gold nanocluster-MOF clusterzyme system [14]. The reason for designing this kind of composite system is based on two challenges of metal nanocluster enzyme-catalysis. First, the surface ligands and their motion hinder the efficient adsorption and activation of substrates on metal nanoclusters. Second, enzyme catalysis occurs under water, while organic ligand-protected metal nanoclusters with atomically precise structures are generally oil-soluble. They cannot be well dispersed in water to catalyze the transformation of substrates. Therefore, Zhao and coworkers [14] conceived that introduction of water soluble γ-cyclodextrin-metal-organic framework (γ-CD-MOF) would be a method to on the one hand restrict the motion of organic ligands on the surface of metal nanoclusters and on the other hand bring them into the water phase for better catalysis (Figure 10.2).

The combination of $Au_{40}$(S-Adm)$_{22}$ nanocluster and γ-CD-MOF was readily achieved by mixing $Au_{40}$(S-Adm)$_{22}$ nanocluster with γ-CD-MOF in a solution and stirring for two hours. The disappearance of the characteristic peaks of γ-CD-MOF at 264 nm suggests the successful combination of two components. They further characterized the composite $Au_{40}$(S-Adm)$_{22}$ nanocluster-γ-CD-MOF clusterzyme system by other techniques such as XPS and confirmed its composition. The composite clusterzyme demonstrated good water solubility. Meanwhile, it can be well detached by simply washing the system with organic solvents such as the DCM. The detached $Au_{40}$(S-Adm)$_{22}$ nanocluster has retained composition and structures, which has been verified by UV–vis and thin layer chromatography (TLC). Further enzymic activity tests revealed the exceptional performance of the composite $Au_{40}$(S-Adm)$_{22}$ nanocluster-γ-CD-MOF clusterzyme system in the HRP-mimicking reaction.

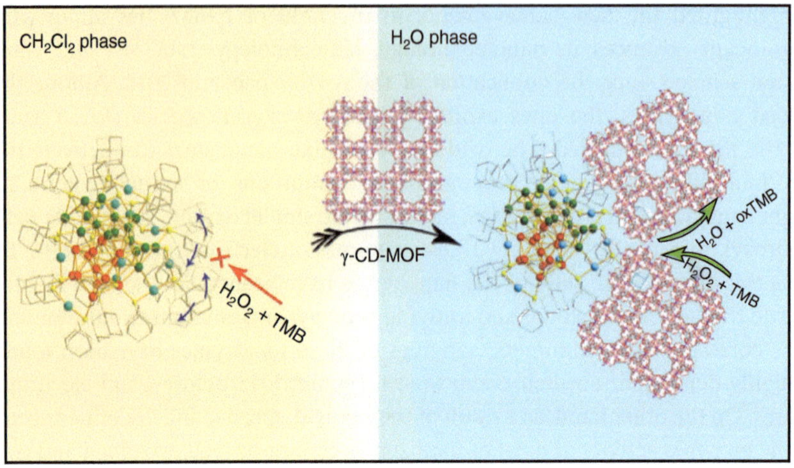


**Figure 10.2** $Au_{40}(S\text{-}Adm)_{22}$ nanocluster-γ-CD-MOF clusterzyme system showing active surface and good water solubility. *Source:* [14]/with permission of American Chemical Society.

This enzyme reaction has been described above. $Au_{40}(S\text{-}Adm)_{22}$ nanocluster-γ-CD-MOF clusterzyme system exhibited much higher activity than $Au_{40}(S\text{-}Adm)_{22}$ nanocluster, γ-CD-MOF, and even the natural HRP enzyme. Results showed that the $Au_{40}(S\text{-}Adm)_{22}$ nanocluster-γ-CD-MOF clusterzyme system has a higher affinity toward the substrate TMB compared to the natural enzyme HRP. The combination strategy of metal nanoclusters with γ-CD-MOF has been verified to be generalizable by Zhikun Wu and coworkers. They prepared the similar composite system by mixing the γ-CD-MOF with other metal nanoclusters, such as $Au_{38}(SCH_2CH_2Ph)_{24}$ and $Au_{44}(SPh\text{-}2,4\text{-}dimethyl)_{26}$. The as-obtained clusterzyme systems all demonstrated highly improved enzymic activity. Based on the atomically precise structure of the clusterzyme, detailed catalytic process was described with the help of theoretical calculations. This work presents the potential enzymic activity of metal nanoclusters for the first time, promoting the development of clusterzyme chemistry.

After one year of this work, Liu and coworkers synthesized a series of single-metal-doped $Au_{24}M(SR)_{18}$ nanoclusters and developed their promising enzymic performance, such as the glutathione peroxidase-like (GPx-like), catalase-like (CAT-like) and superoxide dismutase-like (SOD-like) activities. In this work, they proposed the term "clusterzyme" for the first time [15]. By using water-soluble 3-mercaptopropionic acid (MPA) as the protecting ligand, $HAuCl_4$, $HAuCl_4 + Cu^{2+}$ (molar ratio of Au/Cu = 24/1) or $HAuCl_4 + Cd^{2+}$ (molar ratio of Au/Cd = 24/1) as the metal precursor, they synthesized $Au_{25}(MPA)_{18}$, $Au_{24}Cu(MPA)_{18}$, and $Au_{24}Cd(MPA)_{18}$ nanoclusters. Owing to the high hydrophilicity of the MPA ligand, these nanoclusters do not have the water-solubility issue that commonly exists in organic ligand-protected metal nanoclusters. Meanwhile, based on the extended X-ray absorption fine structure analysis, it is confirmed that the copper and cadmium atom are both located at the outer layer of the metal nanoclusters, thus forming the Cu–S and Cd–S interactions. Bearing good water-solubility and outer surface

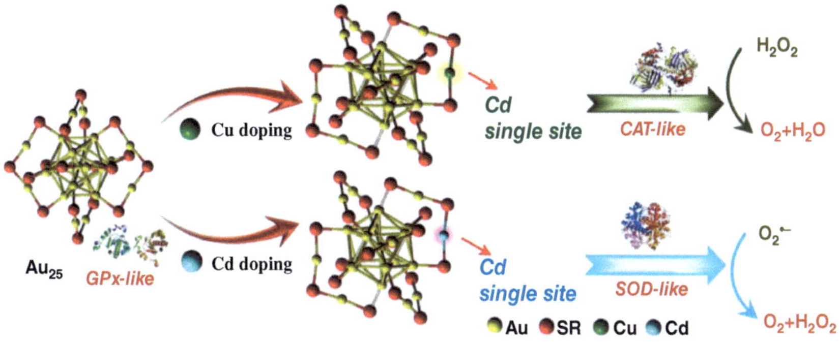


**Figure 10.3** Single-metal-doped $Au_{24}Cu(SR)_{18}$ and $Au_{24}Cd(SR)_{18}$ nanoclusters, and their CAT-like and SOD-like enzymic activities. *Source:* [15]/Springer Nature/CC BY 4.0.

metal catalytic sites, these metal nanoclusters are potentially nanozyme candidates. Liu and coworkers [15] investigated their enzymic activities in depth and revealed that $Au_{25}(MPA)_{18}$, $Au_{24}Cu(MPA)_{18}$, and $Au_{24}Cd(MPA)_{18}$ nanoclusters were all catalytically active. More interestingly, they display different selectivity toward the enzymic reactions. Specifically, $Au_{25}(MPA)_{18}$ demonstrates high GPx-like activity, while $Au_{24}Cu(MPA)_{18}$ and $Au_{24}Cd(MPA)_{18}$ exhibit promising CAT-like and SOD-like activities, respectively (Figure 10.3). Benefiting from their atomically precise structures, their catalytic mechanisms were well established based on control experiments and theoretical calculations. The term "clusterzyme" was formally presented by Xiao-Dong Zhang and coworkers in this work.

The CAT-like and SOD-like activities of $Au_{24}Cu(MPA)_{18}$ and $Au_{24}Cd(MPA)_{18}$ clusterzymes lead to intriguing modulation of neuroinflammation by metal nanoclusters. Bearing the superhigh antioxidant activity (137–160 times higher than the natural trolox), $Au_{24}Cu(MPA)_{18}$ and $Au_{24}Cd(MPA)_{18}$ demonstrate anti-inflammatory property. However, their anti-inflammatory mechanisms are different. $Au_{24}Cu(MPA)_{18}$ clusterzyme reduces neuroinflammation by inhibiting tumor necrosis factor-α(TNFα), showing selectivity against anti-neuroinflammation, whereas $Au_{24}Cd(MPA)_{18}$ clusterzyme significantly mitigate the neuroinflammation via inhibiting levels of interleukin (IL)-1β and IL-6. Additionally, owing to the intrinsic ultrasmall size of these clusterzymes, they can penetrate the kidney barriers and be excreted by renal, avoiding long-term hepatotoxicity and multiorgan injuries, thus being presumably influential as biomedicines, especially in the field of neuroscience.

## 10.2 Catalytic Performance of Metal Nanocluster Enzyme

After the two elegant works on clusterzyme that are reported by Zhikun Wu group and Xiao-Dong Zhang group, numerous works have been represented recently. In this section, we select some typical examples (Figure 10.4) to show the intriguing enzymic activities and potentially practical applications of metal nanoclusters.

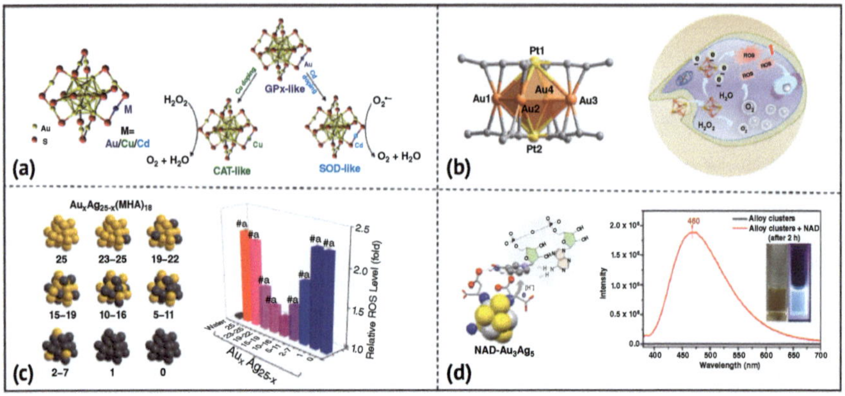


**Figure 10.4** Representatives of clusterzymes and their atomically precise structures and exceptional enzymic activities. *Source:* [16]/Springer Nature.

Employing the similar single atom doping technology (Figure 10.4a), Sun and coworkers synthesized $Au_{24}Ag(MPA)_{18}$ and $Au_{21}Cu(SG)_{18}$ nanoclusters successively [17]. Both nanoclusters are well dissolved in water as a result of the good hydrophilicity of the surface MPA and SG ligands. The silver atom was also identified to locate at the outer surface of $Au_{24}Ag(MPA)_{18}$ nanocluster by X-ray absorption near-edge structure analysis. Owing to the single silver atom substitution, the total antioxidant activity of $Au_{24}Ag(MPA)_{18}$ is much higher (22 time higher) than the undoped nanocluster $Au_{25}(MPA)_{18}$. Moreover, the clusterzyme $Au_{24}Ag(MPA)_{18}$ also showed a 72 times and 4 times higher antioxidant activity than the natural enzymes of Trolox and anthocyanin, respectively. In vivo experimental results demonstrate that $Au_{24}Ag(MPA)_{18}$ shows significant effect in the neuroprotective function via reactive oxygen species-scavenging ability and inflammation reduction. The lattice expansion was verified to be one of the origins of the high activity of the clusterzyme, which displays GPx and CAT activities with the maximum reaction rate reaching to 17.84 and 68.7 μM/min, respectively. The high structural stability and catalytic activity endow the $Au_{24}Ag(MPA)_{18}$ clusterzyme with potential applications in relieving acute inflammation in brain tissues and serum at the early stage.

Ma and coworkers further developed the dual function application of clusterzyme in three-dimensional imaging and acute inflammation inhibition. Their inspiration comes from the previous achievement of high enzymic activity of clusterzymes and the strong fluorescence of the reported $Au_{22}(SG)_{18}$ nanocluster. Similarly, they prepared a single atom-doped $Au_{21}Cu(SG)_{18}$ nanocluster by the in situ synthesis method [18]. They envisioned that the doped copper atom acted as the catalytic sites while the whole nanocluster acted as a fluorescence probe. The achievement of both high catalytic activity and strong fluorescence is very challenging as a result of the conflicts in free electron utilization of other materials. Herein, This strategy worked and the $Au_{21}Cu(SG)_{18}$ nanocluster on the one hand exhibited 18 times higher antioxidant activity, 90 times higher CAT-like activity, and 3 times higher SOD-like activity than the mother nanocluster $Au_{22}(SG)_{18}$ and on the other hand

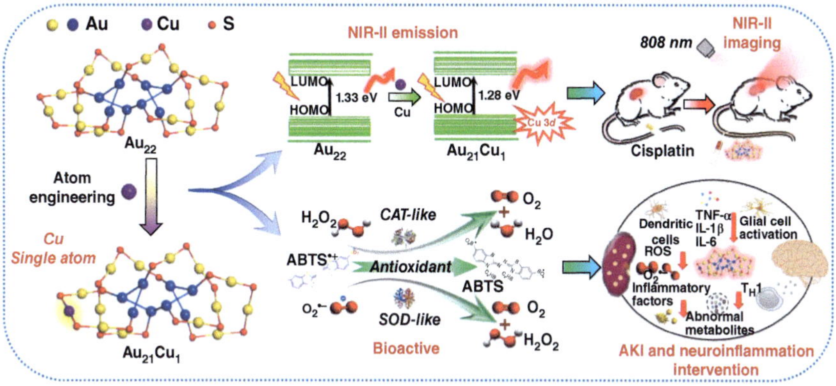


**Figure 10.5** Illustration of the synthesis, structure, and the dual function of $Au_{21}Cu(SG)_{18}$ clusterzyme in three-dimensional bioimaging and acute inflammation inhibition.
*Source:* [18]/with permission of American Association for the Advancement of Science – AAAS.

displayed negligible fluorescence loss compared to the highly fluorescent mother nanocluster (Figure 10.5). Furthermore, as a result of the single copper atom doping, the HOMO-LUMO gap of $Au_{21}Cu(SG)_{18}$ nanocluster decreased from 1.33 eV of the mother nanocluster $Au_{22}(SG)_{18}$ to 1.28 eV, thus endowing $Au_{21}Cu(SG)_{18}$ nanocluster with exceptional near-infrared-II fluorescence that is beneficial for bioimaging. Further in vivo experiments demonstrate that the $Au_{21}Cu(SG)_{18}$ clusterzyme can monitor cisplatin-induced renal injury quickly ranging from 20 minutes to 120 minutes and visualize the injury in three dimensions using NIR-II light-sheet microscopy. Besides, it inhibits oxidative stress and inflammation in the cisplatin-treated mouse model, showing a promising future application.

Employing the alkynyl group of levonorgestrel ($C_{21}H_{27}O_2$) as the protecting group, Jia and coworkers synthesized a levonorgestrel-protected gold nanocluster $Au_8(C_{21}H_{27}O_2)_8$ [19]. This nanocluster demonstrates high photoluminescence with 58.7% quantum yield. Meanwhile, it exhibits promising catalytic activity in converting the oxygen gas to reactive oxygen species under X-ray irradiation. Combined with the satisfactory biocompatibility of the nanocluster originated from the biocompatible levonorgestrel on surface, this nanocluster has been developed as an efficient clusterzyme for cancer therapy. Control experiments on the cell colony formation assay showed that the surviving faction of cells treated by radiotherapy and $Au_8(C_{21}H_{27}O_2)_8$ clusterzyme is significantly smaller than the ones that were treated by only radiotherapy. The ratio of wound healing with $Au_8(C_{21}H_{27}O_2)_8$ clusterzyme and radiotherapy was determined to be as low as 4.9%. The mechanism of radioenhancement was identified to be the generation of reactive oxygen species upon irradiation, which result in irreversible cell apoptosis. In vivo tumor formation experiments give the similar results, and tumors are significantly suppressed after one treatment with $Au_8(C_{21}H_{27}O_2)_8$ clusterzyme and radiotherapy. Compared to the tumors that were treated with X-ray irradiation alone, tumors treated with an X-ray dose of 4 Gy and the $Au_8(C_{21}H_{27}O_2)_8$ clusterzyme exhibited an inhibition ratio of 74.2% in the cancer therapy experiments.

The clusterzyme was found to not only reduce the X-ray dose but also decrease the side effects of radiation in normal tissues.

After this successful attempt, Hua and coworkers further investigated the enzyme activity of foreign metal-doped metal nanoclusters protected by levonorgestrel. By using different metal precursors, they synthesized three doped metal nanoclusters, namely $Au_4Pt_2(C_{21}H_{27}O_2)_8$, $Ag_4Pt_2(C_{21}H_{27}O_2)_8$, and $Cu_4Pt_2(C_{21}H_{27}O_2)_8$ [20]. All three metal nanoclusters are luminescent. $Au_4Pt_2(C_{21}H_{27}O_2)_8$ and $Cu_4Pt_2(C_{21}H_{27}O_2)_8$ exhibit high quantum yields of 37.9% and 39.4%, which are higher than that (2.8%) of $Ag_4Pt_2(C_{21}H_{27}O_2)_8$. Moreover, compared to $Cu_4Pt_2(C_{21}H_{27}O_2)_8$, $Au_4Pt_2(C_{21}H_{27}O_2)_8$ demonstrates much higher catalytic activity toward decomposition of $H_2O_2$ and generation of oxygen gas. Bearing high photoluminescence and the catalase-like activity, $Au_4Pt_2(C_{21}H_{27}O_2)_8$ can be applied for bioimaging as well as cancer radiotherapy (Figure 10.4b). Experimental results verified this assumption and showed that the $Au_4Pt_2(C_{21}H_{27}O_2)_8$ clusterzyme displayed satisfactory catalase-like activity and high radiosensitization enhancement ratio. These properties are in line with the calculated energy level structures and photo mass energy absorption coefficients of the clusterzyme. In vivo experiments further indicate that $Au_4Pt_2(C_{21}H_{27}O_2)_8$ clusterzyme is beneficial for reducing the tumor radioresistance and achieving a promising therapeutic outcome as a result of its catalase-like activity for transforming $H_2O_2$ into $O_2$ in tumor.

Zheng and coworkers systematically investigated the impact of composition of metal kernel and electrical property of surface ligand on the enzyme catalytic activity of metal nanoclusters. Using the one-pot synthetic method, they prepared a series of MHA (MHA = 6-mercaptohexanoic acid)-protected Au–Ag alloyed nanoclusters with different metal kernel compositions, that is, $Au_xAg_{25-x}(x = 25)$, $Au_xAg_{25-x}(x = 23–25)$, $Au_xAg_{25-x}(x = 19–22)$, $Au_xAg_{25-x}(x = 15–19)$, $Au_xAg_{25-x}(x = 10–16)$, $Au_xAg_{25-x}(x = 5–11)$, $Au_xAg_{25-x}(x = 2–7)$, $Au_xAg_{25-x}(x = 1)$, and $Au_xAg_{25-x}(x = 0)$ [21]. These metal nanocluster samples were sent to antimicrobial investigations. Results showed that both $Au_{25}(MHA)_{18}$ and $Ag_{25}(MHA)_{18}$ showed good antimicrobial behavior, while the alloyed metal nanoclusters, especially the $Au_xAg_{25-x}(MHA)_{18}(x = 10–16)$ nanoclusters exhibited low activity. Specifically, $Au_xAg_{25-x}(MHA)_{18}(x = 10–16)$ only showed about 25% of antimicrobial efficacy compared to $Au_{25}(MHA)_{18}$ and $Ag_{25}(MHA)_{18}$. Mechanism studies based on the atomically precise structures of these clusterzymes revealed that the "U-shape" antimicrobial behavior (Figure 10.4c) should be ascribed to the enhanced stability after alloying, which significantly deactivates the capability of $Au_xAg_{25-x}(MHA)_{18}$ clusterzymes in generating reactive oxygen species that play important roles in killing the bacteria. This conclusion is different from that proposed by previous reports that alloying is beneficial for improving catalytic activity of metal nanoclusters.

In addition to the composition of metal kernel, the electrical property of surface ligand also significantly affects the enzymic activity of metal nanoclusters. Shan and coworkers synthesized three similar $Au_{15}$ nanocluster, namely $Au_{15}(NAC)_{13}$ (NAC = *N*-acetyl-L-cysteine), $Au_{15}(MPA)_{13}$ (MPA = 3-mercaptopropionic acid) and $Au_{15}(NMPA)_{13}$ (NMPA = 3-mercapto-2-methylpropanoic acid) [22]. Detailed studies indicate that $Au_{15}(NAC)_{13}$ is not only more stable but also more catalytically active than $Au_{15}(MPA)_{13}$ and $Au_{15}(NMPA)_{13}$ in the transformation of $H_2O_2$, which involves the decomposition of $H_2O_2$ and the oxidation of 3,3′,5,5′-tetramethylbenzidine (TMB). The promising stability and catalytic activity of $Au_{15}(NAC)_{13}$ should be attributed to the electron-withdrawing (NHAc) group on NAC. The NHAc functional groups on the $Au_{15}$ kernel create abundant intramolecular hydrogen bonds, which significantly improve the stability of the clusterzyme. Meanwhile, the electron-withdrawing effect of NHAc endows the gold atoms (which is actually the activation sites for catalytic reduction of $H_2O_2$) on the motif of $Au_{15}$ nanocluster with emptier orbitals, thus facilitating the catalytic oxidation of TMB (Figure 10.6). The enzymatic kinetics data indicates that the peroxidase-like activity of $Au_{15}(NAC)_{13}$ is about 4.3 and 2.7 times higher than that of $Au_{15}(MPA)_{13}$ and $Au_{15}(NMPA)_{13}$ nanoclusters.

Zhao and coworkers further developed the practical application of clusterzymes. They designed and synthesized a diphosphine ligand-protected gild nanocluster $[Au_{14}(dppp)_5I_4]^{2+}$, where dppp represents (1,3-bis(diphenylphosphino) propane) [23]. This nanocluster consists of an icosahedral $Au_{13}$ kernel that protected by four iodine ligands and three dppp ligands. Interestingly, a rarely reported dppp-Au-dppp handle staple was observed on the surface of the metal nanocluster, which provides activation sites for the transformation of oxygen gas into superoxides under visible light irradiation. This kind of oxidase-like activity of $[Au_{14}(dppp)_5I_4]^{2+}$ was applied by Zhikun Wu and coworkers for catalyzing TMB oxidation to give a blue-colored oxidized product oxTMB. This oxidation reaction is highly efficient and the specific activity was determined to be up to 0.182 U/mg. Based on several sequential processes, they established a tandem detection method. Specifically, the addition of acetylcholinesterase would result in disappearance of the blue color of the above mentioned system because the catalytic activity of the clusterzyme is inhibited by acetylcholinesterase. The further addition of organophosphorus pesticides into

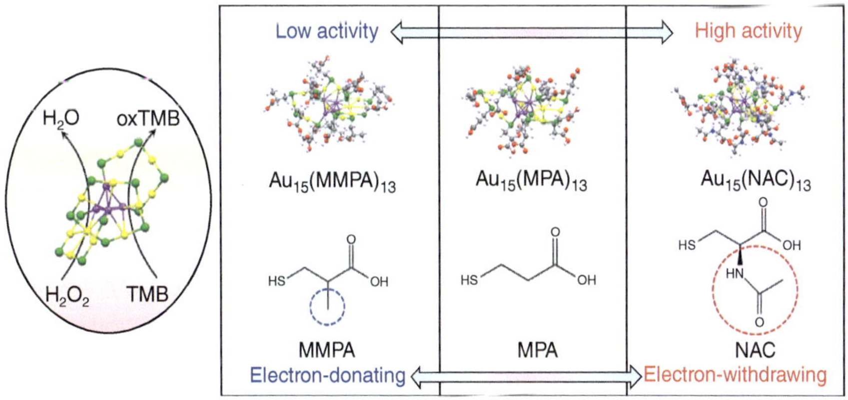


**Figure 10.6** Impact of electrical property of surface ligand on the enzymic activity of metal nanoclusters. *Source:* [22]/with permission of American Chemical Society.

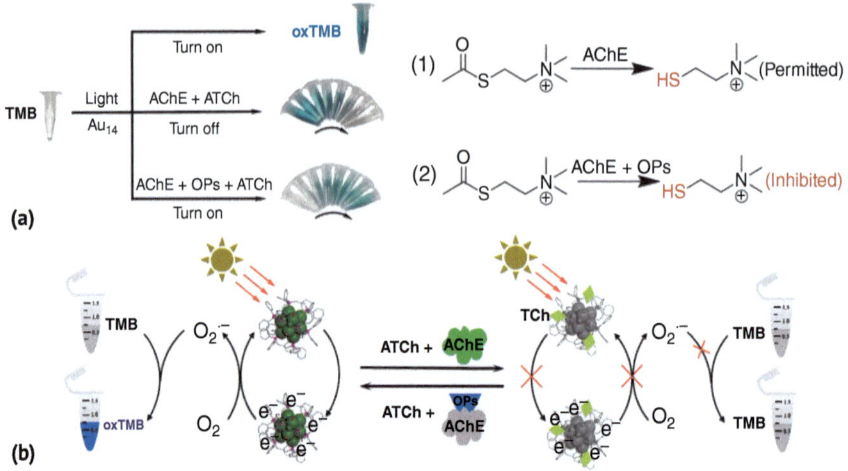


**Figure 10.7** Application of clusterzyme for colorimetrically sense acetylcholinesterase (AChE) and organophosphorus pesticides (OPs). (a) Schematic diagram of the detection of acetylcholinesterase and organophosphorus pesticides. The arrows under the tubes illustrate the color changing. (b) The catalytic process and the possible detection mechanism. *Source:* [23]/with permission of John Wiley & Sons.

the above mixture would then restore the clusterzyme and recover the blue color (Figure 10.7). The detection limits of the clusterzyme for acetylcholinesterase and organophosphorus pesticides reach to 0.04 mU/mL and 0.02 ng/mL, respectively, based on the color turn-on and turn-off mode.

All of the above reports are based on catalytic oxidation reactions of clusterzymes, Chen and coworkers reported the first example of the clusterzyme in catalytic reduction reactions. They synthesized a $[Au_3Ag_5(MSA)_3]^-$ nanocluster, where the MSA represents mercaptosuccinic acid [16]. The redox property of this nanocluster was investigated and it was found that the $[Au_3Ag_5(MSA)_3]^-$ nanocluster could significantly mediate the decomposition of $H_2O_2$ and facilitate the generation of hydroxyl radicals. After verifying the enzymic activity of $[Au_3Ag_5(MSA)_3]^-$, they further found that this clusterzyme was able to mediate the reduction of nicotinamide adenine dinucleotide (NAD) to the reduced form (NADH) of this compound (Figure 10.4d). This observation would be beneficial for developing other types enzymic reactions that catalyzed by clusterzymes.

Even if various clusterzymes have been discovered and their enzyme like activities including the GPx-like, CAT-like, SOD-like, and POD-like activities have been developed, the catalytic enzyme reactions of metal nanoclusters are still very limited. Current reactions are basically based on catalytic oxidation potential of these metal nanoclusters. When we look at the nature, massive types of reactions can be realized by natural enzymes. Therefore, developing new nanocluster structures and their enzyme-like activities apart from the catalytic oxidation activity is still an important and long-term task.

## References


1 B. Yuan, D. Yang, G. Qu, N. J. Turner, Z. Sun, Biocatalytic Reductive Aminations with NAD(P)H-dependent Enzymes: Enzyme Discovery, Engineering and Synthetic Applications. *Chem. Soc. Rev.* **2024**, *53*, 227–262.

2 I. Cho, Z-J. Jia, F. H. Arnold, Site-selective Enzymatic C–H Amidation for Synthesis of Diverse Lactams. *Science* **2019**, *364*, 575–578.

3 J. Piera, J. Bäckvall, Catalytic Oxidation of Organic Substrates by Molecular Oxygen and Hydrogen Peroxide by Multistep Electron Transfer—A Biomimetic Approach. *Angew. Chem. Int. Ed.* **2008**, *47*, 3506–3523.

4 R. Breslow, *Artificial Enzymes*. Wiley-VCH: Weinheim, **2005**.

5 L. Gao, J. Zhuang, L. Nie, J. Zhang, Y. Zhang, N. Gu, T. Wang, J. Feng, D. Yang, S. Perrett, X. Yan, Intrinsic Peroxidase-like Activity of Ferromagnetic Nanoparticles. *Nat. Nanotechnol.* **2007**, *2*, 577–583.

6 Y. Wei, J. Wu, Y. Wu, H. Liu, F. Meng, Q. Liu, A. C. Midgley, X. Zhang, T. Qi, H. Kang, R. Chen, D. Kong, J. Zhuang, X. Yan, X. Huang, Prediction and Design of Nanozymes using Explainable Machine Learning. *Adv. Mater.* **2022**, *34*, 2201736.

7 Z. Yang, M. Yuan, B. Liu, W. Zhang, A. Maleki, B. Guo, P. Ma, Z. Cheng, J. Lin, Conferring $BiVO_4$ Nanorods with Oxygen Vacancies to Realize Enhanced Sonodynamic Cancer Therapy. *Angew. Chem. Int. Ed.* **2022**, *61*, e202209484; *Angew. Chem.* **2022**, *134*, e202209484.

8 B. Chen, C. Zhang, W. Wang, Z. Chu, Z. Zha, X. He, W. Zhou, T. Liu, H. Wang, H. Qian, Ultrastable $AgBiS_2$ Hollow Nanospheres with Cancer Cell-specific Cytotoxicity for Multimodal Tumor Therapy. *ACS Nano*. **2020**, *14*, 14919–14928.

9 P. J. Derry, A. V. Liopo, K. Mouli, E. A. Mchugh, A. T. T. Vo, A. Mckelvey, L. J. Suva, G. Wu, Y. Gao, K. R. Olson, J. M. Tour, T. A. Kent, Oxidation of Hydrogen Sulfide to Polysulfide and Thiosulfate by a Carbon Nanozyme: Therapeutic Implications with an Emphasis on Down Syndrome. *Adv. Mater.* **2023**, *36*, 2211241.

10 F. Gao, T. Shao, Y. Yu, Y. Xiong, L. Yang, Surface-bound Reactive Oxygen Species Generating Nanozymes for Selective Antibacterial Action. *Nat. Commun.* **2021**, *12*, 745.

11 L. Wang, B. Zhu, Y. Deng, T. Li, Q. Tian, Z. Yuan, L. Ma, C. Cheng, Q. Guo, L. Qiu, Biocatalytic and Antioxidant Nanostructures for ROS Scavenging and Biotherapeutics. *Adv. Funct. Mater.* **2021**, *31*, 2101804.

12 H. Wei, E. Wang, Nanomaterials with Enzyme-like Characteristics (Nanozymes): Next-generation Artificial Enzymes. *Chem. Soc. Rev.* **2013**, *42*, 6060–6093.

13 J. Wu, X. Wang, Q. Wang, Z. Lou, S. Li, Y. Zhu, L. Qin, H. Wei, Nanomaterials with Enzyme-like Characteristics (Nanozymes): Next-generation Artificial Enzymes (II). *Chem. Soc. Rev.* **2019**, *48*, 1004–1076.

14 Y. Zhao, S. Zhuang, L. Liao, C. Wang, N. Xia, Z. Gan, W. Gu, J. Li, H. Deng, Z. Wu, A Dual Purpose Strategy to Endow Gold Nanoclusters With Both Catalysis Activity and Water Solubility. *J. Am. Chem. Soc.* **2020**, *142*, 973–977.

**15** H. Liu, Y. Li, S. Sun, Q. Xin, S. Liu, X. Mu, X. Yuan, K. Chen, H. Wang, K. Varga, W. Mi, J. Yang, X.-D. Zhang, Catalytically Potent and Selective Clusterzymes for Modulation of Neuroinflammation through Single-atom Substitutions. *Nat. Commun.* **2021**, *12*, 114.

**16** L. Chen, Y. Du, Y. Lv, D. Fan, J. Wu, L. Wu, M. Cui, H. Yu, M. Zhu, Nicotinamide Adenine Dinucleotide ($NAD^+$) Reduction Enabled by an Atomically Precise Au-Ag Alloy Nanocluster. *Nano Res.* **2023**, *16*, 7770–7776.

**17** S. Sun, H. Liu, Q. Xin, K. Chen, H. Ma, S. Liu, X. Mu, W. Hao, S. Liu, Y. Gao, Y. Wang, J. Pei, R. Zhao, S. Zhang, X. Zhang, H. Wang, Y. Li, X.-D. Zhang, Atomic Engineering of Clusterzyme for Relieving Acute Neuroinflammation Through Lattice Expansion. *Nano Lett.* **2021**, *21*, 2562–2571.

**18** H. Ma, X. Zhang, L. Liu, Y. Huang, S. Sun, K. Chen, Q. Xin, P. Liu, Y. Yan, Y. Wang, Y. Li, H. Liu, R. Zhao, K. Tan, X. Chen, X. Yuan, Y. Li, Y. Liu, H. Dai, C. Liu, H. Wang, X.-D. Zhang, Bioactive NIR-II Gold Clusters for Three-dimensional Imaging and Acute Inflammation Inhibition. *Sci. Adv.* **2023**, *9*, eadh7828.

**19** T.-T. Jia, G. Yang, S.-J. Mo, Z.-Y. Wang, B.-J. Li, W. Ma, Y.-X. Guo, X. Chen, X. Zhao, J.-Q. Liu, S.-Q. Zang, Atomically Precise Gold–Levonorgestrel Nanocluster as a Radiosensitizer for Enhanced Cancer Therapy. *ACS Nano.* **2019**, *13*, 8320–8328.

**20** Y. Hua, J.-H. Huang, Z.-H. Shao, X.-M. Luo, Z.-Y. Wang, J.-Q. Liu, X. Zhao, X. Chen, S.-Q. Zang, Composition-dependent Enzyme Mimicking Activity and Radiosensitizing Effect of Bimetallic Clusters to Modulate Tumor Hypoxia for Enhanced Cancer Therapy. *Adv. Mater.* **2022**, *34*, 2203734.

**21** K. Zheng, J. Xie, Composition-dependent Antimicrobial Ability of Full-spectrum $Au_xAg_{25-x}$ Alloy Nanoclusters. *ACS Nano.* **2020**, *14*, 11533–11541.

**22** H. Shan, J. Shi, T. Chen, Y. Cao, Q. Yao, H. An, Z. Yang, Z. Wu, Z. Jiang, J. Xie, Modulating Catalytic Activity and Stability of Atomically Precise Gold Nanoclusters as Peroxidase Mimics via Ligand Engineering. *ACS Nano.* **2023**, *17*, 2368–2377.

**23** H. Zhao, Q. You, W. Zhu, J. Li, H. Deng, M.-B. Li, Y. Zhao, Z. Wu, Nanoclusterzyme for Dual Colorimetric Sensings: A Case Study on $[Au_{14}(Dppp)_5I_4]^{2+}$. *Small* **2023**, *19*, 2207936.

# 11

# Metal Nanocluster Industrial Catalysis

## 11.1 Metal Nanocluster Large-scale Preparation

Despite the significant progress of fundamental research in catalytic performance of metal nanoclusters, their practical application in industry is still limited. There are several challenging technical issues that are needed to be resolved. Large-scale synthesis of metal nanoclusters constitutes one of the challenges. Concerning the fact that the catalytic activity, selectivity, and even stability of metal nanoclusters are strongly dependent on their size, shape, and structures [1], the first thing we need to consider for metal nanocluster large-scale preparation is how to maintain their uniformity. This kind of heterogeneity problems and impurity artifacts are also key issues in homogeneous metal catalysis [2], because the activity and selectivity of the catalytic reactions would change significantly with even a trace amount of impurity mixing in the catalysts. Compared to the homogeneous metal complex catalysts, keep structural uniformity for metal nanocluster catalysts is much more challenging. Apart from that, newly developed synthetic methods that enable high-yield, mild-condition, and economic preparation of metal nanoclusters are highly desirable.

Considering the synthetic efficiency, the previously developed method for the synthesis of metal nanoclusters might be not applicable for their large-scale synthesis. For example, $[Au_{25}(SR)_{18}]^-$ nanocluster is the most extensively studied gold nanocluster up to now [3], while its synthetic method is limited by the in situ two-step reduction synthesis (Figure 11.1a) and the synthetic yield is quite low. Considering the wide application of this kind of nanoclusters in catalysis, a facile, synthetically easy and low-cost method toward the preparation of $[Au_{25}(SR)_{18}]^-$ nanocluster in high yield and at large scale is desired. Lei and coworkers developed a "cluster from cluster" synthetic method [4], which features quantitative conversion and gram-scale preparation of $[Au_{25}(SR)_{18}]^-$ nanocluster (Figure 11.1b). Based on this approach, various $[Au_{25}(SR)_{18}]^-$ nanoclusters with different thiolate ligands were synthesized efficiently.

*Metal Nanocluster Catalysis: From Fundamentals to Applications*, First Edition. Manzhou Zhu and Man-Bo Li. © 2026 Wiley-VCH GmbH. All rights reserved, including rights for text and data mining and training of artificial intelligence technologies or similar technologies. Published 2026 by Wiley-VCH GmbH.

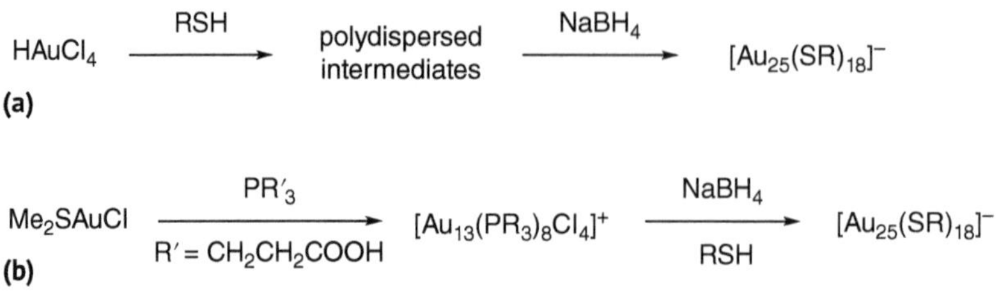


**Figure 11.1** (a) Conventional in situ two-step reduction approach toward synthesis of $[Au_{25}(SR)_{18}]^-$ nanocluster. (b) "Cluster from cluster" synthetic method toward high-yield and large-scale preparation of $[Au_{25}(SR)_{18}]^-$ nanoclusters.

The large-scale synthesis of $[Au_{25}(SR)_{18}]^-$ is based on a careful anatomy of atomically precise structure of the nanocluster. Quan-Ming Wang and coworkers divided the total structure of this nanocluster into two parts, namely an $Au_{13}$ kernel and six $[Au_2(SR)_3]^-$ staples. Therefore, they proposed a synthetic approach initiated from the icosahedral $Au_{13}$ nanocluster. By reacting $[Au_{13}(PR_3)_8Cl_4]^+$ (R = $CH_2CH_2COOCH_3$) with freshly prepared polymeric thiolate-gold complexes under reduction conditions ($NaBH_4$) for 12 hours, they successfully obtained high-quality $[Au_{25}(SR)_{18}]^-$ nanocluster in almost 100% yield. UV–vis and ESI-MS spectra verified the purity of the sample. Next, they also developed a way to quantitative synthesis of the icosahedral $Au_{13}$ nanocluster. Tris(2-carboxyethyl)phosphine hydrochloride was mixed with the gold precursor $Me_2SAuCl$ in a mixed solution of $CH_2Cl_2$ and MeOH under stirring. This mixture was then treated with of $NaBH_4$ ethanol solution, stirred overnight and rotary evaporated. After washing with a mixture of ether and hexane, $[Au_{13}(PR_3)_8Cl_4]^+$ (R = $CH_2CH_2COOCH_3$) was obtained in almost 100% yield. Single crystal structure confirms its atomically precise structure and ESI-MS data confirms that the product is pure. The methyl esterification of the carboxylic sites on phosphine ligands should be attributed to the reaction between tris(2-carboxyethyl)phosphine hydrochloride and methanol under the basic reaction conditions.

This "cluster from cluster" method gives totally 100% yield in two steps. Meanwhile it enables facile gram-scale preparation of $[Au_{25}(SR)_{18}]^-$ nanocluster. This approach has been extended to the synthesis of a variety of $[Au_{25}(SR)_{18}]^-$ nanoclusters in large scale, including $[Au_{25}(S\text{-}C_6H_3\text{-}2,4\text{-}DiF)_{18}]^-$ (2,4-DiF-$C_6H_3$-SH = 2,4-difluorothiophenol), $[Au_{25}(S\text{-}C_6H_3\text{-}2,4\text{-}DiMe)_{18}]^-$ (2,4-DiMe-$C_6H_3$-SH = 2,4-dimethylthiophenol), $[Au_{25}(S\text{-}C_6H_4\text{-}2\text{-}Cl)_{18}]^-$ (2-Cl-$C_6H_4$-SH = 2-chlorothiophenol), $[Au_{25}(S\text{-}C_6H_4\text{-}2\text{-}F)_{18}]^-$ (2-F-$C_6H_4$-SH = 2-fluorothiophenol), $[Au_{25}(S\text{-}C_6H_4\text{-}2\text{-}Me)_{18}]^-$ (2-Me-$C_6H_4$-SH = 2-methylthiophenol), $[Au_{25}(SC_6H_4\text{-}3\text{-}Cl)_{18}]^-$ (3-Cl-$C_6H_4$-SH = 3-chlorothiophenol), $[Au_{25}(S\text{-}C_6H_4\text{-}3\text{-}F)_{18}]^-$ (3-F-$C_6H_4$-SH = 3-fluorothiophenol), $[Au_{25}(S\text{-}C_6H_3\text{-}3,4\text{-}DiF)_{18}]^-$ (3,4-DiF-$C_6H_3$-SH = 3,4-difluorothiophenol) and $[Au_{25}(S\text{-}1\text{-}NAP)_{18}]^-$ (1-NAP-SH = 1-thionaphthol). ESI-MS spectra and single crystal XRD analysis further confirmed their purity and atomically precise structures.

By slightly regulating reaction parameters of the conventional in situ two-step reduction method, large-scale preparation of metal nanoclusters can also be achieved. Shen and coworkers [5] developed an unique reducing agent

$Cu(PPh_3)_2BH_4$ to replace $NaBH_4$ in the synthesis of metal nanoclusters and realized simple and selective synthesis of $[Ag_{25}Cu_4(PhC{\equiv}C)_{12}(PPh_3)_{12}Cl_6H_8]^{3+}$, $[AuCu_{14}(SR)_{12}(PPh_3)_6]^+$ (R = 4-fluorothiophenol), and $[Au_9Cu_2(PPh_3)_8Cl_2]^+$. These metal nanoclusters were obtained in high yields and high purity under the reduction conditions using $Cu(PPh_3)_2BH_4$, which can be applied further for metal nanoclusters large-scale synthesis. Chen and coworkers developed a method to synthesize copper-hydride nanoclusters in large scale [6]. Three nanocluster $AuCu_{24}H_{22}(PPh_3)_{12}$, $Cu_{25}H_{22}((4\text{-}FPh)_3P)_{12}$ and $AuCu_{24}H_{22}((4\text{-}FPh)_3P)_{12}$ that were difficult to be obtained by other methods were synthesized based on their controlled synthetic method, which provides a solution for large-scale preparation of unstable copper nanoclusters. Dong and coworkers also synthesized copper nanocluster $Cu_4I_4(PR_3)_4$ by their in situ synthetic method [7]. Structurally similar nanoclusters including $Cu_4I_4(PPh_3)_4$, $Cu_4I_4(PPh_2Me)_4$, $Cu_4I_4(PPh_2Et)_4$, $Cu_4I_4(PPh_2Pr)_4$ and $Cu_4I_4(PPhMe_2)_4$ were prepared in large scale and sent for reversible crystal-liquid-glass transitions investigation. The successful large-scale preparation of these nanoclusters triggered discovery of a high-performance organic–inorganic hybrid glass. $Cu_4I_4(PPh_2Et)_4$ nanocluster glass exhibits superb optical properties, such as more than 90% transmission in both visible and near-infrared wavelengths, negligible self-absorption, near-unity quantum yield, and high light yield.

With the rapid development of electrocatalysis and photocatalysis, the utilization of light and electric energy also triggers the large-scale synthesis of metal nanoclusters under green reaction conditions. As we mentioned in Section 4.1, photo-induced in situ generation of sulfur ligands or reductants facilitates the formation of metal nanoclusters that are otherwise difficult to be synthesized by conventional methods. Meanwhile, the mild reaction conditions and easy operation of these photo synthetic methods are beneficial for the large-scale preparation of metal nanoclusters (Figure 11.2a). Gram-scale synthesis of copper nanoclusters, gold nanoclusters and silver nanoclusters was achieved by the photo synthetic methods developed by Zhu and coworkers[8–12].

Bearing advantages of avoiding usage of harsh oxidizing or reducing agents, precisely regulating electrical potential, and easy to scale up, electrosynthesis has been widely applied in organo-synthesis and a wide range of organic compounds

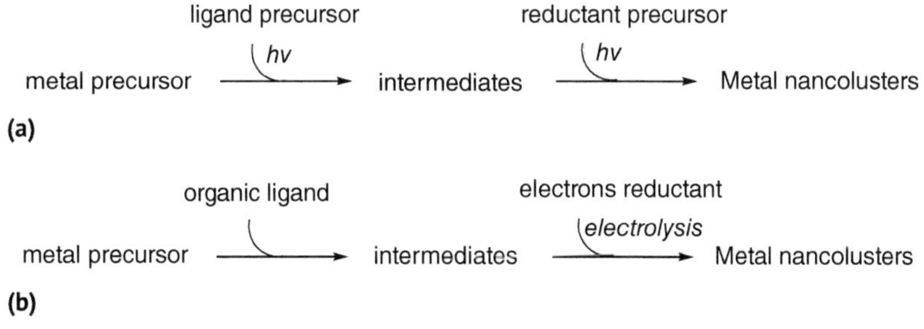


**Figure 11.2** (a) Photo-induced large-scale synthesis of metal nanoclusters. (b) Electro-induced large-scale synthesis of metal nanoclusters.

that were difficult to be obtained have been electrochemically synthesized [13–20]. Dong and coworkers employed the electro-synthetic method for preparing atomically precise gold nanoclusters and successfully revealed their structures by SCXRD [21]. They used a single electrolytic cell to realize the synthesis of metal nanoclusters under a constant potential from simple precursors of $HAuCl_4$ and organic ligands. $[Au_{13}(4\text{-FPhC}{\equiv}C)_2(Dppe)_5]Cl_3$, $[Au_8(2\text{-}CF_3PhC{\equiv}C)_2(Dppp)_4](PF_6)_2$, $[Au_{11}(Dppp)_5]Cl_3$, and $[Au_8(SC_2H_4Ph)_2(Dppp)_4]Cl_2$ were synthesized electrochemically, suggesting the versatility of the strategy. Notably, pulsed electrolysis technology has been utilized in the electrosynthesis of metal nanoclusters (Figure 11.2b). This is crucial for resolving the issues of over-reduction and deposition of gold species on the cathode. Another example of large-scale synthesis of metal nanoclusters by electrolysis comes from Wang and coworkers. They located polymeric activation sites on the organic ligands and realized the large-scale preparation of metal nanoclusters on the surface of electrode by electrochemical polymerization reactions [22]. We have briefly described this case in Section 4.1. Owing to the emergence of continuous flow reactors [23] in photosynthesis and electrosynthesis, it is believed that metal nanocluster synthesis with photo- and electro-continuous flow reactors will be an important supplement for large-scale preparation of metal nanoclusters.

Apart from the in situ synthetic method, post functionalization on the surface of metal nanoclusters by mild organic transformations constitutes another important method for the large-scale preparation of metal nanoclusters. Zhang and coworkers showed the gram-scale functionalization of gold nanoclusters by an amidation reaction [24]. They first synthesized an $Au_{11}$ nanocluster by a conventional synthetic method, using PNHP pincer ligand ($Ph_2PCH_2CH_2NHCH_2CH_2PPh_2$). This nanocluster features a di-coordination mode of PNHP pincer ligand on the surface, which has been shown in Figure 6.15. As a result of the unique coordination mode, the NH group has been exposed on the surface of $[Au_{11}(PNHP)_4Br_2]^+$ nanocluster. The four amino sites enable further functionalization by a mild amidation reaction, which gives $[Au_{11}(PNCORP)_4Br_2]^+$ nanocluster in the presence of RCOCl or RCOOH (Figure 11.3). This strategy is quite efficient, triggering the nanocluster surface functionalization while keeping their structures and coordination modes basically unchanged. A slight sliding of several gold atoms was observed.

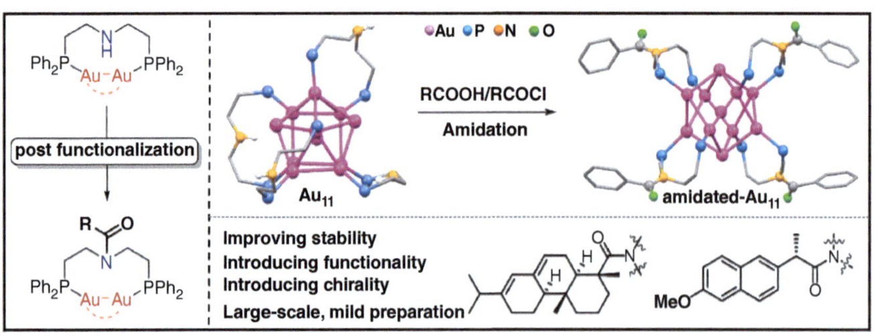


**Figure 11.3** Post functionalization for large-scale synthesis of metal nanoclusters.

This strategy has been applied by Zhang and coworkers [24] for introduction of functionality and chirality in metal nanoclusters. Meanwhile, the stability of the functionalized $Au_{11}$ nanocluster was significantly improved, due to their increased oxidation potential inhibiting nanocluster oxidation under atmosphere oxygen. The amidation functionalization reaction is carried out under mild conditions, thus being beneficial for the scale-up of this method. Gram-scale functionalization of $[Au_{11}(PNHP)_4Br_2]^+$ nanocluster was achieved, manifesting the applicability of this method. He and coworkers also developed a post functionalization method for the large-scale synthesis of silver nanoclusters [25], which were difficult to be achieved by other methods. By applying organic ligand of (4-CHOPhC≡CH, they synthesized a silver nanocluster $\left[Ag_{21}(Ph_2PO_2)_{10}(4\text{-CHOPhC}{\equiv}C)_6\right]^+$. There are six aldehyde groups on the surface of this silver nanocluster, which allow for the surface functionalization by an aldimine condensation reaction. This synthetic method enables introduction of chirality in silver nanoclusters and gram-scale preparation of the functionalized nanoclusters in more than 90% yields.

## 11.2 Metal Nanocluster Recyclability

Although the research of metal nanocluster catalysis would be focused on earth-abundant metal nanoclusters in the future, the efficient metal nanocluster catalysts developed currently mainly fall into noble metals. Concerning the industrial application of these precious metal nanocluster catalysts, recyclability should be another key factor for nanocluster catalyst evaluation. Small portions of reaction in laboratory only require several milligrams of metal nanocluster catalysts, while the amount used in industry might increase to kilograms or tons. The corresponding catalyst cost has to be considered in metal nanocluster catalysis. As a result of the unique size of metal nanoclusters that locates between single metal atoms and nanocrystals, it is hard to distinguish whether metal nanocluster catalysis is homogeneous or heterogeneous. For the immobilized metal nanoclusters (that we discussed in Chapter 3), they are generally easy to be separated from the catalytic system because the supports such as carbon nanomaterial, metals, metallic oxides, polymers, supermolecules, dendrimers, and biomolecules are insoluble. The recyclability of these metal nanocluster catalysts is not a problem and the corresponding reports have also showed that they can be well recovered and recycled after catalysis. For the in situ generated metal nanoclusters and organic ligand-protected metal nanoclusters, the recyclability is a key issue because they are generally soluble in most of solvents.

In fact, developing approaches for separation, recovery and recycling of this kind of homogeneous catalysts has long been studied [26]. Compared to heterogeneous metal catalysts, homogeneous metal catalysts usually exhibit higher catalytic activity and selectivity because they do not have the mass transfer issue during catalysis. However, many high-performance homogeneous metal catalysts cannot be commercialized. One of the most important reasons is the difficulty associated with separating the products from the catalysts. This would case two problems. The first one is the aforementioned high cost of the metal catalysts. The second issue is that the remaining metals would significantly impact the performance of products. Thus, various heterogenization methods have been developed for homogeneous metal catalysts, including the self-supporting, polymerization, immobilization by ordered structure or amorphous silicas [27–30]. Meanwhile, emerging recycling techniques such as temperature-controlled phase-separable catalyst recycling, aqueous biphasic catalytic system and ionic liquid-based catalytic system also endow homogeneous metal catalysts with recyclability. These methods and techniques effectively facilitate the recovery and recycling of metal nanocluster catalysts. Herein, we give a unique example that exclusively belongs to metal nanocluster catalyst recycling.

Organic ligand-protected metal nanoclusters are selectively dissolved in different solvents. For example, the most studied $[Au_{25}(SR)_{18}]^-$ nanocluster can be well dissolved in dichloromethane and toluene, while it is insoluble in hexane, methanol, etc. Based on this feature, we can realize the metal nanocluster homogeneous catalysis and heterogeneous recycling [31–34]. That is, we choose the reaction solvent that allow metal nanocluster dissolve well in it and then add the poor solvent after reaction for metal nanocluster participation and subsequent recycling. This approach has been applied for metal nanocluster catalysis. A metal-doped mixed ligand-protected nanocluster $Au_{17}Cd_2(PNP)_2(SR)_{12}$ showed high catalytic activity in the cascade reaction of ketones, amines and alkynes ($KA^2$ reaction). This reaction was generally catalyzed by copper-based homogeneous or heterogeneous catalysts. Harsh reaction conditions such as high temperature (>100°C) and high catalyst loading (5–30 mol%) were always involved. Herein, the $Au_{17}Cd_2(PNP)_2(SR)_{12}$ nanocluster was able to catalyze the $KA^2$ reaction, affording propargyl amines at room temperature with only 0.01 mol% catalyst loading. Thus, the TON of the metal nanocluster-catalyzed $KA^2$ reaction is as high as around 8000. Apart from the high activity, $Au_{17}Cd_2(PNP)_2(SR)_{12}$ can be well recovered and recycled after catalysis [31]. The catalytic reaction was conducted in toluene, which is a good solvent for $Au_{17}Cd_2(PNP)_2(SR)_{12}$. The homogeneous reaction condition contributes to the high catalytic efficiency because the mass transfer issue is circumvented. After the reaction, the poor solvent methanol was added and the $Au_{17}Cd_2(PNP)_2(SR)_{12}$ nanocluster started to participate out from the reaction system (Figure 11.4). A simple centrifugation enables the separation of the metal nanocluster catalyst from the products. The recovered $Au_{17}Cd_2(PNP)_2(SR)_{12}$ nanocluster catalyst exhibits identical UV–vis and ESI-MS spectra with the freshly prepared one, indicating that its composition and structure were basically retained. The further recycling experiments gave almost the same catalytic results with the freshly prepared nanocluster, demonstrating that the catalytic activity and selectivity were also remained. This kind of metal nanocluster recovery and recycling method features simple operation and high efficiency, which would be highly valuable for industrial application in the future.

Even if the high structural stability and catalytic stability of metal nanoclusters are generally reported in the case of several recycling experiments, it should be

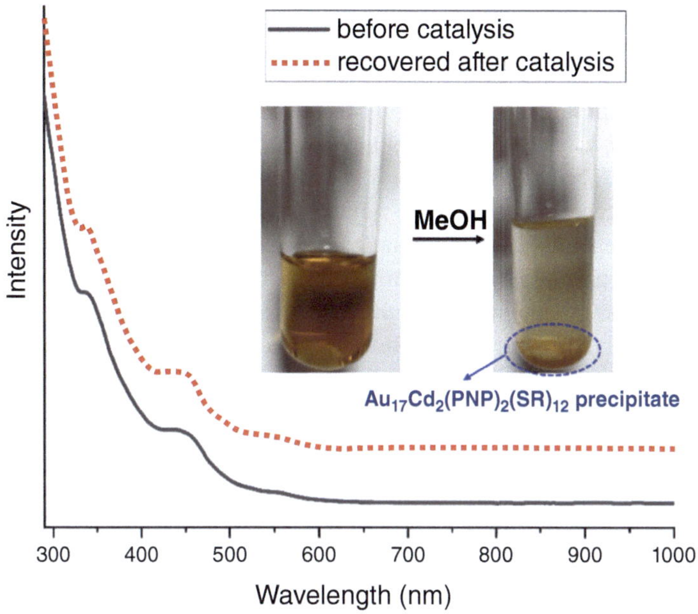


**Figure 11.4** Metal nanocluster homogeneous catalysis and heterogeneous recycling. *Source:* [31]/John Wiley & Sons.

noted that their catalytic performance might be changed after a long run of the reaction [35]. We cannot exclude the possibility that the metal nanocluster catalysts might lose their activity slowly after hundreds or thousands of times recycling experiments. Therefore, a continuous flow device would provide a more convenient and credible method for evaluating the recycling ability of metal nanocluster catalysts. This technique is also attracting for the industrial application of metal nanoclusters.

## 11.3 Metal Nanocluster Catalysis in Environmental and Sustainable Chemistry

In the process of human civilization moving towards sustainable development, the distribution of environmental and resource benefits has always been an important cornerstone for building a harmonious society. Entering the 21st century, with the profound transformation from industrial civilization to ecological civilization, human society is experiencing unprecedented ecological tests. Water, air, and soil are the essential "three elements of life" for human survival. However, at present, this foundation that relies on survival is being eroded by multiple pollutants: heavy metals, recalcitrant organic dyes, and antibiotics are rampant in water bodies; In the air, volatile organic compounds (VOCs), nitrogen oxides, and sulfur dioxide continue to pose a threat to respiratory health; In the soil, pesticide residues and heavy metals continue to accumulate, quietly disrupting the balance of ecological cycles. These pollutions not only shake the ecological foundation of the earth, but also directly endanger the quality of human life and health safety. Therefore, building a solid defense line for pollution prevention and control, and safeguarding the purity of these key resources, has become an urgent issue of the times. As a "small giant" in the field of catalysis, metal nanoclusters have shown important application potential and development prospects in the field of environment and sustainable development due to their unique physical and chemical properties. They primarily function to identify and monitor pollutants, efficiently adsorb, immobilize, or separate them, and convert toxic pollutants into harmless substances (such as $CO_2$, $H_2O$, and innocuous salts) through catalytic reactions, thereby achieving the goal of environmental purification.

### 11.3.1 Identification and Removal of Inorganic Pollutants in Water Bodies

Inorganic pollutants in waste water mainly include heavy metal cations (HMIs) and inorganic salt anions. The pollution of HMIs has attracted much attention due to their high toxicity, strong bioaccumulation, and anti-biodegradation properties. It is worth noting that when organisms are exposed to a complex pollution environment of multiple heavy metal ions, the synergistic effect between ions significantly enhances their toxic effects, which further exacerbates the ecological risks and health hazards of environmental pollution. A simple and effective method needs to be developed for determining the composition of metal ions in waste water, in order to reduce environmental pollution and improve water quality. Among various detection methods, fluorescence technology has advantages such as high sensitivity and low cost, while metal nanoclusters are commonly used as fluorescent probes due to their excellent optical properties. By utilizing the characteristics of multiple probes and combining them with the principle of sensing arrays, multiple analytes can be detected simultaneously.

Wei and coworkers developed two new types of fluorescence sensing systems. The research team first encapsulated glutathione (GSH) modified aggregation induced luminescent gold nanoclusters in a metal organic framework (MOF, named ZIF-8), and significantly improved the luminescence efficiency and stability of the material through the confinement effect of ZIF-8 [36]. The dual functional sensing platform constructed based on GSH-Au NCs@ZIF-8 exhibited unique advantages: its MOF structure could achieve efficient enrichment of $Cu^{2+}$ ions, combined with fluorescence quenching effect to achieve an ultra-low detection limit of 0.016 μM; simultaneously integrating the acetylcholinesterase inhibition reaction system, a highly sensitive detection method for organophosphate pesticides had been developed through the specific coordination between thiocholine and $Cu^{2+}$, breaking through the limitations of traditional enzyme inhibition methods that require complex labeling. In response to the diverse detection needs of environmental pollutants, Meng and coworkers further constructed a fluorescent sensor array based on ligand engineering (Figure 11.5) [37]. A specific response system for seven heavy metal

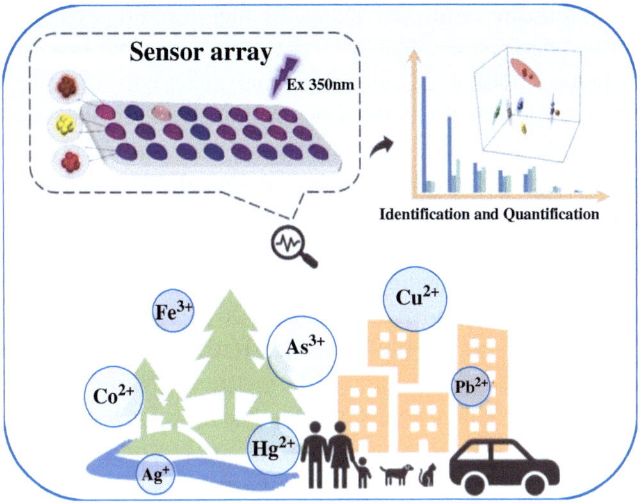


**Figure 11.5** Metal nanocluster-based sensor array for multimetal ion identification in water. *Source:* [37]/with permission of Elsevier.

ions ($Pb^{2+}$, $Fe^{3+}$, $Cu^{2+}$, $Co^{2+}$, $Ag^{+}$, $Hg^{2+}$, $As^{3+}$) was established by designing Au nanoclusters protected by three differentiated ligands: GSH, ovalbumin, and bovine serum albumin (BSA). Each metal nanocluster induced a change in aggregation state through coordination, resulting in characteristic fluorescence enhancement/quenching patterns. By combining linear discriminant analysis and hierarchical clustering analysis, a multidimensional fingerprint database was constructed to achieve synchronous detection of multiple components in untreated environmental water samples. This array-based sensing system combined high-throughput screening and precise identification characteristics, providing a new analytical tool for the field of environmental monitoring. Subsequently, Hu and coworkers continued to expand Au nanoclusters protected by other ligands (mercaptopropionic acid, 2-mercaptobenzoic acid) and embedded them into ZIF-8 through coordination reactions. The developed sensor array could accurately identify eight heavy metal ions ($Ni^{2+}$, $Cr^{3+}$, $Co^{2+}$, $Pb^{2+}$, $Cd^{2+}$, $Ag^{+}$, $Cu^{2+}$, and $Zn^{2+}$) with concentrations ranging from 0.5 to 50 μM [38].

The carrier mainly plays a role in stabilizing metal nanoclusters. In addition to using MOF materials similar to ZIF-8, inorganic nanomaterials such as carbon nitride ($C_3N_4$) and graphene can also be used as carriers. Xiao and coworkers anchored monatomic Pt ($Pt_1$) and $Au_{25}$ nanoclusters with high loading onto graphitic $C_3N_4$ (g-$C_3N_4$) for electrocatalytic detection of As (III) in water [39]. Under neutral conditions at pH 7, its sensitivity to As (III) reached $8.32\,\mu A \cdot ppb^{-1}$. In addition, the detection limit was as low as 0.2 ppb, far below the World Health Organization's standard for As content (10 ppb) in drinking water.

In addition to the Au nanoclusters mentioned above, Ag and Cu NCs also exhibit remarkable performance in the identification and adsorption removal of metal ions in water. TXu and coworkers synthesized different Cu NCs using three functional ligands: polyethylenimine, histidine, and GSH, achieving high-resolution recognition of 12 types of heavy metal ions (including $Pb^{2+}$, $Fe^{3+}$, $Cu^{2+}$, etc.) and 5 categories of dissolved organic matter (DOM, including humic acid, lipids, fatty acids, etc.) [40]. This fluorescence sensing array demonstrated significant environmental applicability: it could effectively identify trace metal ions in tap water and distinguish river water and seawater samples from different regions based on fluorescence response differences, providing a novel technical platform for rapid screening of multicomponent pollutants in complex water bodies. Yuan and coworkers further designed D-penicillamine-modified Ag nanoparticle-NC (Ag NP-NC) composites encapsulated within ZIF-8, constructing AgNP-NC@ZIF-8 composite luminescent materials [41]. Through the confinement effect of MOF, the photoluminescence intensity of AgNP-NC@ZIF-8 was significantly enhanced (66 ± 5 times), and the fluorescence lifetime was extended from (50.9 ± 1.8) ns to (15.9 ± 0.4) μs, while retaining the microporous adsorption properties of ZIF-8. This composite material exhibited unique dual "detection-removal" functions: on one hand, based on the fluorescence response characteristics of AgNP-NCs, highly sensitive detection of Cr(VI) was achieved, showing a good linear relationship in the 40–400 μM concentration range ($R^2 = 0.998$) with a detection limit of (23.5 ± 0.8) μM; on the other hand, effective adsorption and removal of Cr(VI) were realized through the microporous structure of ZIF-8. The system maintained high selectivity for total Cr in complex systems containing interfering ions such as $Pb^{2+}$ and $Fe^{3+}$ (selectivity coefficient >100) and was successfully applied to simultaneous monitoring and remediation of Cr pollution in industrial wastewater and natural water samples.

Beyond heavy metal contamination, the ecological risks posed by other inorganic ion pollutants, such as halides and nitrites, urgently require focused attention to mitigate environmental harm. Liu and coworkers prepared new aluminum based porous nanoclusters with the characteristics of both inorganic molecular sieves and organic macrocycles [42]. Al nanoclusters adapt pure inorganic Archimedean truncated cube as the core, and formed six face centered cup-shaped aromatic macrocyclic cavities on the periphery through precise self-assembly, constructing a unique "core-shell" dual porous structure. It could achieve fast and low concentration capture of iodine species in $I_2$/KI aqueous solution through ion exchange, and achieve visible and efficient adsorption at concentrations as low as 4 ppm. Combined with spectroscopic characterization, the aggregation evolution law of iodine species in the porous structure was revealed. Zhang and coworkers achieved controllable synthesis of highly stable $Ag_{56}$ nanoclusters through precise control of reaction parameters, and used them as near-infrared fluorescence sensors for $Br^-$ ions and $I^-$ ions, with detection limits as low as 85 and 105 nM, respectively [43]. As shown in Figure 11.6, the dichloromethane ($CH_2Cl_2$) solution of $Ag_{56}$ exhibited unique luminescent properties. As the concentration of the solution gradually decreased, the excitation and emission wavelengths of its fluorescence spectrum underwent a blue shift, with the excitation wavelength shifting from 618 to 385 nm and the emission wavelength shifting from 788 to 485 nm. To evaluate the selectivity of $Ag_{56}$ nanoclusters for $Br^-/I^-$ detection, the research team conducted systematic tests on 9 metal ions, 16 anions, 10 amino acids, as well as human blood and urine samples.

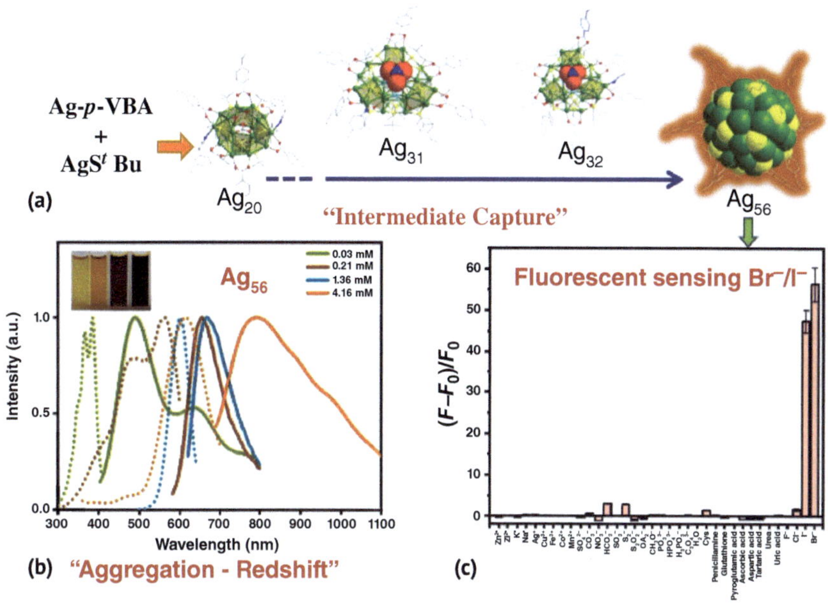


**Figure 11.6** $Ag_{56}$ nanoclusters for specific fluorescence recognition of $Br^-$ and $I^-$. (a) Synthesis evolution diagram of the $Ag_{56}$ nanoclusters; (b) excitation and emission spectra of $Ag_{56}$ in $CH_2Cl_2$ solution, with the excitation spectra shown as a dashed line and the emission spectra as a solid line; (c) fluorescence response $(F\text{-}F_0)/F_0$ of $Ag_{56}$ nanoclusters in the presence of various metal ions and anions at a concentration of 10 μM. *Source:* [43]/ Tsinghua University Press Ltd/CC BY 4.0.

The study systematically investigated their effects on the photoluminescence properties of $Ag_{56}$ nanoclusters, revealing that only $Br^-$ and $I^-$ ions significantly enhanced fluorescence intensity. This observation likely arises from halide-induced partial decomposition of the $Ag_{56}$ nanoclusters through ligand exchange reactions, which creates new emissive species or alters the electronic structure of the original nanoclusters.

Nitrite ($NO_2^-$) is a strong carcinogen that is widely present in groundwater and surface water. Researchers used Au nanoclusters that were protected with BSA as fluorescent probes to detect nitrite [44]. In addition, It was found that [45] the red fluorescence of Au nanoclusters (with excitation/emission peaks at 365/635 nm) could be quenched by trace amounts of $I^-$ etching. Under the catalytic action of $NO_2^-$, $I^-$ were oxidized by $BrO_3^-$ to form free $I_2$. This catalytic process provided a sensitive method for $NO_2^-$ detection. Under optimal conditions, the fluorescence intensity decreased linearly within the range of $NO_2^-$ concentration from 10 nM to 0.8 μM, and the detection limit reached 1.1 nM, which was several orders of magnitude lower than the maximum allowable concentration specified by authoritative institutions. Chen and coworkers [46] used hyperbranched polyethyleneimine (hPEI) stabilized AgNCs for sensitive detection of $NO_2^-$. The detection principle was that $NO_2^-$ reacted with hydrogen peroxide ($H_2O_2$) under acidic conditions to generate peroxynitrite, which could induce the aggregation of hPEI-AgNCs and reduce fluorescence intensity. This sensor exhibited a detection limit of 100 nmol/L and demonstrated promising potential for $NO_2^-$ detection in actual samples such as seawater.

### 11.3.2 Removal of Organic Pollutants from Water Bodies

At present, organic pollutants in waste water mainly refer to dyes, pesticides, antibiotics, etc. These pollutants come from chemical plants and medical related enterprises. Metal nanoclusters, as catalysts, are generally based on the Fenton-like catalytic principle [47], catalyzing various strong oxidants (such as $H_2O_2$, persulfate, etc.) to generate active free radicals or non-free radicals. They can "tear apart" various organic pollutants and ultimately decompose them into nontoxic and harmless $CO_2$ and $H_2O$. Additionally, there are catalytic reaction types that integrate technologies such as light and electricity with Fenton-like systems, which are referred to as photo-Fenton, electro-Fenton, respectively.

Liu and coworkers [48, 49] first reported the integration of Au nanoclusters, specifically $Au_{25}(MHA)_{18}$ (where MHA denotes mercaptohexanoic acid), with carbon nanotubes as Fenton-like reaction catalysts. They discovered that their catalytic function lay in their ability to undergo redox cycling of $Au^+/Au^0$ under an electric field (Figure 11.7). Compared to larger-sized nanoparticles (13.2 and 5.7 nm), the Au nanoclusters (≤2 nm) demonstrated higher efficiency in degrading tetracycline (an antibiotic) and greater reaction kinetic constants. Through experimental research and theoretical calculations, a catalytic mechanism was proposed where $H_2O_2$ molecules adsorbed at the apex sites of the Au nanoclusters tend to directly decompose into hydroxyl radicals (•OH). This work demonstrates the intriguing application of metal nanoclusters in Fenton-like water purification. Notably, the team loaded the metal nanoclusters onto film-forming materials to construct a unique Fenton membrane reactor, which effectively reduced metal nanoclusters loss and enabled high-efficiency catalyst recovery. In addition

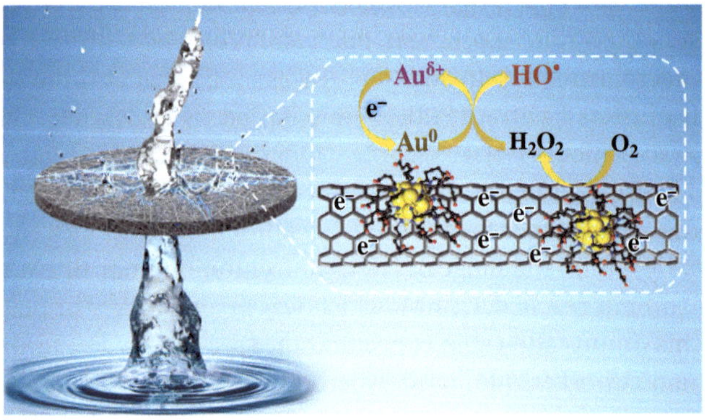


**Figure 11.7** Supported atomically precise Au nanoclusters for enhanced flow-through electro-Fenton reaction. *Source:* [48]/with permission of American Chemical Society.

to the aforementioned carbon nanotubes, they also explored graphene materials. Inspired by this work, Yang and coworkers prepared composite Fenton-like catalysts by loading Cu nanoclusters onto various carriers (such as hydroxyapatite, silica, etc.), achieving 100% degradation of multiple organic dyes [50, 51].

In addition to Fenton-like catalytic systems, metal nanoclusters have also demonstrated significant advantages in the field of photocatalytic degradation of organic pollutants. This type of nanomaterial, with its quantum size effect and adjustable electronic structure, can not only efficiently capture and convert light energy, but also achieve charge directed transfer through precise regulation of band structure, significantly improving pollutant degradation efficiency. Especially in the treatment of typical recalcitrant organic compounds such as azo dyes and antibiotics, their photocatalytic performance shows an order of magnitude improvement compared to traditional semiconductor materials, providing an innovative solution for the treatment of environmental pollutants that is both efficient and sustainable.

Ullahand coworkers [52] constructed Ag nanoclusters protected by natural tripeptide ligands (GSH@Ag NCs) and used them as photocatalytic materials for dye degradation. Under sunlight irradiation, the degradation rate of organic dye Erythrosine B was significantly higher than that of rhodamine B (RhB) (85.1% degradation rate), reaching 94.6%. However, under the illumination of white LED, the degradation efficiency of the above-mentioned dyes showed a decreasing trend, reaching 78.57% and 67.923%, respectively, under the same experimental conditions. GSH@Ag nanoclusters exhibited astonishing high degradation efficiency under sunlight, due to the significant advantage of high irradiance (1370 W) compared to LED light (0.07 W), and the •OH generated on the surface of GSH@Ag NCs initiated degradation reactions.

Another common method is to combine metal nanoclusters with semiconductor carriers, which not only compensates for the insufficient light absorption efficiency of the semiconductor itself, but also effectively stabilizes and protects the metal nanoclusters, avoiding their aggregation during the catalytic process. Titanium dioxide ($TiO_2$) is widely used as a photocatalytic material for environmental applications due to its high activity, chemical stability, resistance to photocorrosion, low toxicity, low cost, and sustainability. However, the rapid electron-hole recombination rate limits its performance. Several research groups have attempted to combine metal nanoclusters with them to achieve ideal photocatalytic degradation of organic pollutants.

Liu and coworkers [53] synthesized a new type of alloy metal nanoclusters, $Au_{23-x}Ag_x(S\text{-}Adm)_{15}$ (where S-Adm represents an adamantanethiol ligand), and loaded them onto the surface of $TiO_2$ carrier. When compared with $TiO_2$ loaded with monometallic nanoclusters ($Au_{23}(S\text{-}Adm)_{16}$), significant differences were observed in the degradation efficiency of RhB and phenol under visible light irradiation, indicating that the Ag-doped alloy metal nanoclusters exhibited markedly improved photocatalytic activity (Figure 11.8).

Cheng and coworkers [54] synthesized a series of thiolate-protected $Ag_4M_2$ nanoclusters (where M represents dopant elements Ni, Pd, and Pt). These metal nanoclusters adopted a distorted hexahedral structural framework, with four Ag atoms located at the midpoints of the side edges and two heterometal atoms positioned at the centers

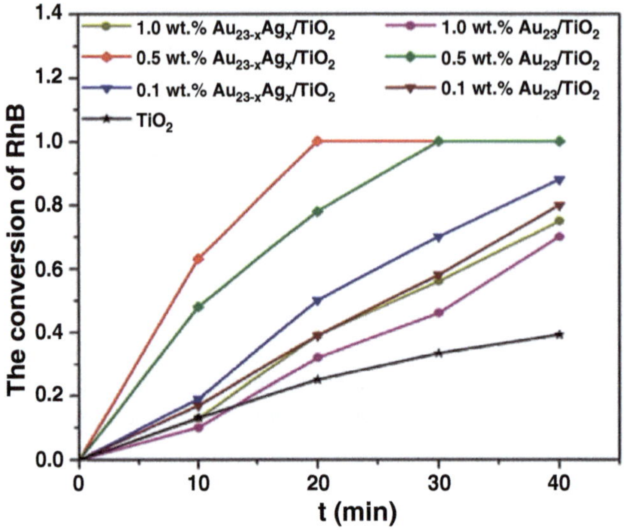


**Figure 11.8** Comparison of photocatalytic degradation of RhB under visible-light irradiation over $Au_{23-x}Ag_x/TiO_2$, $Au_{23}/TiO_2$, and $TiO_2$. *Source:* [53]/with permission of John Wiley & Sons.

of the top and bottom faces. After combining these three metal nanoclusters with $TiO_2$, the photocatalytic performance order for methyl orange and RhB dye degradation was found to be opposite, a phenomenon attributed to two distinct types of intermolecular recombination mechanisms. In both photocatalytic systems, the catalysts and dyes exhibited visible light activity, where intermolecular recombination between photogenerated holes in the catalyst and photogenerated electrons in the dye led to charge separation across the system, including both the catalyst and the dye.

A scientific team from Flinders University in Australia [55, 56] chemically "anchored" $Au_9$ nanoclusters to $TiO_2$, which in turn catalyzed reactions by harvesting ultraviolet energy through energy transfer. This $Au_9$ nanoclusters co-catalyst not only enhanced the photocatalytic effect of $TiO_2$ but also reduced the reaction completion time by sixfold. This heterogeneous semiconductor-mediated photocatalytic system proved highly efficient and sustainable for degrading organic pollutants such as azo dyes. In summary, the application of metal nanoclusters has brought new hope to wastewater treatment. With ongoing research and practice, the prospects for metal nanoclusters technology in the field of water environmental protection are expected to expand significantly.

### 11.3.3 Removal of Bacteria from Water Bodies

Given that natural water bodies are typically exposed to open environments and harbor diverse microbial populations, including potential pathogens that pose significant risks to drinking water safety, establishing an efficient water disinfection system capable of reliably neutralizing these harmful microorganisms becomes a critical component in safeguarding water quality. Liu and coworkers [57] reported an electrochemical water disinfection system using atomically precise $Ag_{28}$ nanoclusters as electrode materials. In this study, ultra-small $Ag_{28}$ nanoclusters were selected as electrocatalytic models and further encapsulated within the ZIF-67 framework to enhance catalytic durability (labeled as Ag NCs@ZIF-67). Subsequently, Ag NCs@ZIF-67 assembled as a capacitor deionized water disinfection (CEWD) anode, efficient water disinfection was achieved through the synergistic effect of in situ generation of residual reactive oxygen species (ROS) and synchronous microbial capture (Figure 11.9). In addition, the design of the system implemented a cyclic process of "alive microbe capture-killing-dead microbe desorption," which could be used for continuous water disinfection. On this basis, the water disinfection system had an efficient killing effect on broad-spectrum microorganisms (antibacterial activity >99.99%), good durability (After 40 cycles, the performance only decreased by 0.75%, and the antibacterial efficiency remained at 99.90% even after continuous operation for five hours), multifunctionality (can be adapted to other metal nanoclusters), scalability (water production rate reached 213.33 $L \cdot h^{-1} \cdot m^{-2}$), energy saving (energy consumption as low as 4.91 $Wh \cdot m^{-3}$; 1.04 $Wh \cdot m^{-3}$ when excluding pumping costs), and was suitable for various practical water samples. Zhu and coworkers designed Au nanoclusters with aggregation induced luminescence (AIE) properties, achieving their antibacterial and deodorizing properties under visible light driving [58]. Thanks to its unique AIE characteristics, visible light absorption advantages, and rich surface chemical properties, this AIE-type Au nanoclusters exhibited unprecedented visible light driven antibacterial activity, showing excellent inhibitory effects on both Gram-positive bacteria (≥98.5%) and Gram-negative bacteria (≥99.94%). In addition, these Au nanoclusters had good biocompatibility and were easy to apply in downstream antibacterial and deodorizing scenarios. For example, daily materials (such as latex, ceramic glaze, organic fibers, and fabrics) modified by them could

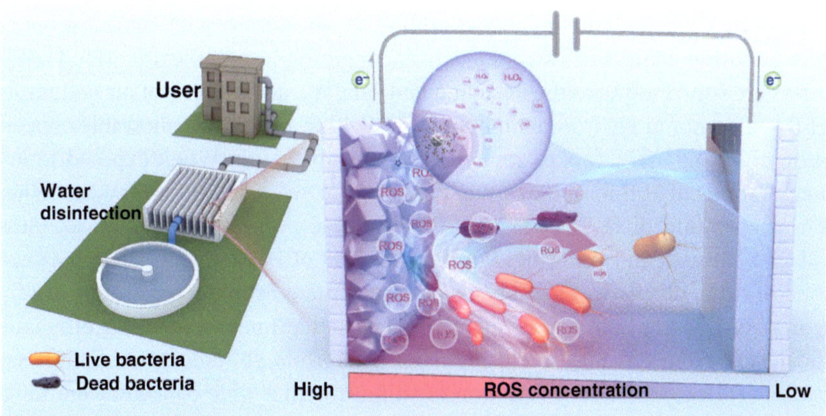


**Figure 11.9** CEWD sterilization application scenarios and working principle schematic diagram. *Source:* [57]/Springer Nature/CC BY 4.0.

achieve up to 99% antibacterial efficiency and >97.9% deodorizing efficiency under visible light irradiation.

In addition, the application research of metal nanoclusters in water environment disinfection was relatively limited at present, but significant accumulation had been formed in their research on bacterial detection and antibacterial performance [59, 60]. For example, Cao and coworkers [61] prepared Ag nanoclusters-porphyrin complexes through self-assembly method, with a structure of $Ag_9$-AgTPyP, which exhibited strong bactericidal properties under visible light. In addition, films containing $Ag_9$-AgTPyP had good killing effects on super resistant bacteria. This material established a novel long-acting charge transfer state (with high affinity for $O_2$) through AgTPyP to adjacent $Ag_9$ nanoclusters, significantly improving the efficiency of ROS generation. Its unique framework structure further accelerated the transmission of ROS. Wang and coworkers [62] prepared a stable $Au_{25}(MHA)_{18}$ nanoclusters by reducing Au (III) to Au (0) using a one pot method. At lower concentrations (100–150 μM) of $Au_{25}(MHA)_{18}$, almost 100% of bacteria could be killed in a short period of time (30–60 minutes), achieving rapid and efficient killing of Gram-negative bacteria.

Although these antibacterial studies are mainly applied in the biological field and have not been attempted in actual complex water environments, they can still serve as a reference for future efficient water disinfection work and indirectly demonstrate the broad application prospects of metal nanoclusters in the antibacterial field.

### 11.3.4 Air Pollution Control: The Ability of Metal Nanoclusters as "Air Purifiers"

Due to the synergistic effects of multiple pollution sources such as industrial pollution (including chemical production, metal smelting, etc.), mobile emissions (transportation exhaust emissions and particulate matter and sulfides generated by coal-fired heating), indoor pollution sources (VOCs from decorative materials and tobacco smoke), and unfavorable meteorological diffusion conditions, the current atmospheric environmental quality continues to deteriorate, with significant composite pollution characteristics, seriously endangering human health. Metal nanoclusters are contributing to the detection and catalytic degradation of air pollutants.

$H_2S$ is a common environmental pollutant, with a maximum allowable concentration of 10 mg · $m^{-3}$ (approximately 6.6 ppm) in the air. When exposed to low concentrations of $H_2S$ for a long time, it may cause eye irritation symptoms. When the concentration of $H_2S$ is too high, it can cause pulmonary edema, dizziness, headache, nausea, and vomiting. If a large amount of $H_2S$ gas is inhaled in a short period of time, people will quickly experience acute poisoning, respiratory paralysis, and even death. Zhang and coworkers reported a fluorescent AuAg bimetallic nanocluster protected by 2-mercapto-5-benzimidazole sulfonic acid sodium salt dihydrate (MBZS) for visual detection of sulfides with a wide concentration range [63]. MBZS-AuAg NCs could emit fluorescence of different colors similar to chameleon like in the presence of sulfides at different concentrations, and the fluorescence intensity changes linearly with the sulfide concentration, allowing for quantitative

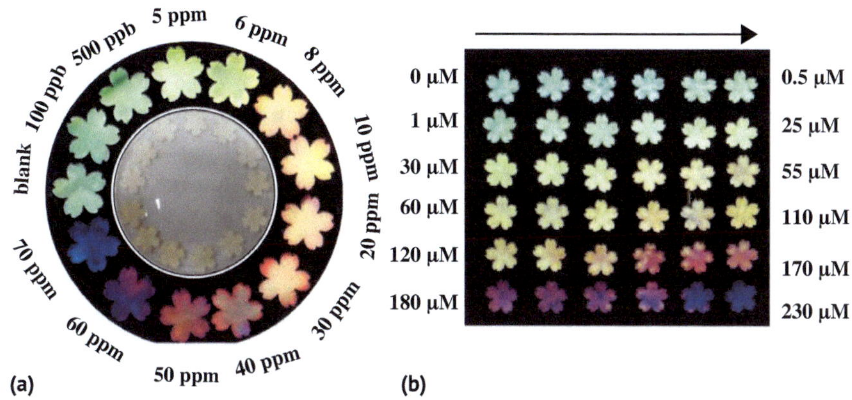


**Figure 11.10** Images of flower-like filter papers immobilized with MBZS-AuAg nanoclusters after exposure to (a) $H_2S$ and (b) $S^{2-}$ at different concentrations. All filter papers were air-dried for enhanced visual contrast. In Figure 11.6b, the $S^{2-}$ concentration increases from left to right, with each filter paper representing a distinct concentration. *Source:* [63]/with permission of American Chemical Society.

and visual detection of sulfides over a wide concentration range (Figure 11.10). In addition, MBZS-AuAg nanoclusters were further fixed on filter paper as test strips for detecting sulfides, which could not only detect the concentration of $S^{2-}$ in water samples, but also detect $H_2S$ in important industrial and mining environments, achieving visual detection and grading warning of gaseous $H_2S$.

VOCs mainly include alkanes, alkenes, alkynes, benzene derivatives, halogenated hydrocarbons, alcohols, aldehydes, ketones, etc. Some substances (such as benzene and formaldehyde [HCHO]) are carcinogenic, and long-term exposure can cause leukemia, liver, and kidney damage. Intake of high concentrations of VOCs may damage the central nervous system and even endanger life.

In 2023, Tian and coworkers [64] invented a Ni/Mn oxide supported $Ag_{30}$ nanocluster catalyst, which was found to have significantly improved catalytic activity and good oxidative degradation effect on HCHO, while $Ag_{25}$ and $Ag_{44}$ nanoclusters did not achieve the same excellent catalytic activity. Perhaps due to the unique structure of $Ag_{30}$ nanoclusters, they had a good activation effect on $O_2$, which could be converted into ROS to decompose HCHO. In the same year, Zhumeng Xu, general manager of Laizhou Jieli Industry and Trade Co., Ltd, invented a $Pt_6$ nanoclusters based catalyst, in which $Pt_6$ nanoclusters were loaded on various transition metal oxide supports ($M_xN_yO_z$, M and N were independently selected from two of Ti, Co, Mn, Fe, Cu, Ni, Mg, and Al; the values of $x$, $y$, and $z$ satisfied the valence equilibrium) with a Pt loading amount of 0.5–3wt%. This $Pt_6$ nanoclusters supported catalyst had excellent HCHO removal catalytic activity, high utilization efficiency of Pt, and stable catalyst structure, which could catalyze the oxidation of HCHO at room temperature without the need for light, thereby converting it into non-toxic and harmless $H_2O$ and $CO_2$. The catalytic reaction proceeded smoothly under mild conditions, exhibiting rapid HCHO degradation and maintaining excellent long-term catalytic activity. The patent was awarded the Third Prize of the 2024 Annual Technological Innovation Award by China Building Materials Circulation Association.

The two invention patents mentioned above had both completed the transformation of achievements and had been officially launched in the market. They were applied in a series of new air purification products, including virus protection HCHO removers, air conditioner (fresh air) inlet/outlet system HCHO removers, metal nanoclusters based HCHO decomposition sheets/hangers, HCHO removal equipment filters, and Jingweikang (a brand name), among others. This series of products could not only remove HCHO, benzene, and benzene series substances but also effectively eliminated active short-carbon-chain gaseous organic compounds (odors). They could effectively maintain the surrounding environment and the health of humans and animals. This series of air purification products with independent intellectual property rights, such as metal nanoclusters-based CHO removal sheets, have been recognized by society and favored by a large number of users since their launch.

In summary, as the "molecular level guardians" in the field of environmental remediation, metal nanoclusters are reshaping the pollution control technology system with their high specific surface area and precise catalytic properties. Their characteristics of high efficiency and sustainability indicate technological innovation in the field of environmental purification. In the near future, these "ultra-small teams" may enter thousands of households and become the "core components" of household water purifiers, urban air purification towers, and other devices.

## References


1 A. T. Bell, The Impact of Nanoscience on Heterogeneous Catalysis. *Science* **2003**, *299*, 1688–1691.

2 R. H. Crabtree, Resolving Heterogeneity Problems and Impurity Artifacts in Operationally Homogeneous Transition Metal Catalysts. *Chem. Rev.* **2012**, *112*, 1536–1554.

3 X. Kang, H. Chong, M. Zhu, $Au_{25}(SR)_{18}$: The Captain of the Great Nanocluster Ship. *Nanoscale* **2018**, *10*, 10758–10834.

4 Z. Lei, J.-J. Li, Z.-A. Nan, Z.-G. Jiang, Q. M. Wang, Cluster from Cluster: A Quantitative Approach to Magic Gold Nanoclusters $[Au_{25}(SR)_{18}]^-$. *Angew. Chem. Int. Ed.* **2021**, *60*, 14415–14419.

5 H. Shen, Y.-Z. Han, Q. Wu, J. Peng, B. K. Teo, N. Zheng, Simple and Selective Synthesis of Copper-containing Metal Nanoclusters Using $(PPh_3)_2CuBH_4$ as Reducing Agent. *Small Methods* **2021**, *5*, 2000603–2000609.

6 A. Chen, X. Kang, S. Jin, W. Du, S. Wang, M. Zhu, Gram-scale Preparation of Stable Hydride M@Cu24 (M = Au/Cu) Nanoclusters. *J. Phys. Chem. Lett.* **2019**, *10*, 6124–6128.

7 C. Dong, X. Song, B. E. Hasanov, Y. Yuan, L. Gutiérrez-Arzaluz, P. Yuan, S. Nematulloev, M. Bayindir, O. F. Mohammed, O. M. Bakr, Organic–Inorganic Hybrid Glasses of Atomically Precise Nanoclusters. *J. Am. Chem. Soc.* **2024**, *146*, 7373–7385.

**8** Z.-M. Zhu, Y. Zhao, H. Zhao, C. Liu, Y. Zhang, W. Fei, H. Bi, M.-B. Li, Photochemical Route for Synthesizing Atomically Precise Metal Nanoclusters from Disulfide. *Nano Lett.* **2023** *23*, 7508–7515.

**9** J.-Q. Fan, K. Cen, H.-J. Xu, H.-Y. Wang, Y. Yang, Z.-M. Zhu, H. Liu, D. Chen, W. Fan, M.-B. Li, Photochemical Synthesis of Group 10 Metal Nanoclusters for Electrocatalysis. *Nanoscale* **2023**, *15*, 19079–19084.

**10** Y. Zhao, Z.-M. Zhu, W. Fan, W. Zhu, J.-J. Yang, Y. Tao, W. Fei, H. Bi, S. Zhang, M.-B. Li, Photosynthesis of $Au_8Cu_6$ Nanocluster for Photocatalysis in Oxidative Functionalization of Alkynes. *Nat. Commun.* **2024**, *15*, 9632.

**11** Y.-X. Wang, J. Zhang, H.-F. Su, X. Cui, C.-Y. Wei, H. Li, X.-M. Zhang, Photochemical Synthesis of Atomically Precise Ag Nanoclusters. *ACS Nano.* **2023,** *17*, 11607–11615.

**12** Y.-X. Wang, F.-Q. Zhang, Z. Qi, X. Zhao, N. Zhang, H. Li, H. Li, X.-M. Zhang, Photochemical Synthesis of $Ag_{12}Cu_7$ Nanocluster with Cuprophilicity-related Long-lived Phosphorescence. *Aggregate* **2025**, *6*, e675.

**13** L. Zeng, J. Wang, D. Wang, H. Yi, A. Lei, Comprehensive Comparisons between Directing and Alternating Current Electrolysis in Organic Synthesis. *Angew. Chem. Int. Ed.* **2023**, *62*, e202309620.

**14** Y. Wan, J. Xu, R. Lv, Heterogeneous Electrocatalysts Design for Nitrogen Reduction Reaction under Ambient Conditions. *Mater. Today*. **2019**, *27*, 69–90.

**15** J. C. Siu, N. Fu, S. Lin, Catalyzing Electrosynthesis: A Homogeneous Electrocatalytic Approach to Reaction Discovery. *Acc. Chem. Res.* **2020**, *53*, 547–560.

**16** P. Li, Y. Wang, H. Zhao, Y. Qiu, Electroreductive Cross-coupling Reactions: Carboxylation, Deuteration, and Alkylation. *Acc. Chem. Res.* **2025**, *58*, 113–129.

**17** S. Zhang, J. Shi, J. Li, M.-B. Li, G. Li, M. Findlater, Catalyst-dependent Direct and Deoxygenative Coupling of Alcohols by Convergent Paired Electrolysis. *CCS Chem.* **2022**, *4*, 1938–1948.

**18** S. Zhang, Y. Liang, K. Liu, X. Zhan, W. Fan, M.-B. Li, M. Findlater, Electrochemically Generated Carbanions Enable Isomerizing Allylation and Allenylation of Aldehydes with Alkenes and Alkynes. *J. Am. Chem. Soc.* 2023, **145**, 14143–14154.

**19** K. Liu, M. Lei, X. Li, X. Zhang, Y. Zhang, W. Fan, M.-B. Li, S. Zhang, Paired Electrocatalysis Unlocks Crossdehydrogenative Coupling of C($sp^3$)-H Bonds Using a Pentacoordinated Cobalt-salen Catalyst. *Nat. Commun.* **2024**, *15*, 2897.

**20** J. Feng, Y. Xia, M. Shen, J. Wang, W. Fan, S. Zhang, M.-B. Li, Electrochemical Allene C–H Functionalization via Carbanion Sampling. *Angew. Chem. Int. Ed.* **2025**, e202508369.

**21** J. Dong, Y. Li, Y. Ding, H.-F. Su, X. Cui, Y.-X. Wang, H. Li, Electrosynthesis of Atomically Precise Au Nanoclusters. *Adv. Sci.* **2025**, *12*, 2414057.

**22** Y.-M. Wang, J. Cai, Q.-Y. Wang, Y. Li, Z. Han, S. Li, C.-H. Gong, S. Wang, S.-Q. Zang, T. C. W. Mak, Electropolymerization of Metal Clusters Establishing a Versatile Platform for Enhanced Catalysis Performance. *Angew. Chem. Int. Ed.* **2022**, *61*, e202114538.

**23** M. B. Plutschack, B. Pieber, K. Gilmore, P. H. Seeberger, The Hitchhiker's Guide to Flow Chemistry. *Chem. Rev.* **2017**, *117*, 11796–11893.

**24** Y. Zhang, S.-R. He, Y. Yang, T.-S. Zhang, Z.-M. Zhu, W. Fei, M.-B. Li, Preorganized Nitrogen Sites for $Au_{11}$ Amidation: A Generalizable Strategy toward Precision Functionalization of Metal Nanoclusters. *J. Am. Chem. Soc.* **2023**, *145*, 12164–12172.

**25** R.-L. He, F. Hu, Z.-J. Guan, Q.-M. Wang, Large Scale Synthesis of a Stable Prefunctionalized Silver Nanocluster. *Angew. Chem. Int. Ed.* **2024**, *63*, e202410827.

**26** D. J. Cole-Hamilton, Homogeneous Catalysis–New Approaches to Catalyst Separation, Recovery, and Recycling. *Science* **2003**, *299*,1702–1706.

**27** M. A. Rivero-Crespo, G. Toupalas, B. Morandi, Preparation of Recyclable and Versatile Porous Poly(aryl thioether)s by Reversible Pd-Catalyzed C–S/C–S Metathesis. *J. Am. Chem. Soc.* **2021**, *143*, 21331–21339.

**28** J. Pritchard, G. A. Filonenko, R. V. Putten, E. J. M. Hensen, E. A. Pidko, Heterogeneous and Homogeneous Catalysis for the Hydrogenation of Carboxylic Acid Derivatives: History, Advances and Future Directions. *Chem. Soc. Rev.* **2015**, *44*, 3808–3833.

**29** A. E. C. Collis, I. T. Horváth, Heterogenization of Homogeneous Catalytic Systems. *Catal. Sci. Technol.* **2011**, *1*, 912–919.

**30** C. Tan, K. Yang, J. Dong, Y. Liu, Y. Liu, J. Jiang, Y. Cui, Boosting Enantioselectivity of Chiral Organocatalysts with Ultrathin Two-dimensional Metal-organic Framework Nanosheets. *J. Am. Chem. Soc.* **2019**, *141*, 17685–17695.

**31** J.-Q. Fan, Y. Yang, C.-B. Tao, M.-B. Li, Cadmium-doped and Pincer Ligand-modified Gold Nanocluster for Catalytic KA2 Reaction. *Angew. Chem. Int. Ed.* **2023**, *62*, e202215741.

**32** W. Zhu, S. Zhang, W. Fan, Y. Yang, H. Zhao, W. Fei, H. Bi, J. He, M.-B. Li, Z. Wu, Atomically Precise Metal Nanoclusters as Single Electron Transferers for Hydroborylation. *Precis. Chem.* **2023**, *1*, 175–182.

**33** Y. Yang, C. Chen, G.-Y. Xu, J. Yuan, S.-F. Ye, L. Chen, Q.-L. Lv, G. Luo, J. Yang, M.-B. Li, Z. Wu, An Efficient Nanocluster Catalyst for Sonogashira Reaction. *J. Catal.* **2021**, *401*, 206–213.

**34** H. Liu, K. Li, F. Li, Y. Zhao, W. Fan, Y. Pei, M.-B. Li, Cobalt-kernelled Icosahedral Gold Nanocluster. *Chin. Chem. Lett.* **2025**, https://doi.org/10.1016/j.cclet.2025.111230.

**35** Á. Molnár, A. Papp, Catalyst Recycling-A Survey of Recent Progress and Current Status. *Coord. Chem. Rev.* **2017**, *349*, 1–65.

**36** D. Wei, M. Li, Y. Wang, N. Zhu, X. Hu, B. Zhao, Z. Zhang, D. Yin, Encapsulating Gold Nanoclusters into Metal-organic Frameworks to Boost Luminescence for Sensitive Detection of Copper Ions and Organophosphorus Pesticides. *J. Hazard. Mater.* **2023**, *441*, 129890.

**37** H. Meng, Y. Wang, R. Wu, Y. Li, D. Wei, M. Li, N. Zhu, F. Zhu, Z. Zhang, H. Zhao, Identification of multi-component metal ion mixtures in complex systems using fluorescence sensor arrays. *J. Hazard. Mater.* **2023**, *455*, 131546.

**38** X. Hu, Z. Mu, Y. Li, L. Bai, M. Qing, Metal-organic Frameworks-scaffold Gold Nanoclusters Enabled Aggregation-induced Enhanced Fluorescent Sensor Array for High-Throughput Detection of Heavy Metal Ions. *Microchem. J.* **2025**, *210*, 113020.

**39** X. Xiao, P.-H. Li, L. Tang, D. Wu, H. Xia, Z.-Y. Song, Y.-H. Zhao, B. Liang, M. Yang, R. Tang, J. Yao, X.-J. Huang, X. Chen, Z. Lei, Stepwise Coordination Engineering of $Pt_1/Au_{25}$ Dual Catalytic Sites with Enhanced Electrochemical Activity and Stability. *Adv. Mater.* **2025**, *37*, 2417900.

**40** J. Xu, H. Zhou, Y. Zhang, Y. Zhao, H. Yuan, X. He, Y. Wu, S. Zhang, Copper Nanoclusters-Based Fluorescent Sensor Array to Identify Metal Ions and Dissolved Organic Matter. *J. Hazard. Mater.* **2022**, *428*, 128158.

**41** H. Yuan, H. Zhou, Y. Zhao, H. Tan, R. Antoine, S. Zhang, Metal-organic Frameworks Encapsulated Ag Nanoparticle-nanoclusters with Enhanced Luminescence For Simultaneous Detection and Removal of Chromium(VI). *Microchem. J.* **2022**, *181*, 107722.

**42** Y.-J. Liu, Y.-F. Sun, S.-H. Shen, S.-T. Wang, Z.-H. Liu, W.-H. Fang, D. S. Wright, J. Zhang, Water-stable Porous $Al_{24}$ Archimedean Solids for Removal of Trace Iodine. *Nat. Commun.* **2022**, *13*, 6632.

**43** Z. Zhang, J. Dai, C.-C. Feng, C. Shi, L. Zhou, X. Yu, C. Yang, X. Zhang, H. Yang, From Intermediate Capture to Functional Cluster Construction: Synthesis of Silver Clusters and Their $Br^-/I^-$ Sensing Applications. *Polyoxometalates*, **2025**, *4*, 9140086.

**44** Q. Yue, L. Sun, T. Shen, X. Gu, S. Zhang, J. Liu, Synthesis of Fluorescent Gold Nanoclusters Directed by Bovine Serum Albumin and Application for Nitrite Detection. *J. Fluoresc.* **2013**, *23*, 1313–1318.

**45** Z. Su, X. Wang, M. Luo, L. Li, Y. Tu, J. Yan, Fluorometric Determination of Nitrite through Its Catalytic Effect on the Oxidation of Iodide and Subsequent Etching of Gold Nanoclusters by Free Iodine. *Microchim. Acta*, **2019**, *186*, 619.

**46** C. Chen, Z. Yuan, H.-T. Chang, F. Lu, Z. Li, C. Lu, Silver Nanoclusters as Fluorescent Nanosensors for Selective and Sensitive Nitrite Detection. *Anal. Methods.* **2016**, *8*, 2628–2633.

**47** B. Huang, Z. Wu, H. Zhou, X. Wang, Y. Liu, H. Zhang, Z. Xiong, B. Lai, The Structure-performance Relationships in Active Center Size-dependent Fenton like Catalysis: From Nanoparticles to Single Atoms. *Appl. Catal. B Environ*, **2024**, *355*, 124157.

**48** F. Liu, Y. Liu, Q. Yao, Y. Wang, X. Fang, C. Shen, F. Li, M. Huang, Z. Wang, W. Sand, J. Xie, Supported Atomically-precise Gold Nanoclusters for Enhanced Flow-through Electro-Fenton. *Environ. Sci. Technol*, **2020**, *54*, 5913–5921.

**49** Y. Liu, Y. Zheng, B. Du, R. R. Nasaruddin, T. Chen, J. Xie, Golden Carbon Nanotube Membrane for Continuous Flow Catalysis. *Ind. Eng. Chem. Res.* **2017**, *56*, 2999–3007.

**50** Y. Yang, Y. Zhang, S. Jiang, X. Shan, X. Guo, H. Liu, L. Shen, W. Chang, Enhancing Fenton-like Degradation of Organic Pollutants at Neutral pH by Multivalent Cu NCs/HAp Nanocatalysts. *Langmuir*, **2023**, *40*, 827–836.

**51** Y. Yang, Y. Ye, H. Chang, Z. Zhang, J. Yang, Q. Wang, P. Pan, X. Fu, C. Xie, L. Yang, W. Chang, Copper Nanoclusters-doped Novel Carrier with Synergistic Adsorption-catalytic Active Sites to Enable High-performance Dye Removal. *J. Colloid. Interface. Sci.* **2024**, *667*, 478–490.

**52** S. Ullah, Q. Li, R. Ullah, S. Anwar, M. F. Hameed, M. Zhu, Facile Synthesis of Water-Soluble Silver Nanoclusters for the Photocatalytic Degradation of Dyes by Multivariate Optimization Approach. *Nanoscale Adv.* **2023**, *5*, 3326–3335.

**53** C. Liu, X. Ren, F. Lin, X. Fu, X. Lin, T. Li, K. Sun, J. Huang, Structure of the $Au_{23-x}Ag_x(S\text{-}Adm)_{15}$ Nanocluster and Its Application for Photocatalytic Degradation of Organic Pollutants. *Angew. Chem. Int. Ed.* **2019**, *58*, 11335–11339.

**54** X. Cheng, X. Sui, J. Xu, X. Liu, M. Chen, Y. Zhu, On the Photocatalysis Evolution of Heteroatom-Doped $Ag_4M_2$ Nanoclusters. *RSC Adv.* **2021**, *11*, 32526–32532.

**55** A. Motamedisade, A. Heydari, Y. Yin, A. S. Alotabi, G. G. Andersson, Enhanced Photocatalytic Degradation of Methyl Orange Using Nitrogen-functionalized Mesoporous $TiO_2$ Decorated with $Au_9$ Nanoclusters. *Sol. RRL.* **2024**, *8*, 2300943.

**56** A. Motamedisade, A. Heydari, D. J. Osborn, A. S. Alotabi, G. G. Andersson, $Au_9$ Clusters Deposited as Co-catalysts on S-modified Mesoporous $TiO_2$ for Photocatalytic Degradation of Methyl Orange. *Appl. Surf. Sci.* **2024**, *655*, 159475.

**57** Y. Liu, L. Wang, Q. Ma, X. Xu, X. Gao, H. Zhu, T. Feng, X. Dou, M. Eguchi, Y. Yamauchi, X. Yuan, Simultaneous Generation of Residue-free Reactive Oxygen Species and Bacteria Capture for Efficient Electrochemical Water Disinfection. *Nat. Commun.* **2024**, *15*, 10175.

**58** H. Zhu, S. Wang, Y. Wang, C. Song, Q. Yao, X. Yuan, J. Xie, Gold Nanocluster with AIE: A Novel Photodynamic Antibacterial and Deodorant Molecule. *Biomaterials* **2022**, *288*, 121695.

**59** Y. Zheng, M. Wei, H. Wu, F. Li, D. Ling, Antibacterial Metal Nanoclusters. *J. Nanobiotechnol.* **2022**, *20*, 328.

**60** S. K. Krishnan, E. Singh, P. Singh, M. Meyyappan, H. S. Nalwa, A Review on Graphene-Based Nanocomposites for Electrochemical and Fluorescent Biosensors. *RSC Adv.* **2019**, *9*, 8778–8881.

**61** M. Cao, S. Wang, J.-H. Hu, B.-H. Lu, Q.-Y. Wang, S.-Q. Zang, Silver Cluster-Porphyrin-assembled Materials as Advanced Bioprotective Materials for Combating Superbacteria. *Adv. Sci.* **2022**, *9*, 2103721.

**62** Y. Wang, M. J. Malkmes, C. Jiang, P. Wang, L. Zhu, H. Zhang, Y. Zhang, H. Huang, L. Jiang, Antibacterial Mechanism and Transcriptome Analysis of Ultra-small Gold Nanoclusters as An Alternative of Harmful Antibiotics Against Gram-negative Bacteria. *J. Hazard. Mater.* **2021**, *416*, 126236.

**63** B. Zhang, Y. Wang, Y. Wang, F. Huo, P. G. Karmaker, L. Chen, X. Yang, B. Zhao, Chameleon-like Response Mechanism of Gold-silver Bimetallic Nanoclusters Stimulated by Sulfur Ions and Their Application in Visual Fluorescence Sensing. *Anal. Chem.* **2024**, *96*, 5029–5036.

**64** S. Tian, D. Wang, S. Jin, A Nickel-manganese Oxide Supported Catalyst and Its Preparation Method and Application, CN116440923B, 2024-06-25.

# 12

# Conclusion and Perspective

## 12.1 Metal Nanocluster Catalysis: Opportunity

Being the key intermediate between single metal atoms and metal nanoparticles, metal nanoclusters bridge the gap of homogeneous catalysts and heterogeneous catalysts. Thus, metal nanocluster catalysis has the potential to combine the high activity/selectivity of homogeneous catalysis and catalyst separation/recycling capacity of heterogeneous catalysis. The research on metal nanocluster catalysis would offer references for the construction of homogeneous–heterogeneous-integrated catalysts that can be further applied in industry. Moreover, atomically precise regulation on the compositions, especially the spatial structures of metal nanoclusters is beneficial for investigating their catalytic mechanism at atomic level, thus providing an ideal model to reveal the relationship between the structure and catalytic performance of heterogeneous catalysis. Concerning both the fundamental research and practical application, metal nanoclusters bring the following opportunities for synthetic chemistry and catalysis science.

1. It has the opportunity to trigger the discovery of dominant catalysts with super high catalytic activity. Metal nanoclusters feature the ultra-small size and much higher specific surface area than conventional metal nanoparticles [1–3]. Therefore, more metal atoms locate at the surface and can be employed as activation sites, endowing metal nanoclusters with high activity. On the other hand, metal nanoclusters allow catalytic synergy of different metal atoms on the surface [4, 5]. This enables more diversified activation modes of metal nanoclusters compared to the single metal atom catalysts, also endowing metal nanoclusters with higher activity under certain catalytic conditions.
2. It has the opportunity to trigger unique catalytic selectivity that otherwise cannot be obtained by other kinds of metal catalysts. The unique selectivity may originate from two aspects. First, size-effect dependent catalytic selectivity has been observed in many reaction conditions, such as the alkene oxidation, hydrogenation, etc. [6–8] The critical size of metal nanoclusters results in their

*Metal Nanocluster Catalysis: From Fundamentals to Applications*, First Edition. Manzhou Zhu and Man-Bo Li. © 2026 Wiley-VCH GmbH. All rights reserved, including rights for text and data mining and training of artificial intelligence technologies or similar technologies. Published 2026 by Wiley-VCH GmbH.

unique selectivity during catalysis. Second, metal kernel–surface ligand and metal kernel–support cooperation also led to different selectivity of metal nanoclusters compared to the single metal atom and metal nanoparticle catalysts. This phenomenon has been frequently observed in multicomponent reactions, such as the $KA^2$ reaction, Ugi reaction, etc. [9–11].

3. It has the opportunity to trigger the discovery of novel activation mode, which would lead to the development of more efficient and selective catalytic processes. The catalytic mechanisms of metal nanoclusters are sometimes distinct from that of metal complexes or metal nanoparticles. For example, an $Au_4Cu_4$ nanocluster has been revealed to be able to catalyze the click reaction of alkynes and azides by $\pi$-activation instead of the conventional deprotonation activation mode [12]. The key intermediate [$Au_4Cu_4$-$\pi$-(R-C≡C-H] has been observed and verified by a series of characterization techniques, including thin-layer chromatography, ESI-MS, etc. This novel activation mode enables click reaction involving internal alkynes that otherwise cannot be achieved by other click reaction catalysts.
4. It has the opportunity to trigger the discovery of new phenomena in structural chemistry and material chemistry. New structures including metal–metal, metal–ligand, and metal–support interactions have been observed in metal nanoclusters [13–16]. Hierarchical chirality [17–21] has been discovered in the metal kernels, metal–organic surface, and cluster–cluster stacking levels [22–27]. Intriguing metal nanocluster-based assembly materials such as the cluster-based metal–organic framework (MOF) have been synthesized [28, 29]. These findings on new structures and materials would stimulate the development of the other research areas concerning synthesis and catalysis, such as information storage, gas separation, biosensors, photoelectric device, etc.

## 12.2 Metal Nanocluster Catalysis: Challenge

Despite of the attractive opportunities of metal nanocluster catalysis, the corresponding fundamental research and practical application face the following challenges.

1. Balancing the structural stability and catalytic activity of metal nanoclusters is challenging. As we discussed in Chapter 2, the in situ generated metal nanoclusters present super high catalytic activity [30]. However, isolation of these highly active species from the reaction system generally results in their aggregation or decomposition. A key scientific issue is developing synthetic methods that on the one hand stabilize the sensitive metal kernels of metal nanoclusters and on the other hand allow the full exposure of metal activation sites on their surface [31–34]. Although the support immobilization (Chapter 3) and organic ligand protection (Chapter 4) strategies enable the construction of such metal nanoclusters in certain conditions, the synthesis of metal nanoclusters bearing both stable structures and high catalytic activity is still a challenging task.

2. Metal nanoclusters demonstrate unique catalytic selectivity, while how to purposely control their selectivity to give the desired products is challenging. It has been reported that metal nanoclusters exhibited intriguing selectivity in some reactions [35–39]. However, these reports are sporadic and the results were generally obtained by chance. On the other hand, stereoselectivity control, especially the enantioselectivity control of metal nanocluster catalysis is still far from satisfactory. Rational design of metal nanocluster structures [40] that solve the certain selectivity problem of reaction processes should be an attractive and challenging task [41]. Concerning on this research, in-depth understanding on catalytic mechanism of metal nanoclusters is the key.
3. As we mentioned above, catalytic mechanism is crucial for positive feedback of metal nanocluster catalysts design and synthesis [42]. However, for most of the metal nanocluster catalytic processes, their activation modes are unclear. This should be partially ascribed to the ill-defined compositions and structures of metal nanoclusters during catalysis. Therefore, development of metal nanoclusters with atomically precise structures is urgent and essential. Organic ligand protected metal nanoclusters that possess precision structures would be an ideal model for the catalysis investigation [43–47]. Actually, natural enzymes can be viewed as metal nanoclusters-protected by organic ligands. Revealing their complex catalytic mechanism and realizing the large-scale synthesis of their analogues would be helpful for developing dominant metal nanocluster catalysts. Thus, mimicking structures and functions of natural enzymes by artificial metal nanoclusters would be an interesting while challenging research area [48].
4. Just like the investigation of metal nanoclusters in other fields, the fundamental research on metal nanocluster catalysis is in full swing. However, industrial application of metal nanoclusters in catalysis remains in infancy. The successful employment of metal nanoclusters in industry is quite limited. Communication between scientists in colleges/research institutes and engineers in companies is very necessary. Scale-up of metal nanocluster catalytic processes from laboratory to factory would face a lot of technical challenges, which can only be well resolved by the seamless cooperation between scientists and engineers. The successful evolution and application of enzyme from nature to industry provides a referenceable case [49].

## References


1 L. Liu, A. Corma, Metal Catalysts for Heterogeneous Catalysis: From Single Atoms to Nanoclusters and Nanoparticles. *Chem. Rev.* **2018**, *118*, 4981–5079.

2 M. Boronat, A. Leyva-Pérez, A. Corma, Theoretical and Experimental Insights into the Origin of the Catalytic Activity of Subnanometric Gold Clusters: Attempts to Predict Reactivity with Clusters and Nanoparticles of Gold. *Acc. Chem. Res.* **2014**, *47*, 834–844.

**3** J. Zhou, D. Astruc, Recent Trends and Perspectives in Palladium Nanocatalysis: From Nanoparticles to Frameworks, Atomically Precise Nanoclusters and Single-atom Catalysts. *J. Inorg. Organomet. Polym. Mater.* **2024**, *34*, 2903–2925.

**4** G. X. Pei, L. Zhang, X. Sun, Recent Advances of Bimetallic Nanoclusters with Atomic Precision for Catalytic Applications. *Coord. Chem. Rev.* **2024**, *506*, 215692.

**5** X. Liu, X. Cai, Y. Zhu, Catalysis Synergism by Atomically Precise Bimetallic Nanoclusters Doped with Heteroatoms. *Acc. Chem. Res.* **2023**, *56*, 1528–1538.

**6** S. Li, X. Du, Z. Liu, Y. Li, Y. Shao, R. Jin, Size Effects of Atomically Precise Gold Nanoclusters in Catalysis. *Precis. Chem.* **2023**, *1*, 14–28.

**7** R. A. T. M. Van Benthem, H. Hiemstra, P. W. N. M. Van Leeuwen, J. W. Geus, W. N. Speckamp, Sulfoxide-stabilized Giant Palladium Clusters in Catalyzed Oxidations. *Angew. Chem. Int. Ed.* **1995**, *34*, 457–460.

**8** D. Pun, T. Diao, S. S. Stahl, Aerobic Dehydrogenation of Cyclohexanone to Phenol Catalyzed by $Pd(TFA)_2$ /2-dimethylaminopyridine: Evidence for the Role of Pd Nanoparticles. *J. Am. Chem. Soc.* **2013**, *135*, 8213–8221.

**9** J.-Q. Fan, Y. Li, W. W. Xu, M.-B. Li, Loading Lewis Acid/base Pair on Metal Nanocluster for Catalytic Ugi Reaction. *Angew. Chem. Int. Ed.* **2025**, *64*, e202413861; *Angew. Chem.* **2025**, *137*, e202413861.

**10** J.-Q. Fan, Y. Yang, C.-B. Tao, M.-B. Li, Cadmium-doped and Pincer Ligand-modified Gold Nanocluster for Catalytic $KA^2$ Reaction. *Angew. Chem. Int. Ed.* **2023**, *62*, e202215741; *Angew. Chem.* **2023**, *135*, e202215741.

**11** Z.-Q. Tian, Y. Zhang, M.-B. Li, Pincer Ligand Facilitated Metal-catalyzed Multicomponent Reactions. *Eur. J. Org. Chem.* **2025**, *28*, e202401469.

**12** Y. Fang, K. Bao, P. Zhang, H. Sheng, Y. Yun, S.-X. Hu, D. Astruc, M. Zhu, Insight into the Mechanism of the CuAAC Reaction by Capturing the Crucial $Au_4 Cu_4$ –π-Alkyne Intermediate. *J. Am. Chem. Soc.* **2021**, *143*, 1768–1772.

**13** Z. Lei, X.-K. Wan, S.-F. Yuan, Z.-J. Guan, Q.-M. Wang, Alkynyl Approach Toward the Protection of Metal Nanoclusters. *Acc. Chem. Res.* **2018**, *51*, 2465–2474.

**14** X. Qi, Y.-L. Xie, J.-Y. Niu, J.-W. Zhao, Y.-M. Li, W.-H. Fang, J. Zhang, Application of Hard and Soft Acid-Base Theory to Construct Heterometallic Materials with Metal-oxo Clusters. *Angew. Chem. Int. Ed.* **2025**, *64*, e202417548; *Angew. Chem.* **2025**, *137*, e202417548.

**15** P. Yang, U. Kortz, Discovery and Evolution of Polyoxopalladates. *Acc. Chem. Res.* **2018**, *51*, 1599–1608.

**16** S.-F. Yuan, W.-D. Liu, C.-Y. Liu, Z.-J. Guan, Q.-M. Wang, Nitrogen Donor Protection for Atomically Precise Metal Nanoclusters. *Chem. Eur. J.* **2022**, *28*, e202104445.

**17** Y. Zhu, J. Guo, X. Qiu, S. Zhao, Z. Tang, Optical Activity of Chiral Metal Nanoclusters. *Acc. Mater. Res.* **2021**, *2*, 21–35.

**18** S. Knoppe, T. Bürgi, Chirality in Thiolate-protected Gold Clusters. *Acc. Chem. Res.* **2014**, *47*, 1318–1326.

**19** J.-H. Huang, X.-Y. Dong, Y.-J. Wang, S.-Q. Zang, Generation and Manipulation of Chiroptical Activities in Coinage-metal Clusters. *Coord. Chem. Rev.* **2022**, *470*, 214729.

**20** Y. Li, T. Higaki, X. Du, R. Jin, Chirality and Surface Bonding Correlation in Atomically Precise Metal Nanoclusters. *Adv. Mater.* **2020**, *32*, 1905488.

**21** B.-W. Zhou, S. Zhang, L. Zhao, Progress in Optical Properties of Chiral Metal Clusters: Circular Dichroism and Circularly Polarized Luminescence. *Mater. Chem. Front.* **2023**, *7*, 6389–6410.

**22** J. B. Patty, S. Havenridge, D. Tietje-Mckinney, M. A. Siegler, K. K. Singh, R. Hajy Hosseini, M. Ghabin, C. M. Aikens, A. Das, Crystal Structure and Optical Properties of a Chiral Mixed Thiolate/stibine-protected $Au_{18}$ Cluster. *J. Am. Chem. Soc.* **2022**, *144*, 478–484.

**23** M. Chi, J. Wei, Z. Zhao, X. Liu, W. Sun, Y. Feng, H. Lv, G.-Y. Yang, Structural Isomerism of $\{Ag_{14}\}^{10+}$ Nanocluster Encapsulated by Bowl-like Polyoxometalates. *Angew. Chem. Int. Ed.* **2025**, *64*, e202424499; *Angew. Chem.* **2025**, *137*, e202424499.

**24** Y. Guo, Z. Zhang, H. Han, Z. Zhou, Chiral Separation of Copper Sulfide [S–$Cu_{36}$] Nanocluster Using a Chiral Adaptive Counterion. *Nano Lett.* **2024**, *24*, 11985–11991.

**25** G. Deng, S. Malola, P. Yuan, X. Liu, B. K. Teo, H. Häkkinen, N. Zheng, Enhanced Surface Ligands Reactivity of Metal Clusters by Bulky Ligands for Controlling Optical and Chiral Properties. *Angew. Chem. Int. Ed.* **2021**, *60*, 12897–12903; *Angew. Chem.* **2021**, *133*, 13007–13013.

**26** L.-Y. Yao, L. Qin, Z. Chen, J. Lam, V. W.-W. Yam, Assembly of Luminescent Chiral Gold(I)-sulfido Clusters via Chiral Self-sorting. *Angew. Chem. Int. Ed.* **2024**, *63*, e202316200; *Angew. Chem.* **2024**, *136*, e202316200.

**27** X. Liu, W. W. Xu, X. Huang, E. Wang, X. Cai, Y. Zhao, J. Li, M. Xiao, C. Zhang, Y. Gao, W. Ding, Y. Zhu, De Novo Design of $Au_{36}(SR)_{24}$ Nanoclusters. *Nat. Commun.* **2020**, *11*, 3349.

**28** R.-W. Huang, Y.-S. Wei, X.-Y. Dong, X.-H. Wu, C.-X. Du, S.-Q. Zang, T. C. W. Mak, Hypersensitive Dual-function Luminescence Switching of a Silver-chalcogenolate Cluster-based Metal–organic Framework. *Nat. Chem.* **2017**, *9*, 689–697.

**29** Y. Jin, X.-Y. Dong, C. Zhang, S. Li, S.-Q. Zang, T. C. W. Mak, Predesigned Cluster-based Spacers for Versatile Luminescent Metallacages. *J. Am. Chem. Soc.* **2023**, *145*, 13514–13519.

**30** J. Zhang, T. Liu, L. Wang, X. Wang, Recent Process in the *in situ* Generated Metal Nanocluster Catalysis. *Chin. J. Org. Chem.* **2022**, *42*, 2331.

**31** A. Sagadevan, K. Murugesan, O. M. Bakr, M. Rueping, Copper Nanoclusters: Emerging Photoredox Catalysts for Organic Bond Formations. *Chem. Commun.* **2024**, *60*, 13858–13866.

**32** P. Kumar, M. Nemiwal, Advanced Functionalized Nanoclusters (Cu, Ag, and Au) as Effective Catalyst for Organic Transformation Reactions. *Chem. Asian J.* **2024**, *19*, e202400062.

**33** J. Ding, H. Yang, S. Zhang, Q. Liu, H. Cao, J. Luo, X. Liu, Advances in the Electrocatalytic Hydrogen Evolution Reaction by Metal Nanoclusters-based Materials. *Small* **2022**, *18*, 2204524.

**34** A. Pal, S. Biswas, B. Sahoo, Y. Negishi, Cu-nanoclusters as Emerging Catalyst in Organic Synthesis. *Synlett* **2025**, *36*, 1123–1129.

**35** W.-L. Mu, L. Li, X.-Z. Cong, X. Chen, P. Xia, Q. Liu, L. Wang, J. Yan, C. Liu, Hierarchical Assembly of High-nuclearity Copper(I) Alkynide Nanoclusters: Highly Effective $CO_2$ Electroreduction Catalyst Toward Hydrocarbons. *J. Am. Chem. Soc.* **2024**, *146*, 28131–28140.

**36** Y.-M. Wang, X.-C. Lin, K.-M. Mo, M. Xie, Y.-L. Huang, G.-H. Ning, D. Li, An Atomically Precise Pyrazolate-protected Copper Nanocluster Exhibiting Exceptional Stability and Catalytic Activity. *Angew. Chem. Int. Ed.* **2023**, *62*, e202218369; *Angew. Chem.* **2023**, *135*, e202218369.

**37** X.-W. Li, F.-X. Zhang, X.-G. Zhao, Q.-Y. Liu, S.-G. He, J.-B. Ma, Dinitrogen Activation and Cyanide Release by the Gas-phase Cluster Anions $InNbC_2^-$: Synergistic Effects of the p-block Indium and the d-block Niobium. *J. Am. Chem. Soc.* **2025**, *147*, 22185–22191.

**38** Q.-Y. Fang, J. Han, M. Qin, W. Li, C. Zhu, J. Xie, Trinuclear Gold-catalyzed 1,2-difunctionalization of Alkenes. *Angew. Chem. Int. Ed.* **2023**, *62*, e202305121; *Angew. Chem.* **2023**, *135*, e202305121.

**39** C. A. Salazar, J. J. Gair, K. N. Flesch, I. A. Guzei, J. C. Lewis, S. S. Stahl, Catalytic Behavior of Mono-*N*-protected Amino-Acid Ligands in Ligand-Accelerated C—H Activation by Palladium(II). *Angew. Chem. Int. Ed.* **2020**, *59*, 10873–10877; *Angew. Chem.* **2019**, *132*, 10965–10969.

**40** W. Jing, H. Shen, R. Qin, Q. Wu, K. Liu, N. Zheng, Surface and Interface Coordination Chemistry Learned from Model Heterogeneous Metal Nanocatalysts: From Atomically Dispersed Catalysts to Atomically Precise Clusters. *Chem. Rev.* **2023**, *123*, 5948–6002.

**41** B. M. Trost, Selectivity: A Key to Synthetic Efficiency. *Science* **1983**, *219*, 245–250.

**42** B. Zhang, C. Chen, W. Chuang, S. Chen, P. Yang, Size Transformation of the $Au_{22}$ $(SG)_{18}$ Nanocluster and Its Surface-sensitive Kinetics. *J. Am. Chem. Soc.* **2020**, *142*, 11514–11520.

**43** Z. Xu, H. Dong, W. Gu, Z. He, F. Jin, C. Wang, Q. You, J. Li, H. Deng, L. Liao, D. Chen, J. Yang, Z. Wu, Lattice Compression Revealed at the ≈1 nm Scale. *Angew. Chem. Int. Ed.* **2023**, *62*, e202308441; *Angew. Chem.* **2023**, *135*, e202308441.

**44** N. de Silva, J.-M. Ha, A. Solovyov, M. M. Nigra, I. Ogino, S. W. Yeh, K. A. Durkin, A. Katz, A Bioinspired Approach for Controlling Accessibility in Calix[4]Arene-bound Metal Cluster Catalysts. *Nat. Chem.* **2010**, *2*, 1062–1068.

**45** S. Wang, S. Jin, S. Yang, S. Chen, Y. Song, J. Zhang, M. Zhu, Total Structure Determination of Surface Doping $[Ag_{46}Au_{24}(SR)_{32}](BPh_4)_2$ Nanocluster and Its Structure-related Catalytic Property. *Sci. Adv.* **2015**, *1*, e1500441.

**46** C. Zhu, Z.-L. Chen, H. Li, L. Lu, X. Kang, J. Xuan, M. Zhu, Rational Design of Highly Phosphorescent Nanoclusters for Efficient Photocatalytic Oxidation. *J. Am. Chem. Soc.* **2024**, *146*, 23212–23220.

**47** S. Wang, L. Tang, B. Cai, Z. Yin, Y. Li, L. Xiong, X. Kang, J. Xuan, Y. Pei, M. Zhu, Ligand Modification of $Au_{25}$ Nanoclusters for Near-infrared Photocatalytic Oxidative Functionalization. *J. Am. Chem. Soc.* **2022**, *144*, 3787–3792.

**48** O. Pàmies, J.-E. Bäckvall, Combination of Enzymes and Metal Catalysts. A Powerful Approach in Asymmetric Catalysis. *Chem. Rev.* **2003**, *103*, 3247–3262.

**49** R. Buller, S. Lutz, R. J. Kazlauskas, R. Snajdrova, J. C. Moore, U. T. Bornscheuer, From Nature to Industry: Harnessing Enzymes for Biocatalysis. *Science* **2023**, *382*, eadh8615.

# Index

Note: Page numbers in *italics* and **bold** refers to figures and tables respectively.




*Metal Nanocluster Catalysis: From Fundamentals to Applications*, First Edition. Manzhou Zhu and Man-Bo Li. © 2026 Wiley-VCH GmbH. All rights reserved, including rights for text and data mining and training of artificial intelligence technologies or similar technologies. Published 2026 by Wiley-VCH GmbH.

## d



## e